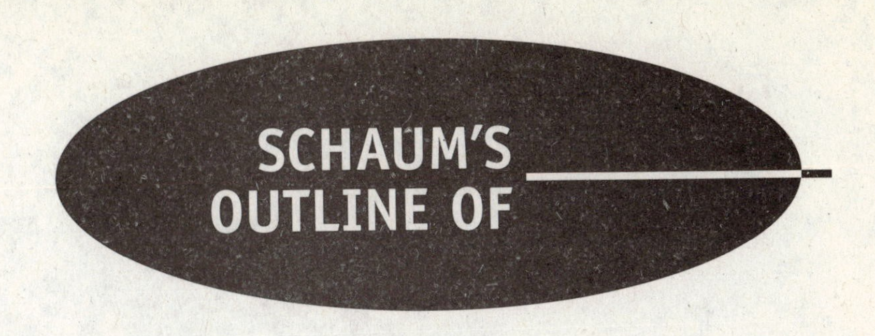

Thermodynamics for Engineers

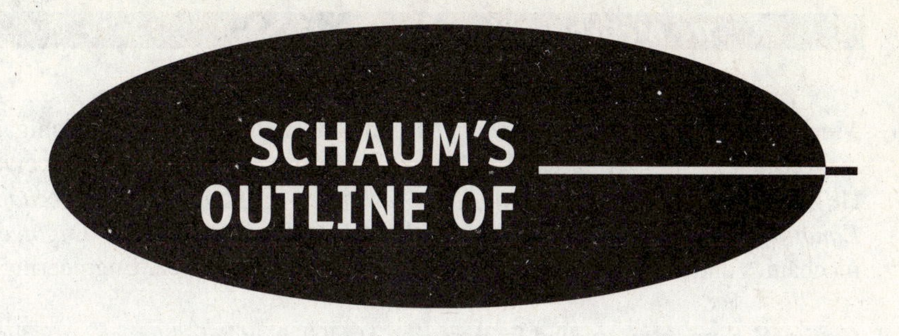

Thermodynamics for Engineers

Second Edition

Merle C. Potter, Ph.D.
Professor Emeritus of Mechanical Engineering
Michigan State University

Craig W. Somerton, Ph.D.
Associate Professor of Mechanical Engineering
Michigan State University

Schaum's Outline Series

New York Chicago San Francisco Lisbon
London Madrid Mexico City Milan New Delhi
San Juan Seoul Singapore Sydney Toronto

Merle C. Potter has a B.S. degree in Mechanical Engineering from Michigan Technological University; his M.S. in Aerospace Engineering and Ph.D. in Engineering Mechanics were received from the University of Michigan. He is the author or coauthor of *The Mechanics of Fluids, Fluid Mechanics, Thermal Sciences, Differential Equations, Advanced Engineering Mathematics, Fundamentals of Engineering,* and numerous papers in fluid mechanics and energy. He is Professor Emeritus of Mechanical Engineering at Michigan State University.

Craig W. Somerton studied Engineering at UCLA, where he was awarded the B.S., M.S., and Ph.D. degrees. He is currently Associate Professor of Mechanical Engineering at Michigan State University. He has published in the *International Journal of Mechcanical Engineering Education* and is a past recipient of the SAE Ralph R. Teetor Education Award.

Schaum's Outline of
THERMODYNAMICS FOR ENGINEERS

6 7 8 9 0 CUS/CUS 0 1 4 3

ISBN 978-0-07-161167-1
MHID 0-07-161167-3

Library of Congress Cataloguing-in-Publication Data is on file with the Library of Congress.

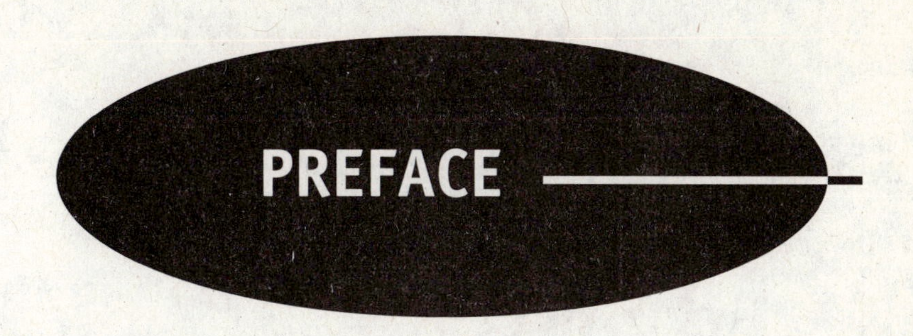

This book is intended for the first course in thermodynamics required by most, if not all, engineering departments. It is designed to supplement the required text selected for the course; it provides a succinct presentation of the material so that the student can more easily determine the major objective of each section of the textbook. If expanded detail is not of importance in this first course, the present Schaum's Outline could itself serve as the required text.

The material presented in a first course in thermodynamics is more or less the same in most engineering schools. Under a quarter system both the first and second laws are covered, with little time left for applications. Under a semester system it is possible to cover some application areas, such as vapor and gas cycles, nonreactive mixtures, or combustion. This book allows such flexibility. In fact, there is sufficient material for a full year of study.

As some U.S. industry continues to avoid the use of SI units, we have written about 20 percent of the examples, solved problems, and supplementary problems in English units. Tables are presented in both systems of units.

The basic thermodynamic principles are liberally illustrated with numerous examples and solved problems that demonstrate how the principles are applied to actual or simulated engineering situations. Supplementary problems that provide students an opportunity to test their problem-solving skills are included at the ends of all chapters. Answers are provided for all these problems at the ends of the chapters. We have also included FE-type questions at the end of most chapters.

In addition, we have included a set of exams that are composed of multiple-choice questions, along with their solutions. The majority of students who take thermodynamics will never see the material again except when they take a national exam (the professional engineers' exams or the GRE/Engineering exam). The national exams are multiple-choice exams with which engineering students are unfamiliar. Thermodynamics provides an excellent opportunity to give these students an experience in taking multiple-choice exams, exams that are typically long and difficult. Studies have shown that grades are independent of the type of exam given, so this may be the course to introduce engineering students to the multiple-choice exam.

The authors wish to thank Mrs. Michelle Gruender for her careful review of the manuscript and Ms. Barbara Gilson for her efficient production of this book.

You, both professors and students, are encouraged to email me at MerleCP@sbcglobal.net if you have comments/corrections/questions or just want to opine.

Merle C. Potter
Craig W. Somerton

CONTENTS

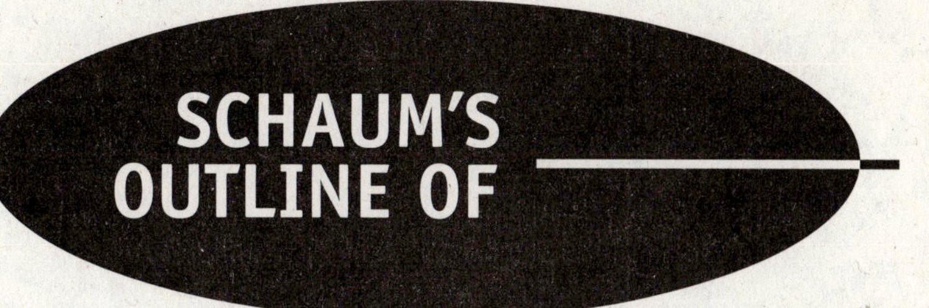

THERMODYNAMICS FOR ENGINEERS

Concepts, Definitions, and Basic Principles

1.1 INTRODUCTION

Thermodynamics is a science in which the storage, the transformation, and the transfer of energy are studied. Energy is *stored* as internal energy (associated with temperature), kinetic energy (due to motion), potential energy (due to elevation), and chemical energy (due to chemical composition); it is *transformed* from one of these species to another; and it is *transferred* across a boundary as either heat or work. In thermodynamics we will develop mathematical equations that relate the transformations and transfers of energy to material properties such as temperature, pressure, or enthalpy. Substances and their properties thus become an important secondary theme. Much of our work will be based on experimental observations that have been organized into mathematical statements, or *laws*; the first and second laws of thermodynamics are the most widely used.

The engineer's objective in studying thermodynamics is most often the analysis or design of a large-scale system—anything from an air-conditioner to a nuclear power plant. Such a system may be regarded as a continuum in which the activity of the constituent molecules is averaged into measurable quantities such as pressure, temperature, and velocity. This outline, then, will be restricted to *macroscopic* or *engineering thermodynamics*. If the behavior of individual molecules is important, a text in *statistical* thermodynamics must be consulted.

1.2 THERMODYNAMIC SYSTEMS AND CONTROL VOLUME

A thermodynamic *system* is a definite quantity of matter contained within some closed surface. The surface is usually an obvious one like that enclosing the gas in the cylinder of Fig. 1-1; however, it may be an imagined boundary like the deforming boundary of a certain amount of mass as it flows through a pump. In Fig. 1-1 the system is the compressed gas, the *working fluid*, and the system boundary is shown by the dotted line.

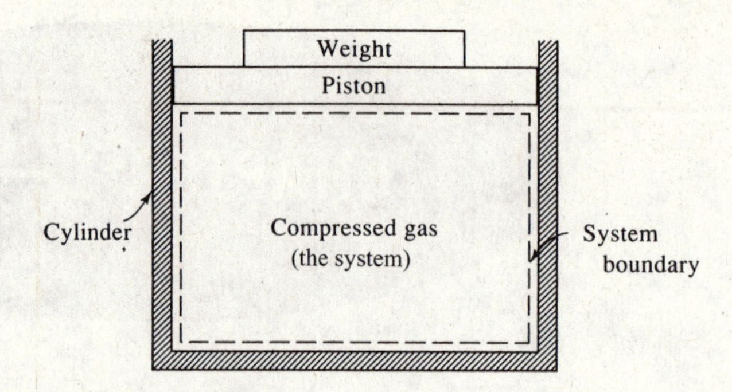

Fig. 1-1 A system.

All matter and space external to a system is collectively called its *surroundings*. Thermodynamics is concerned with the interactions of a system and its surroundings, or one system interacting with another. A system interacts with its surroundings by transferring energy across its boundary. No material crosses the boundary of a given system. If the system does not exchange energy with the surroundings, it is an *isolated* system.

In many cases, an analysis is simplified if attention is focused on a volume in space into which, or from which, a substance flows. Such a volume is a *control volume*. A pump, a turbine, an inflating balloon, are examples of control volumes. The surface that completely surrounds the control volume is called a *control surface*. An example is sketched in Fig. 1-2.

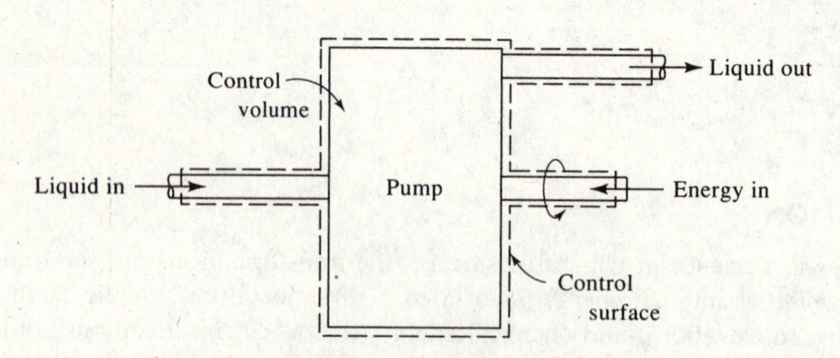

Fig. 1-2 A control volume.

We thus must choose, in a particular problem, whether a system is to be considered or whether a control volume is more useful. If there is mass flux across a boundary of the region, then a control volume is required; otherwise, a system is identified. We will present the analysis of a system first and follow that with a study using the control volume.

1.3 MACROSCOPIC DESCRIPTION

In engineering thermodynamics we postulate that the material in our system or control volume is a *continuum*; that is, it is continuously distributed throughout the region of interest. Such a postulate allows us to describe a system or control volume using only a few measurable properties.

Consider the definition of *density* given by

$$\rho = \lim_{\Delta V \to 0} \frac{\Delta m}{\Delta V} \tag{1.1}$$

where Δm is the mass contained in the volume ΔV, shown in Fig. 1-3. Physically, ΔV cannot be allowed to shrink to zero since, if ΔV became extremely small, Δm would vary discontinuously, depending on the number of molecules in ΔV. So, the zero in the definition of ρ should be replaced by some quantity ε, small, but large enough to eliminate molecular effects. Noting that there are about 3×10^{16} molecules in a cubic millimeter of air at standard conditions, ε need not be very large to contain billions and billions of molecules. For most engineering applications ε is sufficiently small that we can let it be zero, as in (*1.1*).

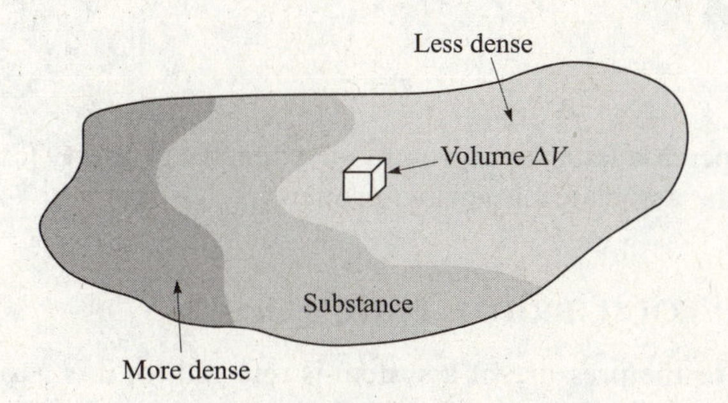

Fig. 1-3 Mass as a continuum.

There are, however, situations where the continuum assumption is not valid, for example, the re-entry of satellites. At an elevation of 100 km the *mean free path*, the average distance a molecule travels before it collides with another molecule, is about 30 mm; the macroscopic approach is already questionable. At 150 km the mean free path exceeds 3 m, which is comparable to the dimensions of the satellite! Under these conditions statistical methods based on molecular activity must be used.

1.4 PROPERTIES AND STATE OF A SYSTEM

The matter in a system may exist in several phases: as a solid, a liquid, or a gas. A *phase* is a quantity of matter that has the same chemical composition throughout; that is, it is homogeneous. Phase boundaries separate the phases, in what, when taken as a whole, is called a *mixture*.

A *property* is any quantity which serves to describe a system. The *state* of a system is its condition as described by giving values to its properties at a particular instant. The common properties are pressure, temperature, volume, velocity, and position; but others must occasionally be considered. Shape is important when surface effects are significant; color is important when radiation heat transfer is being investigated.

The essential feature of a property is that it has a unique value when a system is in a particular state, and this value does not depend on the previous states that the system passed through; that is, it is not a path function. Since a property is not dependent on the path, any change depends only on the initial and final states of the system. Using the symbol ϕ to represent a property, the mathematical equation is

$$\int_{\phi_1}^{\phi_2} d\phi = \phi_2 - \phi_1 \qquad (1.2)$$

This requires that $d\phi$ be an exact differential; $\phi_2 - \phi_1$ represents the change in the property as the system changes from state 1 to state 2. There are quantities which we will encounter, such as work, that are path functions for which an exact differential does not exist.

A relatively small number of *independent properties* suffice to fix all other properties and thus the state of the system. If the system is composed of a single phase, free from magnetic, electrical, and surface effects, the state is fixed when any two properties are fixed; this *simple system* receives most attention in engineering thermodynamics.

Thermodynamic properties are divided into two general types, intensive and extensive. An *intensive property* is one which does not depend on the mass of the system; temperature, pressure, density, and velocity are examples since they are the same for the entire system, or for parts of the system. If we bring two systems together, intensive properties are not summed.

An *extensive property* is one which depends on the mass of the system; volume, momentum, and kinetic energy are examples. If two systems are brought together the extensive property of the new system is the sum of the extensive properties of the original two systems.

If we divide an extensive property by the mass a *specific property* results. The *specific volume* is thus defined to be

$$v = \frac{V}{m} \tag{1.3}$$

We will generally use an uppercase letter to represent an extensive property [exception: m for mass] and a lowercase letter to denote the associated intensive property.

1.5 THERMODYNAMIC EQUILIBRIUM; PROCESSES

When the temperature or the pressure of a system is referred to, it is assumed that all points of the system have the same, or essentially the same, temperature or pressure. When the properties are assumed constant from point to point and when there is no tendency for change with time, a condition of *thermodynamic equilibrium* exists. If the temperature, say, is suddenly increased at some part of the system boundary, spontaneous redistribution is assumed to occur until all parts of the system are at the same temperature.

If a system would undergo a large change in its properties when subjected to some small disturbance, it is said to be in *metastable equilibrium*. A mixture of gasoline and air, or a large bowl on a small table, is such a system.

When a system changes from one equilibrium state to another, the path of successive states through which the system passes is called a *process*. If, in the passing from one state to the next, the deviation from equilibrium is infinitesimal, a *quasiequilibrium* process occurs and each state in the process may be idealized as an equilibrium state. Many processes, such as the compression and expansion of gases in an internal combustion engine, can be approximated by quasiequilibrium processes with no significant loss of accuracy. If a system undergoes a quasiequilibrium process (such as the thermodynamically slow compression of air in a cylinder) it may be sketched on appropriate coordinates by using a solid line, as shown in Fig. 1-4*a*. If the system, however, goes from one equilibrium state to another through a series of nonequilibrium states (as in combustion) a *nonequilibrium process* occurs. In Fig. 1-4*b* the dashed curve represents such a process; between (V_1, P_1) and (V_2, P_2) properties are not uniform throughout the system and thus the state of the system cannot be well defined.

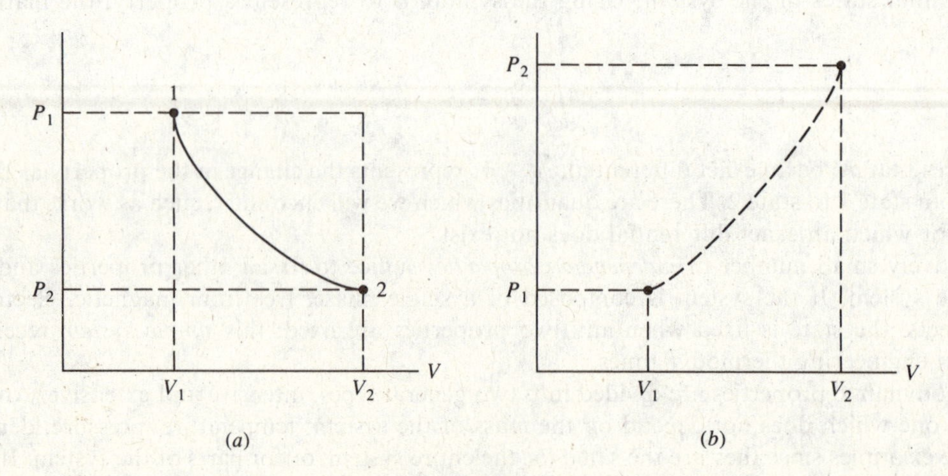

Fig. 1-4 A process.

EXAMPLE 1.1 Whether a particular process may be considered quasiequilibrium or nonequilibrium depends on how the process is carried out. Let us add the weight W to the piston of Fig. 1-5. Explain how W can be added in a nonequilibrium manner and in an equilibrium manner.

Solution: If it is added suddenly as one large weight, as in part (*a*), a nonequilibrium process will occur in the gas, the system. If we divide the weight into a large number of small weights and add them one at a time, as in part (*b*), a quasiequilibrium process will occur.

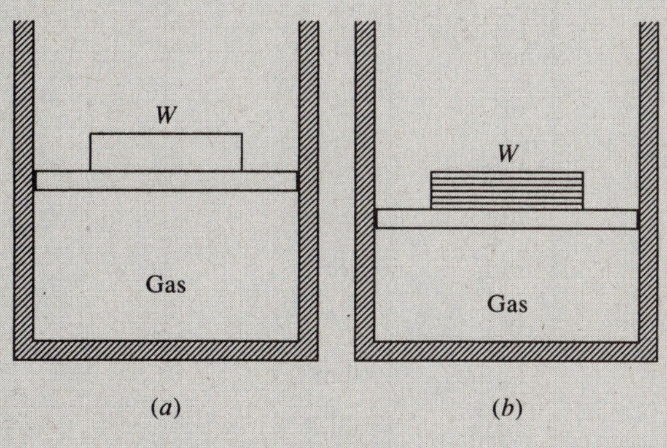

Fig. 1-5

Note that the surroundings play no part in the notion of equilibrium. It is possible that the surroundings do work on the system via friction; for quasiequilibrium it is only required that the properties *of the system* be uniform at any instant during a process.

When a system in a given initial state experiences a series of quasiequilibrium processes and returns to the initial state, the system undergoes a *cycle*. At the end of the cycle the properties of the system have the same values they had at the beginning; see Fig. 1-6.

The prefix *iso-* is attached to the name of any property that remains unchanged in a process. An *isothermal* process is one in which the temperature is held constant; in an *isobaric* process the pressure remains constant; an *isometric* process is a constant-volume process. Note the isobaric and the isometric legs in Fig. 1-6.

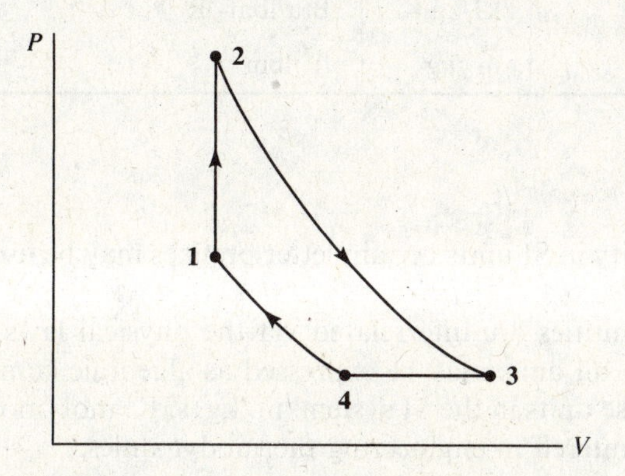

Fig. 1-6 Four processes that make up a cycle.

1.6 UNITS

While the student is undoubtedly most at home with SI (Système International) units, much of the data gathered in the United States is in English units. Therefore, a certain number of examples and problems will be presented in English units. Table 1-1 lists units of the principal thermodynamic quantities. Observe the dual use of W for weight and work; the context and the units will make clear which quantity is intended.

Table 1-1

Quantity	Symbol	SI Units	English Units	To Convert from English to SI Units Multiply by
Length	L	m	ft	0.3048
Mass	m	kg	lbm	0.4536
Time	t	s	sec	—
Area	A	m^2	ft^2	0.09290
Volume	V	m^3	ft^3	0.02832
Velocity	$\mathcal{V}$	m/s	ft/sec	0.3048
Acceleration	a	m/s^2	ft/sec^2	0.3048
Angular velocity	ω	rad/s	sec^{-1}	—
Force, Weight	F, W	N	lbf	4.448
Density	ρ	kg/m^3	lbm/ft^3	16.02
Specific weight	w	N/m^3	lbf/ft^3	157.1
Pressure, Stress	P, τ	kPa	lbf/ft^2	0.04788
Work, Energy	W, E, U	J	ft-lbf	1.356
Heat transfer	Q	J	Btu	1055
Power	$\dot{W}$	W	ft-lbf/sec	1.356
Heat flux	$\dot{Q}$	W or J/s	Btu/sec	1055
Mass flux	$\dot{m}$	kg/s	lbm/sec	0.4536
Flow rate	$\dot{V}$	m^3/s	ft^3/sec	0.02832
Specific heat	C	kJ/kg·K	Btu/lbm-°R	4.187
Specific enthalpy	h	kJ/kg	Btu/lbm	2.326
Specific entropy	s	kJ/kg·K	Btu/lbm-°R	4.187
Specific volume	v	m^3/kg	ft^3/lbm	0.06242

When expressing a quantity in SI units certain letter prefixes may be used to represent multiplication by a power of 10; see Table 1-2.

The units of various quantities are interrelated via the physical laws obeyed by the quantities. It follows that, in either system, all units may be expressed as algebraic combinations of a selected set of *base units*. There are seven base units in the SI system: m, kg, s, K, mol (mole), A (ampere), cd (candela). The last two are rarely encountered in engineering thermodynamics.

EXAMPLE 1.2 Newton's second law, $F = ma$, relates a net force acting on a body to its mass and acceleration. If a force of one newton accelerates a mass of one kilogram at one m/s^2; or, a force of one lbf accelerates 32.2 lbm (1 slug) at a rate of one ft/sec^2, how are the units related?

Solution: The units are related as

$$1\ N = 1\ kg \cdot m/s^2 \qquad \text{or} \qquad 1\ lbf = 32.2\ lbm\text{-}ft/sec^2$$

Table 1-2

Multiplication Factor	Prefix	Symbol
10^{12}	tera	T
10^{9}	giga	G
10^{6}	mega	M
10^{3}	kilo	k
10^{-2}	centi*	c
10^{-3}	milli	m
10^{-6}	micro	μ
10^{-9}	nano	n
10^{-12}	pico	p

*Discouraged except in cm, cm^2, or cm^3.

EXAMPLE 1.3 *Weight* is the force of gravity; by Newton's second law, $W = mg$. How does weight change with elevation?

Solution: Since mass remains constant, the variation of W with elevation is due to changes in the acceleration of gravity g (from about 9.77 m/s^2 on the highest mountain to 9.83 m/s^2 in the deepest ocean trench). We will use the standard value 9.81 m/s^2 (32.2 ft/sec^2), unless otherwise stated.

EXAMPLE 1.4 Express the energy unit J (joule) in terms of SI base units: mass, length, and time.

Solution: Recall that energy or work is force times distance. Hence, by Example 1.2, the energy unit J (joule) is

$$1\ \mathrm{J} = (1\ \mathrm{N})(1\ \mathrm{m}) = (1\ \mathrm{kg{\cdot}m/s^2})(1\ \mathrm{m}) = 1\ \mathrm{kg{\cdot}m^2/s^2}$$

In the English system both the lbf and the lbm are base units. As indicated in Table 1-1, the primary energy unit is the ft-lbf. By Example 1.2,

$$1\ \text{ft-lbf} = 32.2\ \text{lbm-ft}^2/\text{sec}^2 = 1\ \text{slug-ft}^2/\text{sec}^2$$

analogous to the SI relation found above.

1.7 DENSITY, SPECIFIC VOLUME, SPECIFIC WEIGHT

By (*1.1*), density is mass per unit volume; by (*1.3*), specific volume is volume per unit mass. Therefore,

$$v = \frac{1}{\rho} \tag{1.4}$$

Associated with (mass) density is *weight density* or *specific weight* w:

$$w = \frac{W}{V} \tag{1.5}$$

with units N/m^3(lbf/ft^3). [Note that w is volume-specific, not mass-specific.] Specific weight is related to density through $W = mg$ as follows:

$$w = \rho g \tag{1.6}$$

For water, nominal values of ρ and w are, respectively, 1000 kg/m^3 (62.4 lbm/ft^3) and 9810 N/m^3 (62.4 lbf/ft^3). For air at sea level the nominal values are 1.21 kg/m^3 (0.0755 lbm/ft^3) and 11.86 N/m^3 (0.0755 lbf/ft^3).

EXAMPLE 1.5 The mass of air in a room 3 × 5 × 20 m is known to be 350 kg. Determine the density, specific volume, and specific weight.

Solution:

$$\rho = \frac{m}{V} = \frac{350}{(3)(5)(20)} = 1.167 \text{ kg/m}^3 \qquad v = \frac{1}{\rho} = \frac{1}{1.167} = 0.857 \text{ m}^3\text{/kg}$$

$$w = \rho g = (1.167)(9.81) = 11.45 \text{ N/m}^3$$

1.8 PRESSURE

Definition

In gases and liquids it is common to call the effect of a normal force acting on an area the *pressure*. If a force ΔF acts at an angle to an area ΔA (Fig. 1-7), only the normal component ΔF_n enters into the definition of pressure:

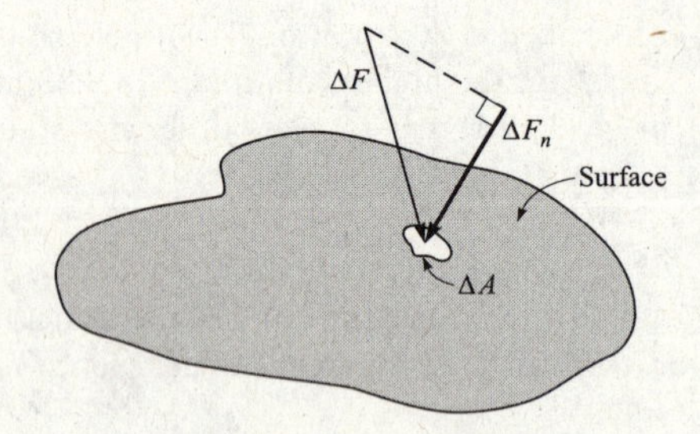

Fig. 1-7 The normal component of a force.

$$P = \lim_{\Delta A \to 0} \frac{\Delta F_n}{\Delta A} \tag{1.7}$$

The SI unit of pressure is the pascal (Pa), where

$$1 \text{ Pa} = 1 \text{ N/m}^2 = 1 \text{ kg/m·s}^2$$

The corresponding English unit is lbf/ft^2 (psf), although lbf/in^2 (psi) is commonly used.

By considering the pressure forces acting on a triangular fluid element of constant depth we can show that the pressure at a point in a fluid in equilibrium (no motion) is the same in all directions; it is a scalar quantity. For gases and liquids in relative motion the pressure may vary with direction at a point; however, this variation is extremely small and can be ignored in most gases and in liquids with low viscosity (e.g., water). We have not assumed in the above discussion that pressure does not vary from point to point, only that at a particular point it does not vary with direction.

Pressure Variation with Elevation

In the atmosphere pressure varies with elevation. This variation can be expressed mathematically by considering the equilibrium of the element of air shown in Fig. 1-8. Summing forces on the element in the vertical direction (up is positive) gives

$$dP = -\rho g \, dz \tag{1.8}$$

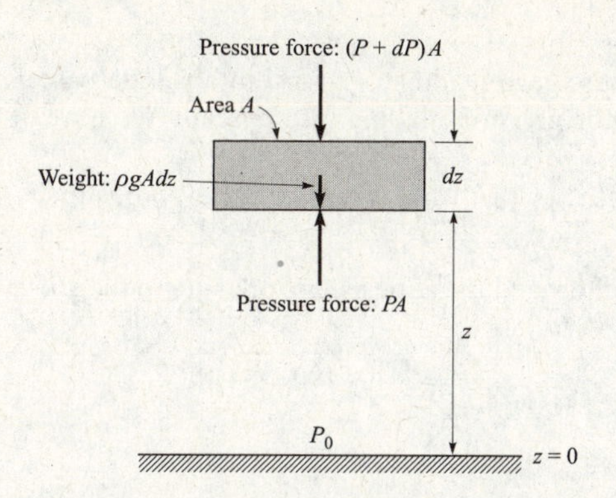

Fig. 1-8 Forces acting on an element of air.

Now if ρ is a known function of z, the above equation can be integrated to give $P(z)$:

$$P(z) - P_0 = -\int_0^z \rho g dz \tag{1.9}$$

For a liquid, ρ is constant. If we write (*1.8*) using $dh = -dz$, we have

$$dP = w\,dh \tag{1.10}$$

where h is measured positive downward. Integrating this equation, starting at a surface where $P = 0$, results in

$$P = wh \tag{1.11}$$

This equation can be used to convert a pressure measured in meters of water or millimeters of mercury to Pascals (Pa).

In most thermodynamic relations *absolute pressure* must be used. Absolute pressure is measured pressure, or *gage* pressure, plus the local atmospheric pressure:

$$P_{abs} = P_{gage} + P_{atm} \tag{1.12}$$

A negative gage pressure is often called a *vacuum*, and gages capable of reading negative pressures are *vacuum gages*. A gage pressure of –50 kPa would be referred to as a vacuum of 50 kPa, with the sign omitted.

Figure 1-9 shows the relationships between absolute and gage pressure.

The word "gage" is generally used in statements of gage pressure; e.g., $P = 200$ kPa gage. If "gage" is not present, the pressure will, in general, be an absolute pressure. Atmospheric pressure is an absolute pressure, and will be taken as 100 kPa (at sea level), unless otherwise stated. It should be noted that atmospheric pressure is highly dependent on elevation; in Denver, Colorado, it is about 84 kPa, and in a mountain city with elevation 3000 m it is only 70 kPa.

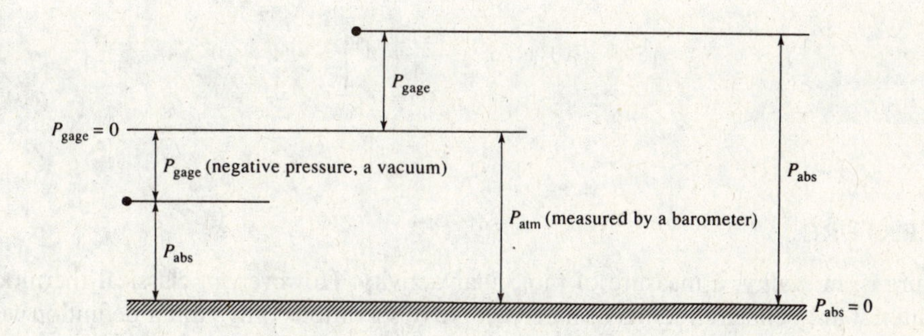

Fig. 1-9 Absolute and gage pressure.

EXAMPLE 1.6 Express a pressure gage reading of 35 psi in absolute pascals.

Solution: First we convert the pressure reading into pascals. We have

$$\left(35\frac{\text{lbf}}{\text{in}^2}\right)\left(144\frac{\text{in}^2}{\text{ft}^2}\right)\left(0.04788\frac{\text{kPa}}{\text{lbf/ft}^2}\right) = 241\text{ kPa gage}$$

To find the absolute pressure we simply add the atmospheric pressure to the above value. Assuming $P_{\text{atm}} = 100$ kPa, we obtain

$$P = 241 + 100 = 341\text{ kPa}$$

EXAMPLE 1.7 The manometer shown in Fig. 1-10 is used to measure the pressure in the water pipe. Determine the water pressure if the manometer reading is 0.6 m. Mercury is 13.6 times heavier than water.

Solution: To solve the manometer problem we use the fact that $P_a = P_b$. The pressure P_a is simply the pressure P in the water pipe plus the pressure due to the 0.6 m of water; the pressure P_b is the pressure due to 0.6 m of mercury. Thus,

$$P + (0.6\text{ m})(9810\text{ N/m}^3) = (0.6\text{ m})(13.6)(9810\text{ N/m}^3)$$

This gives $P = 74\,200$ Pa or 74.2 kPa gage.

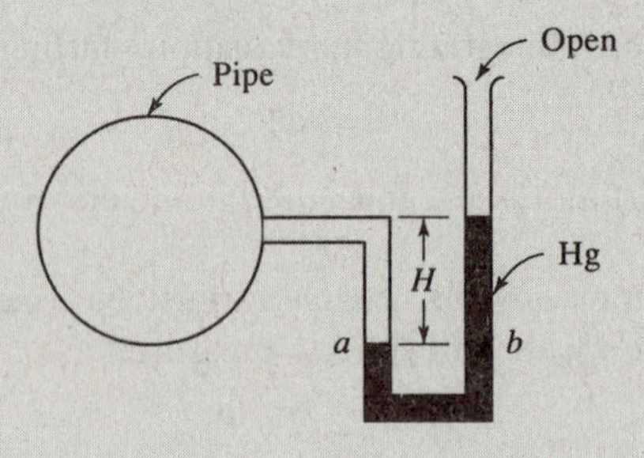

Fig. 1-10

EXAMPLE 1.8 Calculate the force due to the pressure acting on the 1-m-diameter horizontal hatch of a submarine submerged 600 m below the surface.

Solution: The pressure acting on the hatch at a depth of 600 m is found from (*1.11*) as

$$P = \rho g h = (1000\text{ kg/m}^3)(9.81\text{ m/s}^2)(600\text{ m}) = 5.89\text{ MPa gage}$$

The pressure is constant over the area; hence, the force due to the pressure is given by

$$F = PA = (5.89 \times 10^6\text{ N/m}^2)\left[\frac{\pi(1)^2}{4}\text{m}^2\right] = 4.62 \times 10^6\text{ N}$$

1.9 TEMPERATURE

Temperature is, in reality, a measure of molecular activity. However, in classical thermodynamics the quantities of interest are defined in terms of macroscopic observations only, and a definition of temperature using molecular measurements is not useful. Thus we must proceed without actually defining temperature. What we shall do instead is discuss *equality of temperatures*.

Equality of Temperatures

Let two bodies be isolated from the surroundings but placed in contact with each other. If one is hotter than the other, the hotter body will become cooler and the cooler body will become hotter; both bodies will undergo change until all properties (e.g., electrical resistance) of the bodies cease to change. When this occurs, *thermal equilibrium* is said to have been established between the two bodies. Hence, we state that two systems have equal temperatures if no change occurs in any of their properties when the systems are brought into contact with each other. In other words, if two systems are in thermal equilibrium their temperatures are postulated to be equal.

A rather obvious observation is referred to as the *zeroth law of thermodynamics*: if two systems are equal in temperature to a third, they are equal in temperature to each other.

Relative Temperature Scale

To establish a temperature scale, we choose the number of subdivisions, called degrees, between two fixed, easily duplicated points, the ice point and the steam point. The *ice point* exists when ice and water are in equilibrium at a pressure of 101 kPa; the *steam point* exists when liquid water and its vapor are in a state of equilibrium at a pressure of 101 kPa. On the Fahrenheit scale there are 180 degrees between these two points; on the Celsius (formerly called the Centigrade) scale, 100 degrees. On the Fahrenheit scale the ice point is assigned the value of 32 and on the Celsius scale it is assigned the value 0. These selections allow us to write

$$t_F = \frac{9}{5}t_C + 32 \tag{1.13}$$

$$t_C = \frac{5}{9}(t_F - 32) \tag{1.14}$$

Absolute Temperature Scale

The second law of thermodynamics will allow us to define an absolute temperature scale; however, since we do not have the second law at this point and we have immediate use for absolute temperature, an empirical absolute temperature scale will be presented.

The relations between absolute and relative temperatures are

$$T_F = t_F + 459.67 \tag{1.15}$$

$$T_C = t_C + 273.15 \tag{1.16}$$

where the subscript "F" refers to the Fahrenheit scale and the subscript "C" to the Celsius scale. (The values 460 and 273 are used where precise accuracy is not required.) The absolute temperature on the Fahrenheit scale is given in degrees Rankine (°R), and on the Celsius scale it is given in kelvins (K). *Note:* 300 K is read "300 kelvins," not "300 degrees Kelvin." We do not use the degree symbol when temperature is measured in kelvins.

EXAMPLE 1.9 The temperature of a body is 50 °F. Find its temperature in °C, K, and °R.

Solution: Using the conversion equations,

$$t_C = \frac{5}{9}(50 - 32) = 10\,^\circ\text{C} \qquad T_K = 10 + 273 = 283\text{ K} \qquad T_R = 50 + 460 = 510\,^\circ\text{R}$$

Note that T referred to absolute temperature and t to relative temperature.

1.10 ENERGY

A system may possess several different forms of energy. Assuming uniform properties throughout the system, the *kinetic energy* is given by

$$KE = \frac{1}{2}mV^2 \tag{1.17}$$

where V is the velocity of each lump of substance, assumed constant over the entire system. If the velocity is not constant for each lump, then the kinetic energy is found by integrating over the system. The energy that a system possesses due to its elevation h above some arbitrarily selected datum is its *potential energy*; it is determined from the equation

$$PE = mgh \tag{1.18}$$

Other forms of energy include the energy stored in a battery, energy stored in an electrical condenser, electrostatic potential energy, and surface energy. In addition, there is the energy associated with the translation, rotation, and vibration of the molecules, electrons, protons, and neutrons, and the chemical energy due to bonding between atoms and between subatomic particles. These molecular and atomic forms of energy will be referred to as *internal energy* and designated by the letter U. In combustion, energy is released when the chemical bonds between atoms are rearranged; nuclear reactions result when changes occur between the subatomic particles. In thermodynamics our attention will be initially focused on the internal energy associated with the motion of molecules that is influenced by various macroscopic properties such as pressure, temperature, and specific volume. In Chapter 13 the combustion process is studied in some detail.

Internal energy, like pressure and temperature, is a property of fundamental importance. A substance always has internal energy; if there is molecular activity, there is internal energy. We need not know, however, the absolute value of internal energy, since we will be interested only in its increase or decrease.

We now come to an important law, which is often of use when considering isolated systems. The law of *conservation of energy* states that the energy of an isolated system remains constant. Energy cannot be created or destroyed in an isolated system; it can only be transformed from one form to another.

Let us consider the system composed of two automobiles that hit head on and come to rest. Because the energy of the system is the same before and after the collision, the initial KE must simply have been transformed into another kind of energy—in this case, internal energy, primarily stored in the deformed metal.

EXAMPLE 1.10 A 2200-kg automobile traveling at 90 kph (25 m/s) hits the rear of a stationary, 1000-kg automobile. After the collision the large automobile slows to 50 kph (13.89 m/s), and the smaller vehicle has a speed of 88 kph (24.44 m/s). What has been the increase in internal energy, taking both vehicles as the system?

Solution: The kinetic energy before the collision is ($V = 25$ m/s)

$$KE_1 = \frac{1}{2}m_a V_{a1}^2 = \left(\frac{1}{2}\right)(2200)(25^2) = 687\,500 \text{ J}$$

After the collision the kinetic energy is

$$KE_2 = \frac{1}{2}m_a V_{a2}^2 + \frac{1}{2}m_b V_{b2}^2 = \left(\frac{1}{2}\right)(2200)(13.89^2) + \left(\frac{1}{2}\right)(1000)(24.44^2) = 510\,900 \text{ J}$$

The conservation of energy requires that

$$E_1 = E_2 \qquad KE_1 + U_1 = KE_2 + U_2$$

Thus,

$$U_2 - U_1 = KE_1 - KE_2 = 687\,500 - 510\,900$$
$$= 176\,600 \text{ J or } 176.6 \text{ kJ}$$

Solved Problems

1.1 Identify which of the following are extensive properties and which are intensive properties: (*a*) a 10-m^3 volume, (*b*) 30 J of kinetic energy, (*c*) a pressure of 90 kPa, (*d*) a stress of 1000 kPa, (*e*) a mass of 75 kg, and (*f*) a velocity of 60 m/s. (*g*) Convert all extensive properties to intensive properties assuming $m = 75$ kg.

(*a*) Extensive. If the mass is doubled, the volume increases.

(*b*) Extensive. If the mass doubles, the kinetic energy increases.

(*c*) Intensive. Pressure is independent of mass.

(*d*) Intensive. Stress is independent of mass.

(*e*) Extensive. If the mass doubles, the mass doubles.

(*f*) Intensive. Velocity is independent of mass.

(*g*) $\frac{V}{m} = \frac{10}{75} = 0.1333\,\text{m}^3/\text{kg} \qquad \frac{E}{m} = \frac{30}{75} = 0.40\,\text{J/kg} \qquad \frac{m}{m} = \frac{75}{75} = 1.0\,\text{kg/kg}$

1.2 The gas in a cubical volume with sides at different temperatures is suddenly isolated with reference to transfer of mass and energy. Is this system in thermodynamic equilibrium? Why or why not?

It is not in thermodynamic equilibrium. If the sides of the container are at different temperatures, the temperature is not uniform over the entire volume, a requirement of thermodynamic equilibrium. After a period of time elapsed, the sides would all approach the same temperature and equilibrium would eventually be attained.

1.3 Express the following quantities in terms of base SI units (kg, m, and s): (*a*) power, (*b*) kinetic energy, and (*c*) specific weight.

(*a*) Power = (force)(velocity) = (N)(m/s) = $(\text{kg}\cdot\text{m/s}^2)(\text{m/s}) = \text{kg}\cdot\text{m}^2/\text{s}^3$

(*b*) Kinetic energy = mass × velocity^2 = $\text{kg}\cdot\left(\frac{\text{m}}{\text{s}}\right)^2 = \text{kg}\cdot\text{m}^2/\text{s}^2$

(*c*) Specific weight = weight/volume = $\text{N/m}^3 = \text{kg}\cdot\frac{\text{m}}{\text{s}^2}\Big/\text{m}^3 = \text{kg}/(\text{s}^2\cdot\text{m}^2)$

1.4 Determine the force necessary to accelerate a mass of 20 lbm at a rate of 60 ft/sec^2 vertically upward.

A free-body diagram of the mass (Fig. 1-11) is helpful. We will assume standard gravity. Newton's second law, $\sum F = ma$, then allows us to write

$$F - 20 = \left(\frac{20}{32.2}\right)(60)$$

$$\therefore F = 57.3\,\text{lbf}$$

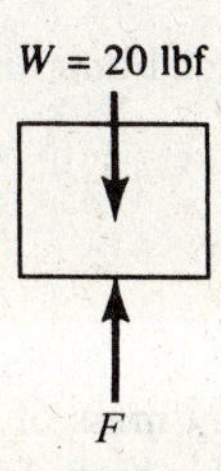

Fig. 1-11

1.5 A cubic meter of water at room temperature has a weight of 9800 N at a location where $g = 9.80$ m/s^2. What is its specific weight and its density at a location where $g = 9.77$ m/s^2?

The mass of the water is

$$m = \frac{W}{g} = \frac{9800}{9.80} = 1000\,\text{kg}$$

Its weight where $g = 9.77$ m/s^2 is $W = mg = (1000)(9.77) = 9770\,\text{N}$.

Specific weight:

$$w = \frac{W}{V} = \frac{9770}{1} = 9770\ \text{N/m}^3$$

Density:

$$\rho = \frac{m}{V} = \frac{1000}{1} = 1000\ \text{kg/m}^3$$

1.6 Assume the acceleration of gravity on a celestial body to be given as a function of altitude by the expression $g = 4 - 1.6 \times 10^{-6} h\ \text{m/s}^2$, where h is in meters above the surface of the planet. A space probe weighed 100 kN on earth at sea level. Determine (*a*) the mass of the probe, (*b*) its weight on the surface of the planet, and (*c*) its weight at an elevation of 200 km above the surface of the planet.

(*a*) The mass of the space probe is independent of elevation. At the surface of the earth we find its mass to be

$$m = \frac{W}{g} = \frac{100\,000}{9.81} = 10\,190\ \text{kg}$$

(*b*) The value of gravity on the planet's surface, with $h = 0$, is $g = 4\ \text{m/s}^2$. The weight is then

$$W = mg = (10\,190)(4) = 40\,760\ \text{N}$$

(*c*) At $h = 200\,000$ m, gravity is $g = 4 - (1.6 \times 10^{-6})(2 \times 10^5) = 3.68\ \text{m/s}^2$. The probe's weight at 200 km is

$$W = mg = (10\,190)(3.68) = 37\,500\ \text{N}$$

1.7 When a body is accelerated under water, some of the water is also accelerated. This makes the body appear to have a larger mass than it actually has. For a sphere at rest this added mass is equal to the mass of one half of the displaced water. Calculate the force necessary to accelerate a 10-kg, 300-mm-diameter sphere which is at rest under water at the rate of 10 m/s^2 in the horizontal direction. Use $\rho_{H_2O} = 1000\ \text{kg/m}^3$.

The added mass is one-half of the mass of the displaced water:

$$m_{\text{added}} = \frac{1}{2}\left(\frac{4}{3}\pi r^3 \rho_{H_2O}\right) = \left(\frac{1}{2}\right)\left(\frac{4}{3}\right)(\pi)\left(\frac{0.3}{2}\right)^3(1000) = 7.069\ \text{kg}$$

The apparent mass of the body is then $m_{\text{apparent}} = m + m_{\text{added}} = 10 + 7.069 = 17.069$ kg. The force needed to accelerate this body from rest is calculated to be

$$F = ma = (17.069)(10) = 170.7\ \text{N}$$

This is 70 percent greater than the force (100 N) needed to accelerate the body from rest in air.

1.8 The force of attraction between two masses m_1 and m_2 having dimensions that are small compared with their separation distance R is given by Newton's third law, $F = km_1m_2/R^2$, where $k = 6.67 \times 10^{-11}\ \text{N·m}^2/\text{kg}^2$. What is the total gravitational force which the sun (1.97×10^{30} kg) and the earth (5.95×10^{24} kg) exert on the moon (7.37×10^{22} kg) at an instant when the earth, moon, and sun form a 90° angle? The earth-moon and sun-moon distances are 380×10^3 and 150×10^6 km, respectively.

A free-body diagram (Fig. 1-12) is very helpful. The total force is the vector sum of the two forces. It is

$$F=\sqrt{F_e^2+F_s^2}=\left\{\left[\frac{(6.67\times 10^{-11})(7.37\times 10^{22})(5.95\times 10^{24})}{(380\times 10^6)^2}\right]^2\right.$$
$$\left.+\left[\frac{(6.67\times 10^{-11})(7.37\times 10^{22})(1.97\times 10^{30})}{(150\times 10^9)^2}\right]^2\right\}^{1/2}$$
$$=(4.10\times 10^{40}+18.5\times 10^{40})^{1/2}=4.75\times 10^{20}\ \text{N}$$

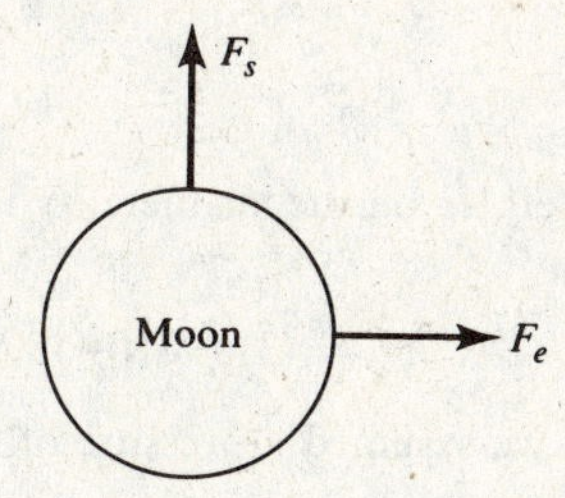

Fig. 1-12

1.9 Calculate the density, specific weight, mass, and weight of a body that occupies 200 ft^3 if its specific volume is 10 ft^3/lbm.

The quantities will not be calculated in the order asked for. The mass is

$$m=\frac{V}{v}=\frac{200}{10}=20\ \text{lbm}$$

The density is

$$\rho=\frac{1}{v}=\frac{1}{10}=0.1\ \text{lbm/ft}^3$$

The weight is, assuming $g=32.2$ ft/sec^2, $W=mg=(20)(32.2/32.2)=20$ lbf. Finally, the specific weight is calculated to be

$$w=\frac{W}{V}=\frac{20}{200}=0.1\ \text{lbf/ft}^3$$

Note that using English units, (*1.6*) would give

$$w=\rho g=\left(\frac{0.1\ \text{lbm/ft}^3}{32.2\ \text{lbm-ft/sec}^2\text{-lbf}}\right)(32.2\ \text{ft/sec}^2)=0.1\ \text{lbf/ft}^3$$

1.10 The pressure at a given point is 50 mmHg absolute. Express this pressure in kPa, kPa gage, and m of H_2O abs if $P_{atm}=80$ kPa. Use the fact that mercury is 13.6 times heavier than water.

The pressure in kPa is found, using (*1.11*), to be

$$P=wh=(9810)(13.6)(0.05)=6671\ \text{Pa or } 6.671\ \text{kPa}$$

The gage pressure is

$$P_{gage}=P_{abs}-P_{atm}=6.671-80=-73.3\ \text{kPa gage}$$

The negative gage pressure indicates that this is a vacuum. In meters of water we have

$$h=\frac{P}{w}=\frac{6671}{9810}=0.68\ \text{m of H}_2\text{O}$$

1.11 A manometer tube which contains mercury (Fig. 1-13) is used to measure the pressure P_A in the air pipe. Determine the gage pressure P_A. $w_{Hg} = 13.6\, w_{H_2O}$.

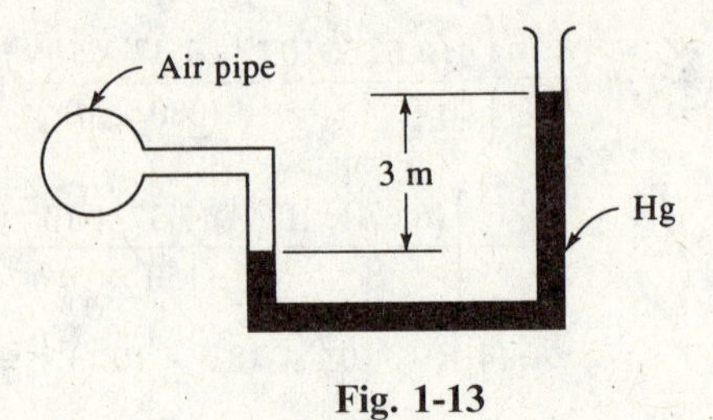

Fig. 1-13

Locate a point a on the left leg on the air-mercury interface and a point b at the same elevation on the right leg. We then have

$$P_a = P_b \qquad P_A = (3)[(9810)(13.6)] = 400\,200 \text{ Pa or } 400.2 \text{ kPa}$$

This is a gage pressure, since we assumed a pressure of zero at the top of the right leg.

1.12 A large chamber is separated into compartments 1 and 2, as shown in Fig. 1-14, which are kept at different pressures. Pressure gage A reads 300 kPa and pressure gage B reads 120 kPa. If the local barometer reads 720 mmHg, determine the absolute pressures existing in the compartments, and the reading of gage C.

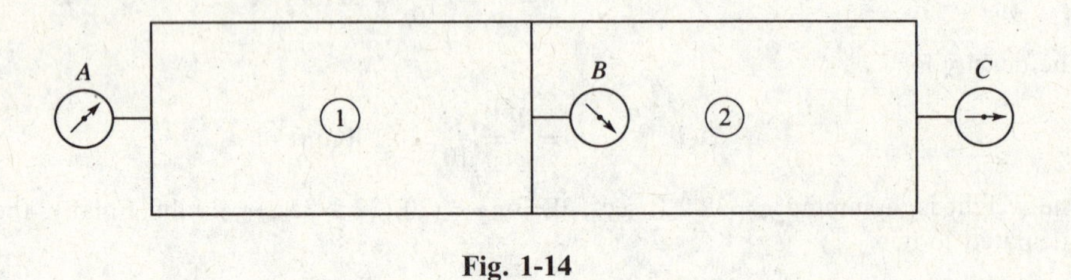

Fig. 1-14

The atmospheric pressure is found from the barometer to be

$$P_{\text{atm}} = (9810)(13.6)(0.720) = 96\,060 \text{ Pa or } 96.06 \text{ kPa}$$

The absolute pressure in compartment 1 is $P_1 = P_A + P_{\text{atm}} = 300 + 96.06 = 396.1$ kPa. If gage C read zero, gage B would read the same as gage A. If gage C read the same as gage A, gage B would read zero. Hence, our logic suggests that

$$P_B = P_A - P_C \quad \text{or} \quad P_C = P_A - P_B = 300 - 120 = 180 \text{ kPa}$$

The absolute pressure in compartment 2 is $P_2 = P_C + P_{\text{atm}} = 180 + 96.06 = 276.1$ kPa.

1.13 A tube can be inserted into the top of a pipe transporting liquids, providing the pressure is relatively low, so that the liquid fills the tube at height h. Determine the pressure in a water pipe if the water seeks a level at height $h = 6$ ft above the center of the pipe.

The pressure is found from (*1.11*) to be

$$P = wh = (62.4)(6) = 374 \text{ lbf/ft}^2 \text{ or } 2.60 \text{ psi gage}$$

1.14 A 10-kg body falls from rest, with negligible interaction with its surroundings (no friction). Determine its velocity after it falls 5 m.

Conservation of energy demands that the initial energy of the system equal the final energy of the system; that is,

$$E_1 = E_2 \qquad \frac{1}{2}m V_1^2 + mgh_1 = \frac{1}{2}m V_2^2 + mgh_2$$

The initial velocity V_1 is zero, and the elevation difference $h_1 - h_2 = 5$ m. Thus, we have

$$mg(h_1 - h_2) = \frac{1}{2}m V_2^2 \qquad \text{or} \qquad V_2 = \sqrt{2g(h_1 - h_2)} = \sqrt{(2)(9.81)(5)} = 9.90 \text{ m/s}$$

1.15 A 0.8-lbm object traveling at 200 ft/sec enters a viscous liquid and is essentially brought to rest before it strikes the side. What is the increase in internal energy, taking the object and the liquid as the system? Neglect the potential energy change.

Conservation of energy requires that the sum of the kinetic energy and internal energy remain constant since we are neglecting the potential energy change. This allows us to write

$$E_1 = E_2 \qquad \frac{1}{2}m V_1^2 + U_1 = \frac{1}{2}m V_2^2 + U_2$$

The final velocity V_2 is zero, so that the increase in internal energy $(U_2 - U_1)$ is given by

$$U_2 - U_1 = \frac{1}{2}m V_1^2 = \left(\frac{1}{2}\right)(0.8 \text{ lbm})(200^2 \text{ ft}^2/\text{sec}^2) = 16{,}000 \text{ lbm-ft}^2/\text{sec}^2$$

We can convert the above units to ft-lbf, the usual units on energy:

$$U_2 - U_1 = \frac{16{,}000 \text{ lbm-ft}^2/\text{sec}^2}{32.2 \text{ lbm-ft/sec}^2\text{-lbf}} = 497 \text{ ft-lbf}$$

Supplementary Problems

1.16 Draw a sketch of the following situations identifying the system or control volume, and the boundary of the system or the control surface. (*a*) The combustion gases in a cylinder during the power stroke, (*b*) the combustion gases in a cylinder during the exhaust stroke, (*c*) a balloon exhausting air, (*d*) an automobile tire being heated while driving, and (*e*) a pressure cooker during operation.

1.17 Which of the following processes can be approximated by a quasiequilibrium process? (*a*) The expansion of combustion gases in the cylinder of an automobile engine, (*b*) the rupturing of a membrane separating a high and low pressure region in a tube, and (*c*) the heating of the air in a room with a baseboard heater.

1.18 A supercooled liquid is a liquid which is cooled to a temperature below that at which it ordinarily solidifies. Is this system in thermodynamic equilibrium? Why or why not?

1.19 Convert the following to SI units: (*a*) 6 ft, (*b*) 4 in^3, (*c*) 2 slugs, (*d*) 40 ft-lbf, (*e*) 200 ft-lbf/sec, (*f*) 150 hp, (*g*) 10 ft^3/sec.

1.20 Determine the weight of a mass of 10 kg at a location where the acceleration of gravity is 9.77 m/s^2.

1.21 The weight of a 10-lb mass is measured at a location where g = 32.1 ft/sec^2 on a spring scale originally calibrated in a region where g = 32.3 ft/sec^2. What will be the reading?

1.22 The acceleration of gravity is given as a function of elevation above sea level by the relation $g = 9.81 - 3.32 \times 10^{-6}h$ m/s^2, with h measured in meters. What is the weight of an airplane at 10 km elevation when its weight at sea level is 40 kN?

1.23 Calculate the force necessary to accelerate a 20,000-lbm rocket vertically upward at the rate of 100 ft/sec^2. Assume $g = 32.2$ ft/sec^2.

1.24 Determine the deceleration of (*a*) a 2200-kg car and (*b*) a 1100-kg car, if the brakes are suddenly applied so that all four tires slide. The coefficient of friction $\eta = 0.6$ on the dry asphalt. ($\eta = F/N$ where N is the normal force and F is the frictional force.)

1.25 The mass which enters into Newton's third law of gravitation (Prob. 1.8) is the same as the mass defined by Newton's second law of motion. (*a*) Show that if g is the gravitational acceleration, then $g = km_e/R^2$, where m_e is the mass of the earth and R is the radius of the earth. (*b*) The radius of the earth is 6370 km. Calculate its mass if the acceleration of gravity is 9.81 m/s^2.

1.26 (*a*) A satellite is orbiting the earth at 500 km above the surface with only the attraction of the earth acting on it. Estimate the speed of the satellite. [*Hint:* The acceleration in the radial direction of a body moving with velocity V in a circular path of radius r is V^2/r; this must be equal to the gravitational acceleration (see Prob. 1.25).]

(*b*) The first earth satellite was reported to have circled the earth at 27 000 km/h and its maximum height above the earth's surface was given as 900 km. Assuming the orbit to be circular and taking the mean diameter of the earth to be 12 700 km, determine the gravitational acceleration at this height using (i) the force of attraction between two bodies, and (ii) the radial acceleration of a moving object.

1.27 Complete the following if $g = 9.81$ m/s^2 and $V = 10$ m^3.

	v (m^3/kg)	ρ (kg/m^3)	w (N/m^3)	m (kg)	W (N)
(*a*)	20				
(*b*)		2			
(*c*)			4		
(*d*)				100	
(*e*)					100

1.28 Complete the following if $P_{atm} = 100$ kPa ($\rho_{Hg} = 13.6\rho_{H_2O}$).

	kPa gage	kPa absolute	mmHg abs	mH_2O gage
(*a*)	5			
(*b*)		150		
(*c*)			30	
(*d*)				30

1.29 Determine the pressure difference between the water pipe and the oil pipe (Fig. 1-15).

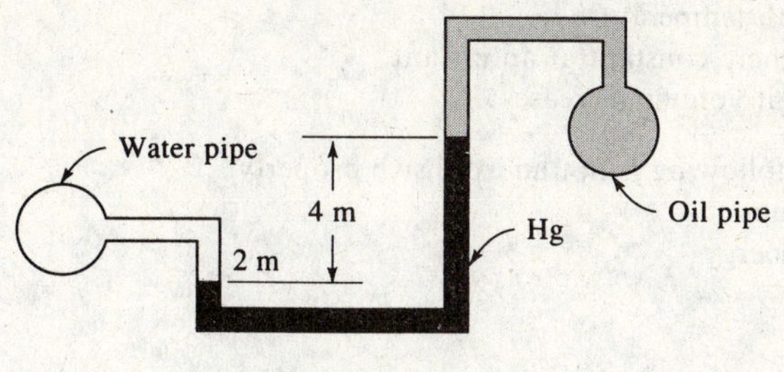

Fig. 1-15

1.30 A bell jar 250 mm in diameter sits on a flat plate and is evacuated until a vacuum of 700 mmHg exists. The local barometer reads 760 mm mercury. Find the absolute pressure inside the jar, and determine the force required to lift the jar off the plate. Neglect the weight of the jar.

1.31 A horizontal 2-m-diameter gate is located in the bottom of a water tank as shown in Fig. 1-16. Determine the force F required to just open the gate.

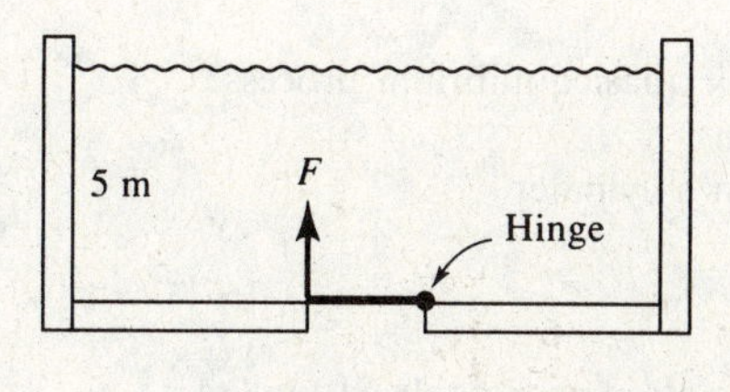

Fig. 1-16

1.32 A temperature of a body is measured to be 26 °C. Determine the temperature in °R, K, and °F.

1.33 The potential energy stored in a spring is given by $\frac{1}{2}Kx^2$, where K is the spring constant and x is the distance the spring is compressed. Two springs are designed to absorb the kinetic energy of a 2000-kg vehicle. Determine the spring constant necessary if the maximum compression is to be 100 mm for a vehicle speed of 10 m/s.

1.34 A 1500-kg vehicle traveling at 60 kph collides head-on with a 1000-kg vehicle traveling at 90 kph. If they come to rest immediately after impact, determine the increase in internal energy, taking both vehicles as the system.

1.35 Gravity is given by $g = 9.81 - 3.32 \times 10^{-6}\ h$ m/s^2, where h is the height above sea level. An airplane is traveling at 900 kph at an elevation of 10 km. If its weight at sea level is 40 kN, determine (*a*) its kinetic energy and (*b*) its potential energy relative to sea level.

Review Questions for the FE Examination

1.1FE Engineering thermodynamics does not include
(A) storage of energy
(B) utilization of energy
(C) transfer of energy
(D) transformation of energy

1.2FE In a quasiequilibrium process, the pressure in a system
(A) remains constant
(B) varies with temperature
(C) is everywhere constant at an instant
(D) increases if volume increases

1.3FE Which of the following is not an extensive property?
(A) Momentum
(B) Kinetic energy
(C) Density
(D) Mass

1.4FE Which of the following would be identified with a control volume?
(A) Compression of air in a cylinder
(B) Filling a tire with air at a service station
(C) Expansion of the gases in a cylinder after combustion
(D) The Goodyear blimp during flight

1.5FE Which of the following is not an intensive property?
(A) Velocity
(B) Pressure
(C) Temperature
(D) Volume

1.6FE Which of the following is a quasiequilibrium process?
(A) Mixing paint in a can
(B) Compression of air in a cylinder
(C) Combustion
(D) A balloon bursting

1.7FE Which of the following is not an acceptable SI unit?
(A) Distance measured in centimeters
(B) Pressure measured in newtons per square meter
(C) Density measured in grams per cubic centimeter
(D) Volume measured in cubic centimeters

1.8FE The unit of joule can be converted to which of the following?
(A) $Pa{\cdot}m^2$
(B) $N{\cdot}kg$
(C) Pa/m^2
(D) $Pa{\cdot}m^3$

1.9FE The standard atmosphere in meters of ammonia ($\rho = 600\ kg/m^3$) is:
(A) 17 m
(B) 19 m
(C) 23 m
(D) 31 m

1.10FE Convert 200 kPa gage to absolute millimeters of mercury ($\rho_{Hg} = 13.6\rho_{water}$).
(A) 1500 mm
(B) 1750 mm
(C) 2050 mm
(D) 2250 mm

1.11FE A gage pressure of 400 kPa acting on a 4-cm-diameter piston is resisted by a spring with a spring constant of 800 N/m. How much is the spring compressed? Neglect the piston weight.
(A) 630 cm
(B) 950 cm
(C) 1320 cm
(D) 1980 cm

1.12FE Calculate the pressure in the 200-mm-diameter cylinder shown in Fig. 1-17. The spring is compressed 40 cm.

(A) 138 kPa
(B) 125 kPa
(C) 110 kPa
(D) 76 kPa

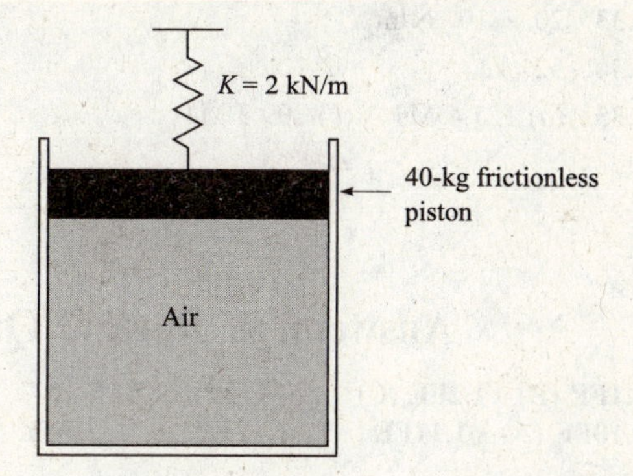

Fig. 1-17

1.13FE What pressure force exists on an 80-cm-diameter horizontal area at a depth below the surface of a lake?
(A) 150 kN
(B) 840 kN
(C) 1480 kN
(D) 5910 kN

1.14FE In thermodynamics our attention is focused on which form of energy?
(A) Kinetic energy
(B) Internal energy
(C) Potential energy
(D) Total energy

Answers to Supplementary Problems

1.16 (*a*) system (*b*) control volume (*c*) control volume (*d*) system (*e*) control volume

1.17 (*a*) can (*b*) cannot (*c*) cannot

1.18 no

1.19 (*a*) 1.829 m (*b*) 65.56 cm^3 (*c*) 29.18 kg (*d*) 54.24 N·m (*e*) 2712 W (*f*) 111.9 kW (*g*) 0.2832 m^3/s

1.20 97.7 N

1.21 9.91 lbf

1.22 39.9 kN

1.23 82,100 lbf

1.24 (*a*) 5.886 m/s^2 (*b*) 5.886 m/s^2

1.25 (*b*) 5.968×10^{24} kg

1.26 (*a*) 8210 m/s
(*b*) (i) 7.55 m/s^2 (ii) 7.76 m/s^2

1.27 (*a*) 0.05, 0.4905, 0.5, 4.905 (*b*) 0.5, 19.62, 20, 196.2 (*c*) 2.452, 0.4077, 4.077, 40 (*d*) 0.1, 10, 98.1, 981
(*e*) 0.981, 1.019, 10, 10.19

1.28 (*a*) 105, 787, 0.5097 (*b*) 50, 1124, 5.097 (*c*) −96, 4, −9.786 (*d*) 294.3, 394.3, 2955

1.29 514 kPa

1.30 8005 Pa, 4584 N

1.31 77.0 kN

1.32 538.8 °R, 299 K, 78.8 °F

1.33 20×10^6 N/m

1.34 521 kJ

1.35 (*a*) 127.4 MJ (*b*) 399.3 MJ

Answers to Review Questions for the FE Examination

1.1FE (B) **1.2FE** (C) **1.3FE** (C) **1.4FE** (B) **1.5FE** (D) **1.6FE** (B) **1.7FE** (C) **1.8FE** (D) **1.9FE** (A) **1.10FE** (A) **1.11FE** (A) **1.12FE** (A) **1.13FE** (C) **1.14FE** (B)

Properties of Pure Substances

2.1 INTRODUCTION

In this chapter the relationships between pressure, specific volume, and temperature will be presented for a pure substance. A pure substance is homogeneous. It may exist in more than one phase, but each phase must have the same chemical composition. Water is a pure substance. The various combinations of its three phases have the same chemical composition. Air is not a pure substance, and liquid air and air vapor have different chemical compositions. In addition, only a *simple compressible substance* will be considered, that is, a substance that is essentially free of magnetic, electrical, or surface tension effects. We will find the pure, simple, compressible substance of much use in our study of thermodynamics. In a later chapter we will include some real effects that cause substances to deviate from the ideal state presented in this chapter.

2.2 THE *P-v-T* SURFACE

It is well known that a substance can exist in three different phases: solid, liquid, and gas. Consider an experiment in which a solid is contained in a piston-cylinder arrangement such that the pressure is maintained at a constant value; heat is added to the cylinder, causing the substance to pass through all the different phases. Our experiment is shown at various stages in Fig. 2-1. We will record the temperature and specific volume during the experiment. Start with the solid at some low temperature; then add heat until it just begins to melt. Additional heat will completely melt the solid, with the temperature remaining constant. After all the solid is melted, the temperature of the liquid again rises until vapor just begins to form; this state is called the *saturated liquid* state. Again, during the phase change from liquid to vapor, often called *boiling*, the temperature remains constant as heat is added. Finally, all the liquid is vaporized and the state of *saturated vapor* exists, after which the temperature again rises with heat addition. This experiment is shown graphically in Fig. 2-2*a*. Note that the specific volume of the solid and liquid are much less than the specific volume of vapor. The scale is exaggerated in this figure so that the differences are apparent.

If the experiment is repeated a number of times using different pressures, a *T-v* diagram results, shown in Fig. 2-2*b*. At pressures that exceed the pressure of the *critical point*, there is no longer a distinction between liquid and vapor; the substance is referred to as a supercritical fluid. Property values of the critical point for various substances are included in Table B-3.

The data obtained in an actual experiment could be presented as a three-dimensional surface with $P = P(v, T)$. Figure 2-3 shows a qualitative rendering of a substance that contracts on freezing. For a substance that expands on freezing, the solid-liquid surface would be at a smaller specific volume than for the solid surface. The regions where only one phase exists are labeled solid, liquid, and vapor. Where two phases exist simultaneously the regions are labeled solid-liquid (S-L), solid-vapor (S-V), and liquid-vapor (L-V). Along the triple line, a line of constant temperature and pressure, all three phases coexist.

The P-v-T surface may be projected unto the P-v plane, the T-v plane, and the P-T plane, thus obtaining the P-v, T-v, and P-T diagrams shown in Fig. 2-4. Again, distortions are made so that the various regions are displayed. Note that when the triple line of Fig. 2-3 is viewed parallel to the v axis it appears to be a point, hence the name *triple point*. A constant pressure line is shown on the T-v diagram and a constant temperature line on the P-v diagram.

Primary practical interest is in situations involving the liquid, liquid-vapor, and vapor regions. A *saturated vapor* lies on the saturated vapor line and a *saturated liquid* on the saturated liquid line. The region to the right of the saturated vapor line is the *superheated vapor region*; the region to the left of the saturated liquid line is the *compressed liquid region* (also called the *subcooled liquid region*). A *supercritical state* is encountered when the pressure and temperature are greater than the critical values.

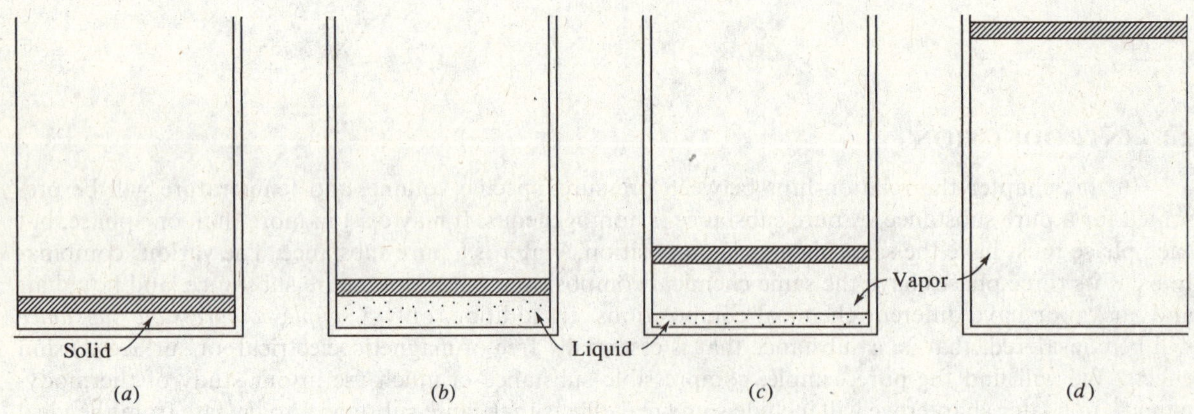

Fig. 2-1 The solid, liquid, and vapor phases.

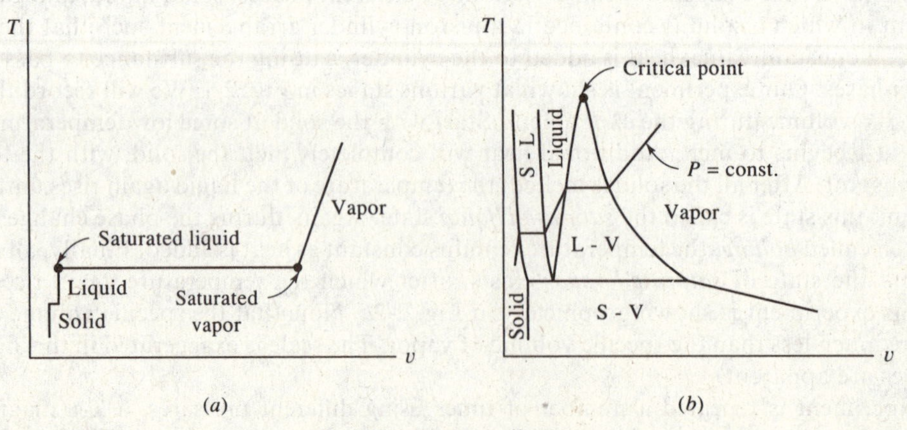

Fig. 2-2 The T-v diagram.

2.3 THE LIQUID-VAPOR REGION

At any state (T, v) between saturated points 1 and 2, shown in Fig. 2-5, liquid and vapor exist as a mixture in equilibrium. Let v_f and v_g represent, respectively, the specific volumes of the saturated liquid and the saturated vapor. Let m be the total mass of a system (such as shown in Fig. 2-1), m_f the amount of mass in the liquid phase, and m_g the amount of mass in the vapor phase. Then for a state of the system represented by (T, v) the total volume of the mixture is the sum of the volume occupied by the liquid and that occupied by the vapor, or

$$mv = m_f v_f + m_g v_g \tag{2.1}$$

The ratio of the mass of saturated vapor to the total mass is called the *quality* of the mixture, designated by the symbol x; it is

$$x = \frac{m_g}{m} \tag{2.2}$$

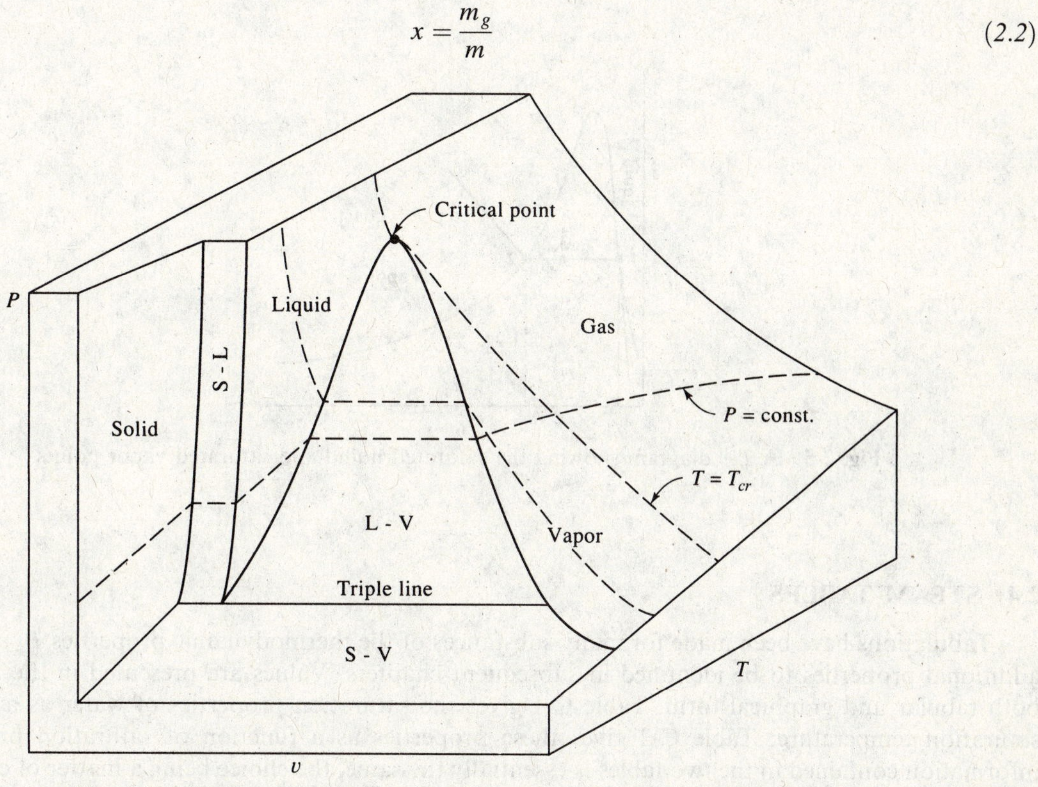

Fig. 2-3 The *P-v-T* rendering of a substance that contracts on freezing.

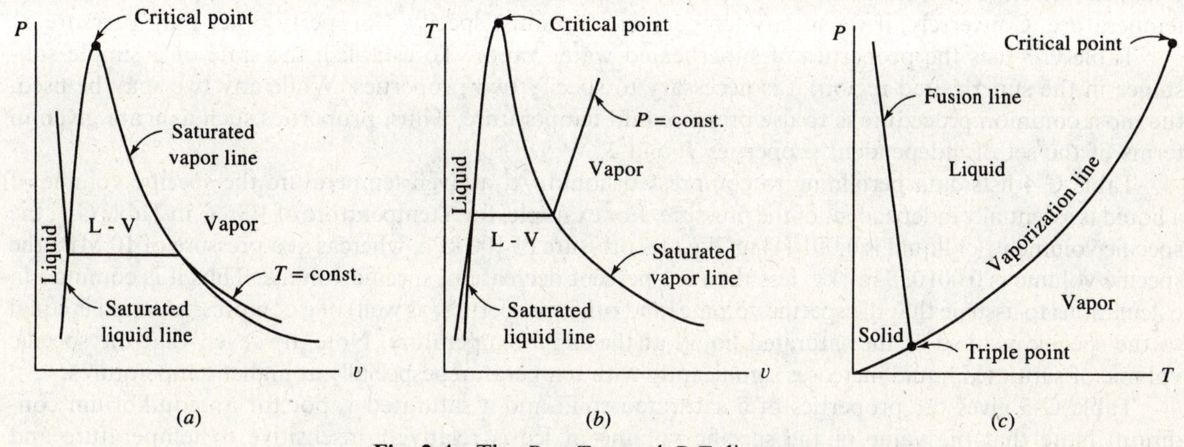

Fig. 2-4 The *P-v*, *T-v*, and *P-T* diagrams.

Recognizing that $m = m_f + m_g$, we may write (*2.1*), using our definition of quality, as

$$v = v_f + x(v_g - v_f) \tag{2.3}$$

Because the difference in saturated vapor and saturated liquid values frequently appears in calculations, we often let the subscript *fg* denote this difference; that is,

$$v_{fg} = v_g - v_f \tag{2.4}$$

Thus, (*2.3*) is

$$v = v_f + xv_{fg} \tag{2.5}$$

Note that the percentage liquid by mass in a mixture is $100(1-x)$ and the percentage vapor is $100x$.

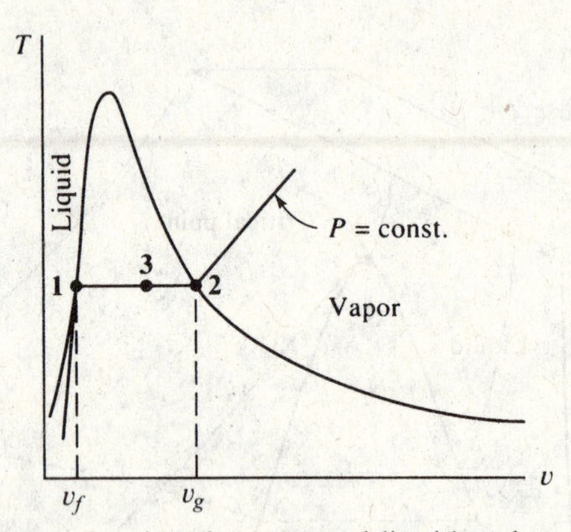

Fig. 2-5 A T-v diagram showing the saturated liquid and saturated vapor points.

2.4 STEAM TABLES

Tabulations have been made for many substances of the thermodynamic properties P, v, and T and additional properties to be identified in subsequent chapters. Values are presented in the appendix in both tabular and graphical form. Table C-1 gives the saturation properties of water as a function of saturation temperature; Table C-2 gives these properties as a function of saturation pressure. The information contained in the two tables is essentially the same, the choice being a matter of convenience. We should note, however, that in the mixture region pressure and temperature are dependent. Thus to establish the state of a mixture, if we specify the pressure, we need to specify a property other than temperature. Conversely, if we specify temperature, we must specify a property other than pressure.

Table C-3 lists the properties of superheated water vapor. To establish the state of a simple substance in the superheated region, it is necessary to specify two properties. While any two may be used, the most common procedure is to use pressure and temperature. Thus, properties such as v are given in terms of the set of independent properties P and T.

Table C-4 lists data pertaining to compressed liquid. At a given temperature the specific volume of a liquid is essentially independent of the pressure. For example, for a temperature of 100 °C in Table C-1, the specific volume v_f of liquid is 0.001044 m^3/kg at a pressure of 100 kPa, whereas at a pressure of 10 MPa the specific volume is 0.001038 m^3/kg, less than a 1 percent decrease in specific volume. Thus it is common in calculations to assume that the specific volume (and other properties, as well) of a compressed liquid is equal to the specific volume of the saturated liquid at the same temperature. Note, however, that the specific volume of saturated liquid increases significantly with temperature, especially at higher temperatures.

Table C-5 gives the properties of a saturated solid and a saturated vapor for an equilibrium condition. Note that the value of the specific volume of ice is relatively insensitive to temperature and pressure for the saturated-solid line. Also, it has a greater value (almost 10 percent greater) than the minimum value on the saturated-liquid line.

EXAMPLE 2.1 Determine the volume change when 1 kg of saturated water is completely vaporized at a pressure of (*a*) 1 kPa, (*b*) 100 kPa, and (*c*) 10 000 kPa.

Solution: Table C-2 provides the necessary values. The quantity being sought is $v_{fg} = v_g - v_f$. Note that P is given in MPa.

(*a*) 1 kPa. Thus, $v_{fg} = 129.2 - 0.001 = 129.2\,\text{m}^3/\text{kg}$.

(*b*) 100 kPa = 0.1 MPa. Again, $v = 1.694 - 0.001 = 1.693\ \text{m}^3/\text{kg}$.

(*c*) 10000 kPa = 10 MPa. Finally, $v_{fg} = 0.01803 - 0.00145 = 0.01658\ \text{m}^3/\text{kg}$.

Notice the large change in specific volume at low pressures compared with the small change as the critical point is approached. This underscores the distortion of the *P-v* diagram in Fig. 2-4.

EXAMPLE 2.2 Four kg of water is placed in an enclosed volume of 1 m^3. Heat is added until the temperature is 150 °C. Find (*a*) the pressure, (*b*) the mass of vapor, and (*c*) the volume of the vapor.

Solution: Table C-1 is used. The volume of 4 kg of saturated vapor at 150 °C is (0.3928)(4) = 1.5712 m^3. Since the given volume is less than this, we assume the state to be in the quality region.

(*a*) In the quality region the pressure is given as $P = 475.8$ kPa.

(*b*) To find the mass of the vapor we must determine the quality. It is found from (*2.3*), using $v = 1/4$ m^3/kg, as

$$0.25 = 0.00109 + x(0.3928 - 0.00109)$$

Thus, $x = 0.2489/0.3917 = 0.6354$. Using (*2.2*), the vapor mass is

$$m_g = mx = (4)(0.6354) = 2.542\,\text{kg}$$

(*c*) Finally, the volume of the vapor is found from

$$V_g = v_g m_g = (0.3928)(2.542) = 0.9985\ \text{m}^3$$

Note that in mixtures where the quality is not very close to zero the vapor phase occupies most of the volume. In this example, with a quality of 63.54 percent it occupies 99.85 percent of the volume.

EXAMPLE 2.3 Four kg of water is heated at a pressure of 220 kPa to produce a mixture with quality $x = 0.8$. Determine the final volume occupied by the mixture.

Solution: Use Table C-2. To determine the appropriate numbers at 220 kPa we linearly interpolate between 0.2 and 0.3 MPa. This provides, at 220 kPa,

$$v_g = \left(\frac{220 - 200}{300 - 200}\right)(0.6058 - 0.8857) + 0.8857 = 0.8297\ \text{m}^3/\text{kg} \qquad v_f = 0.0011\ \text{m}^3/\text{kg}$$

Note that no interpolation is necessary for v_f, since for both pressures v_f is the same to four decimal places. Using (*2.3*), we now find

$$v = v_f + x(v_g - v_f) = 0.0011 + (0.8)(0.8297 - 0.0011) = 0.6640\ \text{m}^3/\text{kg}$$

The total volume occupied by 4 kg is $V = mv = (4\ \text{kg})(0.6640\ \text{m}^3/\text{kg}) = 2.656\ \text{m}^3$.

EXAMPLE 2.4 Two lb of water is contained in a constant-pressure container held at 540 psia. Heat is added until the temperature reaches 700 °F. Determine the final volume of the container.

Solution: Use Table C-3E. Since 540 psia lies between the table entry values, the specific volume is simply

$$v = 1.3040 + (0.4)(1.0727 - 1.3040) = 1.2115 \text{ ft}^3/\text{lbm}$$

The final volume is then $V = mv = (2)(1.2115) = 2.423 \text{ ft}^3$.

2.5 THE IDEAL-GAS EQUATION OF STATE

When the vapor of a substance has relatively low density, the pressure, specific volume, and temperature are related by the specific equation

$$Pv = RT \tag{2.6}$$

where R is a constant for a particular gas and is called the *gas constant*. This *ideal-gas equation* is an *equation of state* in that it relates the state properties P, v, and T; any gas for which this equation is valid is called an *ideal gas* or a *perfect gas*. Note that when using the ideal-gas equation the pressure and temperature must be expressed as absolute quantities.

The gas constant R is related to a *universal gas constant* R_u, which has the same value for all gases, by the relationship

$$R = \frac{R_u}{M} \tag{2.7}$$

where M is the *molar mass*, values of which are tabulated in Tables B-2 and B-3. The *mole* is that quantity of a substance (i.e., that number of atoms or molecules) having a mass which, measured in grams, is numerically equal to the atomic or molecular weight of the substance. In the SI it is convenient to use instead the kilomole (kmol), which amounts to x kilograms of a substance of molecular weight x. For instance, 1 kmol of carbon is a mass of 12 kg (exactly); 1 kmol of molecular oxygen is 32 kg (very nearly). Stated otherwise, $M = 12$ kg/kmol for C, and $M = 32$ kg/kmol for O_2. In the English system one uses the pound-mole (lbmol); for O_2, $M = 32$ lbm/lbmol.

The value of R_u is

$$R_u = 8.314 \text{ kJ/(kmol·K)} = 1545 \text{ ft-lbf/(lbmol-°R)} = 1.986 \text{ Btu/(lbmol-°R)} \tag{2.8}$$

For air M is 28.97 kg/kmol (28.97 lbm/lbmol), so that for air R is 0.287 kJ/kg·K (53.3 ft-lbf/lbm-°R), a value used extensively in calculations involving air.

Other forms of the ideal-gas equation are

$$PV = mRT \qquad P = \rho RT \qquad PV = NR_uT \tag{2.9}$$

where N is the number of moles.

Care must be taken in using this simple convenient equation of state. A low-density ρ can be experienced by either having a low pressure or a high temperature. For air the ideal-gas equation is surprisingly accurate for a wide range of temperatures and pressures; less than 1 percent error is encountered for pressures as high as 3000 kPa at room temperature, or for temperatures as low as −130 °C at atmospheric pressure.

The *compressibility factor* Z helps us in determining whether or not the ideal-gas equation should be used. It is defined as

$$Z = \frac{Pv}{RT} \tag{2.10}$$

and is displayed in Fig. 2-6 for nitrogen. Since air is composed mainly of nitrogen, this figure is acceptable for air also. If $Z = 1$, or very nearly 1, the ideal-gas equation can be used. If Z is not

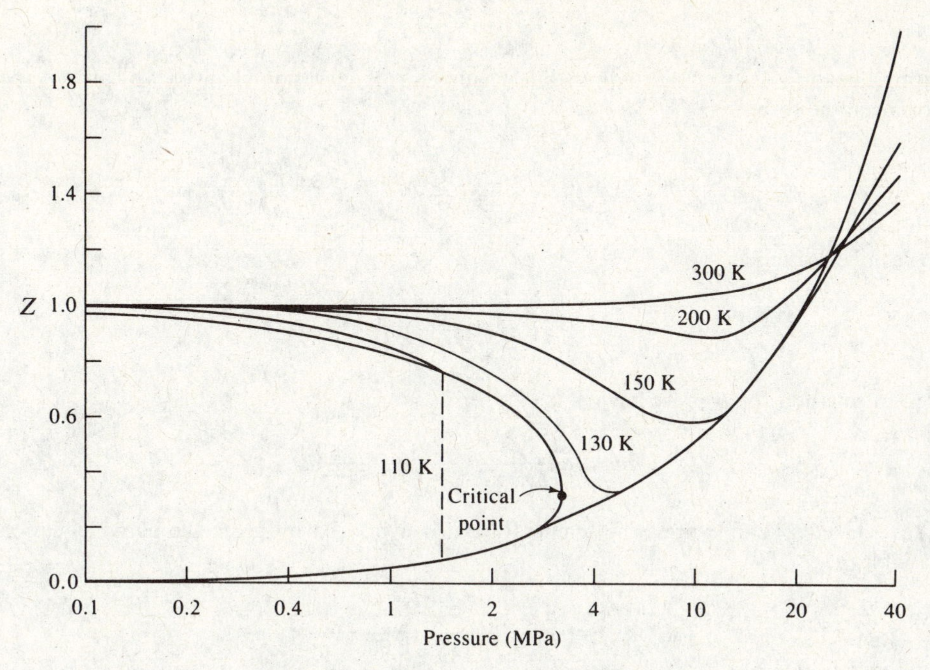

Fig. 2-6 The compressibility factor.

approximately 1, then (*2.10*) may be used. Additional real-gas effects (deviations from ideal-gas behavior) are considered in a subsequent chapter.

The compressibility factor can be determined for any gas by using a generalized compressibility chart presented in Fig. G-1 in the appendix. In the generalized chart the *reduced pressure* P_R and *reduced temperature* T_R must be used. They are calculated from

$$P_R = \frac{P}{P_{cr}} \qquad T_R = \frac{T}{T_{cr}} \qquad v_R = \frac{v}{RT_{cr}/P_{cr}} \tag{2.11}$$

where P_{cr} and T_{cr} are critical-point pressure and temperature, respectively, of Table B-3, and v_R is the pseudo-reduced volume.

EXAMPLE 2.5 An automobile tire with a volume of 0.6 m^3 is inflated to a gage pressure of 200 kPa. Calculate the mass of air in the tire if the temperature is 20 °C.

Solution: Air is assumed to be an ideal gas at the conditions of this example. In the ideal-gas equation, $PV = mRT$, we use absolute pressure and absolute temperature. Thus, using $P_{atm} = 100$ kPa,

$$P = 200 + 100 = 300 \text{ kPa} \qquad \text{and} \qquad T = 20 + 273 = 293 \text{ K}$$

The mass is then calculated to be

$$m = \frac{PV}{RT} = \frac{(300\,000 \text{ N/m}^2)(0.6 \text{ m}^3)}{(287 \text{ N·m/kg·K})(293 \text{ K})} = 2.14 \text{ kg}$$

The units in the above equation should be checked.

EXAMPLE 2.6 The temperature in the atmosphere near the surface of the earth (up to an elevation of 10 000 m) can be approximated by $T(z) = 15 - 0.00651z$ °C. Determine the pressure at an elevation of 3000 m if at $z = 0$, $P = 101$ kPa.

Solution: Equation (*1.8*) relates the pressure change to the elevation change. We can put the ideal-gas equation for air in the form

$$P = 287\rho T = \left(\frac{287}{9.81}\right)(wT) = 29.3\,wT$$

Hence, (*1.8*) can be written as

$$dP = -\frac{P}{29.3T}dz$$

Using the given equation for $T(z)$ we have

$$dP = -\frac{P}{(29.3)(288 - 0.00651z)}dz$$

where we have added 273 to express the temperature in kelvins. To integrate the above equation we must separate variables as

$$\frac{dP}{P} = -\frac{dz}{(29.3)(288 - 0.00651z)}$$

Now integrate between the appropriate limits:

$$\int_{101}^{P}\frac{dP}{P} = -\int_{0}^{3000}\frac{dz}{(29.3)(288 - 0.00651z)} = \left(\frac{1}{29.3}\right)\left(\frac{1}{0.00651}\right)\int_{0}^{3000}\frac{-0.00651dz}{288 - 0.00651z}$$

$$\ln\frac{P}{101} = [5.24\ \ln(288 - 0.00651z)]_{0}^{3000} = -0.368$$

There results $P = (101)(e^{-0.368}) = 69.9$ kPa.

2.6 EQUATIONS OF STATE FOR A NONIDEAL GAS

There are many equations of state that have been recommended for use to account for nonideal-gas behavior. Such behavior occurs where the pressure is relatively high (> 4 MPa for many gases) or when the temperature is near the saturation temperature. There are no acceptable criteria that can be used to determine if the ideal-gas equation can be used or if the nonideal-gas equations of this section must be used. Usually a problem is stated in such a way that it is obvious that nonideal-gas effects must be included; otherwise a problem is solved assuming an ideal gas.

The *van der Waals equation of state* is intended to account for the volume occupied by the gas-molecules and for the attractive forces between molecules. It is

$$P = \frac{RT}{v - b} - \frac{a}{v^2} \qquad (2.12)$$

where the constants a and b are related to the critical-point data of Table B-3 by

$$a = \frac{27R^2T_{cr}^2}{64P_{cr}} \qquad b = \frac{RT_{cr}}{8P_{cr}} \qquad (2.13)$$

These constants are also presented in Table B-8 to simplify calculations.

An improved equation is the Redlich–Kwong equation of state:

$$P = \frac{RT}{v-b} - \frac{a}{v(v+b)\sqrt{T}} \tag{2.14}$$

where the constants are given by

$$a = 0.4275\frac{R^2 T_{cr}^{2.5}}{P_{cr}} \qquad b = 0.0867\frac{RT_{cr}}{P_{cr}} \tag{2.15}$$

and are included in Table B-8.

A *virial equation of state* presents the product Pv as a series expansion. The most common expansion is

$$P = \frac{RT}{v} + \frac{B(T)}{v^2} + \frac{C(T)}{v^3} + \cdots \tag{2.16}$$

where interest is focused on $B(T)$ since it represents the first-order correction to the ideal-gas law. The functions $B(T)$, $C(T)$, etc., must be specified for the particular gas.

EXAMPLE 2.7 Calculate the pressure of steam at a temperature of 500 °C and a density of 24 kg/m³ using (*a*) the ideal-gas equation, (*b*) the van der Waals equation, (*c*) the Redlich–Kwong equation, (*d*) the compressibility factor, and (*e*) the steam table.

Solution:

(*a*) Using the ideal-gas equation, $P = \rho RT = (24)(0.462)(773) = 8570$ kPa, where the gas constant for steam is found in Table B-2.

(*b*) Using values for a and b from Table B-8, the van der Waals equation provides

$$P = \frac{RT}{v-b} - \frac{a}{v^2} = \frac{(0.462)(773)}{\frac{1}{24} - 0.00169} - \frac{1.703}{\left(\frac{1}{24}\right)^2} = 7950\text{ kPa}$$

(*c*) Using values for a and b from Table B-8, the Redlich–Kwong equation gives

$$P = \frac{RT}{v-b} - \frac{a}{v(v+b)\sqrt{T}} = \frac{(0.462)(773)}{\frac{1}{24} - 0.00117} - \frac{43.9}{\left(\frac{1}{24}\right)\left(\frac{1}{24} + 0.00117\right)\sqrt{773}} = 7930\text{ kPa}$$

(*d*) The compressibility factor is found from the generalized compressibility chart of Fig. G-1 in the appendix. To use the chart we must know the reduced temperature and pressure:

$$T_R = \frac{T}{T_{cr}} = \frac{773}{647.4} = 1.19 \qquad P_R = \frac{P}{P_{cr}} = \frac{8000}{22\,100} = 0.362$$

where we have used the anticipated pressure from parts (*a*), (*b*), and (*c*). Using the compressibility chart (it is fairly insensitive to the precise values of T_R and P_R, so estimates of these values are quite acceptable) and (*2.10*), we find

$$P = \frac{ZRT}{v} = \frac{(0.93)(0.462)(773)}{1/24} = 7970\text{ kPa}$$

(*e*) The steam table provides the most precise value for the pressure. Using $T = 500$ °C and $v = 1/24 = 0.0417$ m³/kg, we find $P = 8000$ kPa. Note that the ideal-gas law has an error of 7.1 percent, and the error of each of the other three equations is less than 1 percent.

Solved Problems

2.1 For a specific volume of 0.2 m^3/kg, find the quality of steam if the absolute pressure is (*a*) 40 kPa and (*b*) 630 kPa. What is the temperature of each case?

(*a*) Using information from Table C-2 in (*2.3*), we calculate the quality as follows:

$$v = v_f + x(v_g - v_f) \qquad 0.2 = 0.001 + x(3.993 - 0.001) \qquad \therefore x = 0.04985$$

The temperature is found in Table C-2 next to the pressure entry: $T = 75.9\,°\text{C}$.

(*b*) We must interpolate to find the correct values in Table C-2. Using the values at 0.6 and 0.8 MPa we have

$$v_g = \left(\frac{0.03}{0.2}\right)(0.2404 - 0.3157) + 0.3157 = 0.3044 \qquad v_f = 0.0011$$

Using (*2.3*), we have

$$0.2 = 0.0011 + x(0.3044 - 0.0011) \qquad \therefore x = 0.6558$$

The temperature is interpolated to be

$$T = \left(\frac{0.03}{0.2}\right)(170.4 - 158.9) + 158.9 = 160.6\,°\text{C}$$

2.2 Calculate the specific volume of water at (*a*) 160 °C and (*b*) 221 °C if the quality is 85 percent.

(*a*) Using the entries from Table C-1 and (*2.3*) we find

$$v = v_f + x(v_g - v_f) = 0.0011 + (0.85)(0.3071 - 0.0011) = 0.2612$$

(*b*) We must interpolate to find the values for v_g and v_f. Using entries at 220 °C and 230 °C, we determine

$$v_g = \left(\frac{1}{10}\right)(0.07159 - 0.08620) + 0.08620 = 0.08474 \qquad v_f = 0.00120$$

Using (*2.3*) $v = 0.00120 + (0.85)(0.08474 - 0.00120) = 0.07221\ \text{m}^3/\text{kg}$.

2.3 Ten pounds of steam is contained in a volume of 50 ft^3. Find the quality and the pressure if the temperature is 263 °F.

The temperature is not a direct entry in Table C-1E. We interpolate between temperatures of 260 °F and 270 °F to find

$$v_g = \left(\frac{3}{10}\right)(10.066 - 11.768) + 11.768 = 11.257 \qquad v_f = 0.017$$

From the given information we calculate

$$v = \frac{V}{m} = \frac{50}{10} = 5.0\ \text{ft}^3/\text{lbm}$$

The quality is found from (*2.3*) as follows:

$$5 = 0.017 + x(11.257 - 0.017) \qquad \therefore x = 0.4433$$

The pressure is interpolated to be

$$P = \left(\frac{3}{10}\right)(41.85 - 35.42) + 35.42 = 37.35\ \text{psia}$$

2.4 Saturated water occupies a volume of 1.2 m^3. Heat is added until it is completely vaporized. If the pressure is held constant at 600 kPa, calculate the final volume.

The mass is found, using v_f from Table C-2, to be

$$m = \frac{V}{v_f} = \frac{1.2}{0.0011} = 1091 \text{ kg}$$

When completely vaporized, the specific volume will be v_g, so that

$$V = mv_g = (1091)(0.3157) = 344.4 \text{ m}^3$$

2.5 Water is contained in a rigid vessel of 5 m^3 at a quality of 0.8 and a pressure of 2 MPa. If the pressure is reduced to 400 kPa by cooling the vessel, find the final mass of vapor m_g and mass of liquid m_f.

The initial specific volume is found, using data from Table C-2, to be

$$v = v_f + x(v_g - v_f) = 0.00118 + (0.8)(0.09963 - 0.00118) = 0.07994 \text{ m}^3/\text{kg}$$

Since the vessel is rigid, the specific volume does not change. Hence the specific volume at a pressure of 400 kPa is also 0.07994. We can then find the quality as follows:

$$0.07994 = 0.0011 + x(0.4625 - 0.0011) \qquad \therefore x = 0.1709$$

The total mass of water is

$$m = \frac{V}{v} = \frac{5}{0.07994} = 62.55 \text{ kg}$$

Now (*2.2*) gives the mass of vapor: $m_g = xm = (0.1709)(62.55) = 10.69$ kg. The mass of liquid is then

$$m_f = m - m_g = 62.55 - 10.69 = 51.86 \text{ kg}$$

2.6 Water exists at the critical point in a rigid container. The container and water are cooled until a pressure of 10 psia is reached. Calculate the final quality.

The initial specific volume as found in Table C-2E at a pressure of 3203.6 psia is $v_1 = 0.05053$ ft^3/lbm. Since the container is rigid, the specific volume does not change. Hence, at $P_2 = 10$ psia we have

$$v_2 = 0.05053 = 0.01659 + x_2(38.42 - 0.01659) \qquad \therefore x_2 = 0.000884$$

This shows that the final state is very close to the saturated liquid line.

2.7 Two kilograms of R134a is contained in a piston-cylinder arrangement, as sketched in Fig. 2-1. The 20-mm-dia, 48-kg piston is allowed to rise freely until the temperature reaches 160 °C. Calculate the final volume.

The absolute pressure inside the cylinder results from the atmospheric pressure and the weight of the piston:

$$P = P_{\text{atm}} + \frac{W}{A} = 100\,000 + \frac{(48)(9.81)}{\pi(0.02)^2/4} = 1.60 \times 10^6 \text{ Pa} \quad \text{or } 1.6 \text{ MPa}$$

At this pressure and a temperature of 160 °C, the R134a is superheated. From Table D-3 the specific volume is $v = 0.0217$ m^3/kg. The volume is then

$$V = mv = (2)(0.0217) = 0.0434 \text{ m}^3$$

2.8 A mass of 0.01 kg of steam at a quality of 0.9 is contained in the cylinder shown in Fig. 2-7. The spring just touches the top of the piston. Heat is added until the spring is compressed 15.7 cm. Calculate the final temperature.

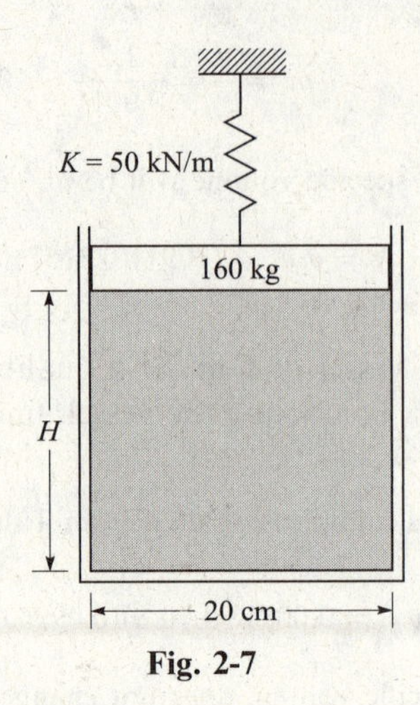

Fig. 2-7

The initial pressure in the cylinder is due to the atmospheric pressure and the weight of the piston:

$$P_1 = P_{\text{atm}} + \frac{W}{A} = 100\,000 + \frac{(160)(9.81)}{\pi(0.2)^2/4} = 150\,000 \text{ Pa} \qquad \text{or } 0.150 \text{ MPa}$$

The initial specific volume is found by interpolating in Table C-2:

$$v_1 = v_f + x(v_g v_f) = 0.0011 + (0.9)(1.164 - 0.0011) = 1.048 \text{ m}^3/\text{kg}$$

The initial volume contained in the cylinder is $V_1 = v_1 m = (1.048)(0.01) = 0.01048 \text{ m}^3$. The height H can now be calculated as follows:

$$V_1 = \frac{\pi d^2}{4} H \qquad 0.01048 = \frac{\pi(0.2)^2}{4} H \qquad \therefore H = 0.334 \text{ m}$$

The final volume is then

$$V_2 = \frac{\pi d^2}{4}(H + 0.157) = \frac{\pi(0.2)^2}{4}(0.334 + 0.157) = 0.01543 \text{ m}^3$$

The final specific volume is

$$v_2 = \frac{V_2}{m} = \frac{0.01543}{0.01} = 1.543 \text{ m}^3/\text{kg}$$

The final pressure is

$$P_2 = P_1 + \frac{Kx}{\pi d^2/4} = 150\,000 + \frac{(50\,000)(0.157)}{\pi(0.2)^2/4} = 400\,000 \text{ Pa} \qquad \text{or } 0.40 \text{ MPa}$$

This pressure and specific volume allow us to determine the temperature. It is obviously greater than the last table entry of 800 °C in the superheat table. We can extrapolate or use the ideal-gas law:

$$T_2 = \frac{P_2 v_2}{R} = \frac{(400)(1.543)}{0.4615} = 1337 \text{ K} \qquad \text{or } 1064\,^\circ\text{C}$$

2.9 Estimate the difference between the weight of air in a room that measures 20 × 100 × 10 ft in the summer when $T = 90\,^\circ\text{F}$ and the winter when $T = 10\,^\circ\text{F}$. Use $P = 14$ psia.

The masses of air in the summer and winter are

$$m_s = \frac{PV}{RT} = \frac{(14)(144)[(20)(100)(10)]}{(53.3)(90+460)} = 1375.4 \text{ lbm}$$

$$m_w = \frac{(14)(144)[(20)(100)(10)]}{(53.3)(10+460)} = 1609.5 \text{ lbm}$$

The difference in the two masses is $\Delta m = 1609.5 - 1375.4 = 234.1$ lbm. Assuming a standard gravity the weight and mass are numerically equal, so that $\Delta W = 234.1$ lbf.

2.10 A pressurized can contains air at a gage pressure of 40 psi when the temperature is 70 °F. The can will burst when the gage pressure reaches 200 psi. At what temperature will the can burst?

We will assume the volume to remain constant as the temperature increases. Using (*2.9*), we can solve for V and write

$$V = \frac{mRT_1}{P_1} = \frac{mRT_2}{P_2}$$

Since m and R are constant,

$$\frac{T_1}{P_1} = \frac{T_2}{P_2}$$

Using absolute values for the pressure and temperature, we find that

$$T_2 = T_1\frac{P_2}{P_1} = (70+460)\frac{(200+14.7)(144)}{(40+14.7)(144)} = 2080\,^\circ\text{R} = 1620\,^\circ\text{F}$$

Supplementary Problems

2.11 Using the steam tables C-1 and C-2 in the appendix, plot to scale the (*a*) P-v, (*b*) P-T, and (*c*) T-v diagrams. Choose either a linear-linear plot or a log-log plot. Note the distortions of the various figures in Sections 2.2 and 2.3. Such distortions are necessary if the various regions are to be displayed.

2.12 Calculate the specific volume for the following situations: (*a*) water at 200 °C, 80% quality; (*b*) R134a at −40 °C, 90% quality.

2.13 If the quality of each of the following substances is 80%, calculate the specific volume: (*a*) Water at 500 psia and (*b*) R134a at 80 psia.

2.14 Five kilograms of steam occupies a volume of 10 m^3. Find the quality and the pressure if the temperature is measured at (*a*) 40 °C and (*b*) 86 °C.

2.15 Determine the final volume of a mixture of water and steam if 3 kg of water is heated at a constant pressure until the quality is 60 percent. The pressure is (*a*) 25 kPa and (*b*) 270 kPa.

2.16 Two kilograms of saturated water at 125 kPa is completely vaporized. Calculate the volume: (*a*) before and (*b*) after.

2.17 The temperature of 10 lb of water is held constant at 205 °F. The pressure is reduced from a very high value until vaporization is complete. Determine the final volume of the steam.

2.18 A rigid vessel with a volume of 10 m^3 contains a water-vapor mixture at 400 kPa. If the quality is 60 percent, find the mass. The pressure is lowered to 300 kPa by cooling the vessel; find m_g and m_f.

2.19 Steam with a quality of 0.85 is contained in a rigid vessel at a pressure of 200 kPa. Heat is then added until the temperature reaches (*a*) 400 °C and (*b*) 140 °C. Determine the final pressures.

2.20 A rigid vessel contains water at 400 °F. Heat is to be added so that the water passes through the critical point. What should the quality be at the temperature of 440 °F?

2.21 R134a is contained in a sealed glass container at 100 °C. As it is cooled, vapor droplets are noted condensing on the sidewalls at 20 °C. Find the original pressure in the container.

2.22 Two kilograms of water is contained in a piston-cylinder arrangement by a 16 000-kg, 2-m-diameter, frictionless piston. See Fig. 2-1. Heat is added until the temperature reaches (*a*) 400 °C, (*b*) 650 °C, and (*c*) 140 °C. Calculate the final volume.

2.23 Two kilograms of steam at a quality of 0.80 is contained in the volume shown (Fig. 2-8). A spring is then brought in contact with the top of the piston and heat is added until the temperature reaches 500 °C. Determine the final pressure. (The force in the spring is Kx, where x is the displacement of the spring. This results in a trial-and-error solution.)

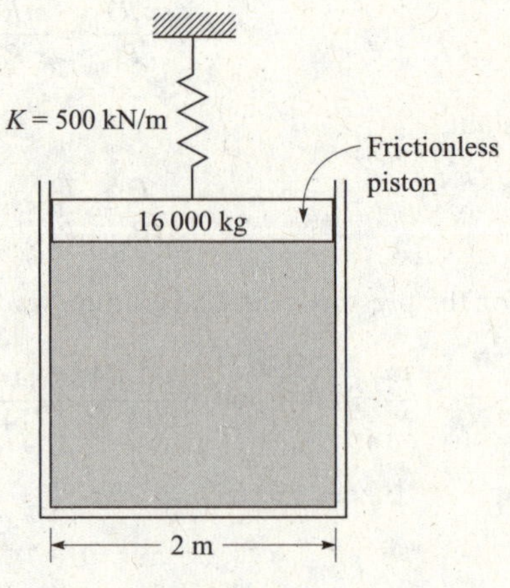

Fig. 2-8

2.24 Determine the volume occupied by 10 kg water at a pressure of 10 MPa and a temperature of (*a*) 5°C, (*b*) 200 °C, (*c*) 400 °C, (*d*) 800 °C, (*e*) 1500 °C, and (*f*) −10 °C.

2.25 For air at 100 psia and 60 °F calculate (*a*) the density, (*b*) the specific volume, (*c*) the specific weight if $g = 32.1$ ft/sec^2, and (*d*) the mass contained in 200 ft^3.

2.26 Provide the missing information for air at an elevation where $g = 9.82$ m/s^2.

	P (kPa)	T (°C)	v (m^3/kg)	ρ (kg/m^3)	w (N/m^3)
(*a*)	100	20			
(*b*)		100	2		
(*c*)	500		0.1		
(*d*)		400			20
(*e*)	200			2	

2.27 Assuming the atmosphere to be isothermal at an average temperature of $-20\,^\circ$C, determine the pressure at elevations of (*a*) 3000 m and (*b*) 10 000 m. Let $P = 101$ kPa at the earth's surface. Compare with measured values of 70.1 kPa and 26.5 kPa, respectively, by calculating the percent error.

2.28 (*a*) Assuming the temperature in the atmosphere to be given by $T = 15 - 0.00651z\,^\circ$C, determine the pressure at an elevation of 10 km. Let $P = 101$ kPa at sea level. (*b*) Compare the result of (*a*) with a measured value of 26.5 kPa by calculating the percent error.

2.29 The gage pressure reading on an automobile tire is 35 psi when the temperature is $0\,^\circ$F. The automobile is driven to a warmer climate where the temperature increases to $150\,^\circ$F on the hot asphalt. Estimate the increased pressure in the tire.

2.30 Nitrogen is contained in a 4-m^3 rigid vessel at a pressure of 4200 kPa. Determine the mass if the temperature is (*a*) $30\,^\circ$C and (*b*) $-120\,^\circ$C.

2.31 Estimate the pressure of nitrogen at a temperature of 220 K and a specific volume of 0.04 m^3/kg using (*a*) the ideal-gas equation, (*b*) the van der Waals equation, (*c*) the Redlich–Kwong equation, and (*d*) the compressibility factor.

2.32 Ten kilograms of $600\,^\circ$C steam is contained in a 182-liter tank. Find the pressure using (*a*) the ideal-gas equation, (*b*) the van der Waals equation, (*c*) the Redlich–Kwong equation, (*d*) the compressibility factor, and (*e*) the steam tables.

2.33 Steam at $300\,^\circ$C has a density of 7.0 kg/m^3. Find the pressure using (*a*) the ideal-gas equation, (*b*) the van der Waals equation, (*c*) the Redlich–Kwong equation, (*d*) the compressibility factor, and (*e*) the steam tables.

Review Questions for the FE Examination

2.1FE The phase change from a liquid to a vapor is referred to as:
(A) vaporization
(B) condensation
(C) sublimation
(D) melting

2.2FE The point that connects the saturated liquid line to the saturated vapor line is called the:
(A) triple point
(B) critical point
(C) superheated point
(D) compressed liquid point

2.3FE Find the volume occupied by 4 kg of $200\,^\circ$C steam.
(A) 0.04 m^3
(B) 0.104 m^3
(C) 0.410 m^3
(D) 4.10 m^3

2.4FE Estimate the volume occupied by 10 kg of water at $200\,^\circ$C and 2 MPa.
(A) 0.099 m^3
(B) 0.012 m^3
(C) 9.4 L
(D) 11.8 L

2.5FE Saturated steam is heated in a rigid tank from 50 °C to 250 °C. Estimate the final pressure.
(A) 20 kPa
(B) 27 kPa
(C) 33 kPa
(D) 44 kPa

2.6FE A 10-L tank holds 0.1 L of liquid water and 9.9 L of steam at 300 °C. What is the quality?
(A) 43%
(B) 61%
(C) 76%
(D) 86%

2.7FE Water at 50 percent quality is heated at constant volume from $P_1 = 100$ kPa to $T_2 = 200$ °C. What is P_2?
(A) 162 kPa
(B) 216 kPa
(C) 1286 kPa
(D) 485 kPa

2.8FE A vertical circular cylinder holds 1 cm of liquid water and 100 cm of vapor. If $P = 200$ kPa, the quality is nearest:
(A) 1.02%
(B) 10.7%
(C) 40.6%
(D) 80.3%

2.9FE Find the quality of steam at 120 °C if the vapor occupies 1200 L and the liquid occupies 2 L.
(A) 42%
(B) 28%
(C) 18%
(D) 2%

2.10FE Select the correct T-v diagram if steam at $v_1 = 0.005$ m^3/kg is heated to $v_2 = 0.5$ m^3/kg while maintaining $P = 500$ kPa. The dots are states 1 and 2 with 1 being on the left.

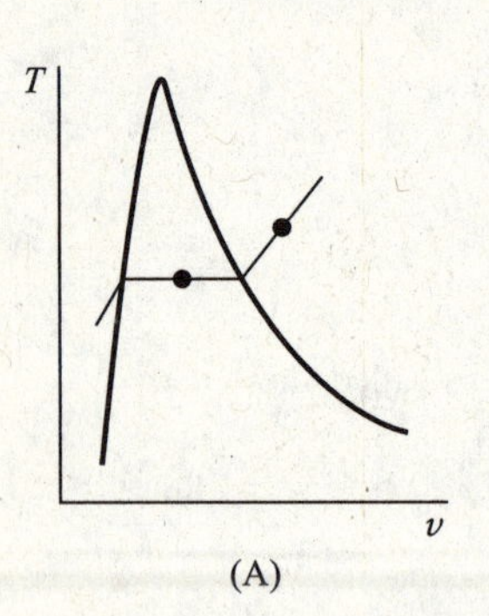

(A)

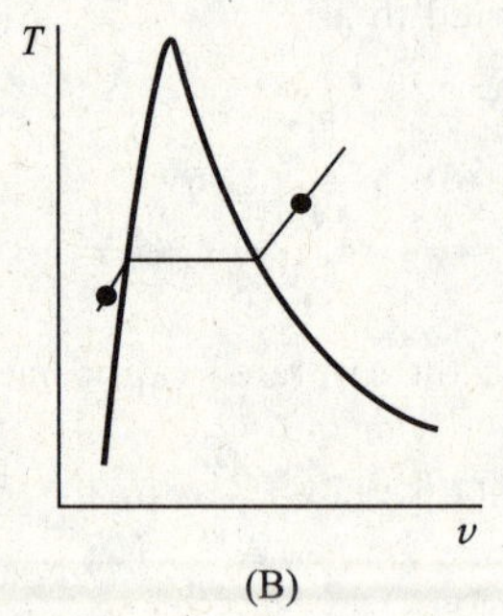

(B)

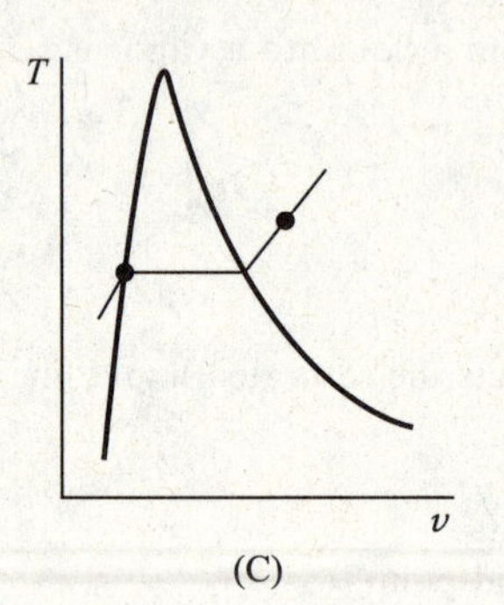

(C)

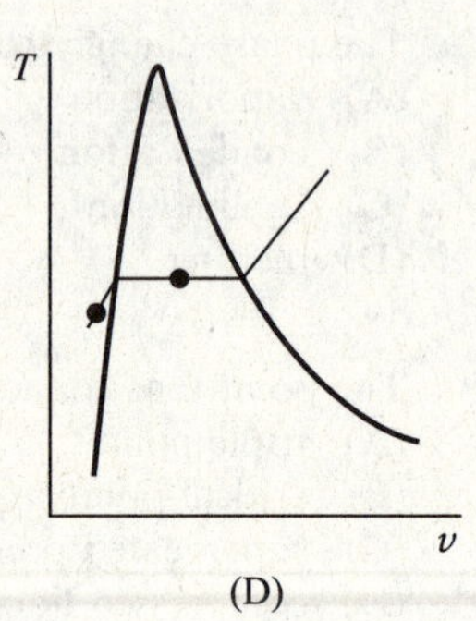

(D)

2.11FE Steam exists in a volume at 288 °C and 2 MPa. Estimate the specific volume.
(A) 0.1221 m^3/kg
(B) 0.1148 m^3/kg
(C) 0.1192 m^3/kg
(D) 0.1207 m^3/kg

2.12FE Calculate the mass of air in a tire with a volume of 0.2 m^3 if its gage pressure is 280 kPa at 25 °C.
(A) 7.8 kg
(B) 0.888 kg
(C) 0.732 kg
(D) 0.655 kg

2.13FE The mass of nitrogen at 150 K and 2 MPa contained in 0.2 m^3 is nearest:
(A) 10.2 kg
(B) 8.89 kg
(C) 6.23 kg
(D) 2.13 kg

2.14FE Estimate the temperature of 2 kg of air contained in a 40-L volume at 2 MPa.
(A) 120 K
(B) 140 K
(C) 160 K
(D) 180 K

2.15FE Find the mass of air in the volume shown in Fig. 2-9 if $T = 20\,°C$ and the frictionless piston has a mass of 75 kg.
(A) 0.0512 kg
(B) 0.0256 kg
(C) 0.0064 kg
(D) 0.0016 kg

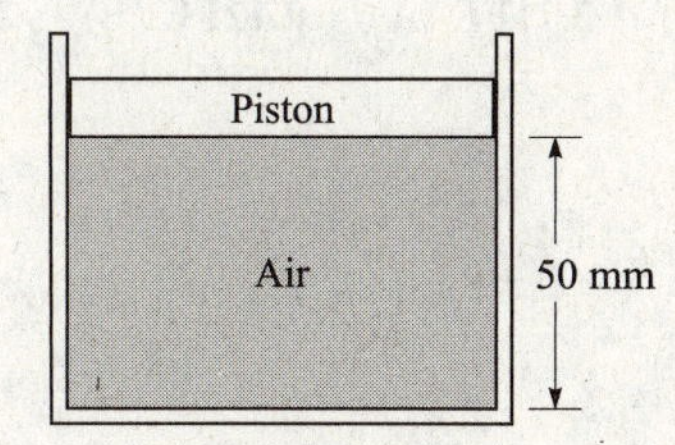

Fig. 2-9

2.16FE Estimate the difference in density between the inside and outside of a house in the winter when $P = 100$ kPa and $T_{inside} = 20\,°C$ and $T_{outside} = -20\,°C$. (This difference leads to air exchange between the inside and outside.)
(A) 0.188 kg/m^3
(B) 0.165 kg/m^3
(C) 0.151 kg/m^3
(D) 0.092 kg/m^3

Answers to Supplementary Problems

2.12 (*a*) 0.1022 m^3/kg (*b*) 0.3213 m^3/kg
2.13 (*a*) 0.7466 ft^3/lbm (*b*) 0.4776 ft^3/lbm
2.14 (*a*) 0.1024, 7.383 kPa (*b*) 0.7312, 60.3 kPa
2.15 (*a*) 11.6 m^3 (*b*) 1.24 m^3
2.16 (*a*) 0.002 m^3 (*b*) 2.76 m^3
2.17 307.2 ft^3
2.18 35.98 kg, 16.47 kg, 19.51 kg
2.19 (*a*) 415 kPa (*b*) 269 kPa
2.20 0.01728
2.21 790 kPa
2.22 (*a*) 4.134 m^3 (*b*) 5.678 m^3 (*c*) 2.506 m^3
2.23 220 kPa
2.24 (*a*) 0.00996 m^3 (*b*) 0.0115 m^3 (*c*) 0.2641 m^3 (*d*) 0.4859 m^3 (*e*) 0.8182 m^3 (*f*) 0.01089 m^3
2.25 (*a*) 0.5196 lbm/ft^3 (*b*) 1.925 ft^3/lbm (*c*) 0.518 lbf/ft^3 (*d*) 103.9 lbm
2.26 (*a*) 0.8409, 1.189, 11.68 (*b*) 53.53, 0.5, 4.91 (*c*) −98.8, 10, 98.2 (*d*) 393.4, 0.491, 2.037 (*e*) 75.4, 0.5, 19.64
2.27 (*a*) 67.3 kPa, −3.99% (*b*) 26.2 kPa, −1.13%
2.28 (*a*) 26.3 kPa (*b*) −0.74%
2.29 51.2 psig

2.30 (*a*) 186.7 kg (*b*) 500 kg

2.31 (*a*) 1630 kPa (*b*) 1580 kPa (*c*) 1590 kPa (*d*) 1600 kPa

2.32 (*a*) 22.2 MPa (*b*) 19.3 MPa (*c*) 19.5 MPa (*d*) 19.5 MPa (*e*) 20 MPa

2.33 (*a*) 1851 kPa (*b*) 1790 kPa (*c*) 1777 kPa (*d*) 1780 kPa (*e*) 1771 kPa

Answers to Review Questions for the FE Examination

2.1FE (A) **2.2FE** (B) **2.3FE** (C) **2.4FE** (C) **2.5FE** (A) **2.6FE** (D) **2.7FE** (C) **2.8FE** (B) **2.9FE** (A)
2.10FE (A) **2.11FE** (A) **2.12FE** (B) **2.13FE** (A) **2.14FE** (C) **2.15FE** (D) **2.16FE** (A)

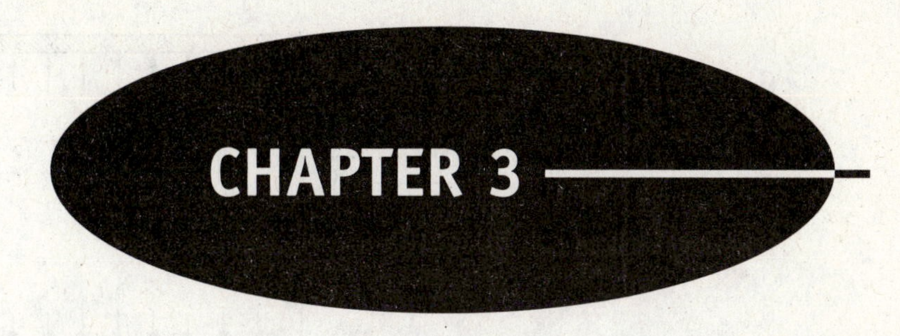

Work and Heat

3.1 INTRODUCTION

In this chapter we will discuss the two quantities that result from energy transfer across the boundary of a system: work and heat. This will lead into a presentation of the first law of thermodynamics. Work will be discussed in detail and will be calculated for several common situations. Heat, however, is a quantity that requires substantial analysis for its calculation. In most engineering programs the subject of heat transfer is covered in a separate course. In thermodynamics, heat is either a given quantity or it is calculated as an unknown in an algebraic equation.

3.2 DEFINITION OF WORK

The term *work* is so broad that we must be very particular in a technical definition. It must include, for example, the work done by expanding exhaust gases after combustion occurs in the cylinder of an automobile engine, as shown in Fig. 3-1. The energy released during the combustion process is transferred to the crankshaft by means of the connecting rod, in the form of work. Thus, in this example, work can be thought of as energy being transferred across the boundary of a system, the system being the gases in the cylinder.

Work, designated W, is often defined as the product of a force and the distance moved in the direction of the force. This is a mechanical definition of work. A more general definition of work is the thermodynamic definition: *Work*, an interaction between a system and its surroundings, is done by a system if the sole external effect on the surroundings could be the raising of a weight. The magnitude of the work is the product of the weight and the distance it could be lifted. Figure 3-2*b* shows that the interaction of Fig. 3-2*a* qualifies as work in the thermodynamic sense.

The convention chosen for positive work is that if the system performs work on the surroundings it is positive. A piston compressing a fluid is doing negative work, whereas a fluid expanding against a piston is doing positive work. The units of work are quickly observed from the units of a force multiplied by a distance: in the SI system, newton-meters (N·m) or joules (J); in the English system, ft-lbf.

The rate of doing work, designated W, is called *power*. In the SI system, power has units joules per second (J/s), or watts (W); in the English system, ft-lbf/sec. We will find occasion to use the unit of horsepower because of its widespread use in rating engines. To convert we simply use 1 hp = 0.746 kW = 550 ft-lbf/sec.

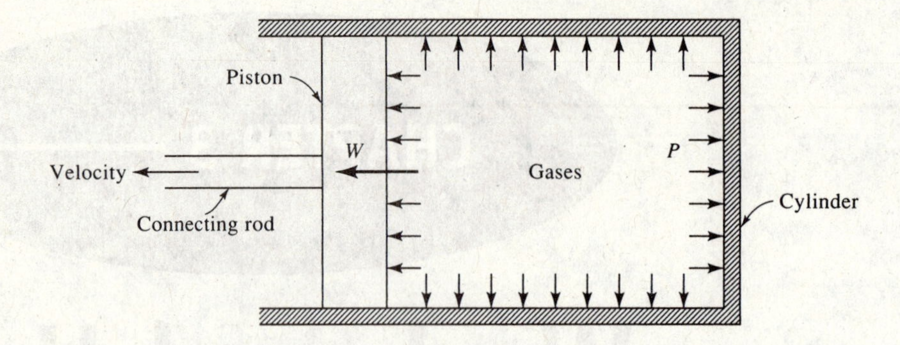

Fig. 3-1 Work being done by expanding gases in a cylinder.

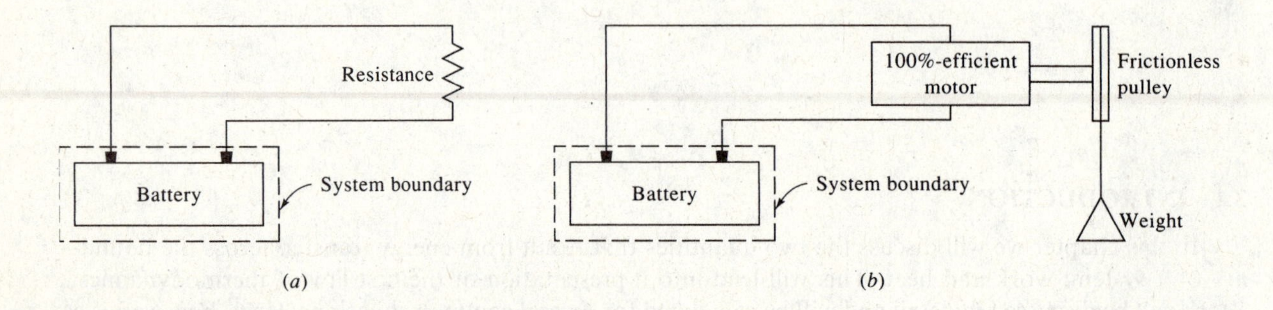

Fig. 3-2 Work being done by electrical means.

The work associated with a unit mass will be designated w (not to be confused with specific weight):

$$w = \frac{W}{m} \tag{3.1}$$

A final general comment concerning work relates to the choice of the system. Note that if the system in Fig. 3-2 included the entire battery-resistor setup in part (*a*), or the entire battery-motor-pulley-weight setup in part (*b*), no energy would cross the system boundary, with the result that no work would be done. The identification of the system is very important in determining work.

3.3 QUASIEQUILIBRIUM WORK DUE TO A MOVING BOUNDARY

There are a number of work modes that occur in various engineering situations. These include the work needed to stretch a wire, to rotate a shaft, to move against friction, to cause a current to flow through a resistor, and to charge a capacitor. Many of these work modes are covered in other courses. In this book we are primarily concerned with the work required to move a boundary against a pressure force.

Consider the piston-cylinder arrangement shown in Fig. 3-3. There is a seal to contain the gas in the cylinder, the pressure is uniform throughout the cylinder, and there are no gravity, magnetic, or electric effects. This assures us of a quasiequilibrium process, one in which the gas is assumed to pass through a series of equilibrium states. Now, allow an expansion of the gas to occur by moving the piston upward a small distance ds. The total force acting on the piston is the pressure times the area of the piston. This pressure is expressed as absolute pressure since pressure is a result of molecular activity; any molecular activity will yield a pressure which will result in work being done when the boundary moves. The infinitesimal work which the system (the gas) does on the surroundings (the piston) is then the force multiplied by the distance:

$$\delta W = PA\, ds \tag{3.2}$$

The symbol δW will be discussed shortly. The quantity $A ds$ is simply dV, the differential volume, allowing (*3.2*) to be written in the form

$$\delta W = P\, dV \tag{3.3}$$

As the piston moves from some position s_1 to another position s_2, the above expression can be integrated to give

$$W_{1-2} = \int_{V_1}^{V_2} P\, dV \tag{3.4}$$

where we assume the pressure is known for each position as the piston moves from volume V_1 to volume V_2. Typical pressure-volume diagrams are shown in Fig. 3-4. The work W_{1-2} is the crosshatched area under the *P-V* curve.

Consideration of the integration process highlights two very important features in (*3.4*). First, as we proceed from state 1 to state 2, the area representing the work is very dependent on the path that we follow. That is, states 1 and 2 in Fig. 3-4(*a*) and (*b*) are identical, yet the areas under the *P-V* curves are very different; in addition to being dependent on the end points, work depends on the actual path that connects the two end points. Thus, work is a path function, as contrasted to a point function, which is dependent only on the end points. The differential of a path function is called an *inexact differential*, whereas the differential of a point function is an *exact differential*. An inexact differential will be denoted with the symbol δ. The integral of δW is W_{1-2}, where the subscript emphasizes that the work is associated with the path as the process passes from state 1 to state 2; the subscript may be

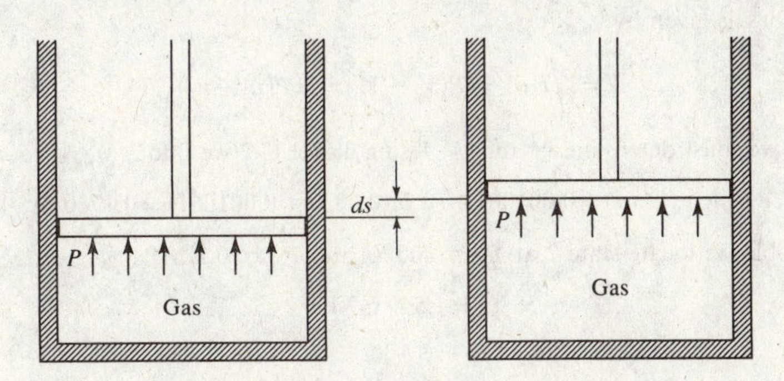

Fig. 3-3 Work due to a moving boundary.

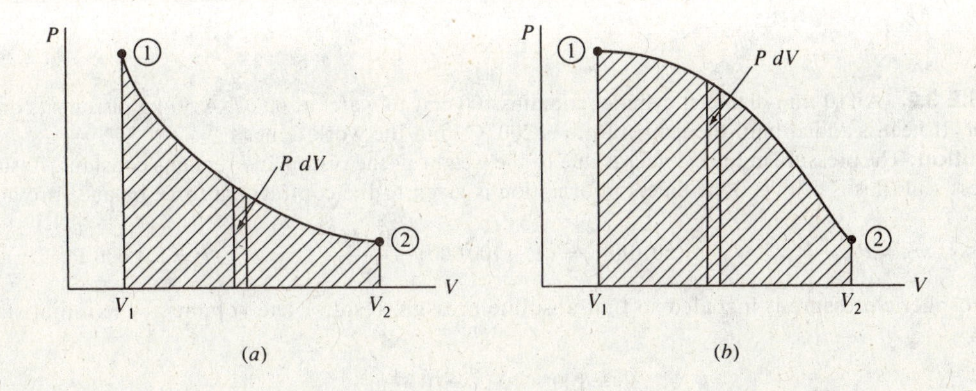

Fig. 3-4 Work depends on the path between two states.

omitted, however, and work done written simply as W. We would never write W_1 or W_2, since work is not associated with a state but with a process. Work is not a property. The integral of an exact differential, for example dT, would be

$$\int_{T_1}^{T_2} dT = T_2 - T_1 \tag{3.5}$$

where T_1 is the temperature at state 1 and T_2 is the temperature at state 2.

The second observation to be made from (*3.4*) is that the pressure is assumed to be constant throughout the volume at each intermediate position. The system passes through each equilibrium state shown in the *P-V* diagrams of Fig. 3-4. An equilibrium state can usually be assumed even though the variables may appear to be changing quite rapidly. Combustion is a very rapid process that cannot be modeled as a quasiequilibrium process. The other processes in the internal combustion engine (expansion, exhaust, intake, and compression) can be assumed to be quasiequilibrium processes; they occur at a slow rate, thermodynamically.

As a final comment regarding work we may now discuss what is meant by a simple system, as defined in Chapter 1. For a system free of surface, magnetic, and electrical effects the only work mode is that due to pressure acting on a moving boundary. For such simple systems only two independent variables are necessary to establish an equilibrium state of the system composed of a homogeneous substance. If other work modes are present, such as a work mode due to an electric field, then additional independent variables would be necessary, such as the electric field intensity.

EXAMPLE 3.1 One kg of steam with a quality of 20 percent is heated at a constant pressure of 200 kPa until the temperature reaches 400 °C. Calculate the work done by the steam.

Solution: The work is given by

$$W = \int P\,dV = P(V_2 - V_1) = mP(v_2 - v_1)$$

To evaluate the work we must determine v_1 and v_2. Using Table C-2 we find

$$v_1 = v_f + x(v_g - v_f) = 0.001061 + (0.2)(0.8857 - 0.001061) = 0.1780\ \text{m}^3/\text{kg}$$

From the superheat table we locate state 2 at $T_2 = 400\,°\text{C}$ and $P_2 = 0.2$ MPa:

$$v_2 = 1.549\ \text{m}^3/\text{kg}$$

The work is then

$$W = (1)(200)(1.549 - 0.1780) = 274.2\ \text{kJ}$$

Note: With the pressure having units of kPa, the result is in kJ.

EXAMPLE 3.2 A 110-mm-diameter cylinder contains 100 cm^3 of water at 60 °C. A 50-kg piston sits on top of the water. If heat is added until the temperature is 200 °C, find the work done.

Solution: The pressure in the cylinder is due to the weight of the piston and remains constant. Assuming a frictionless seal (this is always done unless information is given to the contrary), a force balance provides

$$mg = PA - P_{\text{atm}}A \qquad (50)(9.81) = (P - 100\,000)\frac{\pi(0.110)^2}{4} \qquad \therefore P = 151\,600\ \text{Pa}$$

The atmospheric pressure is included so that absolute pressure results. The volume at the initial state 1 is given as

$$V_1 = 100 \times 10^{-6} = 10^{-4}\ \text{m}^3$$

Using v_1 at 60 °C, the mass is calculated to be

$$m = \frac{V_1}{v_1} = \frac{10^{-4}}{0.001017} = 0.09833 \text{ kg}$$

At state 2 the temperature is 200 °C and the pressure is 0.15 MPa (this pressure is within 1 percent of the pressure of 0.1516 MPa, so it is acceptable). The volume is then

$$V_2 = mv_2 = (0.09833)(1.63) = 0.160 \text{ m}^3$$

Finally, the work is calculated to be

$$W = P(V_2 - V_1) = 151\,600(0.160 - 0.0001) = 24\,300 \text{ J} \qquad \text{or } 24.3 \text{ kJ}$$

EXAMPLE 3.3 Energy is added to a piston-cylinder arrangement, and the piston is withdrawn in such a way that the quantity PV remains constant. The initial pressure and volume are 200 kPa and 2 m^3, respectively. If the final pressure is 100 kPa, calculate the work done by the gas on the piston.

Solution: The work is found from (*3.4*) to be

$$W_{1-2} = \int_2^{V_2} P\,dV = \int_2^{V_2} \frac{C}{V} dV$$

where we have used $PV = C$. To calculate the work we must find C and V_2. The constant C is found from

$$C = P_1 V_1 = (200)(2) = 400 \text{ kJ}$$

To find V_2 we use $P_2 V_2 = P_1 V_1$, which is, of course, the equation that would result from an isothermal process (constant temperature) involving an ideal gas. This can be written as

$$V_2 = \frac{P_1 V_1}{P_2} = \frac{(200)(2)}{100} = 4 \text{ m}^3$$

Finally,

$$W_{1-2} = \int_2^4 \frac{400}{V} dV = 400 \ln \frac{4}{2} = 277 \text{ kJ}$$

This is positive, since work is done during the expansion process by the system (the gas contained in the cylinder).

EXAMPLE 3.4 Determine the horsepower required to overcome the wind drag on a modern car traveling 90 km/h if the drag coefficient C_D is 0.2. The drag force is given by $F_D = \frac{1}{2}\rho V^2 A C_D$, where A is the projected area of the car and V is the velocity. The density ρ of air is 1.23 kg/m^3. Use $A = 2.3$ m^2.

Solution: To find the drag force on a car we must express the velocity in m/s: $V = (90)(1000/3600) = 25$ m/s. The drag force is then

$$F_D = \frac{1}{2}\rho V^2 A C_D = \left(\frac{1}{2}\right)(1.23)(25^2)(2.3)(0.2) = 177 \text{ N}$$

To move this drag force at 25 m/s the engine must do work at the rate

$$W = F_D V = (177)(25) = 4425 \text{ W}$$

The horsepower is then

$$\text{Hp} = \frac{4425 \text{ W}}{746 \text{ W/hp}} = 5.93 \text{ hp}$$

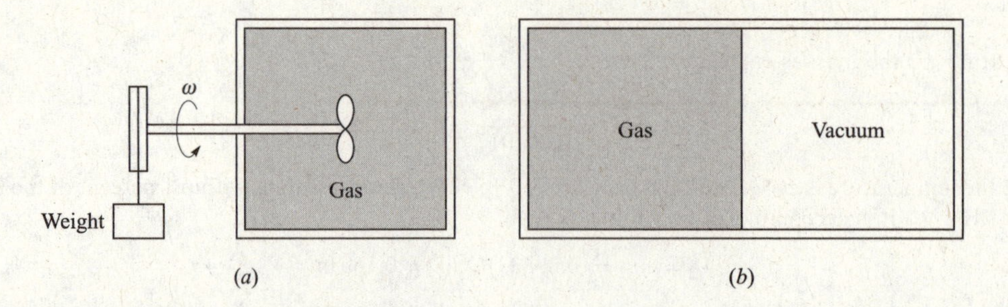

Fig. 3-5 Nonequilibrium work.

3.4 NONEQUILIBRIUM WORK

It must be emphasized that the area on a *P-V* diagram represents the work for a quasiequilibrium process only. For nonequilibrium processes the work cannot be calculated using $\int P\,dV$; either it must be given for the particular process or it must be determined by some other means. Two examples will be given. Consider a system to be formed by the gas in Fig. 3-5. In part (*a*) work is obviously crossing the boundary of the system by means of the rotating shaft; yet the volume does not change. We could calculate the work input by multiplying the weight by the distance it dropped, neglecting friction in the pulley system. This would not, however, be equal to $\int P\,dV$, which is zero. The paddle wheel provides us with a nonequilibrium work mode.

Suppose the membrane in Fig. 3-5*b* ruptures, allowing the gas to expand and fill the evacuated volume. There is no resistance to the expansion of the gas at the moving boundary as the gas fills the volume; hence, there is no work done. Yet there is a change in volume. The sudden expansion is a nonequilibrium process, and again we cannot use $\int P\,dV$ to calculate the work.

EXAMPLE 3.5 A 100-kg mass drops 3 m, resulting in an increased volume in the cylinder of 0.002 m^3 (Fig. 3-6). The weight and the piston maintain a constant gage pressure of 100 kPa. Determine the net work done by the gas on the surroundings. Neglect all friction.

Solution: The paddle wheel does work on the system, the gas, due to the 100-kg mass dropping 3 m. That work is negative and is

$$W = -(F)(d) = -(100)(9.81)(3) = -2940\ \text{J}$$

The work done by the system on this frictionless piston is positive since the system is doing the work. It is

$$W = (PA)(h) = P\Delta V = (200\,000)(0.002) = 400\ \text{J}$$

where absolute pressure has been used. The net work done is thus

$$W_{\text{net}} = -2940 + 400 = -2540\ \text{J}$$

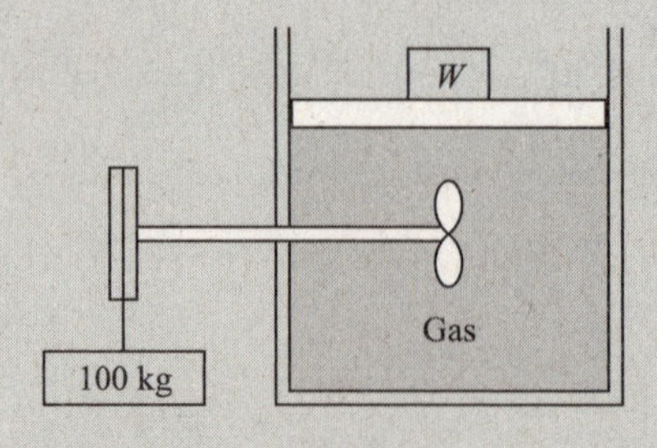

Fig. 3-6

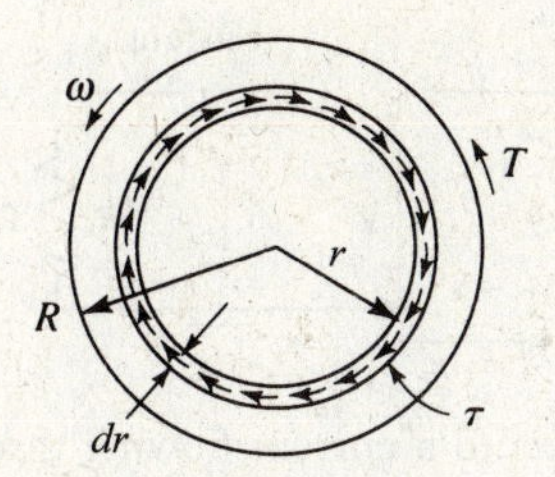

Fig. 3-7 Work due to a rotating shaft transmitting torque.

3.5 OTHER WORK MODES

Work transferred by a rotating shaft (Fig. 3-7) is a common occurrence in mechanical systems. The work results from the shearing forces due to the shearing stress τ, which varies with the radius over the cross-sectional area, moving with angular velocity ω as the shaft rotates. The shearing force is

$$dF = \tau\, dA = \tau(2\pi r\, dr) \tag{3.6}$$

The linear velocity with which this force moves is $r\omega$. Hence, the rate of doing work is

$$\dot{W} = \int_A r\omega\, dF = \int_0^R (r\omega)\tau(2\pi r)\, dr = 2\pi\omega \int_0^R \tau r^2\, dr \tag{3.7}$$

where R is the radius of the shaft. The torque T is found from the shearing stresses by integrating over the area:

$$T = \int_A r\, dF = 2\pi \int_0^R \tau r^2\, dr \tag{3.8}$$

Combining this with (*3.7*) above, the work rate is

$$\dot{W} = T\omega \tag{3.9}$$

To find the work transferred in a given time, we simply multiply (*3.9*) by the number of seconds:

$$W = T\omega\, \Delta t \tag{3.10}$$

Of course, the angular velocity must be expressed in rad/s.

The work necessary to stretch a linear spring (Fig. 3-8) with spring constant K from a length x_1 to x_2 can be found by using the relation

$$F = Kx \tag{3.11}$$

where x is the distance the spring is stretched from the unstretched position. Note that the force is dependent on the variable x. Hence, we must integrate the force over the distance the spring is stretched; this results in

$$W = \int_{x_1}^{x_2} F\, dx = \int_{x_1}^{x_2} Kx\, dx = \frac{1}{2}K(x_2^2 - x_1^2) \tag{3.12}$$

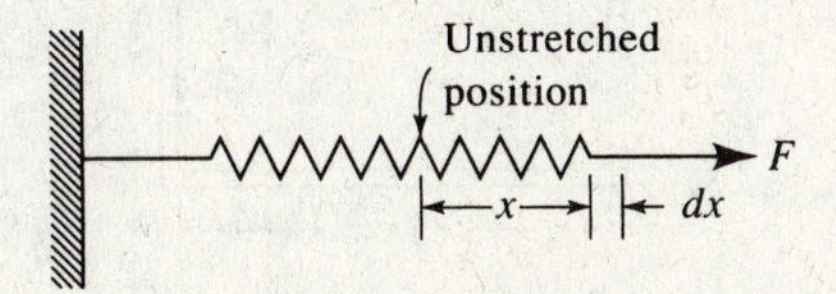

Fig. 3-8 Work needed to stretch a spring.

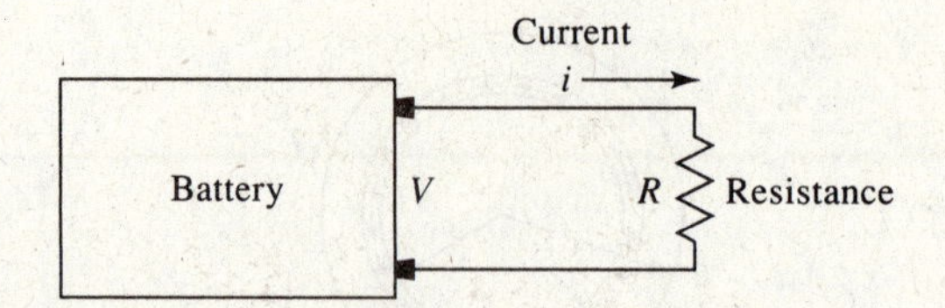

Fig. 3-9 Work due to a current flowing through a resistor.

As a final type let us discuss an electrical work mode, illustrated in Fig. 3-9. The potential difference V across the battery terminals is the "force" that drives the charge q through the resistor during the time increment Δt. The current i is related to the charge by

$$i = \frac{dq}{dt} \tag{3.13}$$

For a constant current the charge is

$$q = i\,\Delta t \tag{3.14}$$

The work, from this nonequilibrium work mode, is then given by the expression

$$W = Vi\,\Delta t \tag{3.15}$$

The power would be the rate of doing work, or

$$\dot{W} = Vi \tag{3.16}$$

This relationship is actually used to define the *electric potential*, the voltage V, since the ampere is a base unit and the watt has already been defined. One volt is one watt divided by one ampere.

EXAMPLE 3.6 The drive shaft in an automobile delivers 100 N·m of torque as it rotates at 3000 rpm. Calculate the horsepower delivered.

Solution: The power is found by using $\dot{W} = T\omega$. This requires ω to be expressed in rad/s:

$$\omega = (3000)(2\pi)\left(\frac{1}{60}\right) = 314.2 \text{ rad/s}$$

Hence

$$\dot{W} = T\omega = (100)(314.2) = 31\,420 \text{ W} \qquad \text{or Hp} = \frac{31\,420}{746} = 42.1 \text{ hp}$$

EXAMPLE 3.7 The air in a circular cylinder (Fig. 3-10) is heated until the spring is compressed 50 mm. Find the work done by the air on the frictionless piston. The spring is initially unstretched, as shown.

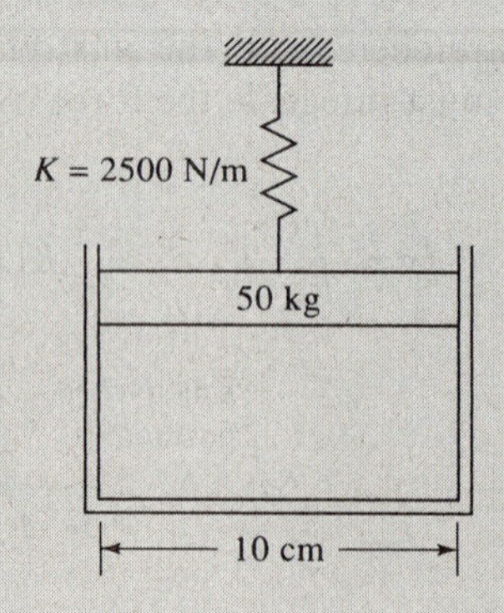

Fig. 3-10

Solution: The pressure in the cylinder is initially found from a force balance:

$$P_1 A_1 = P_{\text{atm}} A + W \qquad P_1 \frac{\pi(0.1)^2}{4} = (100\,000)\frac{\pi(0.1)^2}{4} + (50)(9.81) \qquad \therefore P_1 = 162\,500 \text{ Pa}$$

To raise the piston a distance of 50 mm, without the spring, the pressure would be constant and the work required would be force times distance:

$$W = PA \times d = (162\,500)\frac{\pi(0.1)^2}{4}(0.05) = 63.81 \text{ J}$$

Using (*3.12*), the work required to compress the spring is calculated to be

$$W = \frac{1}{2}K(x_2^2 - x_1^2) = \left(\frac{1}{2}\right)(2500)(0.05^2) = 3.125 \text{ J}$$

The total work is then found by summing the above two values: $W_{\text{total}} = 63.81 + 3.125 = 66.93$ J.

3.6 HEAT

In the preceding section we considered several work modes by which energy is transferred macroscopically to or from a system. Energy can also be transferred microscopically to or from a system by means of interactions between the molecules that form the surface of the system and those that form the surface of the surroundings. If the molecules of the system boundary are more active than those of the boundary of the surroundings, they will transfer energy from the system to the surroundings, with the faster molecules transferring energy to the slower molecules. On this microscopic scale the energy is transferred by a work mode: collisions between particles. A force occurs over an extremely short time span, with work transferring energy from the faster molecules to the slower ones. Our problem is that this microscopic transfer of energy is not observable macroscopically as any of the work modes; we must devise a macroscopic quantity to account for this microscopic transfer of energy.

We have noted that temperature is a property which increases with increased molecular activity. Thus it is not surprising that we can relate microscopic energy transfer to the macroscopic property temperature. This macroscopic transfer of energy that we cannot account for by any of the macroscopic work modes will be called heat. *Heat* is energy transferred across the boundary of a system due to a difference in temperature between the system and the surroundings of the system. A system does not contain heat, it contains energy, and heat is energy in transit. It is often referred to as *heat transfer*.

To illustrate, consider a hot block and a cold block of equal mass. The hot block contains more energy than the cold block due to its greater molecular activity, that is, its higher temperature. When the blocks are brought into contact with each other, energy flows from the hot block to the cold one by means of heat transfer. Eventually, the blocks will attain thermal equilibrium, with both blocks arriving at the same temperature. The heat transfer has ceased, the hot block has lost energy, and the cold block has gained energy.

Heat, like work, is something that *crosses a boundary*. Because a system does not *contain* heat, heat is not a property. Thus, its differential is inexact and is written as δQ, where Q is the heat transfer. For a particular process between state 1 and state 2 the heat transfer could be written as Q_{1-2}, but it will generally be denoted by Q. The *rate* of heat transfer will be denoted by $\dot{Q}$.

By convention, if heat is transferred *to* a system it is considered positive. If it is transferred *from* a system it is negative. This is opposite from the convention chosen for work; if a system performs work on the surroundings it is positive. Positive heat transfer adds energy to a system, whereas positive work subtracts energy from a system. A process in which there is zero heat transfer is called an *adiabatic process*. Such a process is approximated experimentally by insulating the system so that negligible heat is transferred.

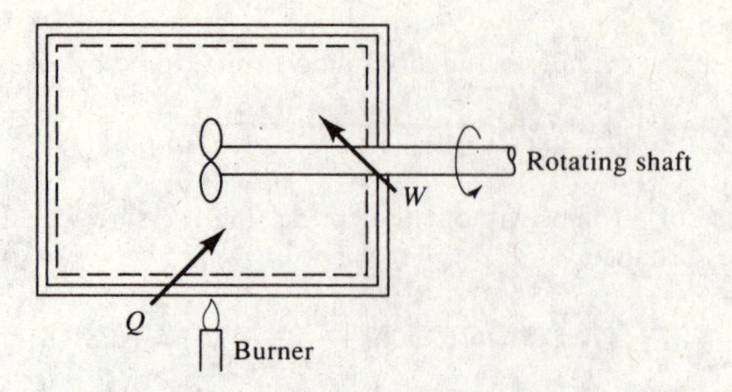

Fig. 3-11 Energy added to a system.

It should be noted that the energy contained in a system may be transferred to the surroundings either by work done by the system or by heat transferred from the system. Thus, heat and work are quantitatively equivalent and are expressed in the same units. An equivalent reduction in energy is accomplished if 100 J of heat is transferred from a system or if 100 J of work is performed by a system. In Fig. 3-11 the burner illustrates heat being added to the system and the rotating shaft illustrates work being done on the system.

It is sometimes convenient to refer to heat transfer per unit mass. Heat transfer per unit mass will be designated q and defined by

$$q = \frac{Q}{m} \tag{3.17}$$

EXAMPLE 3.8 A paddle wheel adds work to a rigid container by rotations caused by dropping a 50-kg weight a distance of 2 m from a pulley. How much heat must be transferred to result in an equivalent effect?

Solution: For this non-quasiequilibrium process the work is given by $W = (mg)(d) = (50)(9.8)(2) = 980$ J. The heat Q that must be transferred equals the work, 980 J.

There are three modes of heat transfer: conduction, convection, and radiation. Often, engineering designs must consider all three modes. *Conduction* heat transfer exists in a material due to the presence of temperature differences within the material. It can exist in all substances but is most often associated with solids. It is expressed mathematically by *Fourier's law of heat transfer*, which for a one-dimensional plane wall takes the form

$$\dot{Q} = kA\frac{\Delta T}{L} \tag{3.18}$$

where k is the *thermal conductivity* with units of W/m·K (Btu/sec-ft-°R), L is the thickness of the wall, ΔT is the temperature difference, and A is the wall area. Often, the heat transfer is related to the common R-factor, *resistivity*, given by $R_{\text{mat}} = L/k$.

Convection heat transfer exists when energy is transferred from a solid surface to a fluid in motion. It is a combination of energy transferred by both conduction and *advection* (energy transfer due to the bulk motion of the fluid); therefore, if there is no fluid motion, there is no convective heat trasfer. Convection is expressed in terms of the temperature difference between the bulk fluid temperature T_∞ and the temperature of the surface T_s. *Newton's law of cooling* expresses this as

$$\dot{Q} = h_c A(T_s - T_\infty) \tag{3.19}$$

where h_c is the *convective heat transfer coefficient*, with units of W/m^2·K (Btu/sec-ft^2-°R), and depends on the properties of the fluid (including its velocity) and the wall geometry. Free convection occurs due to the temperature difference only, whereas forced convection results from the fluid being forced, as with a fan.

Radiation is energy transferred as photons. It can be transferred through a perfect vacuum or through transparent substances such as air. It is calculated using the *Stefan–Boltzmann law* and accounts for the energy emitted and the energy absorbed from the surroundings:

$$\dot{Q} = \varepsilon \sigma A (T^4 - T_{\text{surr}}^4) \tag{3.20}$$

where σ is the *Stefan–Boltzmann constant* (equal to $5.67 \times 10^{-8}\ \text{W/m}^2\cdot\text{K}^4$), ε is the *emissivity* (a number between 0 and 1 where ε is 1 for a *blackbody*, a body that emits the maximum amount of radiation), and T_{surr} is the uniform temperature of the surroundings. The temperatures must be absolute temperatures.

EXAMPLE 3.9 A 10-m-long by 3-m-high wall is composed of an insulation layer with $R = 2\ \text{m}^2\cdot\text{K/W}$ and a wood layer with $R = 0.5\ \text{m}^2\cdot\text{K/W}$. Estimate the heat transfer rate through the wall if the temperature difference is 40 °C.

Solution: The total resistance to heat flow through the wall is

$$R_{\text{total}} = R_{\text{insulation}} + R_{\text{wood}} = 2 + 0.5 = 2.5\ \text{m}^2\cdot\text{K/W}$$

The heat transfer rate is then

$$\dot{Q} = \frac{A}{R_{\text{total}}} \Delta T = \frac{10 \times 3}{2.5} \times 40 = 480\ \text{W}$$

Note that ΔT measured in °C is the same as ΔT measured in kelvins.

EXAMPLE 3.10 The heat transfer from a 2-m-diameter sphere to a 25 °C air stream over a time interval of one hour is 3000 kJ. Estimate the surface temperature of the sphere if the heat transfer coefficient is 10 W/m²K.

Solution: The heat transfer is

$$Q = h_c A (T_s - T_\infty) \Delta t \qquad \text{or } 3 \times 10^6 = 10 \times 4\pi \times 1^2 (T_s - 25) \times 3600$$

The surface temperature is calculated to be

$$T_s = 31.6\,^\circ\text{C}$$

Note that the surface area of a sphere is $4\pi r^2$.

EXAMPLE 3.11 Estimate the rate of heat transfer from a 200 °C sphere which has an emissivity of 0.8 if it is suspended in a cold volume maintained at −20 °C. The sphere has a diameter of 20 cm.

Solution: The rate of heat transfer is given by

$$\dot{Q} = \varepsilon \sigma A (T^4 - T_{\text{surr}}^4) = 0.8 \times 5.67 \times 10^{-8} \times 4\pi \times 0.1^2 (473^4 - 253^4) = 262\ \text{J/s}$$

Solved Problems

3.1 Four kilograms of saturated liquid water is maintained at a constant pressure of 600 kPa while heat is added until the temperature reaches 600 °C. Determine the work done by the water.

The work for a constant-pressure process is $W = \int P\,dV = P(V_2 - V_1) = mP(v_2 - v_1)$. Using entries from Table C-2 and Table C-3, we find

$$W = (4)(600)(0.6697 - 0.0011) = 1605\ \text{kJ}$$

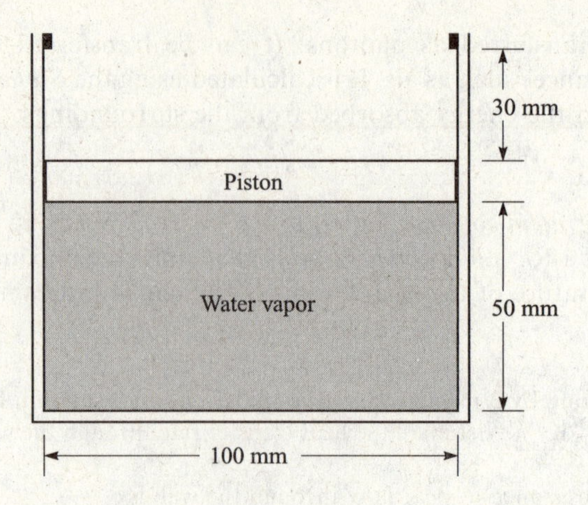

Fig. 3-12

3.2 The frictionless piston shown in Fig. 3-12 has a mass of 16 kg. Heat is added until the temperature reaches 400 °C. If the initial quality is 20 percent, find (*a*) the initial pressure, (*b*) the mass of water, (*c*) the quality when the piston hits the stops, (*d*) the final pressure, and (*e*) the work done on the piston.

(*a*) A force balance on the piston allows us to calculate the initial pressure. Including the atmospheric pressure, which is assumed to be 100 kPa, we have

$$P_1 A = W + P_{\text{atm}} A \qquad P_1 \frac{\pi(0.1)^2}{4} = (16)(9.81) + (100\,000)\frac{\pi(0.1)^2}{4}$$

$$\therefore P_1 = 120\,000 \text{ Pa} \quad \text{or } 120 \text{ kPa}$$

(*b*) To find the mass, we need the specific volume. Using entries from Table C-2, we find

$$v_1 = v_f + x(v_g - v_f) = 0.001 + (0.2)(1.428 - 0.001) = 0.286 \text{ m}^3/\text{kg}$$

The mass is then

$$m = V_1/v_1 = \frac{\pi(0.1)^2}{4}\left(\frac{0.05}{0.286}\right) = 0.001373 \text{ kg}$$

(*c*) When the piston just hits the stops, the pressure is still 120 kPa. The specific volume increases to

$$v_2 = V_2/m = \frac{\pi(0.1)^2}{4}\left(\frac{0.08}{0.001373}\right) = 0.458 \text{ m}^3/\text{kg}$$

The quality is then found as follows, using the entries at 120 °C:

$$0.458 = 0.001 + x_2(1.428 - 0.001) \qquad \therefore x_2 = 0.320 \quad \text{or } 32.0\%$$

(*d*) After the piston hits the stops, the specific volume ceases to change since the volume remains constant. Using $T_3 = 400\,°\text{C}$ and $v_3 = 0.458$, we can interpolate in Table C-3, between pressure 0.6 MPa and 0.8 MPa at 400 °C, to find

$$P_3 = \left(\frac{0.5137 - 0.458}{0.5137 - 0.3843}\right)(0.8 - 0.6) + 0.6 = 0.686 \text{ MPa}$$

(*e*) There is zero work done on the piston after it hits the stops. From the initial state until the piston hits the stops, the pressure is constant at 120 kPa; the work is then

$$W = P(v_2 - v_1)m = (120)(0.458 - 0.286)(0.001373) = 0.0283 \text{ kJ} \quad \text{or } 28.3 \text{ J}$$

3.3 Air is compressed in a cylinder such that the volume changes from 100 to 10 in^3. The initial pressure is 50 psia and the temperature is held constant at 100 °F. Calculate the work.

The work is given by $W = \int P\, dV$. For the isothermal process the equation of state allows us to write

$$PV = mRT = \text{const.}$$

since the mass m, the gas constant R, and the temperature T are all constant. Letting the constant be $P_1 V_1$, the above becomes $P = P_1 V_1 / V$, so that

$$W = P_1 V_1 \int_{V_1}^{V_2} \frac{dV}{V} = P_1 V_1 \ln \frac{V_2}{V_1} = (50)(144)\left(\frac{100}{1728}\right) \ln \frac{10}{100} = -959 \text{ ft-lbf}$$

3.4 Six grams of air is contained in the cylinder shown in Fig. 3-13. The air is heated until the piston raises 50 mm. The spring just touches the piston initially. Calculate (*a*) the temperature when the piston leaves the stops and (*b*) the work done by the air on the piston.

(*a*) The pressure in the air when the piston just raises from the stops is found by balancing the forces on the piston:

$$PA = P_{\text{atm}} A + W \qquad \frac{P\pi(0.2)^2}{4} = (100\,000)\frac{\pi(0.2)^2}{4} + (300)(9.81)$$

$$\therefore P = 193\,700 \text{ Pa} \quad \text{or } 193.7 \text{ kPa}$$

The temperature is found from the ideal-gas law:

$$T = \frac{PV}{mR} = \frac{(193.7)(0.15)(\pi)(0.2)^2/4}{(0.006)(0.287)} = 530 \text{ K}$$

(*b*) The work done by the air is considered to be composed of two parts: the work to raise the piston and the work to compress the spring. The work required to raise the piston a distance of 0.05 m is

$$W = (F)(d) = (P)(A)(d) = (193.7)\frac{\pi(0.2)^2}{4}(0.05) = 0.304 \text{ kJ}$$

The work required to compress the spring is $W = \frac{1}{2}Kx^2 = \frac{1}{2}(400)(0.05^2) = 0.5 \text{ kJ}$. The total work required by the air to raise the piston is

$$W = 0.304 + 0.5 = 0.804 \text{ kJ}$$

3.5 Two kilograms of air experiences the three-process cycle shown in Fig. 3-14. Calculate the net work.

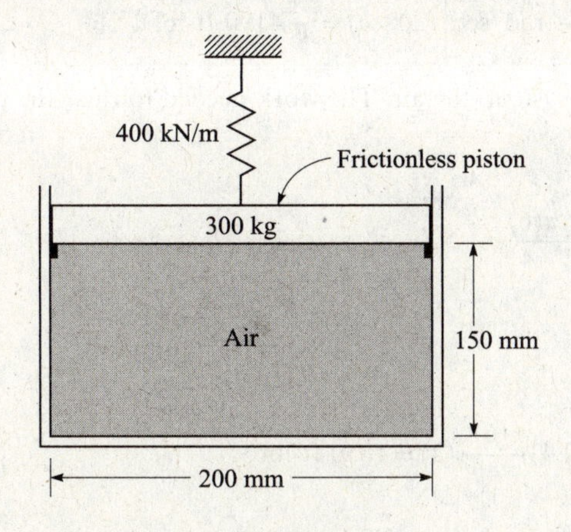

Fig. 3-13

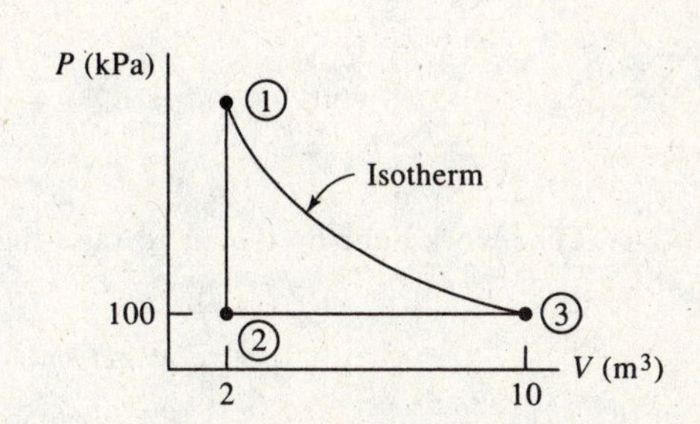

Fig. 3-14

The work for the constant-volume process from state 1 to state 2 is zero since $dV = 0$. For the constant-pressure process the work is

$$W_{2-3} = \int P\,dV = P(V_3 - V_2) = (100)(10 - 2) = 800 \text{ kJ}$$

The work needed for the isothermal process is

$$W_{3-1} = \int P\,dV = \int \frac{mRT}{V}\,dV = mRT \int_{V_3}^{V_1} \frac{dV}{V} = mRT \ln \frac{V_1}{V_3}$$

To find W_{3-1} we need the temperature. It is found from state 3 to be

$$T_3 = \frac{P_3 V_3}{mR} = \frac{(100)(10)}{(2)(0.287)} = 1742\,^\circ\text{R}$$

Thus, the work for the constant-temperature process is

$$W_{3-1} = (2)(0.287)(1742) \ln \frac{2}{10} = -1609 \text{ kJ}$$

Finally, the net work is

$$W_{\text{net}} = W_{1-2}^{0} + W_{2-3} + W_{3-1} = 800 - 1609 = -809 \text{ kJ}$$

The negative sign means that there must be a net input of work to complete the cycle in the order shown above.

3.6 A paddle wheel (Fig. 3-15) requires a torque of 20 ft-lbf to rotate it at 100 rpm. If it rotates for 20 sec, calculate the net work done by the air if the frictionless piston raises 2 ft during this time.

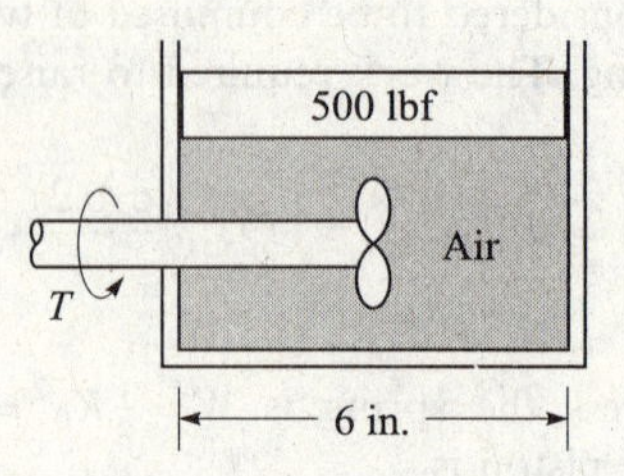

Fig. 3-15

The work input by the paddle wheel is

$$W = -T\omega\,\Delta t = (-20 \text{ ft-lbf})\left[\frac{(100)(2\pi)}{60} \text{ rad/sec}\right](20 \text{ sec}) = -4190 \text{ ft-lbf}$$

The negative sign accounts for work being done on the system, the air. The work needed to raise the piston requires that the pressure be known. It is found as follows:

$$PA = P_{\text{atm}}A + W \qquad P\frac{\pi(6)^2}{4} = (14.7)\frac{\pi(6)^2}{4} + 500 \qquad \therefore P = 32.4 \text{ psia}$$

The work done by the air to raise the piston is then

$$W = (F)(d) = (P)(A)(d) = (32.4)\frac{\pi(6)^2}{4}(2) = 1830 \text{ ft-lbf}$$

and the net work is $W_{\text{net}} = 1830 - 4190 = -2360$ ft-lbf.

3.7 The force needed to compress a nonlinear spring is given by the expression $F = 200x + 30x^2$ N, where x is the displacement of the spring from its unstretched length measured in meters. Determine the work needed to compress the spring a distance of 60 cm.

The work is given by

$$W = \int F\,dx = \int_0^{0.6} (200x + 30x^2)\,dx = (100 \times 0.6^2) + (10 \times 0.6^3) = 38.16 \text{ J}$$

Supplementary Problems

3.8 Two kilograms of saturated steam at 400 kPa is contained in a piston-cylinder arrangement. The steam is heated at constant pressure to 300 °C. Calculate the work done by the steam.

3.9 0.025 kg of steam at a quality of 10 percent and a pressure of 200 kPa is heated in a rigid container until the temperature reaches 200 °C. Find (*a*) the final quality and (*b*) the work done by the steam.

3.10 The frictionless piston shown in equilibrium has a mass of 64 kg (Fig. 3-16). Energy is added until the temperature reaches 220 °C. The atmospheric pressure is 100 kPa. Determine (*a*) the initial pressure, (*b*) the initial quality, (*c*) the quality when the piston just hits the stops, (*d*) the final quality (or pressure if superheat), (*e*) the work done on the piston.

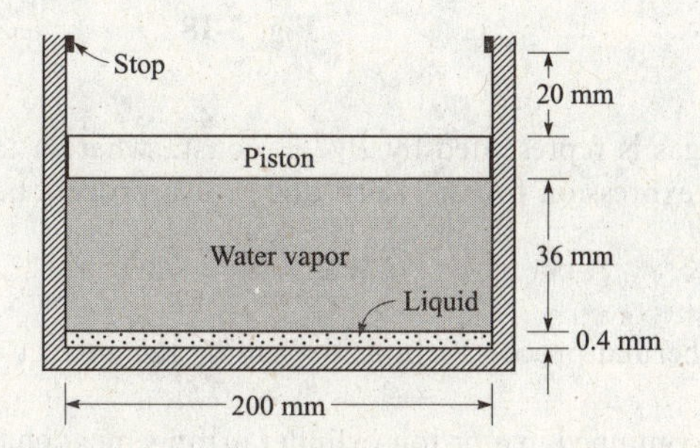

Fig. 3-16

3.11 Saturated water vapor at 180 °C is contained in a piston-cylinder arrangement at an initial volume of 0.1 m^3. Energy is added and the piston withdrawn so that the temperature remains constant until the pressure is 100 kPa.

(*a*) Find the work done. (Since there is no equation that relates p and V, this must be done graphically.)

(*b*) Use the ideal-gas law and calculate the work.

(*c*) What is the percent error in using the ideal-gas law?

3.12 A 75-lb piston and weights rest on a stop (Fig. 3-17). The volume of the cylinder at this point is 40 in^3. Energy is added to the 0.4 lbm of water mixture until the temperature reaches 300 °F. Atmospheric pressure is 14 psia.

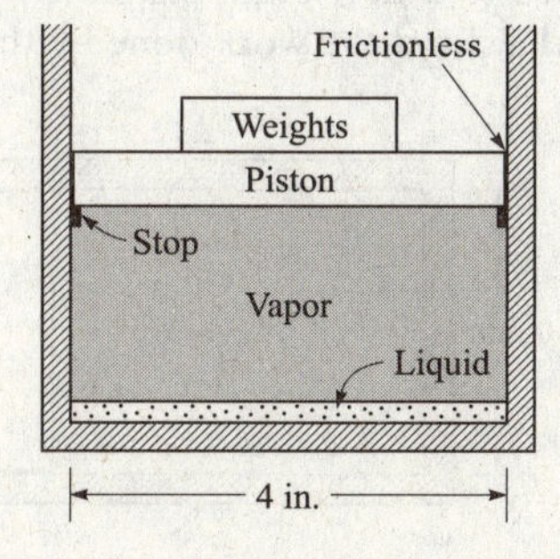

Fig. 3-17

(a) What is the initial specific volume of the mixture of vapor and liquid?

(b) What is the temperature in the cylinder when the piston just lifts off the stop?

(c) Determine the work done during the entire process.

3.13 Air is compressed in a cylinder such that the volume changes from 0.2 to 0.02 m^3. The pressure at the beginning of the process is 200 kPa. Calculate the work if (a) the pressure is constant, and (b) the temperature is constant at 50 °C. Sketch each process on a P-V diagram.

3.14 Air contained in a circular cylinder (Fig. 3-18) is heated until a 100-kg weight is raised 0.4 m. Calculate the work done by the expanding air on the weight. Atmospheric pressure is 80 kPa.

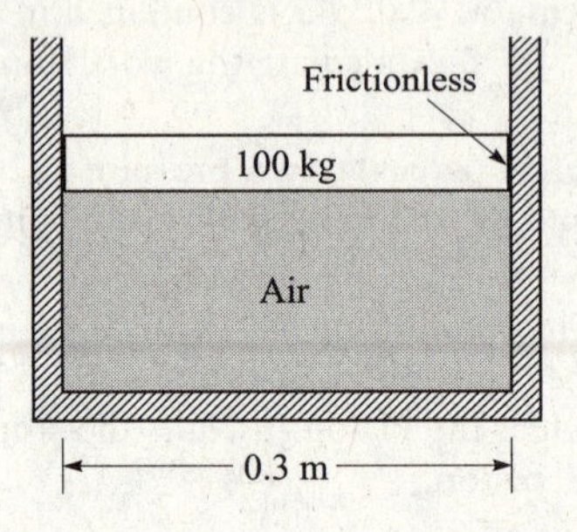

Fig. 3-18

3.15 A process for an ideal gas is represented by PV^n = const., where n takes on a particular value for a given process. Show that the expression for the work done for a process between states 1 and 2 is given by

$$W = \frac{P_2V_2 - P_1V_1}{1-n}$$

Is this valid for an isothermal process? If not, determine the correct expression.

3.16 The pressure in the gas contained in a piston-cylinder arrangement changes according to $P = a + 30/V$ where P is in psi and V is in ft^3. Initially the pressure is 7 psia and the volume is 3 ft^3. Determine the work done if the final pressure is 50 psia. Show the area that represents the work on a P-V diagram.

3.17 Air undergoes a three-process cycle. Find the net work done for 2 kg of air if the processes are

$1 \rightarrow 2$: constant-pressure expansion
$2 \rightarrow 3$: constant volume
$3 \rightarrow 1$: constant-temperature compression

The necessary information is $T_1 = 100\,°C$, $T_2 = 600\,°C$, and $P_1 = 200$ kPa. Sketch the cycle on a P-V diagram.

3.18 An unstretched spring is attached to a horizontal piston (Fig. 3-19). Energy is added to the gas until the pressure in the cylinder is 400 kPa. Find the work done by the gas on the piston. Use $P_{atm} = 75$ kPa.

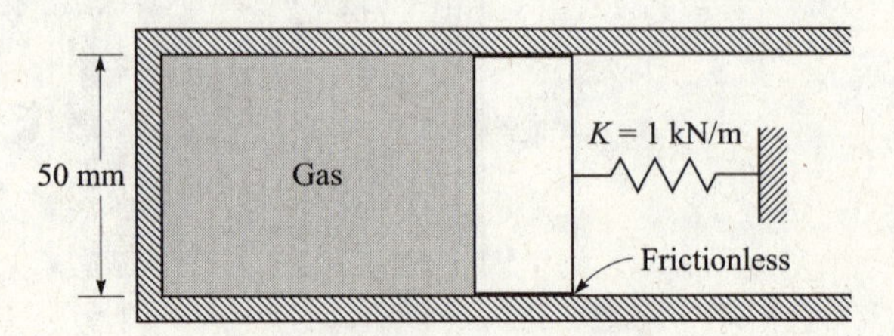

Fig. 3-19

3.19 Air is expanded in a piston-cylinder arrangement at a constant pressure of 200 kPa from a volume of 0.1 m^3 to a volume of 0.3 m^3. Then the temperature is held constant during an expansion of 0.5 m^3. Determine the total work done by the air.

3.20 A 30-ft-diameter balloon is to be filled with helium from a pressurized tank. The balloon is initially empty ($r = 0$) at an elevation where the atmospheric pressure is 12 psia. Determine the work done by the helium while the balloon is being filled. The pressure varies with radius according to $P = 0.04(r - 30)^2 + 12$ where P is in psi.

3.21 Estimate the work necessary to compress the air in an air-compressor cylinder from a pressure of 100 kPa to 2000 kPa. The initial volume is 1000 cm^3. An isothermal process is to be assumed.

3.22 An electric motor draws 3 A from the 12-V battery (Fig. 3-20). Ninety percent of the energy is used to spin the paddle wheel shown. After 50 s of operation the 30-kg piston is raised a distance of 100 mm. Determine the net work done by the gas on the surroundings. Use $P_{atm} = 95$ kPa.

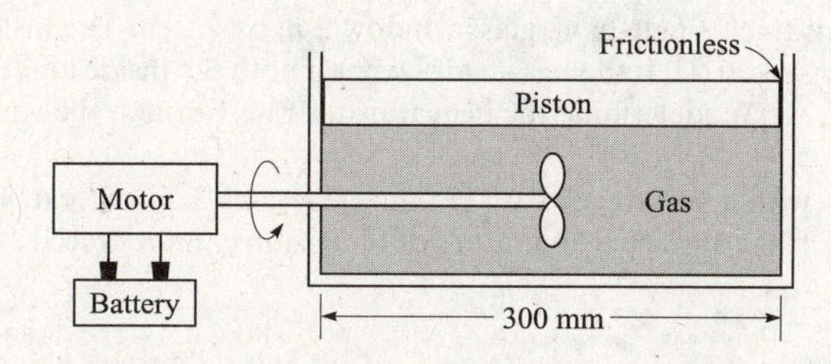

Fig. 3-20

3.23 A torque of 2 N·m is necessary to rotate a paddle wheel at a rate of 20 rad/s. The paddle wheel is located in a rigid vessel containing gas. What is the net work done on the gas during 10 min of operation?

3.24 Estimate the work done by a gas during an unknown process. Data obtained that relates pressure and volume are:

P	200	250	300	350	400	450	500	kPa
V	800	650	550	475	415	365	360	cm^3

3.25 Wind is blowing at 80 kph around a 250-mm-diameter tower that is 100 m high. The drag coefficient is 0.4 (see Example 3.4). Calculate the total force acting on the tower and the rate at which the wind does work on the tower.

3.26 Derive an expression for the work required to stretch an unstretched length of wire a relatively small distance l. The force is related to the amount of stretch x by $F = EAx/L$, where L is the original length of the wire. A is the cross-sectional area, and E is a material constant, and modulus of elasticity.

3.27 A linear spring with a free length of 0.8 ft requires a work input of 4 ft-lbf to extend it to its maximum usable length. If the spring constant is 100 lbf/ft, determine the maximum length of the spring.

3.28 A linear spring requires 20 J of work to compress it from an unstretched length of 100 mm to a length of 20 mm. Find the spring constant.

3.29 The force necessary to compress a nonlinear spring is given by $F = 10x^2$ N, where x is the distance the spring is compressed, measured in meters. Calculate the work needed to compress the spring from 0.2 to 0.8 m.

3.30 An automobile engine develops 100 hp, 96 percent of which is transferred to the driveshaft. Calculate the torque transferred by the driveshaft if it is rotating at 300 rpm.

3.31 A paddle wheel is placed in a small creek in an attempt to generate electricity. The water causes the tip of the 2-ft-radius paddles to travel at 4 ft/sec while a force of 100 lbf acts at an average distance of 1.2 ft from the hub. Determine the maximum continuous amperage output which could be used to charge a bank of 12-V batteries.

3.32 An electrical voltage of 110 V is applied across a resistor with the result that a current of 12 A flows through the resistor. Determine (*a*) the power necessary to accomplish this and (*b*) the work done during a period of 10 min.

3.33 A gasoline engine drives a small generator that is to supply sufficient electrical energy for a motor home. What is the minimum horsepower engine that would be necessary if a maximum of 200 A is anticipated from the 12-V system?

3.34 A house has a 0.5-cm-thick single-pane glass window 2 m by 1.5 m. The inside temperature is 20 °C and the outside temperature is − 20 °C. If there is an air layer on both the inside and the outside of the glass each with an *R*-factor of 0.1 $m^2 \cdot K/W$, determine the heat transfer rate through the window if $k_{glass} = 1.4$ W/m·K.

3.35 An electronic gizmo with a surface area of 75 mm^2 generates 3 W of heat. It is cooled by convection to air maintained at 25 °C. If the surface temperature of the gizmo cannot exceed 120 °C, estimate the heat transfer coefficient needed.

3.36 A 0.3-cm-diameter, 10-m-long copper wire has an emissivity of 0.04. Assuming heat transfer by radiation only, estimate the net rate of heat transferred to the 300-K surroundings if the wire is at 900 K.

Review Questions for the FE Examination

3.1FE A person carries a 200-N weight from one end of a hall to the other end 100 m away. How much work has the person done?
(A) 0 J
(B) 20 J
(C) 20 kJ
(D) 20 kW

3.2FE Which work mode is a nonequilibrium work mode?
(A) Compressing a spring
(B) Transmitting torque with a rotating shaft
(C) Energizing an electrical resistor
(D) Compressing gas in a cylinder

3.3FE Ten kilograms of saturated steam at 800 kPa is heated at constant pressure to 400 °C. The work required is:
(A) 1150 kJ
(B) 960 kJ
(C) 660 kJ
(D) 115 kJ

3.4FE Ten kilograms of air at 800 kPa is heated at constant pressure from 170 °C to 400 °C. The work required is nearest:
(A) 1150 kJ
(B) 960 kJ
(C) 660 kJ
(D) 115 kJ

3.5FE A stop is located 20 mm above the piston shown in Fig. 3-21. If the mass of the frictionless piston is 64 kg, what work must the air do on the piston to increase the pressure in the cylinder to 500 kPa?
(A) 13 J
(B) 28 J
(C) 41 J
(D) 53 J

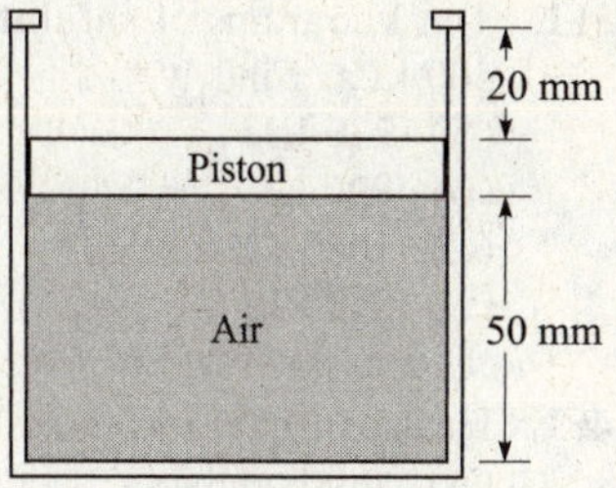

Fig. 3-21

3.6FE Which of the following does not transfer work to or from a system?
(A) A moving piston
(B) The expanding membrane of a balloon
(C) An electrical resistance heater
(D) A membrane that bursts

3.7FE Calculate the work needed to expand a piston in a cylinder at a constant pressure if saturated liquid water is completely vaporized at 200 °C.
(A) 2790 kJ/kg
(B) 1940 kJ/kg
(C) 850 kJ/kg
(D) 196 kJ/kg

3.8FE Which of the following statements about work for a quasiequilibrium process is incorrect?
(A) The differential of work is inexact
(B) Work is the area under a *P-T* diagram
(C) Work is a path function
(D) Work is always zero for a constant-volume process

3.9FE A paddle wheel and an electric heater operate in a system. If the torque is 200 N·m, the rotational speed is 400 rpm, the voltage is 20 V, and the amperage is 10 A, what is the rate of doing work?
(A) − 9200 J
(B) − 9200 kW
(C) − 2040 W
(D) − 2040 kW

3.10FE A 200-mm-diameter piston is lowered by increasing the pressure from 100 kPa to 800 kPa such that the *P-V* relationship is $PV^2 = \text{const}$. What is the work done on the system if $V_1 = 0.1\ \text{m}^3$?
(A) − 18.3 kJ
(B) − 24.2 kJ
(C) − 31.6 kJ
(D) − 42.9 kJ

3.11FE A 120-V electric resistance heater draws 10 A. It operates for 10 minutes in a rigid volume. Calculate the work done on the air in the volume.
(A) 720 000 kJ
(B) 720 kJ
(C) 12 000 kJ
(D) 12 kJ

3.12FE Find the work if air expands from 0.2 m^3 to 0.8 m^3 while the pressure in kPa is $P = 0.2 + 0.4V$.
(A) 0.48 kJ
(B) 0.42 kJ
(C) 0.36 kJ
(D) 0.24 kJ

3.13FE Ten kilograms of saturated liquid water expands until $T_2 = 200\,^\circ\mathrm{C}$ while the pressure remains constant at 500 kPa. Find W_{1-2}.
(A) 230 kJ
(B) 926 kJ
(C) 1080 kJ
(D) 2120 kJ

3.14FE Heat can be transferred by conduction through:
(A) solids only
(B) liquids only
(C) gases only
(D) all of the above

3.15FE Convection heat transfer involves both:
(A) conduction and advection
(B) conduction and radiation
(C) radiation and advection
(D) work and conduction

3.16FE Radiation is emitted by:
(A) liquids only
(B) opaque solids only
(C) gases only
(D) all materials at a finite temperature

Answers to Supplementary Problems

3.8 153.8 kJ
3.9 (*a*)0.7002 (*b*) 0.0
3.10 (*a*)120 kPa (*b*) 0.0619 (*c*) 0.0963 (*d*) 1.52 MPa (*e*) 0.0754 kJ
3.11 (*a*) 252 kJ (*b*) 248 kJ (*c*) −1.6%
3.12 (*a*) 0.05787 $\mathrm{ft^3/lbm}$ (*b*) 228 °F (*c*) 25,700 ft-lbf
3.13 (*a*) −36 kJ (*b*) −92.1 kJ
3.14 2.654 kJ
3.15 No. $P_1 V_1 \ln (V_2/V_1)$
3.16 −6153 ft-lbf
3.17 105 kJ
3.18 0.2976 kJ
3.19 70.65 kJ
3.20 2.54×10^8 ft-lbf
3.21 −0.300 kJ
3.22 −919 J
3.23 −24,000 N·m
3.24 132 J
3.25 3040 N, 0.0
3.26 $EAl^2/2L$
3.27 1.0828 ft
3.28 6250 N/m
3.29 1.68 J

3.30 2280 N·m

3.31 27.1 A

3.32 (*a*) 1320 W (*b*) 792 kJ

3.33 3.22 hp

3.34 590 J/s

3.35 421 W/m·K

3.36 138.5 W

Answers to Review Questions for the FE Examination

3.1FE (A) **3.2FE** (C) **3.3FE** (A) **3.4FE** (C) **3.5FE** (D) **3.6FE** (D) **3.7FE** (D) **3.8FE** (B) **3.9FE** (C) **3.10FE** (A) **3.11FE** (B) **3.12FE** (D) **3.13FE** (D) **3.14FE** (D) **3.15FE** (A) **3.16FE** (D)

CHAPTER 4

The First Law of Thermodynamics

4.1 INTRODUCTION

The first law of thermodynamics is commonly called the conservation of energy. In elementary physics courses, the study of the conservation of energy emphasizes changes in kinetic and potential energy and their relationship to work. A more general form of conservation of energy includes the effects of heat transfer and internal energy changes. This more general form is usually called the *first law of thermodynamics*. Other forms of energy may also be included, such as electrostatic, magnetic, strain, and surface energy. We will present the first law for a system and then for a control volume.

4.2 THE FIRST LAW OF THERMODYNAMICS APPLIED TO A CYCLE

Having discussed the concepts of work and heat, we are now ready to present the first law of thermodynamics. Recall that a law is not derived or proved from basic principles but is simply a statement that we write based on our observations of many experiments. If an experiment shows a law to be violated, either the law must be revised or additional conditions must be placed on the applicability of the law. Historically, the *first law of thermodynamics* was stated for a cycle: the net heat transfer is equal to the net work done for a system undergoing a cycle. This is expressed in equation form by

$$\Sigma W = \Sigma Q \tag{4.1}$$

or

$$\oint \delta W = \oint \delta Q \tag{4.2}$$

where the symbol $\oint$ implies an integration around a complete cycle.

The first law can be illustrated by considering the following experiment. Let a weight be attached to a pulley/paddle-wheel setup, such as that shown in Fig. 4-1a. Let the weight fall a certain distance thereby doing work on the system, contained in the insulated tank shown, equal to the weight multiplied by the distance dropped. The temperature of the system (the fluid in the tank) will rise an amount ΔT.

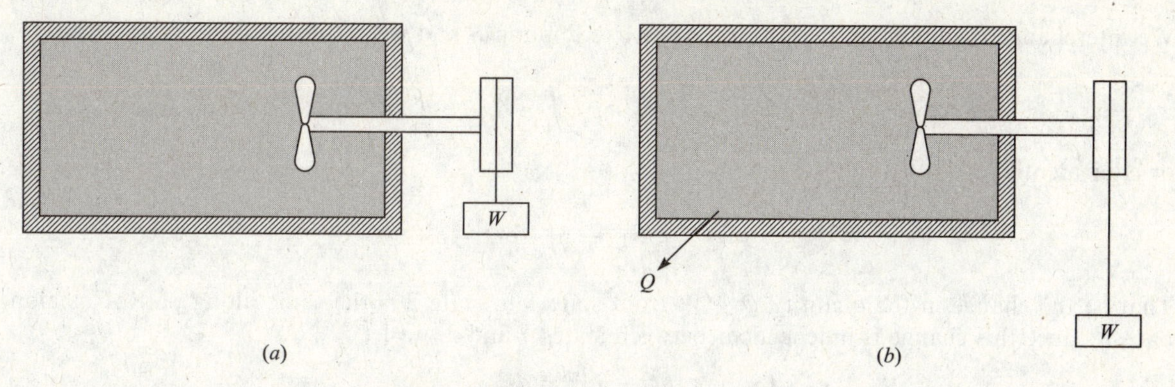

Fig. 4-1 The first law applied to a cycle.

Now, the system is returned to its initial state (the completion of the cycle) by transferring heat to the surroundings, as implied by the Q in Fig. 4-1b. This reduces the temperature of the system to its initial temperature. The first law states that this heat transfer will be exactly equal to the work which was done by the falling weight.

EXAMPLE 4.1 A spring is stretched a distance of 0.8 m and attached to a paddle wheel (Fig. 4-2). The paddle wheel then rotates until the spring is unstretched. Calculate the heat transfer necessary to return the system to its initial state.

Solution: The work done by the spring on the system is given by

$$W_{1-2} = \int_0^{0.8} F\,dx = \int_0^{0.8} 100x\,dx = (100)\left[\frac{(0.8)^2}{2}\right] = 32\,\text{N}\cdot\text{m}$$

Since the heat transfer returns the system to its initial state, a cycle results. The first law then states that $Q_{2-1} = W_{1-2} = 32$ J.

4.3 THE FIRST LAW APPLIED TO A PROCESS

The first law of thermodynamics is often applied to a process as the system changes from one state to another. Realizing that a cycle results when a system undergoes several processes and returns to the initial state, we could consider a cycle composed of the two processes represented by A and B in Fig. 4-3. Applying the first law to this cycle, (*4.2*) takes the form

$$\int_1^2 \delta Q_A + \int_2^1 \delta Q_B = \int_1^2 \delta W_A + \int_2^1 \delta W_B$$

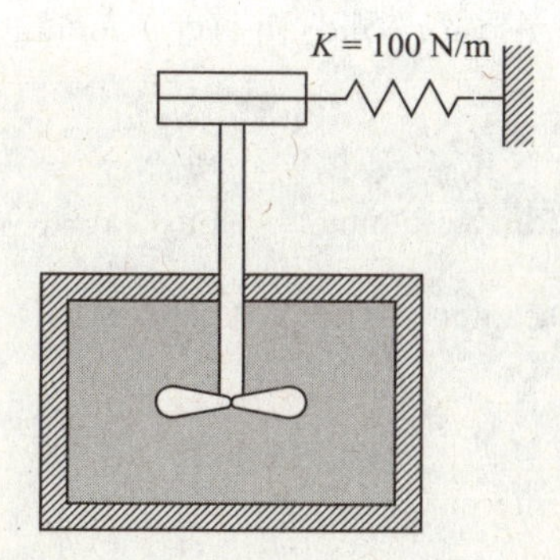

Fig. 4-2

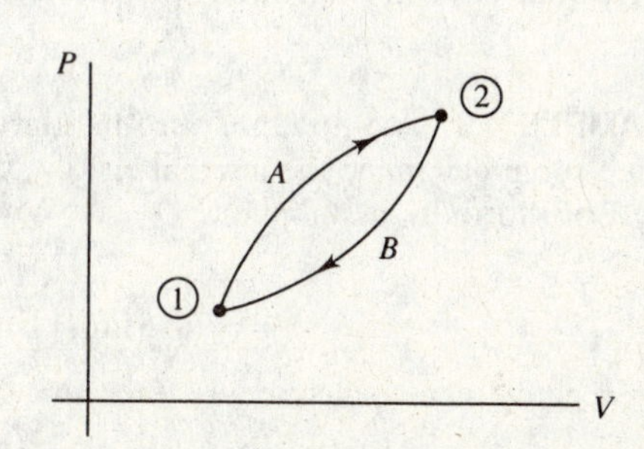

Fig. 4-3 A cycle composed of two processes.

We interchange the limits on the process from 1 to 2 along B and write this as

$$\int_1^2 \delta Q_A - \int_1^2 \delta Q_B = \int_1^2 \delta W_A - \int_1^2 \delta W_B$$

or equivalently

$$\int_1^2 (\delta Q - \delta W)_A = \int_1^2 (\delta Q - \delta W)_B$$

That is, the change in the quantity $Q - W$ from state 1 to state 2 is the same along path A as along path B; since this change is independent between states 1 and 2, we let

$$\delta Q - \delta W = dE \tag{4.3}$$

where dE is an exact differential. The quantity E is an extensive property of the system and can be shown experimentally to represent the energy of the system at a particular state. Equation (*4.3*) can be integrated to yield

$$Q_{1-2} - W_{1-2} = E_2 - E_1 \tag{4.4}$$

where Q_{1-2} is the heat transferred to the system during the process from state 1 to state 2, W_{1-2} is the work done by the system on the surroundings during the process, and E_2 and E_1 are the values of the property E. More often than not the subscripts will be dropped on Q and W when working problems.

The property E represents all of the energy: kinetic energy KE, potential energy PE, and internal energy U which includes chemical energy and the energy associated with the atom. Any other form of energy is also included in the total energy E. Its associated intensive property is designated e.

The first law of thermodynamics then takes the form

$$\begin{aligned} Q_{1-2} - W_{1-2} &= KE_2 - KE_1 + PE_2 - PE_1 + U_2 - U_1 \\ &= \frac{m}{2}(V_2^2 - V_1^2) + mg(z_2 - z_1) + U_2 - U_1 \end{aligned} \tag{4.5}$$

If we apply the first law to an isolated system, one for which $Q_{1-2} = W_{1-2} = 0$, the first law becomes the conservation of energy; that is,

$$E_2 = E_1 \tag{4.6}$$

The internal energy U is an extensive property. Its associated intensive property is the specific internal energy u; that is, $u = U/m$. For simple systems in equilibrium, only two properties are necessary to establish the state of a pure substance, such as air or steam. Since internal energy is a property, it depends only on, say, pressure and temperature; or, for saturated steam, it depends on quality and temperature (or pressure). Its value for a particular quality would be

$$u = u_f + xu_{fg} \tag{4.7}$$

We can now apply the first law to systems involving working fluids with tabulated property values. Before we apply the first law to systems involving substances such as ideal gases or solids, it is convenient to introduce several additional properties that will simplify that task.

EXAMPLE 4.2 A 5-hp fan is used in a large room to provide for air circulation. Assuming a well-insulated, sealed room determine the internal energy increase after 1 h of operation.

Solution: By assumption, $Q = 0$. With $\Delta PE = KE = 0$ the first law becomes $-W = \Delta U$. The work input is

$$W = (-5\,\text{hp})(1\,\text{h})(746\,W/\text{hp})(3600\,\text{s/h}) = -1.343 \times 10^7\,\text{J}$$

The negative sign results because the work is input to the system. Finally, the internal energy increase is

$$\Delta U = -(-1.343 \times 10^7) = 1.343 \times 10^7\,\text{J}$$

EXAMPLE 4.3 A rigid volume contains 6 ft^3 of steam originally at a pressure of 400 psia and a temperature of 900 °F. Estimate the final temperature if 800 Btu of heat is added.

Solution: The first law of thermodynamics, with $\Delta KE = \Delta PE = 0$, is $Q - W = \Delta U$. For a rigid container the work is zero. Thus,

$$Q = \Delta U = m(u_2 - u_1)$$

From the steam tables we find $u_1 = 1324$ Btu/lbm and $v_1 = 1.978$ ft^3/lbm. The mass is then

$$m = \frac{V}{v} = \frac{6}{1.978} = 3.033 \text{ lbm}$$

The energy transferred to the volume by heat is given. Thus,

$$800 = 3.033(u_2 - 1324) \qquad \therefore u_2 = 1588 \text{ Btu/lbm}$$

From Table C-3E we must find the temperature for which $v_2 = 1.978$ ft^3/lbm and $u_2 = 1588$ Btu/lbm. This is not a simple task since we do not know the pressure. At 500 psia if $v = 1.978$ ft^3/lbm, then $u = 1459$ Btu/lbm and $T = 1221$ °F. At 600 psia $v = 1.978$ ft^3/lbm, then $u = 1603$ Btu/lbm and $T = 1546$ °F. Now we linearly interpolate to find the temperature at $u_2 = 1588$ Btu/lbm:

$$T_2 = 1546 - \left(\frac{1603 - 1588}{1603 - 1459}\right)(1546 - 1221) = 1512\,°\text{F}$$

EXAMPLE 4.4 A frictionless piston is used to provide a constant pressure of 400 kPa in a cylinder containing steam originally at 200 °C with a volume of 2 m^3. Calculate the final temperature if 3500 kJ of heat is added.

Solution: The first law of thermodynamics, using $\Delta PE = \Delta KE = 0$, is $Q - W = \Delta U$. The work done during the motion of the piston is

$$W = \int P\,dV = P(V_2 - V_1) = 400(V_2 - V_1)$$

The mass before and after remains unchanged. Using the steam tables, this is expressed as

$$m = \frac{V_1}{v_1} = \frac{2}{0.5342} = 3.744 \text{ kg}$$

The volume V_2 is written as $V_2 = mv_2 = 3.744\,v_2$. The first law is then, finding u_1 from the steam tables,

$$3500 - (400)(3.744v_2 - 2) = (u_2 - 2647) \times (3.744)$$

This requires a trial-and-error process. One plan for obtaining a solution is to guess a value for v_2 and calculate u_2 from the equation above. If this value checks with the u_2 from the steam tables at the same temperature, then the guess is the correct one. For example, guess $v_2 = 1.0\,m^3$/kg. Then the equation gives $u_2 = 3395$ kJ/kg. From the steam tables, with $P = 0.4$ MPa, the u_2 value allows us to interpolate $T_2 = 654$ °C and the v_2 gives $T_2 = 600$ °C. Therefore, the guess must be revised. Try $v_2 = 1.06$ m^3/kg. The equation gives $u_2 = 3372$ kJ/kg. The tables are interpolated to give $T_2 = 640$ °C; for v_2, $T_2 = 647$ °C. The actual v_2 is a little less than 1.06 m^3/kg, with the final temperature being approximately

$$T_2 = 644\,°\text{C}$$

4.4 ENTHALPY

In the solution of problems involving systems, certain products or sums of properties occur with regularity. One such combination of properties can be demonstrated by considering the addition of heat to the constant-pressure situation shown in Fig. 4-4. Heat is added slowly to the system (the gas in the cylinder), which is maintained at constant pressure by assuming a frictionless seal between the piston and

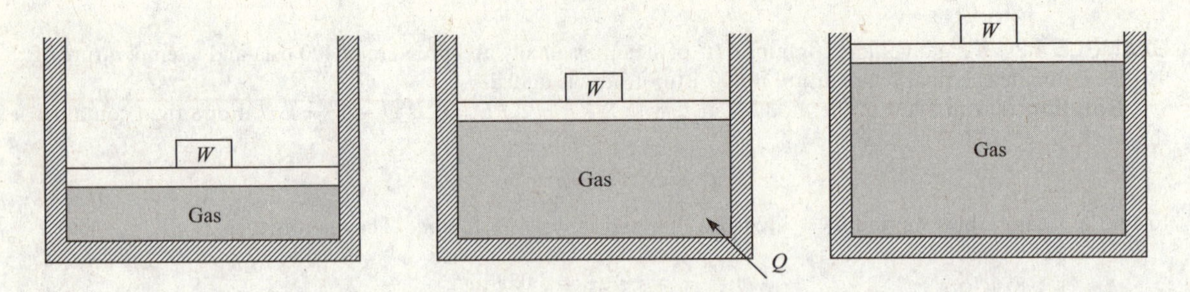

Fig. 4-4 Constant-pressure heat addition.

the cylinder. If the kinetic energy changes and potential energy changes of the system are neglected and all other work modes are absent, the first law of thermodynamics requires that

$$Q - W = U_2 - U_1 \tag{4.8}$$

The work done raising the weight for the constant-pressure process is given by

$$W = P(V_2 - V_1) \tag{4.9}$$

The first law can then be written as

$$Q = (U + PV)_2 - (U + PV)_1 \tag{4.10}$$

The quantity in parentheses is a combination of properties and is thus a property itself. It is called the *enthalpy* H of the system; that is,

$$H = U + PV \tag{4.11}$$

The specific enthalpy h is found by dividing by the mass. It is

$$h = u + Pv \tag{4.12}$$

Enthalpy is a property of a system and is also found in the steam tables. The energy equation can now be written for a constant-pressure equilibrium process as

$$Q_{1-2} = H_2 - H_1 \tag{4.13}$$

The enthalpy was defined using a constant-pressure system with the difference in enthalpies between two states being the heat transfer. For a variable-pressure process, the difference in enthalpy loses its physical significance when considering a system. But enthalpy is still of use in engineering problems; it remains a property as defined by (*4.11*). In a nonequilibrium constant-pressure process ΔH would not equal the heat transfer.

Because only *changes* in enthalpy or internal energy are important, we can arbitrarily choose the datum from which to measure h and u. We choose saturated liquid at 0 °C to be the datum point for water.

EXAMPLE 4.5 Using the concept of enthalpy solve the problem presented in Example 4.4.

Solution: The energy equation for a constant-pressure process is (with the subscript on the heat transfer omitted)

$$Q = H_2 - H_1 \qquad \text{or} \qquad 3500 = (h_2 - 2860)m$$

Using the steam tables as in Example 4.4, the mass is

$$m = \frac{V}{v} = \frac{2}{0.5342} = 3.744 \text{ kg}$$

Thus,

$$h_2 = \frac{3500}{3.744} + 2860 = 3795\,\text{kJ/kg}$$

From the steam tables this interpolates to

$$T_2 = 600 + \left(\frac{92.6}{224}\right)(100) = 641\,^\circ\text{C}$$

Obviously, enthalpy was very useful in solving the constant-pressure problem. Trial and error was unnecessary, and the solution was rather straightforward. We illustrated that the quantity we made up, enthalpy, is not necessary, but it is quite handy. We will use it often in our calculations.

4.5 LATENT HEAT

The amount of energy that must be transferred in the form of heat to a substance held at constant pressure in order that a phase change occur is called the *latent heat*. It is the change in enthalpy of the substance at the saturated conditions of the two phases. The heat that is necessary to melt (or freeze) a unit mass of a substance at constant pressure is the *heat of fusion* and is equal to $h_{if} = h_f - h_i$, where h_i is the enthalpy of saturated solid and h_f is the enthalpy of saturated liquid. The *heat of vaporization* is the heat required to completely vaporize a unit mass of saturated liquid (or condense a unit mass of saturated vapor); it is equal to $h_{fg} = h_g - h_f$. When a solid changes phase directly to a gas, sublimation occurs; the *heat of sublimation* is equal to $h_{ig} = h_g - h_i$.

The heat of fusion and the heat of sublimation are relatively insensitive to pressure or temperature changes. For ice the heat of fusion is approximately 320 kJ/kg (140 Btu/lbm) and the heat of sublimation is about 2040 kJ/kg (880 Btu/lbm). The heat of vaporization of water is included as h_{fg} in Tables C-1 and C-2.

4.6 SPECIFIC HEATS

For a simple system only two independent variables are necessary to establish the state of the system. Consequently, we can consider the specific internal energy to be a function of temperature and specific volume; that is,

$$u = u(T, v) \tag{4.14}$$

Using the chain rule from calculus we express the differential in terms of the partial derivatives as

$$du = \left.\frac{\partial u}{\partial T}\right|_v dT + \left.\frac{\partial u}{\partial v}\right|_T dv \tag{4.15}$$

Since u, v, and T are all properties, the partial derivative is also a property and is called the *constant-volume specific heat* C_v; that is,

$$C_v = \left.\frac{\partial u}{\partial T}\right|_v \tag{4.16}$$

One of the classical experiments of thermodynamics, first performed by Joule in 1843, is illustrated in Fig. 4-5. Pressurize volume A with an ideal gas and evacuate volume B. After equilibrium is attained, open the valve. Even though the pressure and volume of the ideal gas have changed markedly, the temperature does not change. Because there is no change in temperature, there is no net heat transfer to the water. Observing that no work is done we conclude, from the first law, that the internal energy of an ideal gas does not depend on pressure or volume.

For such a gas, which behaves as an ideal gas, we have

$$\left.\frac{\partial u}{\partial v}\right|_T = 0 \tag{4.17}$$

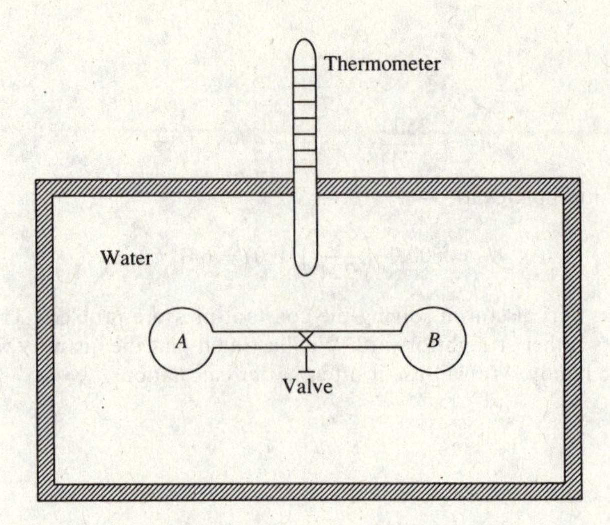

Fig. 4-5 Joule's experiment.

Combining (*4.15*), (*4.16*), and (*4.17*),

$$du = C_v \, dT \tag{4.18}$$

This can be integrated to give

$$u_2 - u_1 = \int_{T_1}^{T_2} C_v \, dT \tag{4.19}$$

For a known $C_v(T)$ this can be integrated to find the change in internal energy over any temperature interval for an ideal gas.

Likewise, considering specific enthalpy to be dependent on the two variables T and P, we have

$$dh = \left.\frac{\partial h}{\partial T}\right|_P dT + \left.\frac{\partial h}{\partial P}\right|_T dP \tag{4.20}$$

The *constant-pressure specific heat* C_p is defined as

$$C_p = \left.\frac{\partial h}{\partial T}\right|_P \tag{4.21}$$

For an ideal gas we have, returning to the definition of enthalpy, (*4.12*),

$$h = u + Pv = u + RT \tag{4.22}$$

where we have used the ideal-gas equation of state. Since u is only a function of T, we see that h is also only a function of T for an ideal gas. Hence, for an ideal gas

$$\left.\frac{\partial h}{\partial P}\right|_T = 0 \tag{4.23}$$

and we have, from (*4.20*),

$$dh = C_p \, dT \tag{4.24}$$

Over the temperature range T_1 to T_2 this is integrated to give

$$h_2 - h_1 = \int_{T_1}^{T_2} C_p \, dT \tag{4.25}$$

for an ideal gas.

It is often convenient to specify specific heats on a per-mole, rather than a per-unit-mass, basis; these *molar specific heats* are $\bar{C}_v$ and $\bar{C}_p$. Clearly, we have the relations

$$\bar{C}_v = MC_v \qquad \text{and} \qquad \bar{C}_p = MC_p$$

where M is the molar mass. Thus values of $\bar{C}_v$ and $\bar{C}_p$ may be simply calculated from the values of C_v and C_p listed in Table B-2. (The "overbar notation" for a molar quantity is used throughout this book.)

The equation for enthalpy can be used to relate, for an ideal gas, the specific heats and the gas constant. In differential form (*4.12*) takes the form

$$dh = du + d(Pv) \tag{4.26}$$

Introducing the specific heat relations and the ideal-gas equation, we have

$$C_p\, dT = C_v\, dT + R\, dT \tag{4.27}$$

which, after dividing by dT, gives

$$C_p = C_v + R \tag{4.28}$$

This relationship—or its molar equivalent $\bar{C}_p = \bar{C}_v + R_u$—allows C_v to be determined from tabulated values or expressions for C_p. Note that the difference between C_p and C_v for an ideal gas is always a constant, even though both are functions of temperature.

The *specific heat ratio* k is also a property of particular interest; it is defined as

$$k = \frac{C_p}{C_v} \tag{4.29}$$

This can be substituted into (*4.28*) to give

$$C_p = R\frac{k}{k-1} \tag{4.30}$$

or

$$C_v = \frac{R}{k-1} \tag{4.31}$$

Obviously, since R is a constant for an ideal gas, the specific heat ratio will depend only on temperature.

For gases, the specific heats slowly increase with increasing temperature. Since they do not vary significantly over fairly large temperature differences, it is often acceptable to treat C_v and C_p as constants. For such situations there results

$$u_2 - u_1 = C_v(T_2 - T_1) \tag{4.32}$$

$$h_2 - h_1 = C_p(T_2 - T_1) \tag{4.33}$$

For air we will use $C_v = 0.717$ kJ/kg·°C (0.171 Btu/lbm-°R) and $C_p = 1.00$ kJ/kg·°C (0.24 Btu/lbm-°R), unless otherwise stated. For more accurate calculations with air, or other gases, one should consult ideal-gas tables, such as those in Appendix E, which tabulate $h(T)$ and $u(T)$, or integrate using expressions for $C_p(T)$ found in Table B-5.

For liquids and solids the specific heat C_p is tabulated in Table B-4. Since it is quite difficult to maintain constant volume while the temperature is changing, C_v values are usually not tabulated for liquids and solids; the difference $C_p - C_v$ is usually quite small. For most liquids the specific heat is relatively insensitive to temperature change. For water we will use the nominal value of 4.18 kJ/kg·°C (1.00 Btu/lbm-°R). For ice the specific heat in kJ/kg·°C is approximately $C_p = 2.1 + 0.0069T$, where T is measured in °C; and in English units of Btu/lbm-°F it is $C_p = 0.47 + 0.001T$, where T is measured in °F. The variation of specific heat with pressure is usually quite slight except for special situations.

EXAMPLE 4.6 The specific heat of superheated steam at approximately 150 kPa can be determined by the equation

$$C_p = 2.07 + \frac{T - 400}{1480} \text{ kJ/kg} \cdot {}^\circ\text{C}$$

(*a*) What is the enthalpy change between 300 °C and 700 °C for 3 kg of steam? Compare with the steam tables. (*b*) What is the average value of C_p between 300 °C and 700 °C based on the equation and based on the tabulated data?

Solution:

(*a*) The enthalpy change is found to be

$$\Delta H = m \int_{T_1}^{T_2} C_p\, dT = 3 \int_{300}^{700} \left(2.07 + \frac{T - 400}{1480}\right) dT = 2565 \text{ kJ}$$

From the tables we find, using $P = 150$ kPa,

$$\Delta H = (3)(3928 - 3073) = 2565 \text{ kJ}$$

(*b*) The average value $C_{p,\text{av}}$ is found by using the relation

$$mC_{p,\text{av}}\, \Delta T = m \int_{T_1}^{T_2} C_p\, dT \quad \text{or}$$

$$(3)(400 C_{p,\text{av}}) = 3 \int_{300}^{700} \left(2.07 + \frac{T - 400}{1480}\right) dT$$

The integral was evaluated in part (*a*); hence, we have

$$C_{p,\text{av}} = \frac{2565}{(3)(400)} = 2.14 \text{ kJ/kg} \cdot {}^\circ\text{C}$$

Using the values from the steam table, we have

$$C_{p,\text{av}} = \frac{\Delta h}{\Delta T} = (3928 - 3073)/400 = 2.14 \text{ kJ/kg} \cdot {}^\circ\text{C}$$

Because the steam tables give the same values as the linear equation of this example, we can safely assume that the $C_p(T)$ relationship for steam over this temperature range is closely approximated by a linear relation. This linear relation would change, however, for each pressure chosen; hence, the steam tables are essential.

EXAMPLE 4.7 Determine the value of C_p for steam at $T = 800\,^\circ$F and $P = 800$ psia.

Solution: To determine C_p we use a finite-difference approximation to (*4.21*). We use the entries at $T = 900\,^\circ$F and $T = 700\,^\circ$F, which gives a better approximation to the slope compared to using the values at 800 °F and 750 °F or at 900 °F and 800 °F. Table C-3E provides us with

$$C_p \cong \frac{\Delta h}{\Delta T} = \frac{1455.6 - 1338.0}{200} = 0.588 \text{ Btu/lbm-}^\circ\text{F}$$

Figure 4-6 shows why it is better to use values on either side of the position of interest. If values at 900 °F and 800 °F are used (a forward difference), C_p is too low. If values at 800 °F and 750 °C are used (a backward difference), C_p is too high. Thus, both a forward and a backward value (a central difference) should be used, resulting in a more accurate estimate of the slope.

EXAMPLE 4.8 Determine the enthalpy change for 1 kg of nitrogen which is heated from 300 to 1200 K by (*a*) using the gas tables, (*b*) integrating $C_p(T)$, and (*c*) assuming constant specific heat. Use $M = 28$ kg/kmol.

Solution:

(*a*) Using the gas table in Appendix E, find the enthalpy change to be

$$\Delta h = 36\,777 - 8723 = 28\,054 \text{ kJ/kmol} \qquad \text{or} \qquad 28\,054/28 = 1002 \text{ kJ/kg}$$

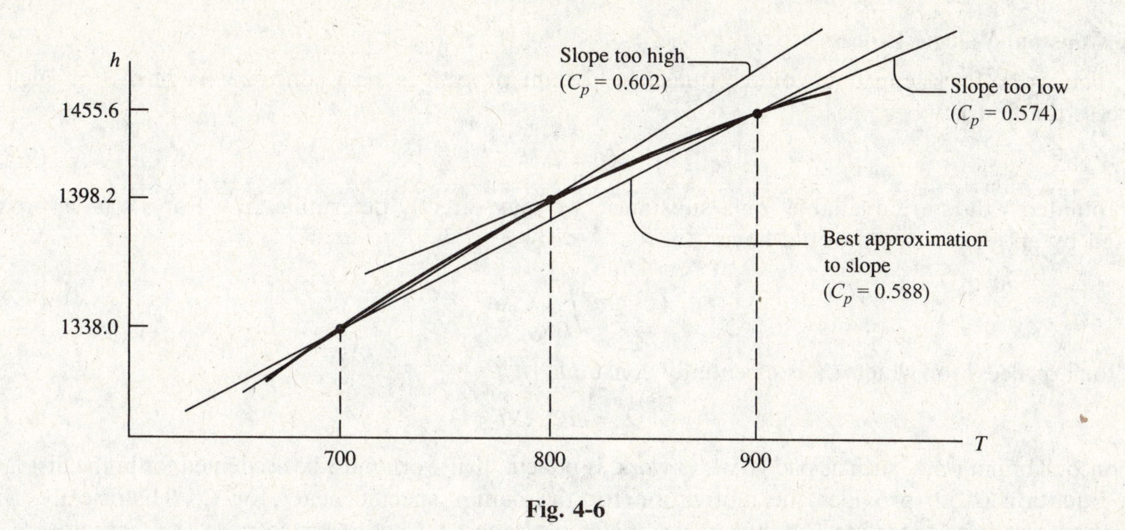

Fig. 4-6

(*b*) The expression for $C_p(T)$ is found in Table B-5. The enthalpy change is

$$\Delta h = \int_{300}^{1200} \left[39.06 - 512.79\left(\frac{T}{100}\right)^{-1.5} + 1072.7\left(\frac{T}{100}\right)^{-2} - 820.4\left(\frac{T}{100}\right)^{-3} \right] dt$$

$$= (39.06)(1200 - 300) - (512.79)\left(\frac{100}{-0.5}\right)(12^{-0.5} - 3^{-0.5})$$

$$+ (1072.7)\left(\frac{100}{-1}\right)(12^{-1} - 3^{-1}) - (820.4)\left(\frac{100}{-2}\right)(12^{-2} - 3^{-2})$$

$$= 28\,093 \text{ kJ/kmol} \quad \text{or } 1003 \text{ kJ/kg}$$

(*c*) Assuming constant specific heat (found in Table B-2) the enthalpy change is found to be

$$\Delta h = C_p \Delta T = (1.042)(1200 - 300) = 938 \text{ kJ/kg}$$

Note the value found by integrating is essentially the same as that found from the gas tables. However, the enthalpy change found by assuming constant specific heat is in error by over 6 percent. If T_2 were closer to 300 K, say 600 K, the error would be much smaller.

4.7 THE FIRST LAW APPLIED TO VARIOUS PROCESSES

The Constant-Temperature Process

For the isothermal process, tables may be consulted for substances for which tabulated values are available. Internal energy and enthalpy vary slightly with pressure for the isothermal process, and this variation must be accounted for in processes involving many substances. The energy equation is

$$Q - W = \Delta U \tag{4.34}$$

For a gas that approximates an ideal gas, the internal energy depends only on the temperature and thus $\Delta U = 0$ for an isothermal process; for such a process

$$Q = W \tag{4.35}$$

Using the ideal-gas equation $PV = mRT$, the work for a quasiequilibrium process can be found to be

$$W = \int_{V_1}^{V_2} P\,dV = mRT \int_{V_1}^{V_2} \frac{dV}{V} = mRT \ln\frac{V_2}{V_1} = mRT \ln\frac{P_1}{P_2} \tag{4.36}$$

The Constant-Volume Process

The work for a constant-volume quasiequilibrium process is zero, since dV is zero. For such a process the first law becomes

$$Q = \Delta U \tag{4.37}$$

If tabulated values are available for a substance, we may directly determine ΔU. For a gas, approximated by an ideal gas, we would have

$$Q = m \int_{T_1}^{T_2} C_v \, dT \tag{4.38}$$

or, for a process for which C_v is essentially constant,

$$Q = mC_v \, \Delta T \tag{4.39}$$

If nonequilibrium work, such as paddle-wheel work, is present, that work must be accounted for in the first law.

Equation (*4.39*) provides the motivation for the name "specific heat" for C_v. Historically, this equation was used to define C_v; thus, it was defined as the heat necessary to raise the temperature of one unit of substance one degree in a constant-volume process. Today scientists prefer the definition of C_v to be in terms of properties only, without reference to heat transfer, as in (*4.16*).

The Constant-Pressure Process

The first law, for a constant-pressure quasiequilibrium process, was shown in Sec. 4.4 to be

$$Q = \Delta H \tag{4.40}$$

Hence, the heat transfer for such a process can easily be found using tabulated values, if available.

For a gas that behaves as an ideal gas, we have

$$Q = m \int_{T_1}^{T_2} C_p \, dT \tag{4.41}$$

For a process involving an ideal gas for which C_p is constant there results

$$Q = mC_p \, \Delta T \tag{4.42}$$

For a nonequilibrium process the work must be accounted for directly in the first law and cannot be expressed as $P(V_2 - V_1)$. For such a process (*4.40*) would not be valid.

The Adiabatic Process

There are numerous examples of processes for which there is no, or negligibly small, heat transfer, e.g., the compression of air in an automobile engine or the exhaust of nitrogen from a nitrogen tank. The study of such processes is, however, often postponed until after the second law of thermodynamics is presented. This postponement is not necessary, and because of the importance of the adiabatic quasiequilibrium process, it is presented here.

The differential form of the first law for the adiabatic process is

$$-\delta w = du \tag{4.43}$$

or, for a quasiequilibrium process, using $\delta w = P \, dv$, thereby eliminating nonequilibrium work modes,

$$du + P \, dv = 0 \tag{4.44}$$

The sum of the differential quantities on the left represents a perfect differential which we shall designate as $d\psi$, ψ being a property of the system. This is similar to the motivation for defining the enthalpy h as a property. Since

$$d\psi = du + P \, dv \tag{4.45}$$

is a property of the system, it is defined for processes other than the adiabatic quasiequilibrium process.

Let us investigate this adiabatic quasiequilibrium process for an ideal gas with constant specific heats. For such a process, (*4.44*) takes the form

$$C_v\,dT + \frac{RT}{v}dv = 0 \tag{4.46}$$

Rearranging, we have

$$\frac{C_v}{R}\frac{dT}{T} = -\frac{dv}{v} \tag{4.47}$$

This is integrated, assuming constant C_v, between states 1 and 2 to give

$$\frac{C_v}{R}\ln\frac{T_2}{T_1} = -\ln\frac{v_2}{v_1} \tag{4.48}$$

which can be put in the form

$$\frac{T_2}{T_1} = \left(\frac{v_1}{v_2}\right)^{R/C_v} = \left(\frac{v_1}{v_2}\right)^{k-1} \tag{4.49}$$

referring to (*4.31*). Using the ideal-gas law, this can be written as

$$\frac{T_2}{T_1} = \left(\frac{P_2}{P_1}\right)^{(k-1)/k} \qquad \frac{P_2}{P_1} = \left(\frac{v_1}{v_2}\right)^k \tag{4.50}$$

Finally, the above three relations can be put in general forms, without reference to particular points. For the adiabatic quasiequilibrium process involving an ideal gas with constant C_p and C_v, we have

$$Tv^{k-1} = \text{const.} \qquad TP^{(1-k)/k} = \text{const.} \qquad Pv^k = \text{const.} \tag{4.51}$$

For a substance that does not behave as an ideal gas, we must utilize tables. For such a process we return to (*4.45*) and recognize that $d\psi = 0$, or $\psi = \text{const.}$ We do not assign the property ψ a formal name, but, as we shall show in Chap. 7, the ψ function is constant whenever the quantity denoted by s, *the entropy*, is constant. Hence, when using the tables, an adiabatic quasiequilibrium process between states 1 and 2 requires $s_1 = s_2$.

The Polytropic Process

A careful inspection of the special quasiequilibrium processes presented in this chapter suggests that each process can be expressed as

$$PV^n = \text{const.} \tag{4.52}$$

The work is calculated

$$\begin{aligned} W &= \int_{V_1}^{V_2} P\,dV = P_1V_1^n \int_{V_1}^{V_2} V^{-n}\,dV \\ &= \frac{P_1V_1^n}{1-n}(V_2^{1-n} - V_1^{1-n}) = \frac{P_2V_2 - P_1V_1}{1-n} \end{aligned} \tag{4.53}$$

except (*4.36*) is used if $n = 1$. The heat transfer follows from the first law.

Each quasiequilibrium process is associated with a particular value for n as follows:

Isothermal: $n = 1$

Constant-volume: $n = \infty$

Constant-pressure: $n = 0$

Adiabatic: $n = k$

The processes are displayed on a (ln P) vs. (ln V) plot in Fig. 4-7. The slope of each straight line is the exponent on V in (*4.52*). If the slope is none of the values ∞, k, 1, or zero, then the process can be

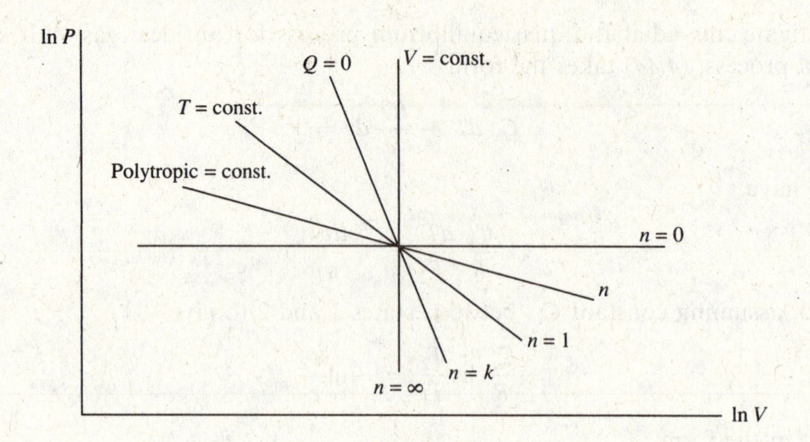

Fig. 4-7 Polytropic exponents for various processes.

referred to as a *polytropic process*. For such a process any of the equations (*4.49*), (*4.50*), or (*4.51*) can be used with k simply replaced by n; this is convenient in processes in which there is some heat transfer but which do not maintain temperature, pressure, or volume constant.

EXAMPLE 4.9 Determine the heat transfer necessary to increase the pressure of 70 percent quality steam from 200 to 800 kPa, maintaining the volume constant at 2 m^3. Assume a quasiequilibrium process.

Solution: For the constant-volume quasiequilibrium process the work is zero. The first law reduces to $Q = m(u_2 - u_1)$. The mass is found to be

$$m = \frac{V}{v} = \frac{2}{0.0011 + (0.7)(0.8857 - 0.0011)} = \frac{2}{0.6203} = 3.224 \text{ kg}$$

The internal energy at state 1 is

$$u_1 = 504.5 + (0.7)(2529.5 - 504.5) = 1922 \text{ kJ/kg}$$

The constant-volume process demands that $v_2 = v_1 = 0.6203$ m^3/kg. From the steam tables at 800 kPa we find, by extrapolation, that

$$u_2 = \left(\frac{0.6203 - 0.6181}{0.6181 - 0.5601}\right)(3661 - 3476) = 3668 \text{ kJ/kg}$$

Note that extrapolation was necessary since the temperature at state 2 exceeds the highest tabulated temperature of 800 °C. The heat transfer is then

$$Q = (3.224)(3668 - 1922) = 5629 \text{ kJ}$$

EXAMPLE 4.10 A piston-cylinder arrangement contains 0.02 m^3 of air at 50 °C and 400 kPa. Heat is added in the amount of 50 kJ and work is done by a paddle wheel until the temperature reaches 700 °C. If the pressure is held constant how much paddle-wheel work must be added to the air? Assume constant specific heats.

Solution: The process cannot be approximated by a quasiequilibrium process because of the paddle-wheel work. Thus, the heat transfer is not equal to the enthalpy change. The first law may be written as

$$Q - W_{\text{paddle}} = m(h_2 - h_1) = mC_p(T_2 - T_1)$$

To find m we use the ideal-gas equation. It gives us

$$m = \frac{PV}{RT} = \frac{(400\,000)(0.02)}{(287)(273 + 50)} = 0.0863 \text{ kg}$$

From the first law the paddle-wheel work is found to be

$$W_{\text{paddle}} = Q - mC_p(T_2 - T_1) = 50 - (0.0863)(1.00)(700 - 50) = -6.095 \text{ kJ}$$

Note: We could have used the first law as $Q - W_{\text{net}} = m(u_2 - u_1)$ and then let $W_{\text{paddle}} = W_{\text{net}} - P(V_2 - V_1)$. We would then need to calculate V_2.

EXAMPLE 4.11 Calculate the work necessary to compress air in an insulated cylinder from a volume of 6 ft^3 to a volume of 1.2 ft^3. The initial temperature and pressure are 50 °F and 30 psia, respectively.

Solution: We will assume that the compression process is approximated by a quasiequilibrium process, which is acceptable for most compression processes, and that the process is adiabatic due to the presence of the insulation. The first law is then written as

$$-W = m(u_2 - u_1) = mC_v(T_2 - T_1)$$

The mass is found from the ideal-gas equation to be

$$m = \frac{PV}{RT} = \frac{[(30)(144)](6)}{(53.3)(460 + 50)} = 0.9535 \text{ lbm}$$

The final temperature T_2 is found for the adiabatic quasiequilibrium process from (*4.49*); it is

$$T_2 = T_1\left(\frac{V_1}{V_2}\right)^{k-1} = (510)\left(\frac{6.0}{1.2}\right)^{1.4-1} = 970.9\,°\text{R}$$

Finally, $W = (-0.9535 \text{ lbm})(0.171 \text{ Btu/lbm-°R})(970.9 - 510)\,°\text{R} = -75.1 \text{ Btu}$.

4.8 GENERAL FORMULATION FOR CONTROL VOLUMES

In the application of the various laws we have thus far restricted ourselves to systems, with the result that no mass has crossed the system boundaries. This restriction is acceptable for many problems of interest and may, in fact, be imposed on the power plant schematic shown in Fig. 4-8. However, if the first law is applied to this system, only an incomplete analysis can be accomplished. For a more complete analysis we must relate W_{in}, Q_{in}, W_{out}, and Q_{out} to the pressure and temperature changes for the pump, boiler, turbine, and condenser, respectively. To do this we must consider each device of the power plant as a control volume into which and from which a fluid flows. For example, water flows into the pump at a low pressure and leaves the pump at a high pressure; the work input into the pump is obviously related

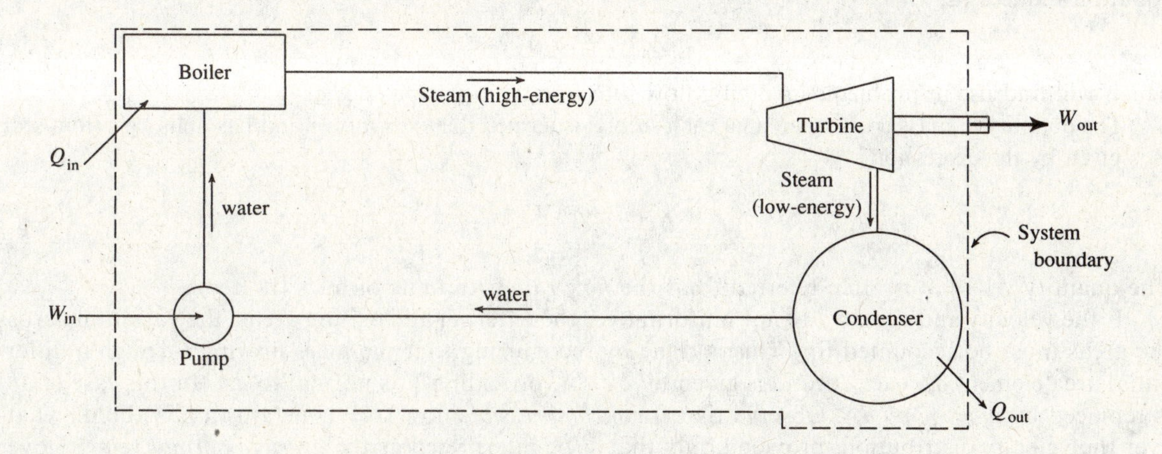

Fig. 4-8 A schematic for a power plant.

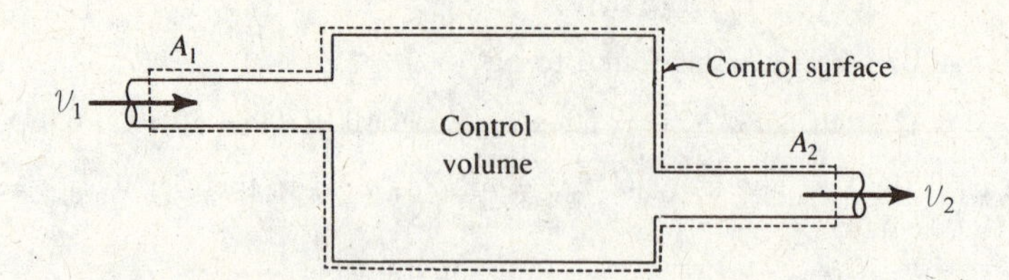

Fig. 4-9 Mass entering and leaving a control volume.

to this pressure rise. We must formulate equations that allow us to make this necessary calculation. For most applications that we will consider it will be acceptable to assume both a *steady flow* (the flow variables do not change with time) and a *uniform flow* (the velocity, pressure, and density are constant over the cross-sectional area). Fluid mechanics treats the more general unsteady, nonuniform situations in much greater depth.

The Continuity Equation

Consider a general control volume with an area A_1 where fluid enters and an area A_2 where fluid leaves, as shown in Fig. 4-9. It could have any shape and any number of entering and exiting areas, but we will derive the continuity equation using the geometry shown. *Conservation of mass* requires that

$$\begin{pmatrix}\text{Mass entering}\\ \text{control volume}\end{pmatrix} - \begin{pmatrix}\text{Mass leaving}\\ \text{control volume}\end{pmatrix} = \begin{pmatrix}\text{Change in mass}\\ \text{within control volume}\end{pmatrix} \tag{4.54}$$
$$m_1 \qquad - \qquad m_2 \qquad = \qquad \Delta m_{\text{c.v.}}$$

The mass that crosses an area A over a time increment Δt can be expressed as $\rho A V\,\Delta t$, where $V\,\Delta t$ is the distance the mass particles travel and $AV\,\Delta t$ is the volume swept out by the mass particles. Equation (*4.54*) can thus be put in the form

$$\rho_1 A_1 V_1 \Delta t - \rho_2 A_2 V_2 \Delta t = \Delta m_{\text{c.v.}} \tag{4.55}$$

where the velocities V_1 and V_2 are perpendicular to the areas A_1 and A_2, respectively. We have assumed the velocity and density to be uniform over the two areas.

If we divide by Δt and let $\Delta t \to 0$, the derivative results and we have the *continuity equation*,

$$\rho_1 A_1 V_1 = \rho_2 A_2 V_2 = \frac{dm_{\text{c.v.}}}{dt} \tag{4.56}$$

For the steady-flow situation, in which the mass in the control volume remains constant, the continuity equation reduces to

$$\rho_1 A_1 V_1 = \rho_2 A_2 V_2 \tag{4.57}$$

which will find use in problems involving flow into and from various devices.

The quantity of mass crossing an area each second is termed the *mass flux* $\dot{m}$ and has units kg/s (lbm/sec). It is given by the expression

$$\dot{m} = \rho A V \tag{4.58}$$

The quantity $AV = \dot{V}$ is often referred to as the *flow rate* with units of m^3/s (ft^3/sec).

If the velocity and density are not uniform over the entering and exiting areas, the variation across the areas must be accounted for. This is done by recognizing that the mass flowing through a differential area element dA each second is given by $\rho V\,dA$, providing V is normal to dA. In this case (*4.58*) is replaced by $\dot{m} = \int_A \rho V\,dA$. Observe that for *incompressible* flow ($\rho =$ constant), (*4.58*) holds whatever the velocity distribution, provided only that V be interpreted as the *average normal velocity* over the area A.

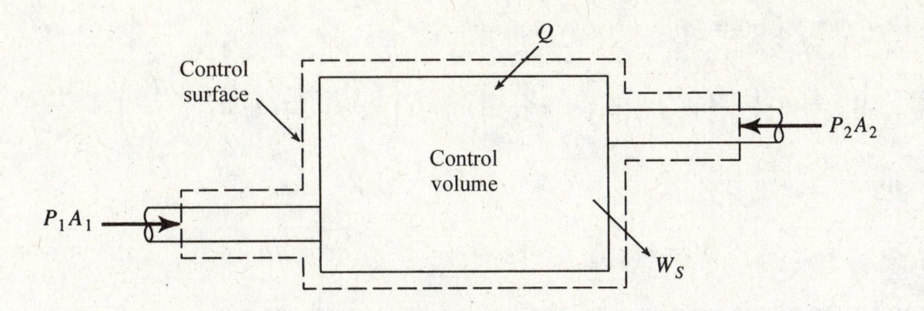

Fig. 4-10 The control volume used for an energy balance.

EXAMPLE 4.12 Water is flowing in a pipe that changes diameter from 20 to 40 mm. If the water in the 20-mm section has a velocity of 40 m/s, determine the velocity in the 40-mm section. Also calculate the mass flux.

Solution: The continuity equation (*4.57*) is used. There results, using $\rho_1 = \rho_2$,

$$A_1 V_1 = A_2 V_2 \qquad \left[\frac{\pi(0.02)^2}{4}\right](40) = \frac{\pi(0.04)^2}{4} V_2 \qquad \therefore V_2 = 10 \text{ m/s}$$

The mass flux is found to be

$$\dot{m} = \rho A_1 V_1 = (1000)\left(\frac{\pi(0.02)^2}{4}\right)(40) = 12.57 \text{ kg/s}$$

where $\rho = 1000 \text{ kg/m}^3$ is the standard value for water.

The Energy Equation

Consider again a general control volume as sketched in Fig. 4-10. The first law of thermodynamics for this control volume can be stated as

$$\begin{pmatrix}\text{Net energy}\\ \text{transferred to}\\ \text{the c.v.}\end{pmatrix} + \begin{pmatrix}\text{Energy}\\ \text{entering}\\ \text{the c.v.}\end{pmatrix} - \begin{pmatrix}\text{Energy}\\ \text{leaving}\\ \text{the c.v.}\end{pmatrix} = \begin{pmatrix}\text{Change of}\\ \text{energy in}\\ \text{the c.v.}\end{pmatrix}$$
$$Q - W \quad + \quad E_1 \quad - \quad E_2 \quad = \quad \Delta E_{\text{c.v.}} \tag{4.59}$$

The work W is composed of two parts: the work due to the pressure needed to move the fluid, sometimes called *flow work*, and the work that results from a rotating shaft, called *shaft work* W_S. This is expressed as

$$W = P_2 A_2 V_2 \Delta t - P_1 A_1 V_1 \Delta t + W_S \tag{4.60}$$

where PA is the pressure force and $V \Delta t$ is the distance it moves during the time increment Δt. The negative sign results because the work done on the system is negative when moving the fluid into the control volume.

The energy E is composed of kinetic energy, potential energy, and internal energy. Thus,

$$E = \frac{1}{2} m V^2 + mgz + mu \tag{4.61}$$

The first law can now be written as

$$\begin{aligned} Q - W_S - P_2 A_2 V_2 \,\Delta t + P_1 A_1 V_1 \,\Delta t + \rho_1 A_1 V_1 \left(\frac{V_1^2}{2} + gz_1 + u_1\right) \Delta t \\ - \rho_2 A_2 V_2 \left(\frac{V_2^2}{2} + gz_2 + u_2\right) \Delta t = \Delta E_{\text{c.v.}} \end{aligned} \tag{4.62}$$

Divide through by Δt to obtain *the energy equation*:

$$\dot{Q} - \dot{W}_S = \dot{m}_2\left(\frac{V_2^2}{2} + gz_2 + u_2 + \frac{P_2}{\rho_2}\right) - \dot{m}_1\left(\frac{V_1^2}{2} + gz_1 + u_1 + \frac{P_1}{\rho_1}\right) + \frac{dE_{\text{c.v.}}}{dt} \tag{4.63}$$

where we have used

$$\dot{Q} = \frac{Q}{\Delta t} \qquad \dot{W}_S = \frac{W}{\Delta t} \qquad \dot{m} = \rho A V \tag{4.64}$$

For steady flow, a very common situation, the energy equation becomes

$$\dot{Q} - \dot{W}_S = \dot{m}\left[h_2 - h_1 + g(z_2 - z_1) + \frac{V_2^2 - V_1^2}{2}\right] \tag{4.65}$$

where the enthalpy of (*4.12*) has been introduced. This is the form most often used when a gas or a vapor is flowing.

Quite often the kinetic energy and potential energy changes are negligible. The first law then takes the simplified form

$$\dot{Q} - \dot{W}_S = \dot{m}(h_2 - h_1) \tag{4.66}$$

or

$$q - w_S = h_2 - h_1 \tag{4.67}$$

where $q = \dot{Q}/\dot{m}$ and $w_S = \dot{W}_S/\dot{m}$. This simplified form of the energy equation has a surprisingly large number of applications.

For a control volume through which a liquid flows, it is most convenient to return to (*4.63*). For a steady flow with $\rho_2 = \rho_1 = \rho$, neglecting the heat transfer and changes in internal energy, the energy equation takes the form

$$-\dot{W}_S = \dot{m}\left[\frac{P_2 - P_1}{\rho} + \frac{V_2^2 - V_1^2}{2} + g(z_2 - z_1)\right] \tag{4.68}$$

This is the form to use for a pump or a hydroturbine. If $\dot{Q}$ and Δu are not zero, simply include them.

4.9 APPLICATIONS OF THE ENERGY EQUATION

There are several points that must be considered in the analysis of most problems in which the energy equation is used. As a first step, it is very important to identify the control volume selected in the solution of a problem; dotted lines are used to outline the control surface. If at all possible, the control surface should be chosen so that the flow variables are uniform or known functions over the areas where the fluid enters or exits the control volume. For example, in Fig. 4-11 the area could be chosen as in part (*a*), but the velocity and the pressure are certainly not uniform over the area. In part (*b*), however, the control surface is chosen sufficiently far downstream from the abrupt area change that the exiting velocity and pressure can be approximated by uniform distributions.

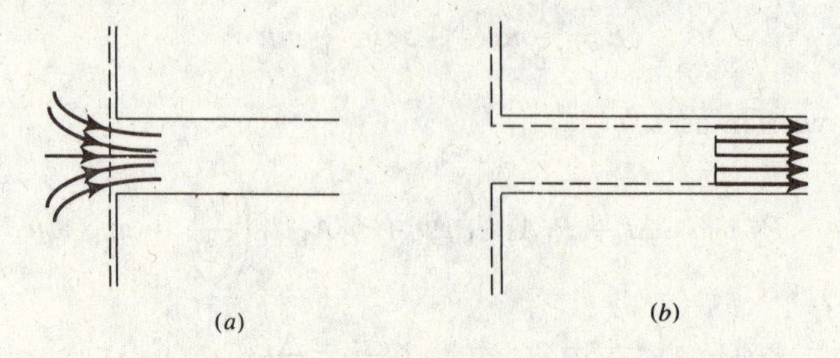

Fig. 4-11 The control surface at an entrance.

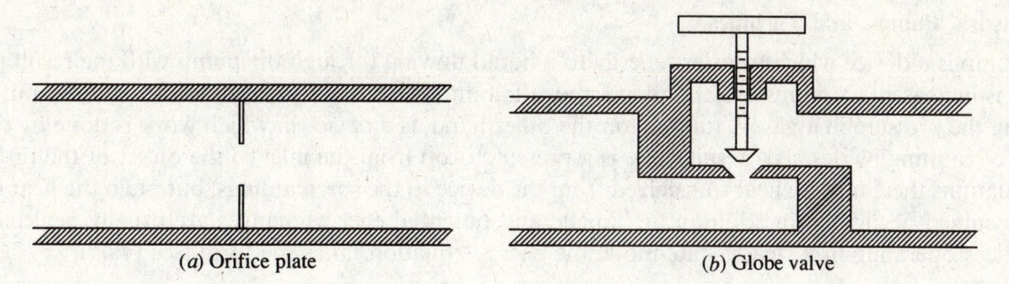

Fig. 4-12 Throttling devices.

It is also necessary to specify the process by which the flow variables change. Is it incompressible? isothermal? constant-pressure? adiabatic? A sketch of the process on a suitable diagram is often of use in the calculations. If the working substance behaves as an ideal gas, then the appropriate equations can be used; if not, tabulated values must be used, such as those provided for steam. For real gases that do not behave as ideal gases, specialized equations may be available for calculations; some of these equations will be presented in a later chapter.

Often heat transfer from a device or an internal energy change across a device, such as flow through a pump, is not desired. For such situations, the heat transfer and internal energy change may be lumped together as *losses*. In a pipeline losses occur because of friction; in a centrifugal pump, losses occur because of poor fluid motion around the rotating blades. For many devices the losses are included as an efficiency of the device. Examples will illustrate.

Kinetic energy or potential energy changes can often be neglected in comparison with other terms in the energy equation. Potential energy changes are usually included only in situations where liquid is involved and where the inlet and exit areas are separated by a large vertical distance. The following applications will illustrate many of the above points.

Throttling Devices

A throttling device involves a steady-flow adiabatic process that provides a sudden pressure drop with no significant potential energy or kinetic energy changes. The process occurs relatively rapidly, with the result that negligible heat transfer occurs. Two such devices are sketched in Fig. 4-12. If the energy equation is applied to such a device, with no work done and neglecting kinetic and potential energy changes, we have, for this adiabatic non-quasiequilibrium process [see (*4.67*)],

$$h_1 = h_2 \tag{4.69}$$

where section 1 is upstream and section 2 is downstream. Most valves are throttling devices, for which the energy equation takes the form of (*4.69*). They are also used in many refrigeration units in which the sudden drop in pressure causes a change in phase of the working substance. The throttling process is analogous to the sudden expansion of Fig. 3-5*b*.

EXAMPLE 4.13 Steam enters a throttling valve at 8000 kPa and 300 °C and leaves at a pressure of 1600 kPa. Determine the final temperature and specific volume of the steam.

Solution: The enthalpy of the steam as it enters is found from the superheat steam table to be $h_1 = 2785$ kJ/kg. This must equal the exiting enthalpy as demanded by (*4.69*). The exiting steam is in the quality region, since at 1600 kPa $h_g = 2794$ kJ/kg. Thus the final temperature is $T_2 = 201.4\,°\mathrm{C}$.

To find the specific volume we must know the quality. It is found from

$$h_2 = h_f + x_2 h_{fg} \qquad 2785 = 859 + 1935x_2 \qquad x_2 = 0.995$$

The specific volume is then $v_2 = 0.0012 + (0.995)(0.1238 - 0.0012) = 0.1232\ \mathrm{m^3/kg}$.

Compressors, Pumps, and Turbines

A pump is a device which transfers energy to a liquid flowing through the pump with the result that the pressure is increased. Compressors and blowers also fall into this category but have the primary purpose of increasing the pressure in a gas. A turbine, on the other hand, is a device in which work is done by the fluid on a set of rotating blades; as a result there is a pressure drop from the inlet to the outlet of the turbine. In some situations there may be heat transferred from the device to the surroundings, but often the heat transfer can be assumed negligible. In addition the kinetic and potential energy changes are usually neglected. For such devices operating in a steady-state mode the energy equation takes the form [see (*4.66*)]

$$-\dot{W}_S = \dot{m}(h_2 - h_1) \qquad \text{or} \qquad -w_S = h_2 - h_1 \tag{4.70}$$

where $\dot{W}_S$ is negative for a compressor and positive for a gas or steam turbine. In the event that heat transfer does occur, from perhaps a high-temperature working fluid, it must, of course, be included in the above equation.

For liquids, such as water, the energy equation (*4.68*), neglecting kinetic and potential energy changes, becomes

$$-w_S = \frac{P_2 - P_1}{\rho} \tag{4.71}$$

EXAMPLE 4.14 Steam enters a turbine at 4000 kPa and 500 °C and leaves as shown in Fig. 4-13. For an inlet velocity of 200 m/s calculate the turbine power output. (*a*) Neglect any heat transfer and kinetic energy change. (*b*) Show that the kinetic energy change is negligible.

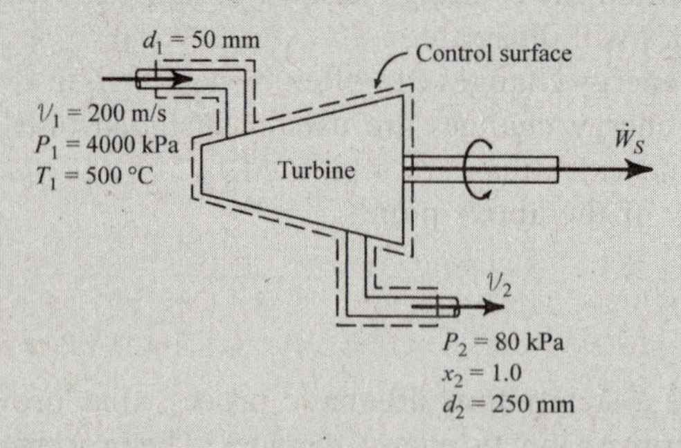

Fig. 4-13

Solution:

(*a*) The energy equation in the form of (*4.70*) is $-\dot{W}_T = (h_2 - h_1)\dot{m}$. We find $\dot{m}$ as follows:

$$\dot{m} = \rho_1 A_1 V_1 = \frac{1}{v_1} A_1 V_1 = \frac{\pi(0.025)^2(200)}{0.08643} = 4.544 \text{ kg/s}$$

The enthalpies are found from Table C-3 to be

$$h_1 = 3445.2 \text{ kJ/kg} \qquad h_2 = 2665.7 \text{ kJ/kg}$$

The maximum power output is then $\dot{W}_T = -(2665.7 - 3445.2)(4.544) = 3542 \text{ kJ/s}$ or 3.542 MW.

(*b*) The exiting velocity is found to be

$$V_2 = \frac{A_1 V_1 \rho_1}{A_2 \rho_2} = \frac{\pi(0.025)^2(200/0.08643)}{\pi(0.125)^2/2.087} = 193 \text{ m/s}$$

The kinetic energy change is then

$$\Delta KE = \dot{m}\left(\frac{V_2^2 - V_1^2}{2}\right) = (4.544)\left(\frac{193^2 - 200^2}{2}\right) = -6250 \text{ J/s} \quad \text{or } -6.25 \text{ kJ/s}$$

This is less than 0.1 percent of the enthalpy change and is indeed negligible. Kinetic energy changes are usually omitted in the analysis of a turbine.

EXAMPLE 4.15 Determine the maximum pressure increase across the 10-hp pump shown in Fig. 4-14. The inlet velocity of the water is 30 ft/sec.

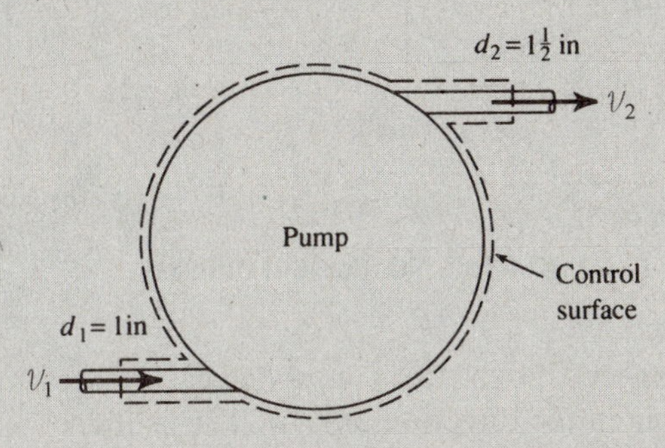

Fig. 4-14

Solution: The energy equation (*4.68*) is used. By neglecting the heat transfer and assuming no increase in internal energy, we establish the maximum pressure rise. Neglecting the potential energy change, the energy equation takes the form

$$-\dot{W}_S = \dot{m}\left(\frac{P_2 - P_1}{\rho} + \frac{V_2^2 - V_1^2}{2}\right)$$

The velocity V_1 is given, and V_2 is found from the continuity equation as follows:

$$\rho A_1 V_1 = \rho A_2 V_2 \qquad \left[\frac{\pi(1)^2}{4}\right](30) = \frac{\pi(1.5)^2}{4} V_2 \qquad \therefore V_2 = 13.33 \text{ ft/sec}$$

The mass flux, needed in the energy equation, is then, using $\rho = 62.4$ lbm/ft^3,

$$\dot{m} = \rho A_1 V_1 = (62.4)\left[\frac{\pi(1)^2}{(4 \times 144)}\right](30) = 10.21 \text{ lbm/sec}$$

Recognizing that the pump work is negative, the energy equation is

$$-(-10)(550) \text{ ft-lbf/sec} = (10.21 \text{ lbm/sec})\left[\frac{(P_2 - P_1)\text{ lbf/ft}^2}{62.4 \text{ lbm/ft}^3} + \frac{(13.33^2 - 30^2)\text{ ft}^2/\text{sec}^2}{(2)(32.2 \text{ lbm-ft/sec}^2\text{-lbf})}\right]$$

where the factor 32.2 lbm-ft/sec^2-lbf is needed to obtain the correct units on the kinetic energy term. This predicts a pressure rise of

$$P_2 - P_1 = (62.4)\left[\frac{5500}{10.21} - \frac{13.33^2 - 30^2}{(2)(32.2)}\right] = 34{,}310 \text{ lbf/ft}^2 \quad \text{or } 238.3 \text{ psi}$$

Note that in this example the kinetic energy terms are retained because of the difference in inlet and exit areas; if they were omitted, only a 2 percent error would result. In most applications the inlet and exit areas will be equal so that $V_2 = V_1$; but even with different areas, as in this example, kinetic energy changes are usually ignored in a pump or turbine and (*4.71*) is used.

Nozzles and Diffusers

A nozzle is a device that is used to increase the velocity of a flowing fluid. It does this by reducing the pressure. A diffuser is a device that increases the pressure in a flowing fluid by reducing the velocity. There is no work input into the devices and usually negligible heat transfer. With the additional assumptions of negligible internal energy and potential energy changes, the energy equation takes the form

$$0 = \frac{V_2^2}{2} - \frac{V_1^2}{2} + h_2 - h_1 \tag{4.72}$$

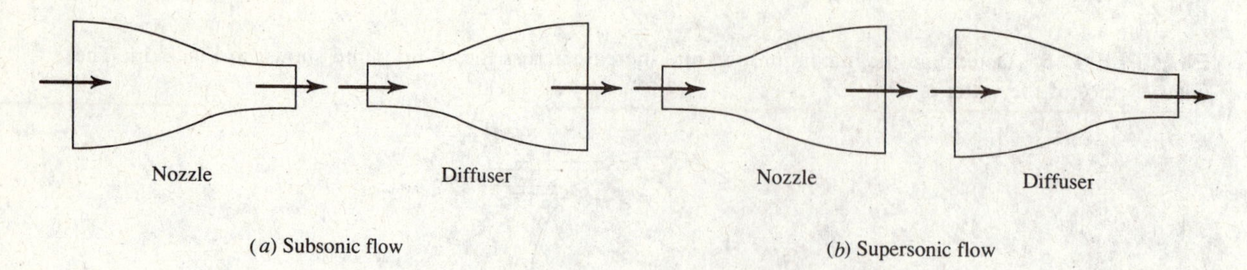

Fig. 4-15 Nozzles and diffusers.

Based on our intuition we expect a nozzle to have a decreasing area in the direction of flow and a diffuser to have an increasing area in the direction of flow. This is indeed the case for a subsonic flow in which $V < \sqrt{kRT}$. For a supersonic flow in which $V > \sqrt{kRT}$ the opposite is true: a nozzle has an increasing area and a diffuser has a decreasing area. This is shown in Fig. 4-15.

Three equations may be used for nozzle and diffuser flow; energy, continuity, and a process equation, such as for an adiabatic quasiequilibrium flow. Thus, we may have three unknowns at the exit, given the entering conditions. There may also be shock waves in supersonic flows or "choked" subsonic flows. These more complicated flows are included in a fluid mechanics course. Only the more simple situations will be included here.

EXAMPLE 4.16 Air flows through the supersonic nozzle shown in Fig. 4-16. The inlet conditions are 7 kPa and 420 °C. The nozzle exit diameter is adjusted such that the exiting velocity is 700 m/s. Calculate (*a*) the exit temperature, (*b*) the mass flux, and (*c*) the exit diameter. Assume an adiabatic quasiequilibrium flow.

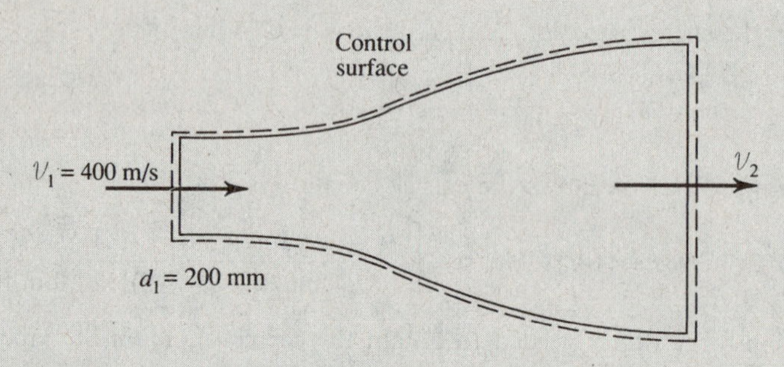

Fig. 4-16

Solution:

(*a*) To find the exit temperature the energy equation (*4.72*) is used. It is, using $\Delta h = C_p \Delta T$,

$$\frac{V_1^2}{2} + C_p T_1 = \frac{V_2^2}{2} + C_p T_2$$

We then have, using $C_p = 1000$ J/kg·K,

$$T_2 = \frac{V_1^2 - V_2^2}{2C_p} + T_1 = \frac{400^2 - 700^2}{(2)(1000)} + 420 = 255\,^\circ\text{C}$$

(*b*) To find the mass flux we must find the density at the entrance. From the inlet conditions we have

$$\rho_1 = \frac{P_1}{RT_1} = \frac{7000}{(287)(693)} = 0.03520 \text{ kg/m}^3$$

The mass flux is then $\dot{m} = \rho_1 A_1 V_1 = (0.0352)(\pi)(0.1)^2(400) = 0.4423$ kg/s.

(c) To find the exit diameter we would use the continuity equation $\rho_1 A_1 V_1 = \rho_2 A_2 V_2$. This requires the density at the exit. It is found by assuming adiabatic quasiequilibrium flow. Referring to (*4.49*), we have

$$\rho_2 = \rho_1 \left(\frac{T_2}{T_1}\right)^{1/(k-1)} = (0.0352)\left(\frac{528}{693}\right)^{1/(1.4-1)} = 0.01784 \text{ kg/m}^3$$

Hence,

$$d_2^2 = \frac{\rho_1 d_1^2 V_1}{\rho_2 V_2} = \frac{(0.0352)(0.2^2)(400)}{(0.01784)(700)} = 0.0451 \qquad \therefore d_2 = 0.212 \text{ m} \quad \text{or } 212 \text{ mm}$$

Heat Exchangers

An important device that has many applications in engineering is the heat exchanger. Heat exchangers are used to transfer energy from a hot body to a colder body or to the surroundings by means of heat transfer. Energy is transferred from the hot gases after combustion in a power plant to the water in the pipes of the boiler and from the hot water that leaves an automobile engine to the atmosphere, and electrical generators are cooled by water flowing through internal flow passages.

Many heat exchangers utilize a flow passage into which a fluid enters and from which the fluid exits at a different temperature. The velocity does not normally change, the pressure drop through the passage is usually neglected, and the potential energy change is assumed zero. The energy equation then results in

$$\dot{Q} = (h_2 - h_1)\dot{m} \tag{4.73}$$

since no work occurs in the heat exchanger.

Energy may be exchanged between two moving fluids, as shown schematically in Fig. 4-17. For a control volume including the combined unit, which is assumed to be insulated, the energy equation, as applied to the control volume of Fig. 4-17*a*, would be

$$0 = \dot{m}_A(h_{A2} - h_{A1}) + \dot{m}_B(h_{B2} - h_{B1}) \tag{4.74}$$

The energy that leaves fluid A is transferred to fluid B by means of the heat transfer $\dot{Q}$. For the control volumes shown in Fig. 4-17*b* we have

$$\dot{Q} = \dot{m}_B(h_{B2} - h_{B1}) \qquad -\dot{Q} = \dot{m}_A(h_{A2} - h_{A1}) \tag{4.75}$$

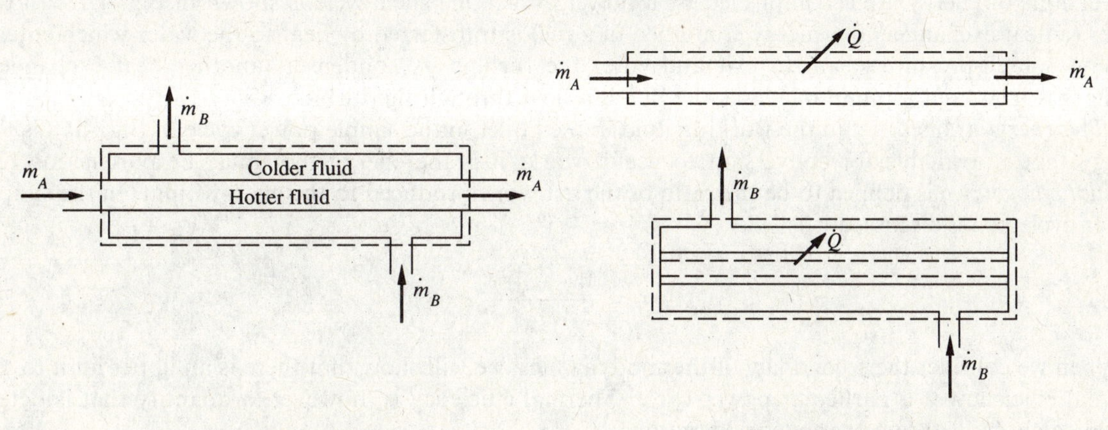

Fig. 4-17 A heat exchanger.

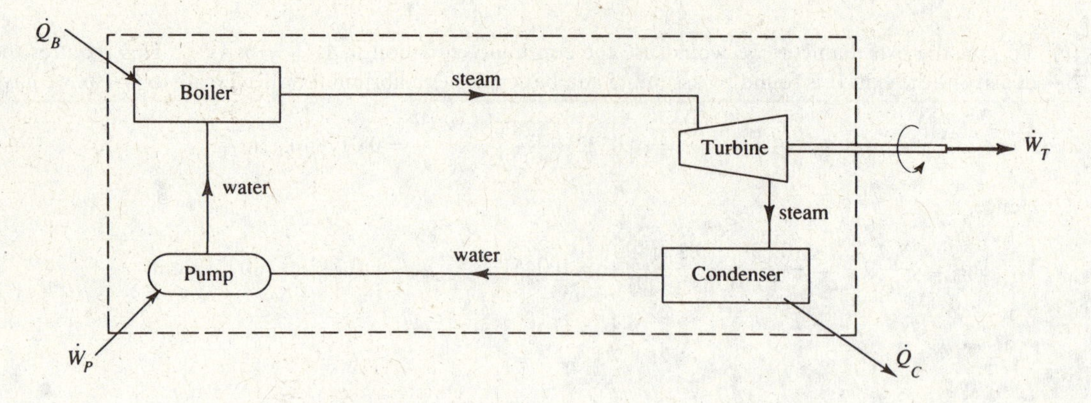

Fig. 4-18 A simple power schematic.

EXAMPLE 4.17 A liquid, flowing at 100 kg/s, enters a heat exchanger at 450 °C and exits at 350 °C. The specific heat of the liquid is 1.25 kJ/kg · °C. Water enters at 5000 kPa and 20 °C. Determine the minimum mass flux of the water so that the water does not completely vaporize. Neglect the pressure drop through the exchanger. Also, calculate the rate of heat transfer.

Solution: The energy equation (*4.74*) is used as $\dot{m}_s(h_{s1} - h_{s2}) = \dot{m}_w(h_{w2} - h_{w1})$, or

$$\dot{m}_s C_p(T_{s1} - T_{s2}) = \dot{m}_w(h_{w2} - h_{w1})$$

Using the given values, we have (use Table C-4 to find h_{w1})

$$(100)(1.25) \times (450 - 350) = \dot{m}_w(2792.8 - 88.7) \qquad \therefore \dot{m}_w = 4.623 \text{ kg/s}$$

where we have assumed a saturated vapor state for the exiting steam to obtain the maximum allowable exiting enthalpy. The heat transfer is found using the energy equation (*4.75*) applied to one of the separate control volumes.

$$\dot{Q} = \dot{m}_w(h_{w2} - h_{w1}) = (4.623)(2792.8 - 88.7) = 12\,500 \text{ kW} \qquad \text{or } 12.5 \text{ MW}$$

Power and Refrigeration Cycles

When energy in the form of heat is transferred to a working fluid, energy in the form of work may be extracted from the working fluid. The work may be converted to an electrical form of energy, such as is done in a power plant, or to a mechanical form, such as is done in an automobile. In general, such conversions of energy are accomplished by a power cycle. One such cycle is shown in Fig. 4-18. In the boiler (a heat exchanger) the energy contained in a fuel is transferred by heat to the water which enters, causing a high-pressure steam to exit and enter the turbine. A condenser (another heat exchanger) discharges heat, and a pump increases the pressure lost through the turbine.

The energy transferred to the working fluid in the boiler in the simple power cycle of Fig. 4-18 is the energy that is available for conversion to useful work; it is the energy that must be purchased. The *thermal efficiency* η is defined to be the ratio of the net work produced to the energy input. In the simple power cycle being discussed it is

$$\eta = \frac{\dot{W}_T - \dot{W}_P}{\dot{Q}_B} \tag{4.76}$$

When we consider the second law of thermodynamics, we will show that there is an upper limit to the thermal efficiency of a particular power cycle. Thermal efficiency is, however, a quantity that is determined solely by first-law energy considerations.

Other components can be combined in an arrangement like that shown in Fig. 4-19, resulting in a refrigeration cycle. Heat is transferred to the working fluid (the refrigerant) in the evaporator (a heat

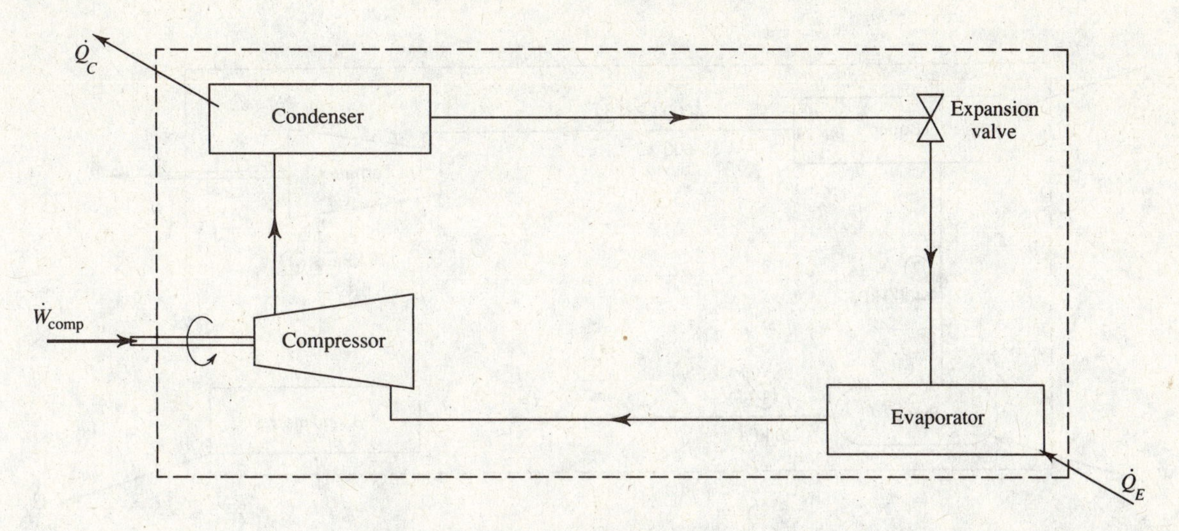

Fig. 4-19 A simple refrigeration schematic.

exchanger). The working fluid is then compressed by the compressor. Heat is transferred from the working fluid in the condenser, and then its pressure is suddenly reduced in the expansion valve. A refrigeration cycle may be used to add energy to a body (heat transfer $\dot{Q}_C$) or it may be used to extract energy from a body (heat transfer $\dot{Q}_E$).

It is not useful to calculate the thermal efficiency of a refrigeration cycle since the objective is not to do work but to accomplish heat transfer. If we are extracting energy from a body, our purpose is to cause maximum heat transfer with minimum work input. To measure this, we define a *coefficient of performance* (abbreviated COP) as

$$\text{COP} = \frac{\dot{Q}_E}{\dot{W}_{\text{comp}}} = \frac{\dot{Q}_E}{\dot{Q}_C - \dot{Q}_E} \tag{4.77}$$

If we are adding energy to a body, our purpose is, again, to do so with a minimum work input. In this case the coefficient of performance is defined as

$$\text{COP} = \frac{\dot{Q}_C}{\dot{W}_{\text{comp}}} = \frac{\dot{Q}_C}{\dot{Q}_C - \dot{Q}_E} \tag{4.78}$$

A device which can operate with this latter objective is called a *heat pump*; if it operates with the former objective only it is a *refrigerator*.

It should be apparent from the definitions that thermal efficiency can never be greater than unity but that the coefficient of performance can be greater than unity. Obviously, the objective of the engineer is to maximize either one in a particular design. The thermal efficiency of a power plant is around 35 percent; the thermal efficiency of an automobile engine is around 20 percent. The coefficient of performance for a refrigerator or a heat pump ranges from 2 to 6, with a heat pump having the greater values.

EXAMPLE 4.18 Steam leaves the boiler of a simple steam power cycle at 4000 kPa and 600 °C. It exits the turbine at 20 kPa as saturated steam. It then exits the condenser as saturated water. (See Fig. 4-20.) Determine the thermal efficiency if there is no loss in pressure through the condenser and the boiler.

Solution: To determine the thermal efficiency we must calculate the heat transferred to the water in the boiler, the work done by the turbine, and the work required by the pump. We will make the calculations for 1 kg of steam since the mass is unknown. The boiler heat transfer is, neglecting kinetic and potential energy changes, $q_B = h_3 - h_2$. To find h_2 we assume that the pump simply increases the pressure [see (*4.71*)]:

$$w_p = (P_2 - P_1)v = (4000 - 20)(0.001) = 3.98 \text{ kJ/kg}$$

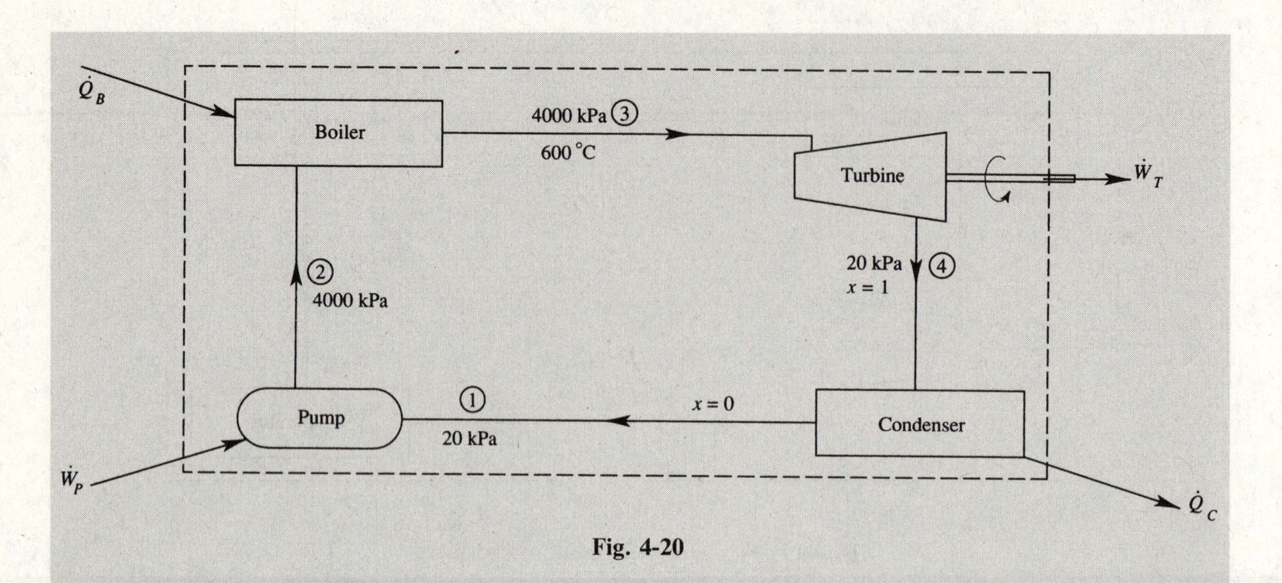

Fig. 4-20

The enthalpy h_2 is thus found to be, using (*4.70*),

$$h_2 = w_p + h_1 = 3.98 + 251.4 = 255.4\,\text{kJ/kg}$$

where h_1 is assumed to be that of saturated water at 20 kPa. From the steam tables we find $h_3 = 3674\,\text{kJ/kg}$. There results

$$q_B = 3674 - 255.4 = 3420\,\text{kJ/kg}$$

The work output from the turbine is $w_T = h_3 - h_4 = 3674 - 2610 = 1064\,\text{kJ/kg}$. Finally, the thermal efficiency is

$$\eta = \frac{w_T - w_P}{q_B} = \frac{1064 - 4}{3420} = 0.310 \quad \text{or } 31.0\%$$

Note that the pump work could have been neglected with no significant change in the results.

Transient Flow

If the steady-flow assumption of the preceding sections is not valid, then the time dependence of the various properties must be included. The filling of a rigid tank with a gas and the release of gas from a pressurized tank are examples that we will consider.

The energy equation is written as

$$\dot{Q} - \dot{W}_S = \frac{dE_{\text{c.v.}}}{dt} + \dot{m}_2\left(\frac{V_2^2}{2} + gz_2 + h_2\right) - \dot{m}_1\left(\frac{V_1^2}{2} + gz_1 + h_1\right) \tag{4.79}$$

We will consider the kinetic energy and potential energy terms to be negligible so that $E_{\text{c.v.}}$ will consist of internal energy only. The first problem we wish to study is the filling of a rigid tank, as sketched in Fig. 4-21. In the tank, there is only an entrance. With no shaft work present the energy equation reduces to

$$\dot{Q} = \frac{d}{dt}(um) - \dot{m}_1 h_1 \tag{4.80}$$

where m is the mass in the control volume. If we multiply this equation by dt and integrate from an initial time t_i, to some final time t_f, we have

$$Q = u_f m_f - u_i m_i - m_1 h_1 \tag{4.81}$$

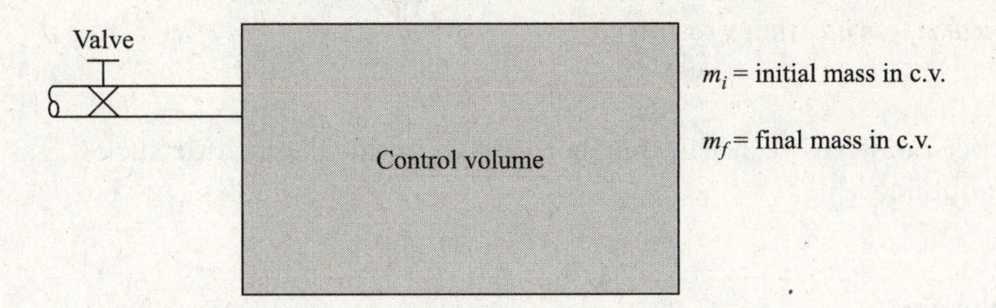

Fig. 4-21 The filling of a rigid tank.

where

m_1 = mass that enters
m_f = final mass in control volume
m_i = initial mass in control volume

In addition, for the filling process the enthalpy h_1 is assumed constant over the time interval.

The continuity equation for the unsteady-flow situation may be necessary in the solution process. Since the final mass is equal to the initial mass plus the mass that entered, this is expressed as

$$m_f = m_i + m_1 \tag{4.82}$$

Now consider the discharge of a pressurized tank. This problem is more complicated than the filling of a tank in that the properties at the exiting area are not constant over the time interval of interest; we must include the variation of the variables with time. We will assume an insulated tank, so that no heat transfer occurs, and again neglect kinetic energy and potential energy. The energy equation becomes, assuming no shaft work,

$$0 = \frac{d}{dt}(um) + \dot{m}_2(P_2 v_2 + u_2) \tag{4.83}$$

where m is the mass in the control volume. From the continuity equation,

$$\frac{dm}{dt} = -\dot{m}_2 \tag{4.84}$$

If this is substituted into (*4.83*), we have

$$d(um) = (P_2 v_2 + u_2)dm \tag{4.85}$$

We will assume that the gas escapes through a small valve opening, as shown in Fig. 4-22. Just upstream of the valve is area A_2 with properties P_2, v_2, and u_2. The velocity at this exiting area is assumed to be quite small so that P_2, v_2, and u_2 are approximately the same as the respective quantities in the control volume. With this assumption (*4.85*) becomes

$$d(um) = (Pv + u)\,dm \tag{4.86}$$

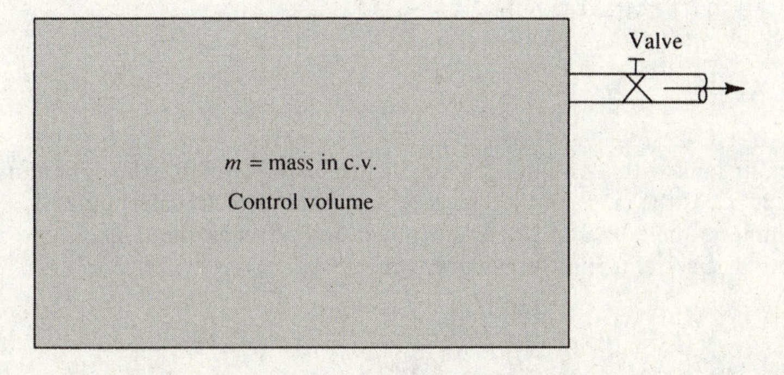

Fig. 4-22 The discharge of a pressurized tank.

Letting $d(um) = u\,dm + m\,du$, there results

$$m\,du = Pv\,dm \tag{4.87}$$

Now we will restrict ourselves to a gas that behaves as an ideal gas. For such a gas $du = C_v\,dT$ and $Pv = RT$, and we obtain

$$mC_v\,dT = RT\,dm \tag{4.88}$$

This is put in the form

$$\frac{C_v}{R}\frac{dT}{T} = \frac{dm}{m} \tag{4.89}$$

which can be integrated from the initial state, signified by the subscript i, to the final state, signified by the subscript f. There results

$$\frac{C_v}{R}\ln\frac{T_f}{T_i} = \ln\frac{m_f}{m_i} \qquad \text{or} \qquad \frac{m_f}{m_i} = \left(\frac{T_f}{T_i}\right)^{1/(k-1)} \tag{4.90}$$

where we have used $C_v/R = 1/(k-1)$; see (*4.31*). In terms of the pressure ratio, (*4.50*) allows us to write

$$\frac{m_f}{m_i} = \left(\frac{P_f}{P_i}\right)^{1/k} \tag{4.91}$$

Remember that these equations are applicable if there is no heat transfer from the volume; the process is quasistatic in that the properties are assumed uniformly distributed throughout the control volume (this requires a relatively slow discharge velocity, say 100 m/s or less); and the gas behaves as an ideal gas.

EXAMPLE 4.19 A completely evacuated, insulated, rigid tank with a volume of 300 ft^3 is filled from a steam line transporting steam at 800 °F and 500 psia. Determine (*a*) the temperature of steam in the tank when its pressure is 500 psia and (*b*) the mass of steam that flowed into the tank.

Solution:

(*a*) The energy equation used is (*4.81*). With $Q = 0$ and $m_i = 0$, we have $u_f m_f = m_i h_1$. The continuity equation (*4.82*) allows us to write $m_f = m_1$, which states that the final mass m_f in the tank is equal to the mass m_1 that entered the tank. Thus, there results $u_f = h_1$. From Table C3-E, h_1 is found, at 800 °F and 500 psia, to be 1412.1 Btu/lbm. Using $P_4 = 500$ psia as the final tank pressure, we can interpolate for the temperature, using $u_f = 1412.1$ Btu/lbm, and find

$$T_f = \left(\frac{1412.1 - 1406.0}{1449.2 - 1406.0}\right)(100) + 1100 = 1114.1\,°\text{F}$$

(*b*) We recognize that $m_1 = m_f = V_{\text{tank}}/v_f$. The specific volume of the steam in the tank at 500 psia and 1114.1 °F is

$$v_f = \left(\frac{1114.1 - 1100}{100}\right)(1.9518 - 1.8271) + 1.8271 = 1.845\ \text{ft}^3/\text{lbm}$$

This gives $m_f = 300/1.845 = 162.6$ lbm.

EXAMPLE 4.20 An air tank with a volume of 20 m^3 is pressurized to 10 MPa. The tank eventually reaches room temperature of 25 °C. If the air is allowed to escape with no heat transfer until $P_f = 200$ kPa, determine the mass of air remaining in the tank and the final temperature of air in the tank.

Solution: The initial mass of air in the tank is found to be

$$m_i = \frac{P_i V}{RT_i} = \frac{10 \times 10^6(20)}{(287)(298)} = 2338\ \text{kg}$$

Equation (*4.91*) gives, using $k = 1.4$,

$$m_f = m_i\left(\frac{P_f}{P_i}\right)^{1/k} = (2338)\left(\frac{2\times 10^5}{10\times 10^6}\right)^{1/1.4} = 143.0 \text{ kg}$$

To find the final temperature (*4.90*) is used:

$$T_f = T_i\left(\frac{m_f}{m_i}\right)^{k-1} = (298)(143/2338)^{0.4} = 97.46 \text{ K} \quad \text{or} -175.5\,^\circ\text{C}$$

A person who accidently comes in contact with a flow of gas from a pressurized tank faces immediate freezing (which is treated just like a burn).

Solved Problems

4.1 A 1500-kg automobile traveling at 30 m/s is brought to rest by impacting a shock absorber composed of a piston with small holes that moves in a cylinder containing water. How much heat must be removed from the water to return it to its original temperature?

As the piston moves through the water, work is done due to the force of impact moving with the piston. The work that is done is equal to the kinetic energy change; that is,

$$W = \frac{1}{2}m V^2 = \left(\frac{1}{2}\right)(1500)(30)^2 = 675\,000 \text{ J}$$

The first law for a cycle requires that this amount of heat must be transferred from the water to return it to its original temperature; hence, $Q = 675$ kJ.

4.2 A piston moves upward a distance of 5 cm while 200 J of heat is added (Fig. 4-23). Calculate the change in internal energy of the vapor if the spring is originally unstretched.

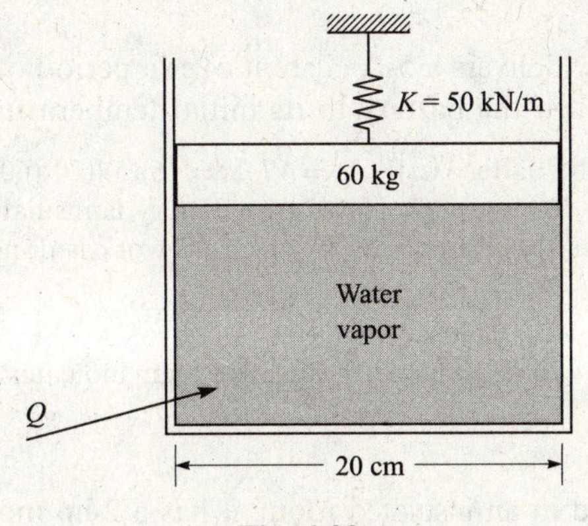

Fig. 4-23

The work needed to raise the weight and compress the spring is

$$W = (mg)(h) + \frac{1}{2}Kx^2 + (P_{atm})(A)(h)$$

$$= (60)(9.81)(0.05) + \left(\frac{1}{2}\right)(50\,000)(0.05)^2 + (100\,000)\left[\frac{\pi(0.2)^2}{4}\right](0.05) = 250 \text{ J}$$

The first law for a process without kinetic or potential energy changes is

$$Q - W = \Delta U$$

Thus, we have $\Delta U = 200 - 250 = -50$ J.

4.3 A system undergoes a cycle consisting of the three processes listed in the table. Compute the missing values. All quantities are in kJ.

Process	Q	W	ΔE
$1 \to 2$	a	100	100
$2 \to 3$	b	-50	c
$3 \to 1$	100	d	-200

Use the first law in the form $Q - W = \Delta E$. Applied to process $1 \to 2$, we have

$$a - 100 = 100 \qquad \therefore a = 200 \text{ kJ}$$

Applied to process $3 \to 1$, there results

$$100 - d = -200 \qquad \therefore d = 300 \text{ kJ}$$

The net work is then $\Sigma W = W_{1-2} + W_{2-3} + W_{3-1} = 100 - 50 + 300 = 350$ kJ. The first law for a cycle demands that

$$\Sigma Q = \Sigma W \qquad 200 + b + 100 = 350 \qquad \therefore b = 50 \text{ kJ}$$

Finally, applying the first law to process $2 \to 3$ provides

$$50 - (-50) = c \qquad \therefore c = 100 \text{ kJ}$$

Note that, for a cycle, $\Sigma\Delta E = 0$; this, in fact, could have been used to determine the value of c:

$$\Sigma\Delta E = 100 + c - 200 = 0 \qquad \therefore c = 100 \text{ kJ}$$

4.4 A 6-V insulated battery delivers a 5-A current over a period of 20 min. Calculate the heat transfer that must occur to return the battery to its initial temperature.

The work done by the battery is $W_{1-2} = VI\,\Delta t = (6)(5)[(20)(60)] = 36$ kJ. According to the first law, this must equal $-(U_2 - U_1)$ since $Q_{1-2} = 0$ (the battery is insulated). To return the battery to its initial state, the first law, for this second process in which no work is done, gives

$$Q_{2-1} - \cancel{W}^0_{2-1} = \Delta U = U_1 - U_2$$

Consequently, $Q_{2-1} = +36$ kJ, where the positive sign indicates that heat must be transferred to the battery.

4.5 A refrigerator is situated in an insulated room; it has a 2-hp motor that drives a compressor. Over a 30-minute period of time it provides 5300 kJ of cooling to the refrigerated space and 8000 kJ of heating from the coils on the back of the refrigerator. Calculate the increase in internal energy in the room.

In this problem we consider the insulated room as the system. The refrigerator is nothing more than a component in the system. The only transfer of energy across the boundary of the system is via the electrical wires of the refrigerator. For an insulated room ($Q = 0$) the first law provides

$$\cancel{Q}^0 - W = \Delta U$$

Hence, $\Delta U = -(-2 \text{ hp})(0.746 \text{ kW/hp})\ (1800 \text{ s}) = 2686$ kJ.

4.6 A 2-ft^3 rigid volume contains water at 120 °F with a quality of 0.5. Calculate the final temperature if 8 Btu of heat is added.

The first law for a process demands that $Q - W = m\,\Delta u$. To find the mass, we must use the specific volume as follows:

$$v_1 = v_f + x(v_g - v_f) = 0.016 + (0.5)(203.0 - 0.016) = 101.5\ \text{ft}^3/\text{lbm}$$

$$\therefore m = \frac{V}{v} = \frac{2}{101.5} = 0.0197\ \text{lbm}$$

For a rigid volume the work is zero since the volume does not change. Hence, $Q = m\,\Delta u$. The value of the initial internal energy is

$$u_1 = u_f + xu_{fg} = 87.99 + (0.5)(961.9) = 568.9\ \text{Btu/lbm}$$

The final internal energy is then calculated from the first law:

$$8 = 0.0197(u_2 - 568.9) \qquad \therefore u_2 = 975\ \text{Btu/lbm}$$

This is less than u_g; consequently, state 2 is in the wet region with $v_2 = 101.5\ \text{ft}^3/\text{lbm}$. This requires a trial-and-error procedure to find state 2:

At $T = 140\ °\text{F}$:

$$101.5 = 0.016 + x_2(122.9 - 0.016) \qquad \therefore x_2 = 0.826$$

$$975 = 108 + 948.2x_2 \qquad \therefore x_2 = 0.914$$

At $T = 150\ °\text{F}$:

$$v_g = 96.99 \qquad \therefore \text{slightly superheat}$$

$$975 = 118 + 941.3x_2 \qquad \therefore x_2 = 0.912$$

Obviously, state 2 lies between 140 °F and 150 °F. Since the quality is insensitive to the internal energy, we find T_2 such that $v_g = 101.5\ \text{ft}^3/\text{lbm}$:

$$T_2 = 150 - \left(\frac{101.5 - 96.99}{122.88 - 96.99}\right)(10) = 148\ °\text{F}$$

A temperature slightly less than this provides us with $T_2 = 147\ °\text{F}$.

4.7 A frictionless piston provides a constant pressure of 400 kPa in a cylinder containing R134a with an initial quality of 80 percent. Calculate the final temperature if 80 kJ/kg of heat is transferred to the cylinder.

The original enthalpy is found, using values from Table D-2, to be

$$h_1 = h_f + x_1 h_{fg} = 62.0 + (0.8)(190.32) = 214.3\ \text{kJ/kg}$$

For this constant-pressure process, the first law demands that

$$q = h_2 - h_1 \qquad 80 = h_2 - 214.3 \qquad \therefore h_2 = 294.3\ \text{kJ/kg}$$

Using $P_2 = 400$ kPa and $h_2 = 294.3$ kJ/kg, we interpolate in Table D-3 to find

$$T_2 = \left(\frac{294.3 - 291.8}{301.5 - 291.8}\right)(10) + 50 = 52.6\ °\text{C}$$

4.8 A piston-cylinder arrangement contains 2 kg of steam originally at 200 °C and 90 percent quality. The volume triples while the temperature is held constant. Calculate the heat that must be transferred and the final pressure.

The first law for this constant-temperature process is $Q - W = m(u_2 - u_1)$. The initial specific volume and specific internal energy are, respectively,

$$v_1 = 0.0012 + (0.9)(0.1274 - 0.0012) = 0.1148 \text{ m}^3/\text{kg}$$
$$u_1 = 850.6 + (0.9)(2595.3 - 850.6) = 2421 \text{ kJ/kg}$$

Using $T_2 = 200\,°\text{C}$ and $v_2 = (3)(0.1148) = 0.3444 \text{ m}^3/\text{kg}$, we interpolate in Table C-3 and find the final pressure P_2 to be

$$P_2 = 0.8 - \left(\frac{0.3444 - 0.2608}{0.3520 - 0.2608}\right)(0.2) = 0.617 \text{ MPa}$$

We can also interpolate to find that the specific internal energy is

$$u_2 = 2638.9 - (2638.9 - 2630.6)\left(\frac{0.617 - 0.6}{0.8 - 0.6}\right) = 2638.2 \text{ kJ/kg}$$

To find the heat transfer we must know the work W. It is estimated using graph paper by plotting P vs. v and graphically integrating (counting squares). The work is twice this area since $m = 2$ kg. Doing this, we find

$$W = (2)(228) = 456 \text{ kJ}$$

Thus $Q = W + m(u_2 - u_1) = 456 + (2)(2638.2 - 2421) = 890 \text{ kJ}$.

4.9 Estimate the constant-pressure specific heat and the constant-volume specific heat for R134a at 30 psia and 100 °F.

We write the derivatives in finite-difference form and, using values on either side of 100 °F for greatest accuracy, we find

$$C_p \cong \frac{\Delta h}{\Delta T} = \frac{126.39 - 117.63}{120 - 80} = 0.219 \text{ Btu/lbm-°F}$$

$$C_v \cong \frac{\Delta u}{\Delta T} = \frac{115.47 - 107.59}{120 - 80} = 0.197 \text{ Btu/lbm-°F}$$

4.10 Calculate the change in enthalpy of air which is heated from 300 K to 700 K if

(*a*) $C_p = 1.006$ kJ/kg·°C.
(*b*) $C_p = 0.946 + 0.213 \times 10^{-3}T - 0.031 \times 10^{-6}T^2$ kJ/kg·°C.
(*c*) The gas tables are used.
(*d*) Compare the calculations of (*a*) and (*b*) with (*c*).

(*a*) Assuming the constant specific heat, we find that

$$\Delta h = C_p(T_2 - T_1) = (1.006)(700 - 300) = 402.4 \text{ kJ/kg}$$

(*b*) If C_p depends on temperature, we must integrate as follows:

$$\Delta h = \int_{T_1}^{T_2} C_p\, dT = \int_{300}^{700} (0.946 + 0.213 \times 10^{-3}T - 0.031 \times 10^{-6}T^2)\, dT = 417.7 \text{ kJ/kg}$$

(*c*) Using Table E-1, we find $\Delta h = h_2 - h_1 = 713.27 - 300.19 = 413.1$ kJ/kg.

(*d*) The assumption of constant specific heat results in an error of -2.59 percent; the expression for C_p produces an error of $+1.11$ percent. All three methods are acceptable for the present problem.

4.11 Sixteen ice cubes, each with a temperature of $-10\,°\text{C}$ and a volume of 8 milliliters, are added to 1 liter of water at 20 °C in an insulated container. What is the equilibrium temperature? Use $(C_p)_{\text{ice}} = 2.1$ kJ/kg·°C.

Assume that all of the ice melts. The ice warms up to 0 °C, melts at 0 °C, and then warms up to the final temperature T_2. The water cools from 20 °C to the final temperature T_2. The mass of ice is calculated to be

$$m_i = \frac{V}{v_i} = \frac{(16)(8 \times 10^{-6})}{0.00109} = 0.1174 \text{ kg}$$

where v_i is found in Table C-5. If energy is conserved, we must have

$$\text{Energy gained by ice} = \text{energy lost by water}$$

$$m_i[(C_p)_i \Delta T + h_i f + (C_p)_w \Delta T] = m_w (C_p)_w \Delta T$$
$$0.1174[(2.1)(10) + 320 + (4.81)(T_2 - 0)] = (1000 \times 10^{-3})(4.18)(20 - T_2)$$
$$T_2 = 9.33\,°\text{C}$$

4.12 A 5-kg block of copper at 300 °C is submerged in 20 liters of water at 0 °C contained in an insulated tank. Estimate the final equilibrium temperature.

Conservation of energy requires that the energy lost by the copper block is gained by the water. This is expressed as

$$m_c (C_p)_c (\Delta T)_c = m_w (C_p)_w (\Delta T)_w$$

Using average values of C_p from Table B-4, this becomes

$$(5)(0.39)(300 - T_2) = (0.02)(1000)(4.18)(T_2 - 0) \qquad \therefore T_2 = 6.84\,°\text{C}$$

4.13 Two pounds of air is compressed from 20 psia to 200 psia while maintaining the temperature constant at 100 °F. Calculate the heat transfer needed to accomplish this process.

The first law, assuming air to be an ideal gas, requires that

$$Q = W + \cancel{\Delta U}^0 = mRT \ln\frac{P_1}{P_2} = (2 \text{ lbm})\left(53.3 \frac{\text{ft-lbf}}{\text{lbm-°R}}\right)(560\,°\text{R})\left(\frac{1 \text{ Btu}}{778 \text{ ft-lbf}}\right)\ln\frac{20}{200}$$
$$= -176.7 \text{ Btu}$$

4.14 Helium is contained in a 2-m^3 rigid volume at 50 °C and 200 kPa. Calculate the heat transfer needed to increase the pressure to 800 kPa.

The work is zero for this constant-volume process. Consequently, the first law gives

$$Q = m\Delta u = mC_v \,\Delta T = \frac{PV}{RT} C_v (T_2 - T_1)$$

The ideal-gas law, $PV = mRT$, allows us to write

$$\frac{P_1}{T_1} = \frac{P_2}{T_2} \qquad \frac{200}{323} = \frac{800}{T_2} \qquad \therefore T_2 = 1292 \text{ K}$$

The heat transfer is then, using values from Table B-2,

$$Q = \frac{(200)(2)}{(2.077)(323)}(3.116)(1292 - 323) = 1800 \text{ kJ}$$

4.15 The air in the cylinder of an air compressor is compressed from 100 kPa to 10 MPa. Estimate the final temperature and the work required if the air is initially at 100 °C.

Since the process occurs quite fast, we assume an adiabatic quasiequilibrium process. Then

$$T_2 = T_1\left(\frac{P_2}{P_1}\right)^{(k-1)/k} = (373)\left(\frac{10\,000}{100}\right)^{(1.4-1)/1.4} = 1390 \text{ K}$$

The work is found by using the first law with $Q = 0$:

$$w = -\Delta u = -C_v(T_2 - T_1) = -(0.717)(1390 - 373) = -729 \text{ kJ/kg}$$

The work per unit mass is calculated since the mass (or volume) was not specified.

4.16 Nitrogen at 100 °C and 600 kPa expands in such a way that it can be approximated by a polytropic process with $n = 1.2$ [see (*4.52*)]. Calculate the work and the heat transfer if the final pressure is 100 kPa.

The final temperature is found to be

$$T_2 = T_1\left(\frac{P_2}{P_1}\right)^{(n-1)/n} = (373)\left(\frac{100}{600}\right)^{(1.2-1)/1.2} = 276.7 \text{ K}$$

The specific volumes are

$$v_1 = \frac{RT_1}{P_1} = \frac{(0.297)(373)}{600} = 0.1846 \text{ m}^3/\text{kg} \qquad v_2 = \frac{RT_2}{P_2} = \frac{(0.297)(276.7)}{100} = 0.822 \text{ m}^3/\text{kg}$$

The work is then [or use (*4.53*)]

$$w = \int P dv = P_1 v_1^n \int v^{-n} dv = (600)(0.1846)^{1.2}\left(\frac{1}{-0.2}\right)(0.822^{-0.2} - 0.1846^{-0.2}) = 143 \text{ kJ/kg}$$

The first law provides us with the heat transfer:

$$q - w = \Delta u = C_v(T_2 - T_1) \qquad q - 143 = (0.745)(276.7 - 373) \qquad \therefore q = 71.3 \text{kJ/kg}$$

4.17 How much work must be input by the paddle wheel in Fig. 4-24 to raise the piston 5 in? The initial temperature is 100 °F.

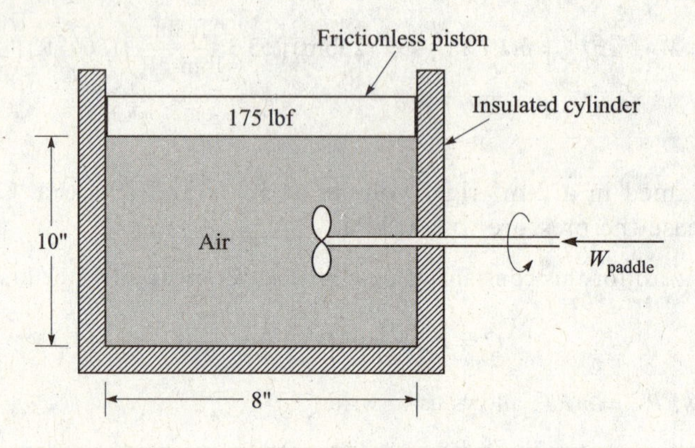

Fig. 4-24

The first law, with $Q = 0$, is

$$W = \Delta U \qquad \text{or} \qquad -PA\Delta h - W_{\text{paddle}} = mC_v(T_2 - T_1)$$

The pressure is found from a force balance on the piston:

$$P = 14.7 + \frac{175}{\pi(4)^2} = 18.18 \text{ psia}$$

The mass of the air is found from the ideal-gas law:

$$m = \frac{PV}{RT} = \frac{(18.18)(144)(\pi)(4)^2(10)/1728}{(53.3)(560)} = 0.0255 \text{ lbm}$$

The temperature T_2 is

$$T_2 = \frac{PV_2}{mR} = \frac{(18.18)(144)(\pi)(4)^2(15)/1728}{(0.0255)(53.3)} = 840\,^\circ\text{R}$$

Finally, the paddle-wheel work is found to be

$$W_{\text{paddle}} = -PA\Delta h - mC_v(T_2 - T_1) = -(18.18)(\pi)(4)^2(5/12) - (0.0255)(0.171)(778)(840 - 560)$$
$$= -1331 \text{ ft-lbf}$$

4.18 For the cycle in Fig. 4-25 find the work output and the net heat transfer if the 0.1 kg of air is contained in a piston-cylinder arrangement.

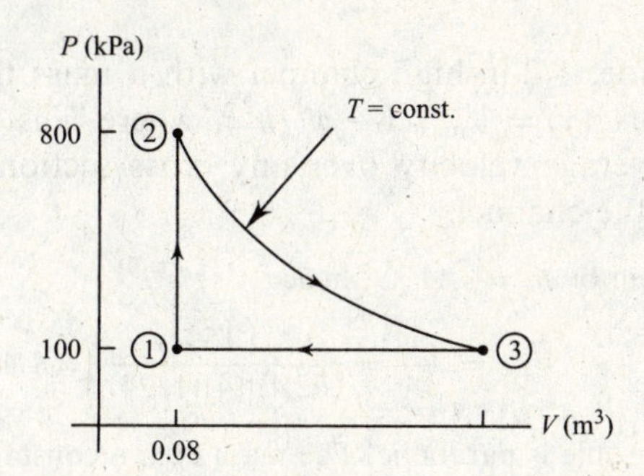

Fig. 4-25

The temperatures and V_3 are

$$T_1 = \frac{P_1V_1}{mR} = \frac{(100)(0.08)}{(0.1)(0.287)} = 278.7 \text{ K} \qquad T_2 = T_3 = \frac{(800)(0.08)}{(0.1)(0.287)} = 2230 \text{ K}$$

$$V_3 = \frac{P_2V_2}{P_3} = \frac{(800)(0.08)}{100} = 0.64 \text{ m}^3$$

Using the definition of work for each process, we find

$$W_{1-2} = 0 \qquad W_{2-3} = mRT \ln\frac{p_2}{p_3} = (0.1)(0.287)(2230)\ln\frac{800}{100} = 133.1 \text{ kJ}$$
$$W_{3-1} = P(V_1 - V_3) = (100)(0.08 - 0.64) = -56 \text{ kJ}$$

The work output is then $W_{\text{net}} = 0 + 133.1 - 56.0 = 77.1$ kJ. Since this is a complete cycle, the first law for a cycle provides us with

$$Q_{\text{net}} = W_{\text{net}} = 77.1 \text{ kJ}$$

4.19 Water enters a radiator through a 4-cm-diameter hose at 0.02 kg/s. It travels down through all the rectangular passageways on its way to the water pump. The passageways are each 10 × 1 mm and there are 800 of them in a cross section. How long does it take water to traverse from the top to the bottom of the 60-cm-high radiator?

The average velocity through the passageways is found from the continuity equation, using $\rho_{\text{water}} = 1000 \text{ kg/m}^3$:

$$\dot{m} = \rho_1 \mathcal{V}_1 A_1 = \rho_2 \mathcal{V}_2 A_2 \qquad \therefore \mathcal{V}_2 = \frac{\dot{m}}{\rho_2 A_2} = \frac{0.02}{(1000)[(800)(0.01)(0.001)]} = 0.0025 \text{ m/s}$$

The time to travel 60 cm at this constant velocity is

$$t = \frac{L}{\mathcal{V}} = \frac{0.60}{0.0025} = 240 \text{ s or 4 min}$$

4.20 A 10-m^3 tank is being filled with steam at 800 kPa and 400 °C. It enters the tank through a 10-cm-diameter pipe. Determine the rate at which the density in the tank is varying when the velocity of the steam in the pipe is 20 m/s.

The continuity equation with one inlet and no outlets is [see (*4.56*)]:

$$\rho_1 A_1 V_1 = \frac{dm_{\text{c.v.}}}{dt}$$

Since $m_{\text{c.v.}} = \rho V$, where V is the volume of the tank, this becomes

$$V\frac{d\rho}{dt} = \frac{1}{v_1}A_1 V_1 \qquad 10\frac{d\rho}{dt} = \left(\frac{1}{0.3843}\right)(\pi)(0.05)^2(20) \qquad \frac{d\rho}{dt} = 0.04087\ \text{kg/m}^3\cdot\text{s}$$

4.21 Water enters a 4-ft-wide, 1/2-in-high channel with a mass flux of 15 lbm/sec. It leaves with a parabolic distribution $V(y) = V_{\max}(1 - y^2/h^2)$, where h is half the channel height. Calculate $V_{\max}$ and V_{avg}, the average velocity over any cross section of the channel. Assume that the water completely fills the channel.

The mass flux is given by $\dot{m} = \rho A V_{\text{avg}}$; hence,

$$V_{\text{avg}} = \frac{\dot{m}}{\rho A} = \frac{15}{(62.4)[(4)(1/24)]} = 1.442\ \text{ft/sec}$$

At the exit the velocity profile is parabolic. The mass flux, a constant, then provides us with

$$\dot{m} = \int_A \rho V dA$$

$$15 = \rho\int_{-h}^{h} V_{\max}\left(1 - \frac{y^2}{h^2}\right)4dy = (62.4)(4V_{\max})\left[y - \frac{y^3}{3h^2}\right]^h - h = (62.4)(4V_{\max})\left[\frac{(4)(1/48)}{3}\right]$$

$$\therefore V_{\max} = 2.163\text{ft/sec}$$

4.22 R134a enters a valve at 800 kPa and 30 °C. The pressure downstream of the valve is measured to be 60 kPa. Calculate the internal energy downstream.

The energy equation across the valve, recognizing that heat transfer and work are zero, is $h_1 = h_2$. The enthalpy before the valve is that of compressed liquid. The enthalpy of a compressed liquid is essentially equal to that of a saturated liquid at the same temperature. Hence, at 30 °C in Table D-1, $h_1 = 91.49$ kJ/kg. Using Table D-2 at 60 kPa we find

$$h_2 = 91.49 = h_f + x_2 h_{fg} = 3.46 + 221.27x_2 \qquad \therefore x_2 = 0.398$$

The internal energy is then

$$u_2 = u_f + x_2(u_g - u_f) = 3.14 + 0.398[(206.12 - 3.14)] = 83.9\ \text{kJ/kg}$$

4.23 The pressure of 200 kg/s of water is to be increased by 4 MPa. The water enters through a 20-cm-diameter pipe and exits through a 12-cm-diameter pipe. Calculate the minimum horsepower required to operate the pump.

The energy equation (*4.68*) provides us with

$$-\dot{W}_p = \dot{m}\left(\frac{\Delta P}{\rho} + \frac{V_2^2 - V_1^2}{2}\right)$$

The inlet and exit velocities are calculated as follows:

$$V_1 = \frac{\dot{m}}{\rho A_1} = \frac{200}{(1000)(\pi)(0.1)^2} = 6.366\ \text{m/s} \qquad V_2 = \frac{\dot{m}}{\rho A_2} = \frac{200}{(1000)(\pi)(0.06)^2} = 17.68\ \text{m/s}$$

The energy equation then gives

$$\dot{W}_P = -200\left[\frac{4\,000\,000}{1000} + \frac{(17.68)^2 - (6.366)^2}{2}\right] = -827\,200\text{ W} \quad \text{or } 1109\text{ hp}$$

Note: The above power calculation provides a minimum since we have neglected any internal energy increase. Also, the kinetic energy change represents only a 3 percent effect on $\dot{W}_P$ and could be neglected.

4.24 A hydroturbine operates on a stream in which 100 kg/s of water flows. Estimate the maximum power output if the turbine is in a dam with a distance of 40 m from the surface of the reservoir to the surface of the backwater.

The energy equation (*4.68*), neglecting kinetic energy changes, takes the form $-\dot{W}_T = \dot{m}g(z_2 - z_1)$, where we have assumed the pressure to be atmospheric on the water's surface above and below the dam. The maximum power output is then

$$\dot{W}_T = -(100)(9.81)(-40) = 39\,240\text{ W} \quad \text{or } 39.24\text{ kW}$$

4.25 A turbine accepts superheated steam at 800 psia and 1200 °F and rejects it as saturated vapor at 2 psia (Fig. 4-26). Predict the horsepower output if the mass flux is 1000 lbm/min. Also, calculate the velocity at the exit.

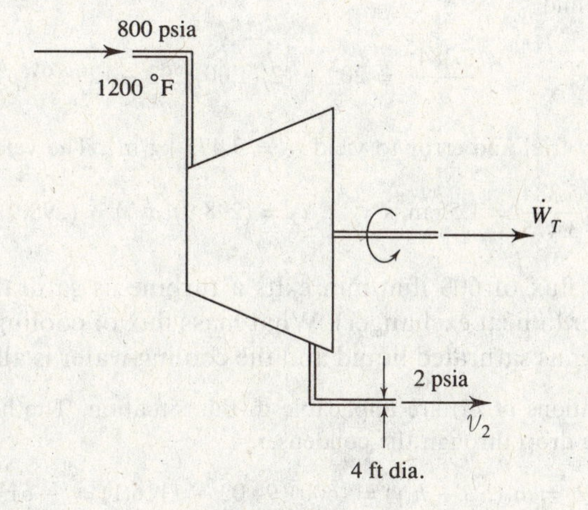

Fig. 4-26

Assuming zero heat transfer, the energy equation (*4.66*) provides us with

$$-\dot{W}_T = \dot{m}(h_2 - h_1) = \left(\frac{1000}{60}\right)(1116.1 - 1623.8) = -8462\text{ Btu/sec} \quad \text{or } 11\,970\text{ hp}$$

where Tables C-3E and C-2E have provided the enthalpies. By (*4.58*),

$$V_2 = \frac{v\dot{m}}{A} = \frac{(173.75)(1000/60)}{\pi(2)^2} = 230\text{ ft/sec}$$

4.26 Air enters a compressor at atmospheric conditions of 20 °C and 80 kPa and exits at 800 kPa and 200 °C. Calculate the rate of heat transfer if the power input is 400 kW. The air exits at 20 m/s through an exit diameter of 10 cm.

The energy equation, neglecting kinetic and potential energy changes, is $\dot{Q} - \dot{W}_S = \dot{m}C_p(T_2 - T_1)$; the mass flux is calculated to be

$$\dot{m} = \rho A V = \frac{P}{RT} A V = \frac{800}{(0.287)(473)}(\pi)(0.05)^2(20) = 0.9257\text{ kg/s}$$

Hence $\dot{Q} = (0.9257)(1.00)(200 - 20) + (-400) = -233.4$ kW. Note that the power input is negative, and a negative heat transfer implies that the compressor is losing heat.

4.27 Air travels through the 4 × 2 m test section of a wind tunnel at 20 m/s. The gage pressure in the test section is measured to be −20 kPa and the temperature 20 °C. After the test section, a diffuser leads to a 6-m-diameter exit pipe. Estimate the velocity and temperature in the exit pipe.

The energy equation (*4.72*) for air takes the form

$$\mathcal{V}_2^2 = \mathcal{V}_1^2 + 2C_p(T_1 - T_2) = 20^2 + (2)(1.00)(293 - T_2)$$

The continuity equation, $\rho_1 A_1 \mathcal{V}_1 = \rho_2 A_2 \mathcal{V}_2$, yields

$$\frac{P_1}{RT_1} A_1 \mathcal{V}_1 = \rho_2 A_2 \mathcal{V}_2 \qquad \therefore \rho_2 \mathcal{V}_2 = \left[\frac{80}{(0.287)(293)}\right]\left[\frac{8}{\pi(3)^2}\right](20) = 5.384 \text{ kg/m}^2\cdot\text{s}$$

The best approximation to the actual process is the adiabatic quasiequilibrium process. Using (*4.49*), letting $\rho = 1/v$, we have

$$\frac{T_2}{T_1} = \left(\frac{\rho_2}{\rho_1}\right)^{k-1} \qquad \text{or} \qquad \frac{T_2}{\rho_2^{0.4}} = \frac{293}{[80/(0.287)(293)]^{0.4}} = 298.9$$

The above three equations include the three unknowns T_2, $\mathcal{V}_2$, and ρ_2. Substitute for T_2 and $\mathcal{V}_2$ back into the energy equation and find

$$\frac{5.384^2}{\rho_2^2} = 20^2 + (2)(1.00)[293 - (298.9)(\rho_2^{0.4})]$$

This can be solved by trial and error to yield $\rho_2 = 3.475$ kg/m³. The velocity and temperature are then

$$\mathcal{V}_2 = \frac{5.384}{\rho_2} = \frac{5.384}{3.475} = 1.55 \text{ m/s} \qquad T_2 = (298.9)(\rho_2^{0.4}) = (298.9)(3.475)^{0.4} = 492 \quad \text{or } 219\,^\circ\text{C}$$

4.28 Steam with a mass flux of 600 lbm/min exits a turbine as saturated steam at 2 psia and passes through a condenser (a heat exchanger). What mass flux of cooling water is needed if the steam is to exit the condenser as saturated liquid and the cooling water is allowed a 15 °F temperature rise?

The energy equations (*4.75*) are applicable to this situation. The heat transfer rate for the steam is, assuming no pressure drop through the condenser,

$$\dot{Q}_s = \dot{m}_s(h_{s2} - h_{s1}) = (600)(94.02 - 1116.1) = -613{,}200 \text{ Btu/min}$$

This energy is gained by the water. Hence,

$$\dot{Q}_w = \dot{m}_w(h_{w2} - h_{w1}) = \dot{m}_w C_p(T_{w2} - T_{w1}) \qquad 613{,}200 = \dot{m}_w(1.00)(15) \qquad \dot{m}_w = 40{,}880 \text{ lbm/min}$$

4.29 A simple steam power plant operates on 20 kg/s of steam, as shown in Fig. 4-27. Neglecting losses in the various components, calculate (*a*) the boiler heat transfer rate, (*b*) the turbine power output, (*c*) the condenser heat transfer rate, (*d*) the pump power requirement, (*e*) the velocity in the boiler exit pipe, and (*f*) the thermal efficiency of the cycle.

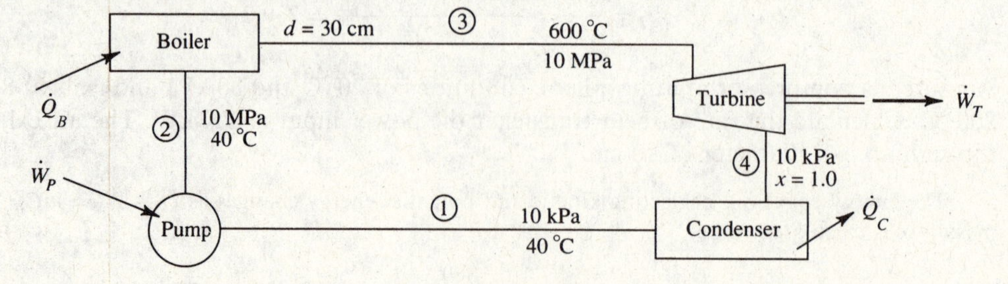

Fig. 4-27

(*a*) $\dot{Q}_B = \dot{m}(h_3 - h_2) = (20)(3625.3 - 167.5) = 69.15$ MW, where we have taken the enthalpy h_2 to be h_f at 40 °C.

(*b*) $\dot{W}_T = \dot{m}(h_4 - h_3) = -(20)(2584.6 - 3625.3) = 20.81$ MW.

(*c*) $\dot{Q}_C = \dot{m}(h_1 - h_4) = (20)(167.57 - 2584.7) = -48.34$ MW.

(*d*) $\dot{W}_P = \dot{m}(P_2 - P_1)/\rho = (20)(10\,000 - 10/1000) = 0.2$ MW.

(*e*) $V = \dot{m}v/A = (20)(0.03837)/\pi(0.15)^2 = 10.9$ m/s.

(*f*) $\eta = (\dot{W}_T - \dot{W}_P)/\dot{Q}_B = (20.81 - 0.2)/69.15 = 0.298$ or 29.8%.

4.30 An insulated 4-m^3 evacuated tank is connected to a 4-MPa 600 °C steam line. A valve is opened and the steam fills the tank. Estimate the final temperature of the steam in the tank and the final mass of the steam in the tank.

From (*4.81*), with $Q = 0$ and $m_i = 0$, there results $u_f = h_1$, since the final mass m_f is equal to the mass m_1 that enters. We know that across a valve the enthalpy is constant; hence,

$$h_1 = h_{\text{line}} = 3674.4 \text{ kJ/kg}$$

The final pressure in the tank is 4 MPa, achieved when the steam ceases to flow into the tank. Using $P_f = 4$ MPa and $u_f = 3674.4$ kJ/kg, we find the temperature in Table C-3 to be

$$T_f = \left(\frac{3674.4 - 3650.1}{3650.1 - 3555.5}\right)(500) + 800 = 812.8 \text{ °C}$$

The specific volume at 4 MPa and 812.8 °C is

$$v_f = \left(\frac{812.8 - 800}{50}\right)(0.1229 - 0.1169) + 0.1229 = 0.1244 \text{ ft}^3/\text{lbm}$$

The mass of steam in the tank is then

$$m_f = \frac{V_f}{v_f} = \frac{4}{0.1244} = 32.15 \text{ kg}$$

Supplementary Problems

4.31 An unknown mass is attached by a pulley to a paddle wheel which is inserted in a volume of water. The mass is then dropped a distance of 3 m. If 100 J of heat must be transferred from the water in order to return the water to its initial state, determine the mass in kilograms.

4.32 While 300 J of heat is added to the air in the cylinder of Fig. 4-28, the piston raises a distance of 0.2 m. Determine the change in internal energy.

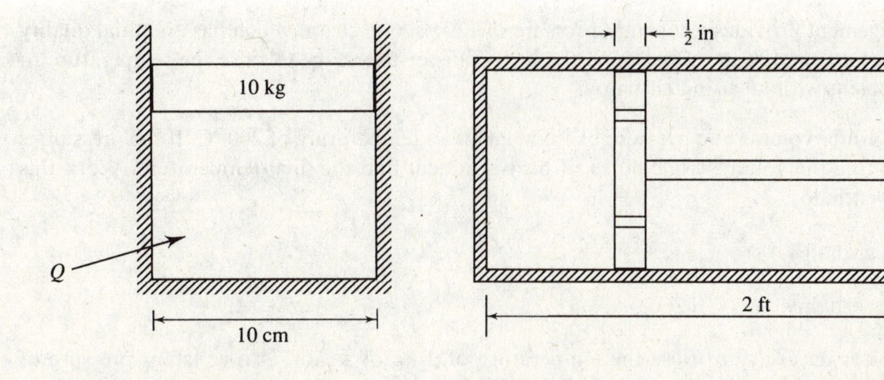

Fig. 4-28 **Fig. 4-29**

4.33 A constant force of 600 lbf is required to move the piston shown in Fig. 4-29. If 2 Btu of heat is transferred from the cylinder when the piston moves the entire length, what is the change in internal energy?

4.34 Each of the letters (*a*) through (*e*) in the accompanying table represents a process. Supply the missing values, in kJ.

	Q	W	ΔE	E_2	E_1
(*a*)	20	5			7
(*b*)		−3	6		8
(*c*)	40			30	15
(*d*)	−10		20	10	
(*e*)		10		−8	6

4.35 A system undergoes a cycle consisting of four processes. Some of the values of the energy transfers and energy changes are given in the table. Fill in all the missing values. All units are kJ.

Process	Q	W	ΔU
$1 \to 2$	−200	(*a*)	0
$2 \to 3$	800	(*b*)	(*c*)
$3 \to 4$	(*d*)	600	400
$4 \to 1$	0	(*e*)	−1200

4.36 A 12-V battery is charged by supplying 3 A over a period of 6 h. If a heat loss of 400 kJ occurs from the battery during the charging period, what is the change in energy stored within the battery?

4.37 A 12-V battery delivers a current of 10 A over a 30-min time period. The stored energy decreases by 300 kJ. Determine the heat lost during the time period.

4.38 A 110-V heater draws 15 A while heating a particular air space. During a 2-h period the internal energy in the space increases by 8000 Btu. Calculate the amount of heat lost in Btu.

4.39 How much heat must be added to a 0.3-m^3 rigid volume containing water at 200 °C in order that the final temperature be raised to 800 °C? The initial pressure is 1 MPa.

4.40 A 0.2-m^3 rigid volume contains steam at 600 kPa and a quality of 0.8. If 1000 kJ of heat is added, determine the final temperature.

4.41 A piston-cylinder arrangement provides a constant pressure of 120 psia on steam which has an initial quality of 0.95 and an initial volume of 100 in^3. Determine the heat transfer necessary to raise the temperature to 1000 °F. Work this problem without using enthalpy.

4.42 Steam is contained in a 4-liter volume at a pressure of 1.4 MPa and a temperature of 200 °C. If the pressure is held constant by expanding the volume while 40 kJ of heat is added, find the final temperature. Work this problem without using enthalpy.

4.43 Work Prob. 4.41 using enthalpy.

4.44 Work Prob. 4.42 using enthalpy.

4.45 Calculate the heat transfer necessary to raise the temperature of 2 kg of steam, at a constant pressure of 100 kPa (*a*) from 50 °C to 400 °C and (*b*) from 400 °C to 750 °C.

4.46 Steam is contained in a 1.2-m^3 volume at a pressure of 3 MPa and a quality of 0.8. The pressure is held constant. What is the final temperature if (*a*) 3 MJ and (*b*) 30 MJ of heat is added? Sketch the process on a T-v diagram.

4.47 Estimate the constant-pressure specific heat for steam at 400 °C if the pressure is (*a*) 10 kPa, (*b*) 100 kPa, and (*c*) 30 000 kPa.

4.48 Determine approximate values for the constant-volume specific heat for steam at 800 °F if the pressure is (*a*) 1 psia, (*b*) 14.7 psia, and (*c*) 3000 psia.

4.49 Calculate the change in enthalpy of 2 kg of air which is heated from 400 K to 600 K if (*a*) $C_p = 1.006$ kJ/kg·K, (*b*) $C_p = 0.946 + 0.213 \times 10^{-3}T - 0.031 \times 10^{-6}T^2$ kJ/kg·K, and (*c*) the gas tables are used.

4.50 Compare the enthalpy change of 2 kg of water for a temperature change from 10 °C to 60 °C with that of 2 kg of ice for a temperature change from −60 °C to −10 °C.

4.51 Two MJ of heat is added to 2.3 kg of ice held at a constant pressure of 200 kPa, at (*a*) − 60 °C and (*b*) 0 °C. What is the final temperature? Sketch the process on a T-v diagram.

4.52 What is the heat transfer required to raise the temperature of 10 lbm of water from 0 °F (ice) to 600 °F (vapor) at a constant pressure of 30 psia? Sketch the process on a T-v diagram.

4.53 Five ice cubes (4 × 2 × 2 cm) at −20 °C are added to an insulated glass of cola at 20 °C. Estimate the final temperature (if above 0 °C) or the percentage of ice melted (if at 0 °C) if the cola volume is (*a*) 2 liters and (*b*) 0.25 liters. Use $\rho_{ice} = 917$ kg/m^3.

4.54 A 40-lbm block of copper at 200 °F is dropped in an insulated tank containing 3 ft^3 of water at 60 °F. Calculate the final equilibrium temperature.

4.55 A 50-kg block of copper at 0 °C and a 100-kg block of iron at 200 °C are brought into contact in an insulated space. Predict the final equilibrium temperature.

4.56 Determine the enthalpy change and the internal energy change for 4 kg of air if the temperature changes from 100 °C to 400 °C. Assume constant specific heats.

4.57 For each of the following quasiequilibrium processes supply the missing information. The working fluid is 0.4 kg of air in a cylinder.

	Process	Q (kJ)	W (kJ)	ΔU (kJ)	ΔH (kJ)	T_2 (°C)	T_1 (°C)	P_2 (kPa)	P_1 (kPa)	V_2 (m^3)	V_1 (m^3)
(*a*)	$T = C$	60				100		50			
(*b*)	$V = C$				80	300		200			
(*c*)	$P = C$	100					200		500		
(*d*)	$Q = 0$						50			0.1	0.48

4.58 For each of the quasiequilibrium processes presented in the table in Prob. 4.57, supply the missing information if the working fluid is 0.4 kg of steam. [Note: for process (*a*) it is necessary to integrate graphically.]

4.59 One thousand Btu of heat is added to 2 lbm of steam maintained at 60 psia. Calculate the final temperature if the initial temperature of the steam is (*a*) 600 °F and (*b*) 815 °F.

4.60 Fifty kJ of heat is transferred to air maintained at 400 kPa with an initial volume of 0.2 m^3. Determine the final temperature if the initial temperature is (*a*) 0 °C and (*b*) 200 °C.

4.61 The initial temperature and pressure of 8000 cm^3 of air are 100 °C and 800 kPa, respectively. Determine the necessary heat transfer if the volume does not change and the final pressure is (*a*) 200 kPa and (*b*) 3000 kPa.

4.62 Calculate the heat transfer necessary to raise the temperature of air, initially at 10 °C and 100 kPa, to a temperature of 27 °C if the air is contained in an initial volume with dimensions 3 × 5 × 2.4 m. The pressure is held constant.

4.63 Heat is added to a fixed 0.15-m^3 volume of steam initially at a pressure of 400 kPa and a quality of 0.5. Determine the final pressure and temperature if (*a*) 800 kJ and (*b*) 200 kJ of heat is added. Sketch the process on a *P-v* diagram.

4.64 Two hundred Btu of heat is added to a rigid air tank which has a volume of 3 ft^3. Find the final temperature if initially (*a*) $P = 60$ psia and $T = 30\,°F$ and (*b*) $P = 600$ psia and $T = 820\,°F$. Use the air tables.

4.65 A system consisting of 5 kg of air is initially at 300 kPa and 20 °C. Determine the heat transfer necessary to (*a*) increase the volume by a factor of two at constant pressure, (*b*) increase the pressure by a factor of two at constant volume, (*c*) increase the pressure by a factor of two at constant temperature, and (*d*) increase the absolute temperature by a factor of 2 at constant pressure.

4.66 Heat is added to a container holding 0.5 m^3 of steam initially at a pressure of 400 kPa and a quality of 80 percent (Fig. 4-30). If the pressure is held constant, find the heat transfer necessary if the final temperature is (*a*) 500 °C and (*b*) 675 °C. Also determine the work done. Sketch the process on a *T-v* diagram.

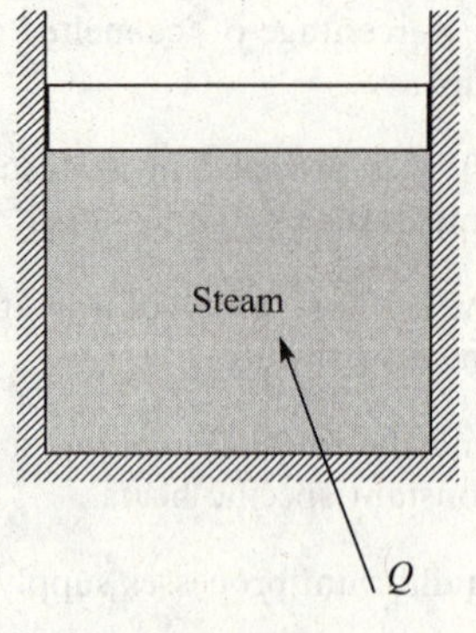

Fig. 4-30

4.67 A rigid 1.5-m^3 tank at a pressure of 200 kPa contains 5 liters of liquid and the remainder steam. Calculate the heat transfer necessary to (*a*) completely vaporize the water, (*b*) raise the temperature to 400 °C, and (*c*) raise the pressure to 800 kPa.

4.68 Ten Btu of heat is added to a rigid container holding 4 lbm of air in a volume of 100 ft^3. Determine ΔH.

4.69 Eight thousand cm^3 of air in a piston-cylinder arrangement is compressed isothermally at 30 °C from a pressure of 200 kPa to a pressure of 800 kPa. Find the heat transfer.

4.70 Two kilograms of air is compressed in an insulated cylinder from 400 kPa to 15 000 kPa. Determine the final temperature and the work necessary if the initial temperature is (*a*) 200 °C and (*b*) 350 °C.

4.71 Air is compressed in an insulated cylinder from the position shown in Fig. 4-31 so that the pressure increases to 5000 kPa from atmospheric pressure of 100 kPa. What is the required work if the mass of the air is 0.2 kg?

4.72 The average person emits approximately 400 Btu of heat per hour. There are 1000 people in an unventilated room 10 × 75 × 150 ft. Approximate the increase in temperature after 15 min, assuming (*a*) constant pressure and (*b*) constant volume. (*c*) Which assumption is the more realistic?

4.73 Two hundred kJ of work is transferred to the air by means of a paddle wheel inserted into an insulated volume (Fig. 4-32). If the initial pressure and temperature are 200 kPa and 100 °C, respectively, determine the final temperature and pressure.

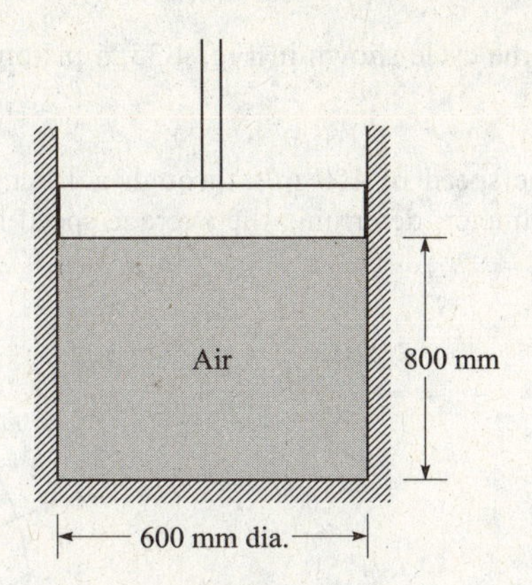

Fig. 4-31

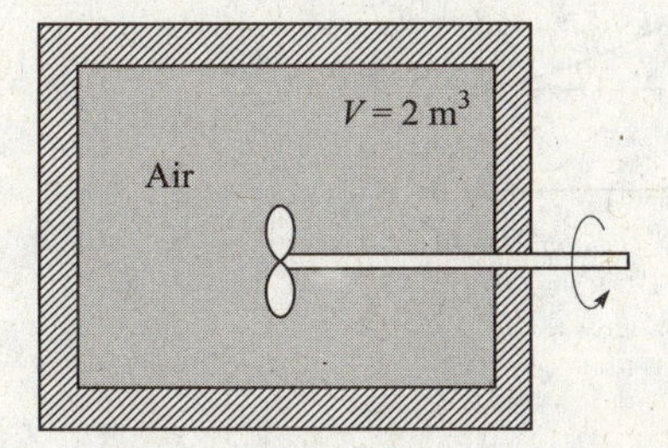

Fig. 4-32

4.74 A 2-kg rock falls from 10 m and lands in a 10-liter container of water. Neglecting friction during the fall, calculate the maximum temperature increase in the water.

4.75 A torque of 10 N·m is required to turn a paddle wheel at the rate of 100 rad/s. During a 45-s time period a volume of air, in which the paddle wheel rotates, is increased from 0.1 to 0.4 m^3. The pressure is maintained constant at 400 kPa. Determine the heat transfer necessary if the initial temperature is (*a*) 0 °C and (*b*) 300 °C.

4.76 For the cycle shown in Fig. 4-33 find the work output and the net heat transfer, if 0.8 lbm of air is contained in a cylinder with $T_1 = 800\,°F$, assuming the process from 3 to 1 is (*a*) an isothermal process and (*b*) an adiabatic process.

4.77 For the cycle shown in Fig. 4-34 find the net heat transfer and work output if steam is contained in a cylinder.

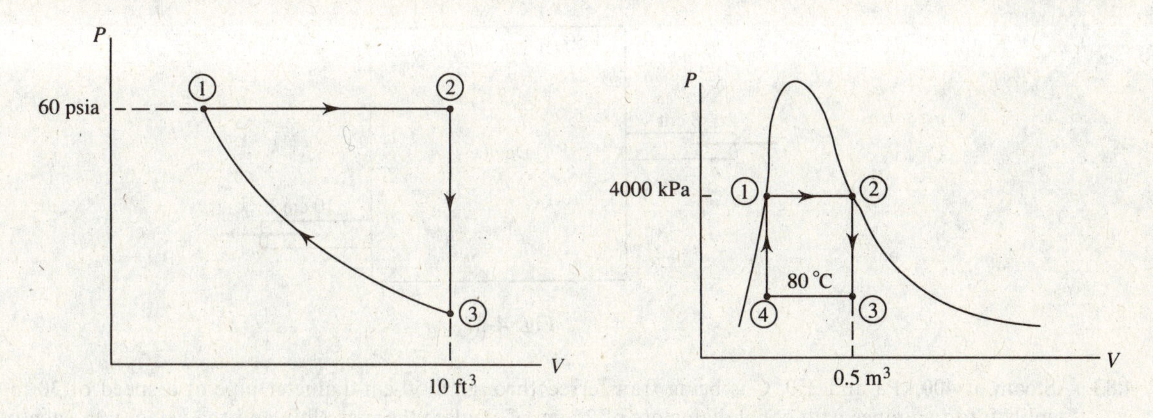

Fig. 4-33

Fig. 4-34

4.78 If 0.03 kg of air undergoes the cycle shown in Fig. 4-35, a piston-cylinder arrangement, calculate the work output.

4.79 Air is flowing at an average speed of 100 m/s through a 10-cm-diameter pipe. If the pipe undergoes an enlargement to 20 cm in diameter, determine the average speed in the enlarged pipe.

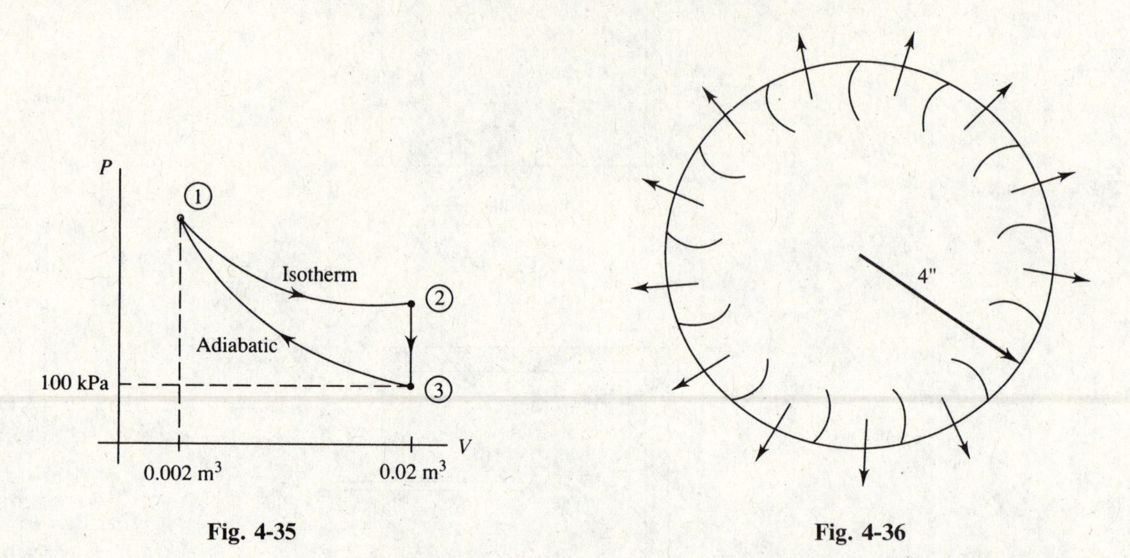

Fig. 4-35 **Fig. 4-36**

4.80 Air enters a vacuum cleaner through a 2-in-diameter pipe at a speed of 150 ft/sec. It passes through a rotating impeller (Fig. 4-36), of thickness 0.5 in., through which the air exits. Determine the average velocity exiting normal to the impeller.

4.81 Air enters a device at 4 MPa and 300 °C with a velocity of 150 m/s. The inlet area is 10 cm^2 and the outlet area is 50 cm^2. Determine the mass flux and the outlet velocity if the air exits at 0.4 MPa and 100 °C.

4.82 Air enters the device shown in Fig. 4-37 at 2 MPa and 350 °C with a velocity of 125 m/s. At one outlet area the conditions are 150 kPa and 150 °C with a velocity of 40 m/s. Determine the mass flux and the velocity at the second outlet for conditions of 0.45 MPa and 200 °C.

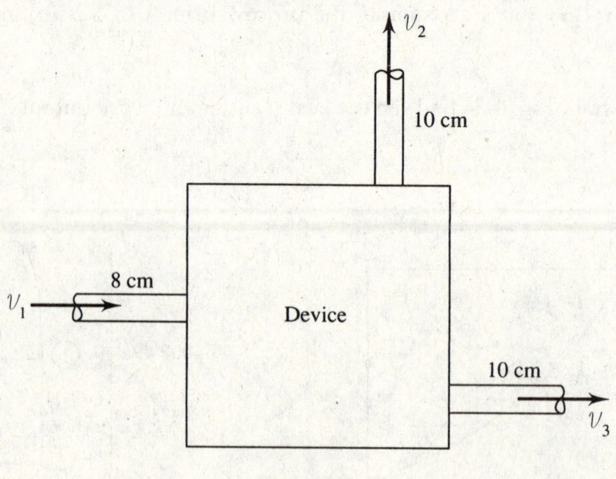

Fig. 4-37

4.83 Steam at 400 kPa and 250 °C is being transferred through a 50-cm-diameter pipe at a speed of 30 m/s. It splits into two pipes with equal diameters of 25 cm. Calculate the mass flux and the velocity in each of the smaller pipes if the pressure and temperature are 200 kPa and 200 °C, respectively.

4.84 Steam enters a device through a 2-in^2 area at 500 psia and 600 °F. It exits through a 10-in^2 area at 20 psia and 400 °F with a velocity of 800 ft/sec. What are the mass flux and the entering velocity?

4.85 Steam enters a 10-m^3 tank at 2 MPa and 600 °C through an 8-cm-diameter pipe with a velocity of 20 m/s. It leaves at 1 MPa and 400 °C through a 12-cm-diameter pipe with a velocity of 10 m/s. Calculate the rate at which the density in the tank is changing.

4.86 Water flows into a 1.2-cm-diameter pipe with a uniform velocity of 0.8 m/s. At some distance down the pipe a parabolic velocity profile is established. Determine the maximum velocity in the pipe and the mass flux. The parabolic profile can be expressed as $V(r) = V_{max}(1 - r^2/R^2)$, where R is the radius of the pipe.

4.87 Water enters the contraction shown in Fig. 4-38 with a parabolic profile $V(r) = 2(1 - r^2)$ m/s, where r is measured in centimeters. The exiting profile after the contraction is essentially uniform. Determine the mass flux and the exit velocity.

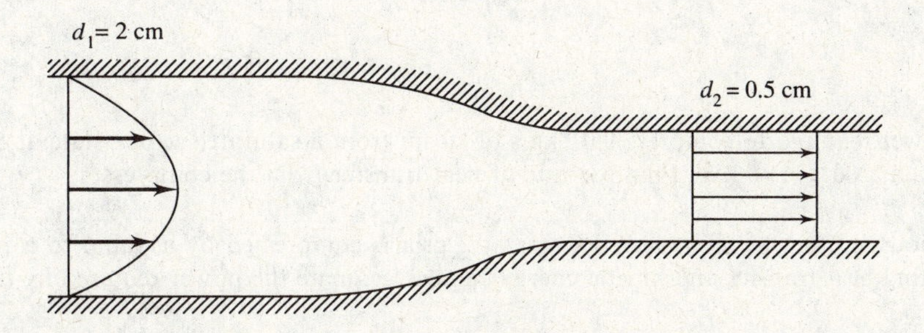

Fig. 4-38

4.88 Air enters a 4-in. constant-diameter pipe at 100 ft/sec with a pressure of 60 psia and a temperature of 100 °F. Heat is added to the air, causing it to pass a downstream area at 70 psia, 300 °F. Calculate the downstream velocity and the heat transfer rate.

4.89 Water at 9000 kPa and 300 °C flows through a partially open valve. The pressure immediately after the valve is measured to be 600 kPa. Calculate the specific internal energy of the water leaving the valve. Neglect kinetic energy changes. (*Note:* The enthalpy of slightly compressed liquid is essentially equal to the enthalpy of saturated liquid at the same temperature.)

4.90 Steam at 9000 kPa and 600 °C passes through a throttling process so that the pressure is suddenly reduced to 400 kPa. (*a*) What is the expected temperature after the throttle? (*b*) What area ratio is necessary for the kinetic energy change to be zero?

4.91 Water at 70 °F flows through the partially open valve shown in Fig. 4-39. The area before and after the valve is the same. Determine the specific internal energy downstream of the valve.

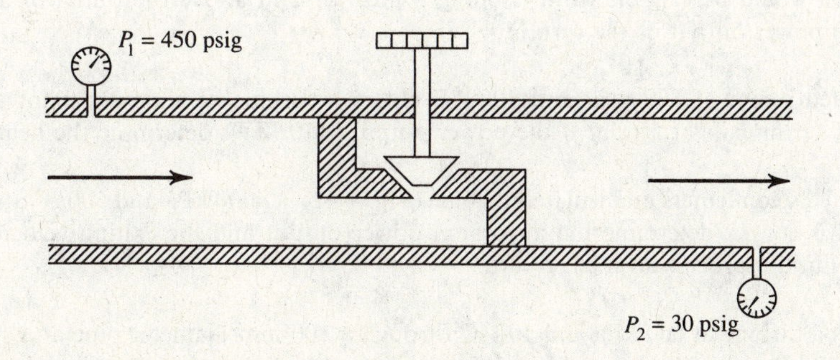

Fig. 4-39

4.92 The inlet conditions on an air compressor are 50 kPa and 20 °C. To compress the air to 400 kPa, 5 kW of energy is needed. Neglecting heat transfer and kinetic and potential energy changes, estimate the mass flux.

4.93 The air compressor shown in Fig. 4-40 draws air from the atmosphere and discharges it at 500 kPa. Determine the minimum power required to drive the insulated compressor. Assume atmospheric conditions of 25 °C and 80 kPa.

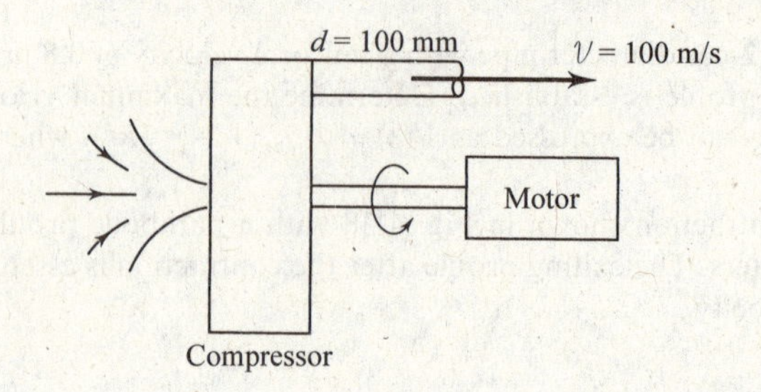

Fig. 4-40

4.94 The power required to compress 0.01 kg/s of steam from a saturated vapor state at 50 °C to a pressure of 800 kPa at 200 °C is 6 kW. Find the rate of heat transfer from the compressor.

4.95 Two thousand lbm/hr of saturated water at 2 psia is compressed by a pump to a pressure of 2000 psia. Neglecting heat transfer and kinetic energy change, estimate the power required by the pump.

4.96 The pump in Fig. 4-41 increases the pressure in the water from 200 to 4000 kPa. What is the minimum horsepower motor required to drive the pump for a flow rate of 0.1 m^3/s?

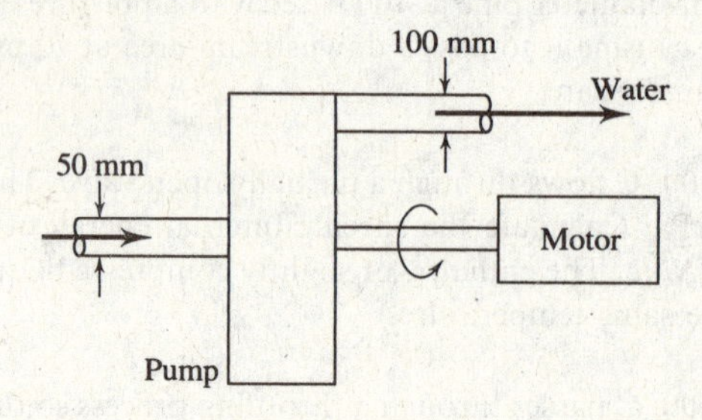

Fig. 4-41

4.97 A turbine at a hydroelectric plant accepts 20 m^3/s of water at a gage pressure of 300 kPa and discharges it to the atmosphere. Determine the maximum power output.

4.98 Water flows in a creek at 1.5 m/s. It has cross-sectional dimensions of 0.6 × 1.2 m upstream of a proposed dam which would be capable of developing a head of 2 m above the outlet of a turbine. Determine the maximum power output of the turbine.

4.99 Superheated steam at 800 psia and 1000 °F enters a turbine at a power plant at the rate of 30 lb/sec. Saturated steam exits at 5 psia. If the power output is 10 MW, determine the heat transfer rate.

4.100 Superheated steam enters an insulated turbine (Fig. 4-42) at 4000 kPa and 500 °C and leaves at 20 kPa. If the mass flux is 6 kg/s, determine the maximum power output and the exiting velocity. Assume an adiabatic quasiequilibrium process so that $s_2 = s_1$.

4.101 Air enters a turbine at 600 kPa and 100 °C through a 100-mm-diameter pipe at a speed of 100 m/s. The air exits at 140 kPa and 20 °C through a 400-mm-diameter pipe. Calculate the power output, neglecting heat transfer.

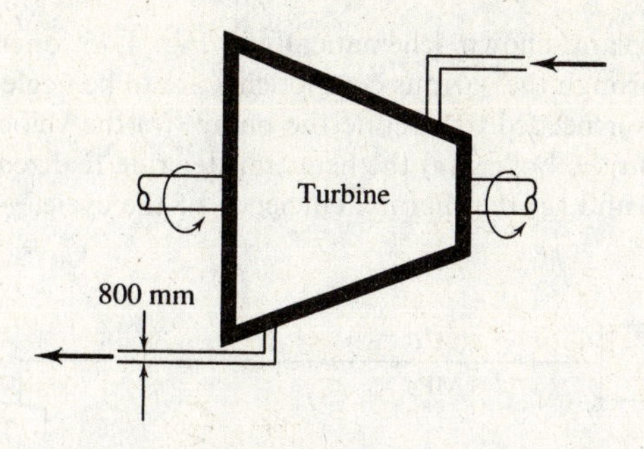

Fig. 4-42

4.102 A turbine delivers 500 kW of power by extracting energy from air at 450 kPa and 100 °C flowing in a 120-mm-diameter pipe at 150 m/s. For an exit pressure of 120 kPa and a temperature of 20 °C determine the heat transfer rate.

4.103 Water flows through a nozzle that converges from 4 in. to 0.8 in. in diameter. For a mass flux of 30 lbm/sec calculate the upstream pressure if the downstream pressure is 14.7 psia.

4.104 Air enters a nozzle like that shown in Fig. 4-43 at a temperature of 195 °C and a velocity of 100 m/s. If the air exits to the atmosphere where the pressure is 85 kPa, find (*a*) the exit temperature, (*b*) the exit velocity, and (*c*) the exit diameter. Assume an adiabatic quasiequilibrium process.

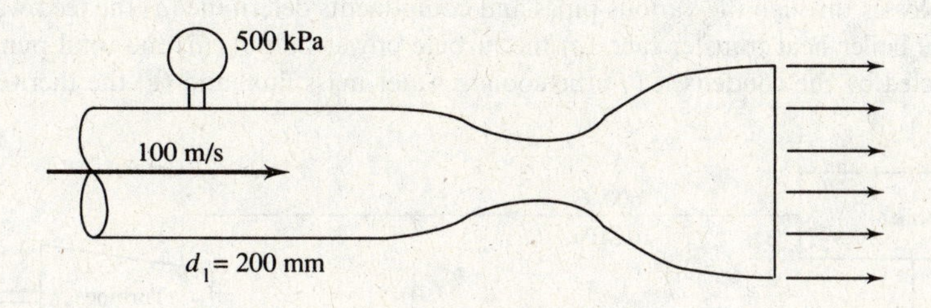

Fig. 4-43

4.105 Nitrogen enters a diffuser at 200 m/s with a pressure of 80 kPa and a temperature of −20 °C. It leaves with a velocity of 15 m/s at an atmospheric pressure of 95 kPa. If the inlet diameter is 100 mm, calculate (*a*) the mass flux and (*b*) the exit temperature.

4.106 Steam enters a diffuser as a saturated vapor at 220 °F with a velocity of 600 ft/sec. It leaves with a velocity of 50 ft/sec at 20 psia. What is the exit temperature?

4.107 Water is used in a heat exchanger (Fig. 4-44) to cool 5 kg/s of air from 400 °C to 200 °C. Calculate (*a*) the minimum mass flux of the water and (*b*) the quantity of heat transferred to the water each second.

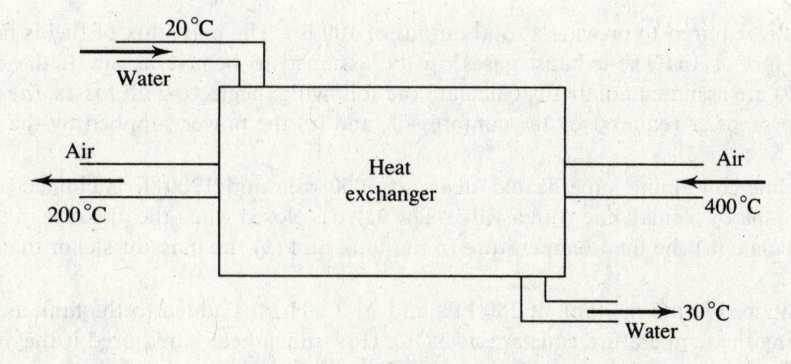

Fig. 4-44

4.108 A simple steam power plant, shown schematically in Fig. 4-45, operates on 8 kg/s of steam. Losses in the connecting pipes and through the various components are to be neglected. Calculate (*a*) the power output of the turbine, (*b*) the power needed to operate the pump, (*c*) the velocity in the pump exit pipe, (*d*) the heat transfer rate necessary in the boiler, (*e*) the heat transfer rate realized in the condenser, (*f*) the mass flux of cooling water required, and (*g*) the thermal efficiency of the cycle.

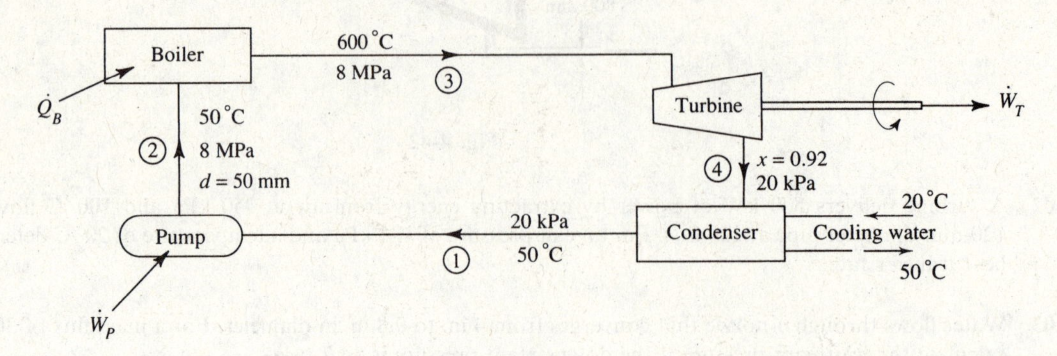

Fig. 4-45

4.109 A feed water heater is used to preheat water before it enters a boiler, as shown schematically in Fig. 4-46. A mass flux of 30 kg/s flows through the system and 7 kg/s is withdrawn from the turbine for the feed water heater. Neglecting losses through the various pipes and components determine (*a*) the feed water heater outlet temperature, (*b*) the boiler heat transfer rate, (*c*) the turbine power output, (*d*) the total pump power required, (*e*) the energy rejected by the condenser, (*f*) the cooling water mass flux, and (*g*) the thermal efficiency of the cycle.

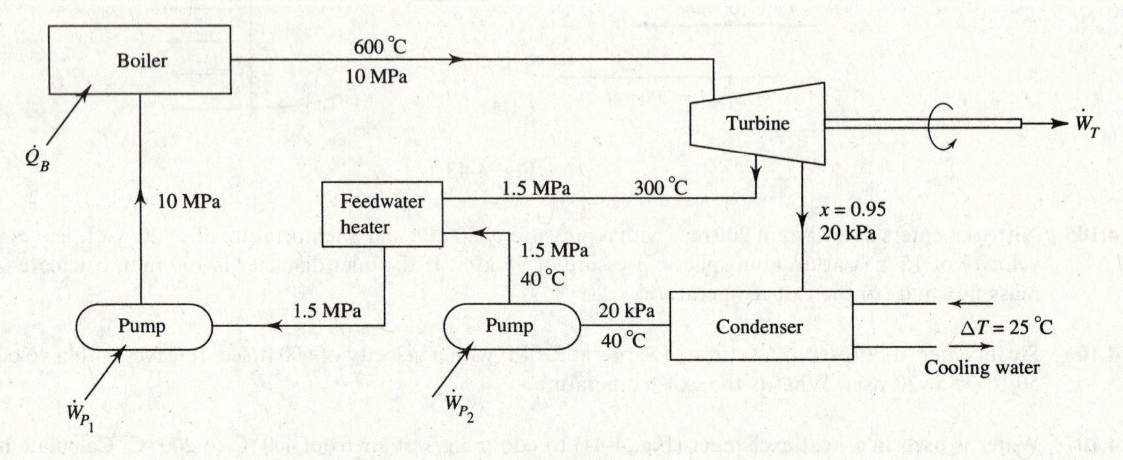

Fig. 4-46

4.110 A turbine is required to provide a total output of 100 hp. The mass flux of fuel is negligible compared with the mass flux of air. The exhaust gases can be assumed to behave as air. If the compressor and turbine (Fig. 4-47) are assumed adiabatic, calculate the following, neglecting all losses: (*a*) the mass flux of the air, (*b*) the horsepower required by the compressor, and (*c*) the power supplied by the fuel.

4.111 A steam line containing superheated steam at 1000 psia and 1200 °F is connected to a 50-ft^3 evacuated insulated tank by a small line with a valve. The valve is closed when the pressure in the tank just reaches 800 psia. Calculate (*a*) the final temperature in the tank and (*b*) the mass of steam that entered the tank.

4.112 Air is contained in a 3-m^3 tank at 250 kPa and 25 °C. Heat is added to the tank as the air escapes, thereby maintaining the temperature constant at 25 °C. How much heat is required if the air escapes until the final pressure is atmospheric? Assume $P_{atm} = 80$ kPa.

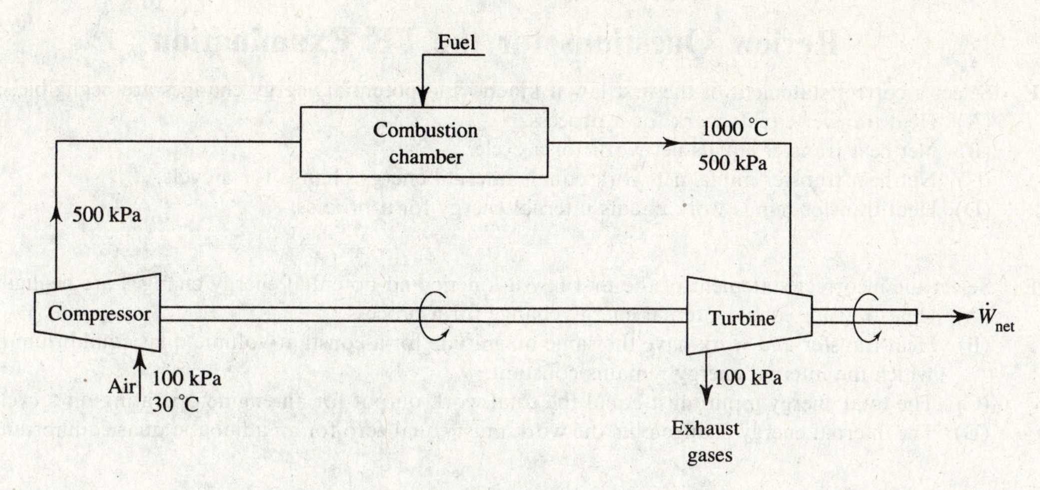

Fig. 4-47

4.113 An air line carries air at 800 kPa (Fig. 4-48). An insulated tank initially contains 20 °C air at atmospheric pressure of 90 kPa. The valve is opened and air flows into the tank. Determine the final temperature of the air in the tank and the mass of air that enters the tank if the valve is left open.

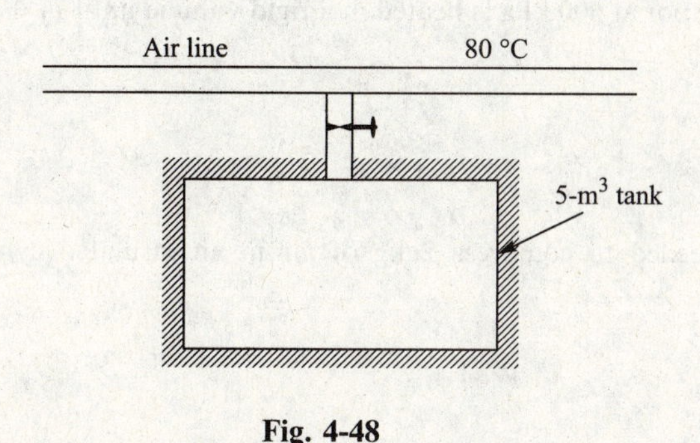

Fig. 4-48

4.114 An insulated tank is evacuated. Air from the atmosphere at 12 psia and 70 °F is allowed to flow into the 100-ft^3 tank. Calculate (*a*) the final temperature and (*b*) the final mass of air in the tank just after the flow ceases.

4.115 (*a*) An insulated tank contains pressurized air at 2000 kPa and 30 °C. The air is allowed to escape to the atmosphere ($P_{atm} = 95$ kPa, $T_{atm} = 30$ °C) until the flow ceases. Determine the final temperature in the tank. (*b*) Eventually, the air in the tank will reach atmospheric temperature. If a valve was closed after the initial flow ceased, calculate the pressure that is eventually reached in the tank.

4.116 An insulated tank with a volume of 4 m^3 is pressurized to 800 kPa at a temperature of 30 °C. An automatic valve allows the air to leave at a constant rate of 0.02 kg/s. (*a*) What is the temperature after 5 min? (*b*) What is the pressure after 5 min? (*c*) How long will it take for the temperature to drop to −20 °C?

4.117 A tank with a volume of 2 m^3 contains 90 percent liquid water and 10 percent water vapor by volume at 100 kPa. Heat is transferred to the tank at 10 kJ/min. A relief valve attached to the top of the tank allows vapor to discharge when the gage pressure reaches 600 kPa. The pressure is maintained at that value as more heat is transferred. (*a*) What is the temperature in the tank at the instant the relief valve opens? (*b*) How much mass is discharged when the tank contains 50 percent vapor by volume? (*c*) How long does it take for the tank to contain 75 percent vapor by volume?

Review Questions for the FE Examination

4.1FE Select a correct statement of the first law if kinetic and potential energy changes are negligible.
(A) Heat transfer equals work for a process.
(B) Net heat transfer equals net work for a cycle.
(C) Net heat transfer minus net work equals internal energy change for a cycle.
(D) Heat transfer minus work equals internal energy for a process.

4.2FE Select the incorrect statement of the first law if kinetic and potential energy changes are negligible.
(A) Heat transfer equals internal energy change for a process.
(B) Heat transfer and work have the same magnitude for a constant-volume quasiequilibrium process in which the internal energy remains constant.
(C) The total energy input must equal the total work output for an engine operating on a cycle.
(D) The internal energy change plus the work must equal zero for an adiabatic quasiequilibrium process.

4.3FE Ten kilograms of hydrogen is contained in a rigid, insulated tank at 20 °C. Estimate the final temperature if a 400-W resistance heater operates in the hydrogen for 40 minutes.
(A) 116 °C
(B) 84 °C
(C) 29 °C
(D) 27 °C

4.4FE Saturated water vapor at 400 kPa is heated in a rigid volume until $T_2 = 400\,°\text{C}$. The heat transfer is nearest:
(A) 407 kJ/kg
(B) 508 kJ/kg
(C) 604 kJ/kg
(D) 702 kJ/kg

4.5FE Find the work needed to compress 2 kg of air in an insulated cylinder from 100 kPa to 600 kPa if $T_1 = 20\,°\text{C}$.
(A) −469 kJ
(B) −390 kJ
(C) −280 kJ
(D) −220 kJ

4.6FE Find the temperature rise after 5 minutes in the volume of Fig. 4-49.
(A) 423 °C
(B) 378 °C
(C) 313 °C
(D) 287 °C

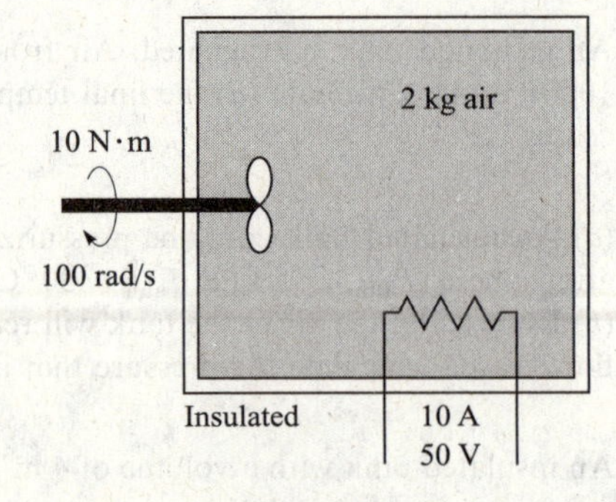

Fig. 4-49

4.7FE One kilogram of air is compressed at $T = 100\,°\text{C}$ until $V_1 = 2V_2$. How much heat is rejected?
(A) 42 kJ
(B) 53 kJ
(C) 67 kJ
(D) 74 kJ

4.8FE Energy is added to 5 kg of air with a paddle wheel until $\Delta T = 100\,°\text{C}$. Find the paddle wheel work if the rigid volume is insulated.
(A) 524 kJ
(B) 482 kJ
(C) 412 kJ
(D) 358 kJ

4.9FE Initially $P_1 = 400$ kPa and $T_1 = 400\,^\circ\text{C}$, as shown in Fig. 4-50. What is T_2 when the frictionless piston hits the stops?
(A) 315 °C
(B) 316 °C
(C) 317 °C
(D) 318 °C

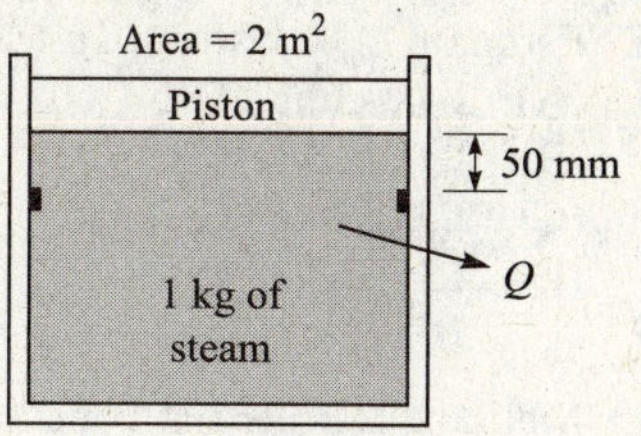

Area = 2 m^2
Piston
50 mm
Q
1 kg of steam

Fig. 4-50

4.10FE What heat is released during the process of Question 4.9FE?
(A) 190 kJ
(B) 185 kJ
(C) 180 kJ
(D) 175 kJ

4.11FE After the piston of Fig. 4-50 hits the stops, how much additional heat is released before $P_3 = 100$ kPa?
(A) 1580 kJ
(B) 1260 kJ
(C) 930 kJ
(D) 730 kJ

4.12FE The pressure of 10 kg of air is increased isothermally at 60 °C from 100 kPa to 800 kPa. Estimate the rejected heat.
(A) 1290 kJ
(B) 1610 kJ
(C) 1810 kJ
(D) 1990 kJ

4.13FE Saturated water is heated at constant pressure of 400 kPa until $T_2 = 400\,^\circ\text{C}$. Estimate the heat removal.
(A) 2070 kJ/kg
(B) 2370 kJ/kg
(C) 2670 kJ/kg
(D) 2870 kJ/kg

4.14FE One kilogram of steam in a cylinder requires 170 kJ of heat transfer while the pressure remains constant at 1 MPa. Estimate the temperature T_2 if $T_1 = 320\,^\circ\text{C}$.
(A) 420 °C
(B) 410 °C
(C) 400 °C
(D) 390 °C

4.15FE Estimate the work required for the process of Question 4.14FE.
(A) 89 kJ
(B) 85 kJ
(C) 45 kJ
(D) 39 kJ

4.16FE The pressure of steam at 400 °C and $u = 2949$ kJ·kg is nearest:
(A) 2000 kPa
(B) 1900 kPa
(C) 1800 kPa
(D) 1700 kPa

4.17FE The enthalpy of steam at $P = 500$ kPa and $v = 0.7\ \text{m}^3/\text{kg}$ is nearest:
(A) 3480 kJ/kg
(B) 3470 kJ/kg
(C) 3460 kJ/kg
(D) 3450 kJ/kg

4.18FE Estimate C_p for steam at 4 MPa and 350 °C.
(A) 2.48 kJ/kg °C
(B) 2.71 kJ/kg °C
(C) 2.53 kJ/kg °C
(D) 2.31 kJ/kg °C

4.19FE Methane is heated at constant pressure of 200 kPa from 0 °C to 300 °C. How much heat is needed?
(A) 731 kJ/kg
(B) 692 kJ/kg
(C) 676 kJ/kg
(D) 623 kJ/kg

4.20FE Estimate the equilibrium temperature if 20 kg of copper at 0 °C and 10 L of water at 30 °C are placed in an insulated container.
(A) 27.2 °C
(B) 25.4 °C
(C) 22.4 °C
(D) 20.3 °C

4.21FE Estimate the equilibrium temperature if 10 kg of ice at 0 °C is mixed with 60 kg of water at 20 °C in an insulated container.
(A) 12 °C
(B) 5.8 °C
(C) 2.1 °C
(D) 1.1 °C

4.22FE The table shows a three-process cycle; determine c.
(A) 140
(B) 100
(C) 80
(D) 40

Process	Q	W	ΔU
$1 \rightarrow 2$	100	a	0
$2 \rightarrow 3$	b	60	40
$3 \rightarrow 1$	40	c	d

4.23FE Find w_{1-2} for the process of Fig. 4-51.
(A) 219 kJ/kg
(B) 166 kJ/kg
(C) 113 kJ/kg
(D) 53 kJ/kg

4.24FE Find w_{3-1} for the process of Fig. 4-51.
(A) −219 kJ/kg
(B) −166 kJ/kg
(C) −113 kJ/kg
(D) −53 kJ/kg

4.25FE Find q_{cycle} for the processes of Fig. 4-51.
(A) 219 kJ/kg
(B) 166 kJ/kg
(C) 113 kJ/kg
(D) 53 kJ/kg

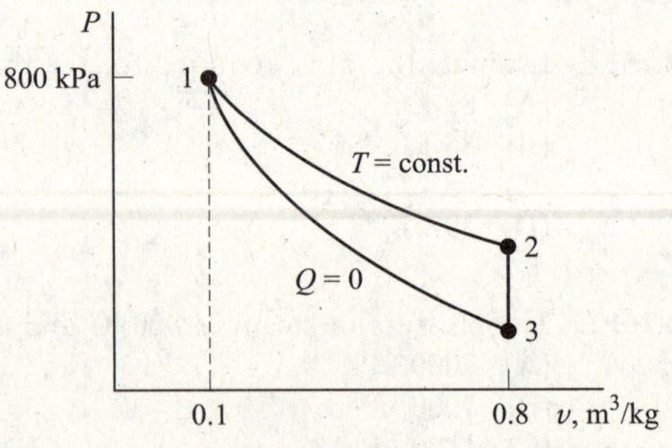

Fig. 4-51

4.26FE Clothes are hung on a clothesline to dry on a freezing winter day. The clothes dry due to:
(A) sublimation
(B) evaporation
(C) vaporization
(D) melting

4.27FE Air is compressed adiabatically from 100 kPa and 20 °C to 800 kPa. T_2 is nearest:
(A) 440 °C
(B) 360 °C
(C) 290 °C
(D) 260 °C

4.28FE The work required to compress 2 kg of air in an insulated cylinder from 100 °C and 100 kPa to 600 kPa is nearest:
(A) 460 kJ
(B) 360 kJ
(C) 280 kJ
(D) 220 kJ

4.29FE One hundred people are in a 10 m × 20 m × 3 m meeting room when the air conditioning fails. Estimate the temperature increase if it is off for 15 min. Each person emits 400 kJ/hr of heat and the lights add 300 W of energy. Neglect all other forms of energy input.
(A) 15 °C
(B) 18 °C
(C) 21 °C
(D) 25 °C

4.30FE Air undergoes a three-process cycle with a $P =$ const. process, a $T =$ const. process, and a $V =$ const. process. Select the correct statement for a piston-cylinder arrangement.
(A) $W = 0$ for the $P =$ const. process
(B) $Q = 0$ for the $V =$ const. process
(C) $Q = 0$ for the $T =$ const. process
(D) $W = 0$ for the $V =$ const. process

4.31FE The term $\dot{m}\Delta h$ in a control volume equation $\dot{Q} - \dot{W}_s = \dot{m}\Delta h$:
(A) Accounts for the rate of change in energy of the control volume.
(B) Represents the rate of change of energy between the inlet and outlet.
(C) Is often neglected in control-volume applications.
(D) Includes the work rate due to the pressure forces.

4.32FE Select an assumption that is made when deriving the continuity equation $\rho_1 A_1 V_1 = \rho_2 A_2 V_2$.
(A) Incompressible flow
(B) Steady flow
(C) Uniform flow
(D) Isothermal flow

4.33FE A nozzle accelerates air from 20 m/s to 200 m/s. What temperature change is expected?
(A) 40 °C
(B) 30 °C
(C) 20 °C
(D) 10 °C

4.34FE Steam enters a valve at 10 MPa and 550 °C and exits at 0.8 MPa. The exiting temperature is nearest:
(A) 590 °C
(B) 535 °C
(C) 520 °C
(D) 510 °C

4.35FE Air enters an insulated compressor at 100 kPa and 20 °C and exits at 800 kPa. The exiting temperature is nearest:
(A) 530 °C
(B) 462 °C
(C) 323 °C
(D) 258 °C

4.36FE If $\dot{m} = 2$ kg/s for the compressor of Question 4.35FE and $d_1 = 20$ cm, calculate V_1.
(A) 62 m/s
(B) 53 m/s
(C) 41 m/s
(D) 33 m/s

4.37FE 10 kg/s of saturated steam at 10 kPa is to be completely condensed using 400 kg/s of cooling water. Estimate the temperature change of the cooling water.
(A) 32 °C
(B) 24 °C
(C) 18 °C
(D) 14 °C

4.38FE 100 kg/min of air enters a relatively short, constant-diameter tube at 25 °C and leaves at 20 °C. Estimate the heat loss.
(A) 750 kJ/min
(B) 670 kJ/min
(C) 500 kJ/min
(D) 360 kJ/min

4.39FE The minimum power needed by a water pump that increases the pressure of 4 kg/s from 100 kPa to 6 MPa is:
(A) 250 kW
(B) 95 kW
(C) 24 kW
(D) 6 kW

4.40FE A key concept in analyzing the filling of an evacuated tank is:
(A) The mass flow rate into the tank remains constant.
(B) The enthalpy across a valve remains constant.
(C) The internal energy in the tank remains constant.
(D) The temperature in the tank remains constant.

4.41FE A given volume of material, initially at 100 °C, cools to 60 °C in 40 seconds. Assuming no phase change and only convective cooling to air at 20 °C, how long would it take the same material to cool to 60 °C if the heat transfer coefficient were doubled?
(A) 3 s
(B) 4 s
(C) 20 s
(D) 80 s

Answers to Supplementary Problems

4.31 3.398 kg

4.32 123.3 J

4.33 0.49 Btu

4.34 (*a*) 15, 22 (*b*) 3, 14 (*c*) 25, 15 (*d*) −30, −10 (*e*) −4, −14

4.35 (*a*) −200 (*b*) 0 (*c*) 800 (*d*) 1000 (*e*) 1200

4.36 378 kJ

4.37 84 kJ

4.38 3260 Btu

4.39 1505 kJ

4.40 686 °C

4.41 6.277 Btu

4.42 833 °C

4.43 6.274 Btu

4.44 834 °C

4.45 (*a*) 6140 kJ (*b*) 1531 kJ

4.46 (*a*) 233.9 °C (*b*) 645 °C

4.47 (*a*) 2.06 kJ/kg·°C (*b*) 2.07 kJ/kg·°C (*c*) 16.5 kJ/kg·°C

4.48 (*a*) 0.386 Btu/lbm-°F (*b*) 0.388 Btu/lbm-°F (*c*) 1.96 Btu/lbm-°F

4.49 (*a*) 402 kJ (*b*) 418 kJ (*c*) 412 kJ

4.50 418 kJ vs. 186 kJ

4.51 (*a*) 104 °C (*b*) 120.2 °C

4.52 14,900 Btu

4.53 (*a*) 16.2 °C (*b*) 76.4%

4.54 62.7 °F

4.55 139.5 °C

4.56 1200 kJ, 860 kJ

4.57 (*a*) 60, 0, 0, 100, 203, 0.856, 0.211; (*b*) 43, 0, 43, 150, 148, 0.329, 0.329; (*c*) 28.4, 71.6, 100, 450, 500, 0.166, 0.109; (*d*) 0, −81, 81, 113, 706, 695, 77.3

4.58 (*a*) 49.4, 10.2, 11.8, 100, 100, 1.37, 0.662; (*b*) 45, 0, 60, 220, 172, 0.526, 0.526; (*c*) 23.5, 100, 320, 500, 0.217, 0.170; (*d*) 0, −190, 190, 245, 550, 1500, 200

4.59 (*a*) 1551 °F (*b*) 1741 °F

4.60 (*a*) 49.0 °C (*b*) 249 °C

4.61 (*a*) −12.0 kJ (*b*) 44.0 kJ

4.62 753 kJ

4.63 (*a*) 1137 kPa, 314 °C (*b*) 533 kPa, 154 °C

4.64 (*a*) 1135 °F (*b*) 1195 °F

4.65 (*a*) 1465 kJ (*b*) 1050 kJ (*c*) −291 kJ (*d*) 1465 kJ

4.66 (*a*) 1584 kJ (*b*) 2104 kJ

4.67 (*a*) 9.85 MJ (*b*) 12.26 MJ (*c*) 9.53 MJ

4.68 14.04 Btu

4.69 −2.22 kJ

4.70 (*a*) −1230 kJ (*b*) −1620 kJ

4.71 −116 kJ

4.72 (*a*) 49.4 °F (*b*) 69.4 °F (*c*) constant pressure

4.73 174.7 °C, 240.1 kPa

4.74 4.69 °C

4.75 (*a*) 373 kJ (*b*) 373 kJ

4.76 (*a*) 7150 ft-lbf, 9.19 Btu (*b*) 9480 ft-lbf, 12.2 Btu

4.77 1926 kJ, 1926 kJ

4.78 4.01 kJ

4.79 25 m/s

4.80 37.5 ft/sec

4.81 3.65 kg/s, 195.3 m/s

4.82 6.64 kg/s, 255 m/s

4.83 4.95 kg/s, 109 m/s

4.84 2.18 lbm/sec, 182.2 ft/sec
4.85 0.01348 $kg/m^3 \cdot s$
4.86 1.6 m/s, 0.0905 kg/s
4.87 0.314 kg/s, 16 m/s
4.88 116.3 ft/sec, 121.2 Btu/sec
4.89 1282 kJ/kg
4.90 (*a*) 569 °C (*b*) 22.3
4.91 39.34 Btu/lbm
4.92 0.021 kg/s
4.93 571 kW
4.94 3.53 kW
4.95 4.72 hp
4.96 346 hp
4.97 6 MW
4.98 21.19 kW
4.99 −1954 Btu/sec
4.100 6.43 MW, 82.2 m/s
4.101 373 kW
4.102 −70.5 kW
4.103 142.1 psia
4.104 (*a*) −4 °C (*b*) 731 m/s (*c*) 147 mm
4.105 (*a*) 1.672 kg/s (*b*) −0.91°C
4.106 238 °F
4.107 (*a*) 23.9 kg/s (*b*) 1 MJ
4.108 (*a*) 9.78 MW (*b*) 63.8 kW (*c*) 4.07 m/s (*d*) 27.4 MW (*e*) 17.69 MW (*f*) 141 kg/s (*g*) 35.5%
4.109 (*a*) 197 °C (*b*) 83.4 MW (*c*) 30.2 MW (*d*) 289 kW (*e*) 53.5 MW (*f*) 512 kg/s (*g*) 35.9%
4.110 (*a*) 0.1590 kg/s (*b*) 37.7 hp (*c*) 126.1 kW
4.111 (*a*) 1587 °F (*b*) 33.1 lbm
4.112 503 kJ
4.113 184 °C, 25.1 kg
4.114 (*a*) 284 °F (*b*) 4.36 lbm
4.115 (*a*) −146 °C (*b*) 227 kPa
4.116 (*a*) 9.2 °C (*b*) 624 kPa (*c*) 11.13 min
4.117 (*a*) 158.9 °C (*b*) 815 kg (*c*) 11.25 h

Answers to Review Questions for the FE Examination

4.1FE (B) **4.2FE** (A) **4.3FE** (C) **4.4FE** (A) **4.5FE** (C) **4.6FE** (C) **4.7FE** (D) **4.8FE** (D) **4.9FE** (A)
4.10FE (D) **4.11FE** (A) **4.12FE** (D) **4.13FE** (C) **4.14FE** (C) **4.15FE** (D) **4.16FE** (D) **4.17FE** (C)
4.18FE (C) **4.19FE** (C) **4.20FE** (B) **4.21FE** (B) **4.22FE** (C) **4.23FE** (B) **4.24FE** (C) **4.25FE** (D)
4.26FE (A) **4.27FE**(D) **4.28FE** (B) **4.29FE** (A) **4.30FE** (D) **4.31FE** (D) **4.32FE** (B) **4.33FE** (C)
4.34FE (D) **4.35FE** (D) **4.36FE** (B) **4.37FE** (D) **4.38FE** (C) **4.39FE** (C) **4.40FE** (B) **4.41FE** (C)

The Second Law of Thermodynamics

5.1 INTRODUCTION

Water flows down a hill, heat flows from a hot body to a cold one, rubber bands unwind, fluid flows from a high-pressure region to a low-pressure region, and we all get old! Our experiences in life suggest that processes have a definite direction. The first law of thermodynamics relates the several variables involved in a physical process but does not give any information as to the direction of the process. It is the second law of thermodynamics which helps us establish the direction of a particular process.

Consider, for example, the situation illustrated in Fig. 5-1. Here, the first law states that the work done by the falling weight is converted to internal energy of the air contained in the fixed volume, provided the volume is insulated so that $Q=0$. It would not be a violation of the first law if we postulated that an internal energy decrease of the air is used to turn the paddle and raise the weight. This, however, would be a violation of the second law of thermodynamics and would thus be an impossibility.

In this chapter we will state the second law as it applies to a cycle. Several devices will be analyzed. This will be followed in Chapter 6 by a statement of the second law as it applies to a process.

5.2 HEAT ENGINES, HEAT PUMPS, AND REFRIGERATORS

We refer to a device operating on a cycle as a heat engine, a heat pump, or a refrigerator, depending on the objective of the particular device. If the objective of the device is to perform work it is a *heat engine*; if its objective is to supply energy to a body it is a *heat pump*; if its objective is to extract energy from a body it is a *refrigerator*. A schematic diagram of a simple heat engine is shown in Fig. 5-2.

The net work produced by the engine would be equal to the net heat transfer, a consequence of the first law:

$$W = Q_H - Q_L \tag{5.1}$$

where Q_H and Q_L are the heat transfers from the high- and low-temperature reservoirs, respectively.

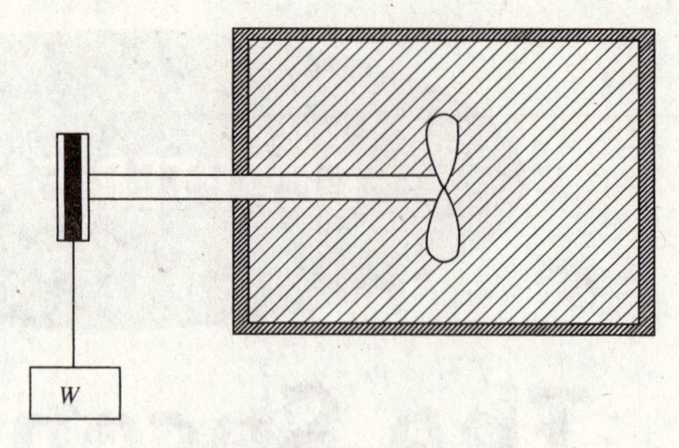

Fig. 5-1 Paddle-wheel work.

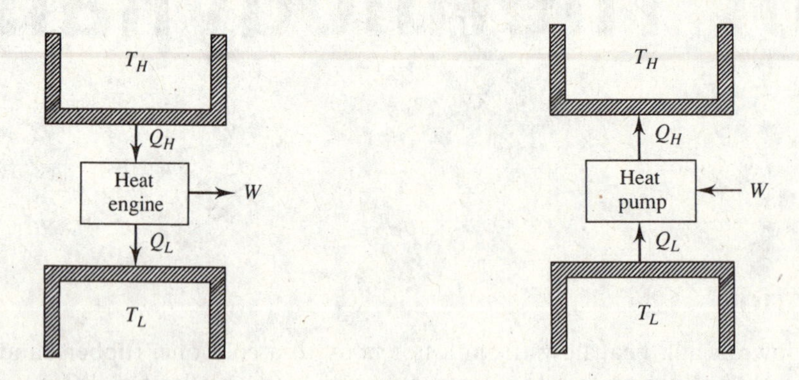

Fig. 5-2 A heat engine. **Fig. 5-3** A heat pump or a refrigerator.

If the cycle of Fig. 5-2 were reversed, a net work input would be required, as shown in Fig. 5-3. A heat pump would provide energy as heat Q_H to the warmer body (e.g., a house), and a refrigerator would extract energy as heat Q_L from the cooler body (e.g., a freezer). The work would also be given by *(5.1)*. Here we use magnitudes only.

Note that an engine or refrigerator operates between two *thermal energy reservoirs*, entities which are capable of providing or accepting heat without changing temperatures. The atmosphere or a lake serves as *heat sinks*; furnaces, solar collectors, or burners serve as *heat sources*. Temperatures T_H and T_L identify the respective temperatures of a source and a sink.

The thermal efficiency of the heat engine and the coefficients of performance of the refrigerator and the heat pump are as defined in Sec. 4.9:

$$\eta = \frac{W}{Q_H} \qquad \text{COP}_{\text{refrig}} = \frac{Q_L}{W} \qquad \text{COP}_{\text{h.p.}} = \frac{Q_H}{W} \tag{5.2}$$

The second law of thermodynamics will place limits on the above measures of performance. The first law would allow a maximum of unity for the thermal efficiency and an infinite coefficient of performance. The second law, however, establishes limits that are surprisingly low, limits that cannot be exceeded regardless of the cleverness of proposed designs.

One additional note concerning heat engines is appropriate. There are devices that we will refer to as heat engines which do not strictly meet our definition; they do not operate on a thermodynamic cycle but instead exhaust the working fluid and then intake new fluid. The internal combustion engine is an example. Thermal efficiency, as defined above, remains a quantity of interest for such devices.

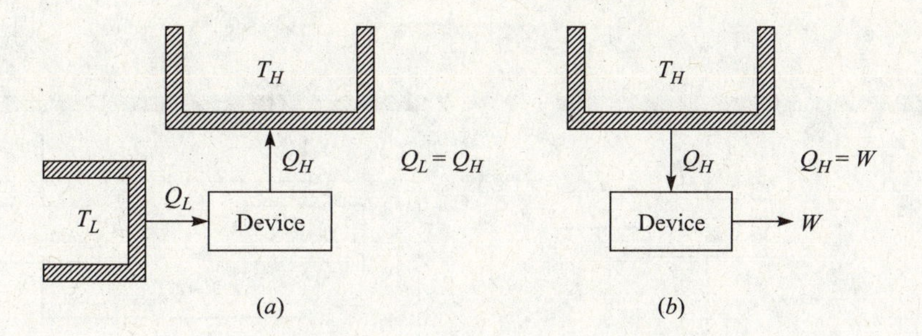

Fig. 5-4 Violations of the second law.

5.3 STATEMENTS OF THE SECOND LAW OF THERMODYNAMICS

As with the other basic laws presented, we do not derive a basic law but merely observe that such a law is never violated. The second law of thermodynamics can be stated in a variety of ways. Here we present two: the *Clausius statement* and the *Kelvin–Planck statement*. Neither is presented in mathematical terms. We will, however, provide a property of the system, entropy, which can be used to determine whether the second law is being violated for any particular situation. The first statement of the second law is:

Clausius Statement It is impossible to construct a device which operates on a cycle and whose sole effect is the transfer of heat from a cooler body to a hotter body.

This statement relates to a refrigerator (or a heat pump). It states that it is impossible to construct a refrigerator that transfers energy from a cooler body to a hotter body without the input of work; this violation is shown in Fig. 5-4*a*.

The second statement of the second law takes the following form:

Kelvin–Planck Statement It is impossible to construct a device which operates on a cycle and produces no other effect than the production of work and the transfer of heat from a single body.

In other words, it is impossible to construct a heat engine that extracts energy from a reservoir, does work, and does not transfer heat to a low-temperature reservoir. This rules out any heat engine that is 100 percent efficient, like the one shown in Fig. 5-4*b*.

Note that the two statements of the second law are negative statements. Neither has ever been proved; they are expressions of experimental observations. No experimental evidence has ever been obtained that violates either statement of the second law. It should also be noted that the two statements are equivalent. This will be demonstrated with an example.

EXAMPLE 5.1 Show that the Clausius and Kelvin–Planck statements of the second law are equivalent.

Solution: We will show that a violation of the Clausius statement implies a violation of the Kelvin–Planck statement, and vice versa, demonstrating that the two statements are equivalent. Consider the system shown in Fig. 5-5*a*. The device on the left transfers heat and violates the Clausius statement, since it has no work input. Let the heat engine transfer the same amount of heat Q_L. Then Q'_H is greater than Q_L by the amount W. If we simply transfer the heat Q_L directly from the engine to the device, as shown in Fig. 5-5*b*, there is no need for the low-temperature reservoir and the net result is a conversion of energy $(Q'_H - Q_H)$ from the high-temperature reservoir into an equivalent amount of work, a violation of the Kelvin–Planck statement of the second law.

Conversely (Problem 5.13), a violation of the Kelvin–Planck is equivalent to a violation of the Clausius statement.

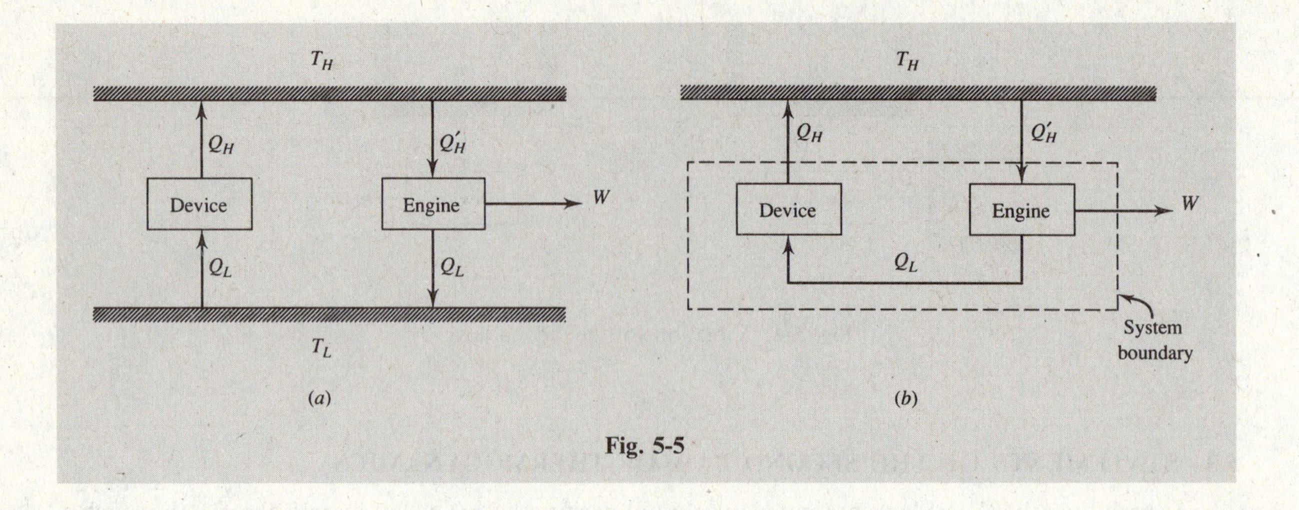

Fig. 5-5

5.4 REVERSIBILITY

In our study of the first law we made use of the concept of equilibrium and we defined equilibrium, or quasi equilibrium, with reference to the system only. We must now introduce the concept of *reversibility* so that we can discuss the most efficient engine that can possibly be constructed, an engine that operates with reversible processes only. Such an engine is called a *reversible engine*.

A *reversible process* is defined as a process which, having taken place, can be reversed and in so doing leaves no change in either the system or the surroundings. Observe that our definition of a reversible process refers to both the system and the surroundings. The process obviously has to be a quasiequilibrium process; additional requirements are:

1. No friction is involved in the process.
2. Heat transfer occurs due to an infinitesimal temperature difference only.
3. Unrestrained expansion does not occur.

The mixing of different substances and combustion also lead to irreversibilities.

To illustrate that friction makes a process irreversible consider the system of block plus an inclined plane, shown in Fig. 5-6*a*. Weights are added until the block is raised to the position shown in part (*b*). Now, to return the system to its original state some weight must be removed so that the block will slide back down the plane, as shown in part (*c*). Note that the surroundings have experienced a significant change; the weights that were removed must be raised, which requires a work input. Also, the block and plane are at a higher temperature due to the friction, and heat must be transferred to the surroundings to return the system to its original state. This will also change the surroundings. Because there has been a

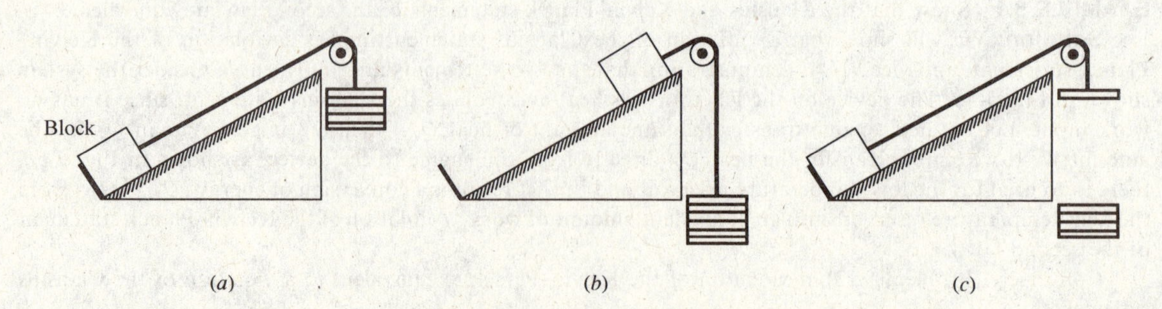

Fig. 5-6 Irreversibility due to friction.

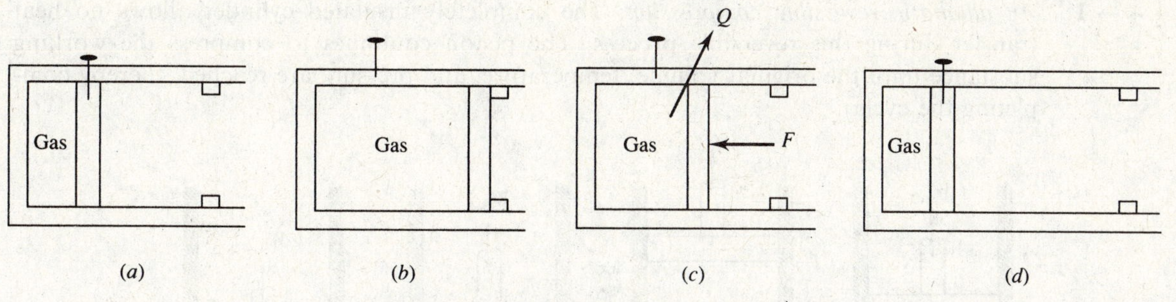

Fig. 5-7 Unrestrained expansion.

change in the surroundings as a result of the process and the reversed process, we conclude that the process was irreversible. A reversible process requires that no friction be present.

To demonstrate the fact that heat transfer across a finite temperature difference makes a process irreversible, consider a system composed of two blocks, one at a higher temperature than the other. Bringing the blocks together results in a heat transfer process; the surroundings are not involved in this process. To return the system to its original state, we must refrigerate the block that had its temperature raised. This will require a work input, demanded by the second law, resulting in a change in the surroundings. Hence, the heat transfer across a finite temperature difference is an irreversible process.

For an example of unrestrained expansion, consider the high-pressure gas contained in the cylinder of Fig. 5-7*a*. Pull the pin and let the piston suddenly move to the stops shown. Note that the only work done by the gas on the surroundings is to move the piston against atmospheric pressure. Now, to reverse this process it is necessary to exert a force on the piston. If the force is sufficiently large, we can move the piston to its original position, shown in part (*d*). This will demand a considerable amount of work, to be supplied by the surroundings. In addition, the temperature will increase substantially, and this heat must be transferred to the surroundings to return the temperature to its original value. The net result is a significant change in the surroundings, a consequence of irreversibility. Unrestrained expansion cannot occur in a reversible process.

5.5 THE CARNOT ENGINE

The heat engine that operates the most efficiently between a high-temperature reservoir and a low-temperature reservoir is the *Carnot engine*. It is an ideal engine that uses reversible processes to form its cycle of operation; thus it is also called a reversible engine. We will determine the efficiency of the Carnot engine and also evaluate its reverse operation. The Carnot engine is very useful, since its efficiency establishes the maximum possible efficiency of any real engine. If the efficiency of a real engine is significantly lower than the efficiency of a Carnot engine operating between the same limits, then additional improvements may be possible.

The cycle associated with the Carnot engine is shown in Fig. 5-8, using an ideal gas as the working substance. It is composed of the following four reversible processes:

1 → 2: *An isothermal expansion.* Heat is transferred reversibly from the high-temperature reservoir at the constant temperature T_H. The piston in the cylinder is withdrawn and the volume increases.

2 → 3: *An adiabatic reversible expansion.* The cylinder is completely insulated so that no heat transfer occurs during this reversible process. The piston continues to be withdrawn, with the volume increasing.

3 → 4: *An isothermal compression.* Heat is transferred reversibly to the low-temperature reservoir at the constant temperature T_L. The piston compresses the working substance, with the volume decreasing.

4 → 1: *An adiabatic reversible compression.* The completely insulated cylinder allows no heat transfer during this reversible process. The piston continues to compress the working substance until the original volume, temperature, and pressure are reached, thereby completing the cycle.

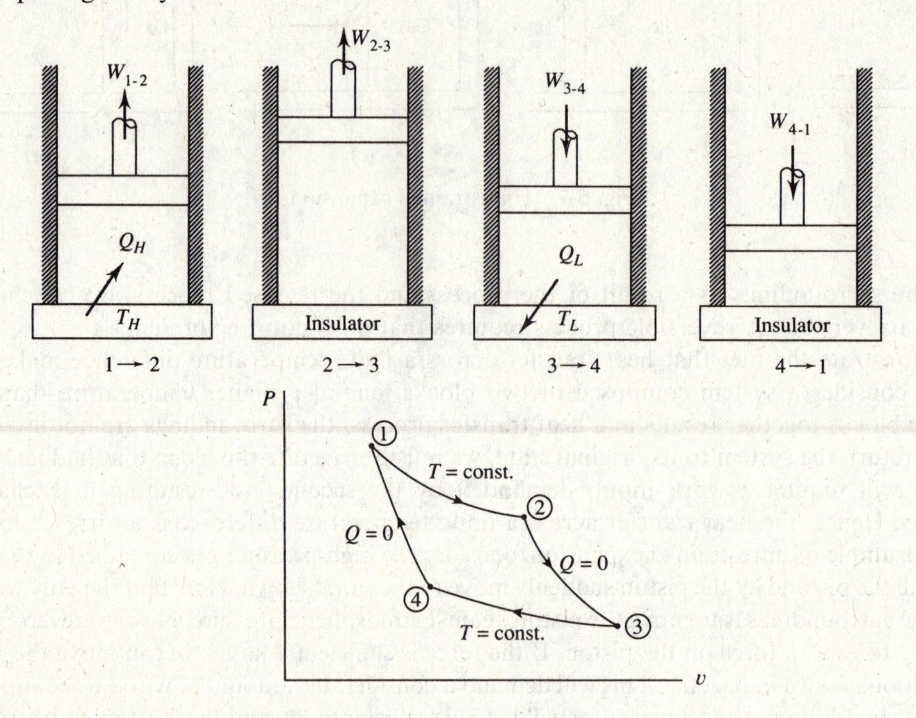

Fig. 5-8 The Carnot cycle.

Applying the first law to the cycle, we note that

$$Q_H - Q_L = W_{\text{net}} \tag{5.3}$$

where Q_L is assumed to be a positive value for the heat transfer to the low-temperature reservoir. This allows us to write the thermal efficiency [see (*4.76*)] for the Carnot cycle as

$$\eta = \frac{Q_H - Q_L}{Q_H} = 1 - \frac{Q_L}{Q_H} \tag{5.4}$$

The following examples will be used to prove two of the following three postulates:

Postulate 1 It is impossible to construct an engine, operating between two given temperature reservoirs, that is more efficient than the Carnot engine.

Postulate 2 The efficiency of a Carnot engine is not dependent on the working substance used or any particular design feature of the engine.

Postulate 3 All reversible engines, operating between two given temperature reservoirs, have the same efficiency as a Carnot engine operating between the same two temperature reservoirs.

EXAMPLE 5.2 Show that the efficiency of a Carnot engine is the maximum possible efficiency.

Solution: Assume that an engine exists, operating between two reservoirs, that has an efficiency greater than that of a Carnot engine; also, assume that a Carnot engine operates as a refrigerator between the same two reservoirs, as sketched in Fig. 5-9*a*. Let the heat transferred from the high-temperature reservoir to the engine be equal to the heat rejected by the refrigerator; then the work produced by the engine will be greater than the work required by the refrigerator (that is, $Q'_L < Q_L$) since the efficiency of the engine is greater than that of a Carnot engine. Now, our system can be organized as shown in Fig. 5-9*b*. The engine drives the refrigerator using the rejected heat from the refrigerator. But, there is some net work $(W' - W)$ that leaves the system. The net result is the conversion of energy from a single reservoir into work, a violation of the second law. Thus, the Carnot engine is the most efficient engine operating between two particular reservoirs.

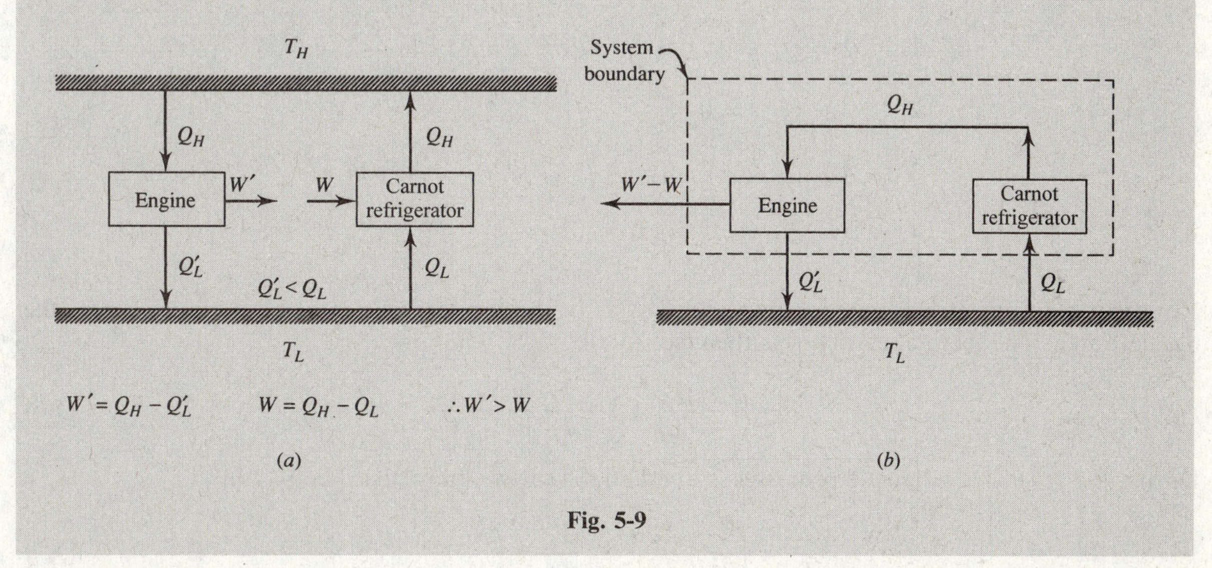

Fig. 5-9

EXAMPLE 5.3 Show that the efficiency of a Carnot engine operating between two reservoirs is independent of the working substance used by the engine.

Solution: Suppose that a Carnot engine drives a Carnot refrigerator as shown in Fig. 5-10*a*. Let the heat rejected by the engine be equal to the heat required by the refrigerator. Suppose the working fluid in the engine results in Q_H being greater than Q'_H; then W would be greater than W' (a consequence of the first law) and we would have the equivalent system shown in Fig. 5-10*b*. The net result is a transfer of heat $(Q_H - Q'_H)$ from a single reservoir and the production of work, a clear violation of the second law. Thus, the efficiency of a Carnot engine is not dependent on the working substance.

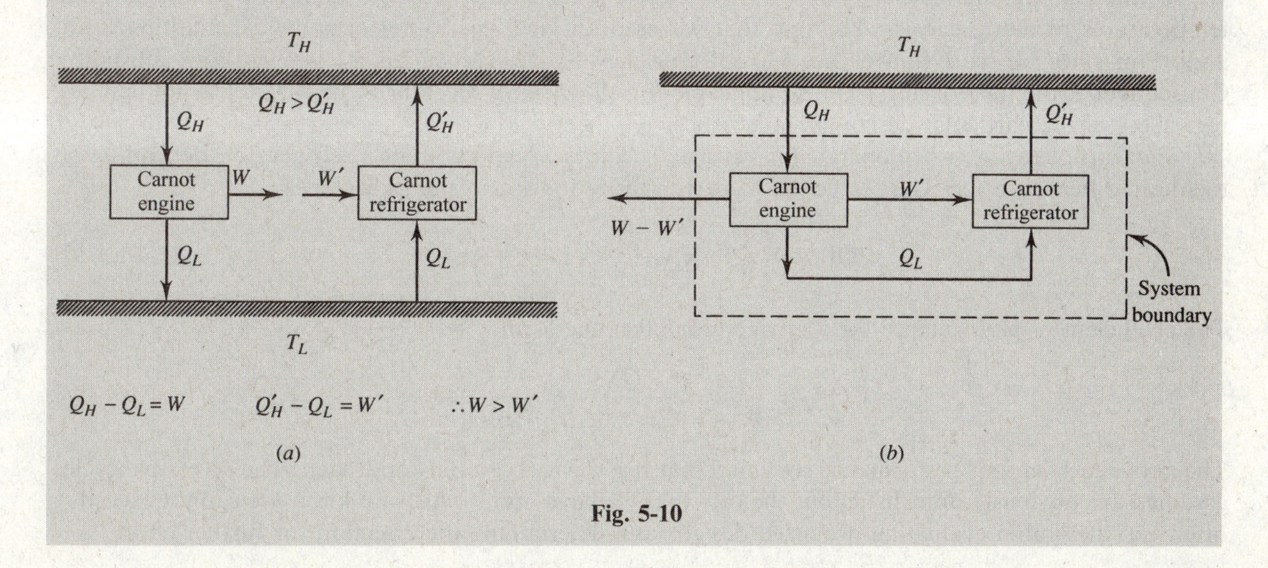

Fig. 5-10

5.6 CARNOT EFFICIENCY

Since the efficiency of a Carnot engine is dependent only on the two reservoir temperatures, the objective of this article will be to determine that relationship. We will assume the working substance to be an ideal gas (see Example 5.3) and simply perform the required calculations for the four processes of Fig. 5-8.

The heat transfer for each of the four processes is as follows:

$$\begin{aligned}
&\mathbf{1 \rightarrow 2:} \quad Q_H = W_{1-2} = \int_{V_1}^{V_2} P\,dV = mRT_H \ln \frac{V_2}{V_1} \\
&\mathbf{2 \rightarrow 3:} \quad Q_{2-3} = 0 \\
&\mathbf{3 \rightarrow 4:} \quad Q_L = -W_{3-4} = -\int_{V_3}^{V_4} P\,dV = -mRT_L \ln \frac{V_4}{V_3} \\
&\mathbf{4 \rightarrow 1:} \quad Q_{4-1} = 0
\end{aligned} \tag{5.5}$$

Note that we want Q_L to be a positive quantity, as in the thermal efficiency relationship; hence, the negative sign. The thermal efficiency is then [see (*5.4*)]

$$\eta = 1 - \frac{Q_L}{Q_H} = 1 + \frac{T_L}{T_H} \frac{\ln V_4/V_3}{\ln V_2/V_1} \tag{5.6}$$

During the reversible adiabatic processes $2 \rightarrow 3$ and $4 \rightarrow 1$, we know that [see (*4.49*)]

$$\frac{T_L}{T_H} = \left(\frac{V_2}{V_3}\right)^{k-1} \qquad \frac{T_L}{T_H} = \left(\frac{V_1}{V_4}\right)^{k-1} \tag{5.7}$$

Thus, we see that

$$\frac{V_3}{V_2} = \frac{V_4}{V_1} \qquad \text{or} \qquad \frac{V_4}{V_3} = \frac{V_1}{V_2} \tag{5.8}$$

Substituting into (*5.6*), we obtain the result

$$\eta = 1 - \frac{T_L}{T_H} \tag{5.9}$$

We have simply replaced Q_L/Q_H with T_L/T_H. We can do this for all reversible engines or refrigerators. We see that the thermal efficiency of a Carnot engine is dependent only on the high and low absolute temperatures of the reservoirs. The fact that we used an ideal gas to perform the calculations is not important since we have shown that Carnot efficiency to be independent of the working substance. Consequently, the relationship (*5.9*) is applicable for all working substances, or for all Carnot engines, regardless of the particular design characteristics.

The Carnot engine, when operated in reverse, becomes a heat pump or a refrigerator, depending on the desired heat transfer. The coefficient of performance for a heat pump becomes

$$\text{COP}_{HP} = \frac{Q_H}{W_{\text{net}}} = \frac{Q_H}{Q_H - Q_L} = \frac{1}{1 - T_L/T_H} \tag{5.10}$$

The coefficient of performance for a refrigerator takes the form

$$\text{COP}_R = \frac{Q_L}{W_{\text{net}}} = \frac{Q_L}{Q_H - Q_L} = \frac{1}{T_H/T_L - 1} \tag{5.11}$$

The above measures of performance set limits that real devices can only approach. The reversible cycles assumed are obviously unrealistic, but the fact that we have limits which we know we cannot exceed is often very helpful in evaluating proposed designs and determining the direction for further effort.

Rather than listing the COP of refrigerators and air-conditioners, manufacturers often list the EER (the *energy efficiency ratio*). It has the same definition as the COP (i.e., Q/W) but the units on Q are Btu's whereas the units on W are in watt-hours. So, it is the Btu's removed divided by the watt-hours of electricity consumed. Since there are 3.412 Btu's per watt-hour, we see that EER = 3.412 COP.

EXAMPLE 5.4 A Carnot engine operates between two temperature reservoirs maintained at 200 °C and 20 °C, respectively. If the desired output of the engine is 15 kW, as shown in Fig. 5-11, determine the heat transfer from the high-temperature reservoir and the heat transfer to the low-temperature reservoir.

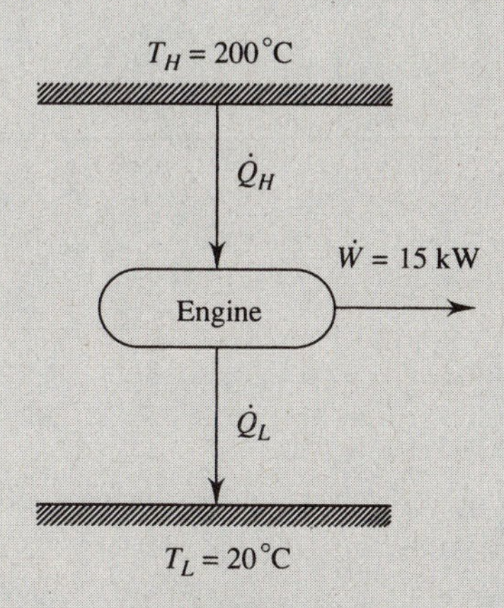

Fig. 5-11

Solution: The efficiency of a Carnot engine is given by

$$\eta = \frac{\dot{W}}{\dot{Q}_H} = 1 - \frac{T_L}{T_H}$$

This gives, converting the temperatures to absolute temperatures,

$$\dot{Q}_H = \frac{\dot{W}}{1 - T_L/T_H} = \frac{15}{1 - 293/473} = 39.42 \text{ kW}$$

Using the first law, we have

$$\dot{Q}_L = \dot{Q}_H - \dot{W} = 39.42 - 15 = 24.42 \text{ kW}$$

EXAMPLE 5.5 A refrigeration unit is cooling a space to −5 °C by rejecting energy to the atmosphere at 20 °C. It is desired to reduce the temperature in the refrigerated space to −25 °C. Calculate the minimum percentage increase in work required, by assuming a Carnot refrigerator, for the same amount of energy removed.

Solution: For a Carnot refrigerator we know that

$$\text{COP}_R = \frac{Q_L}{W} = \frac{1}{T_H/T_L - 1}$$

For the first situation we have $W_1 = Q_L(T_H/T_L - 1) = Q_L(293/268 - 1) = 0.0933Q_L$. For the second situation there results $W_2 = Q_L(293/248 - 1) = 0.181Q_L$. The percentage increase in work is then

$$\frac{W_2 - W_1}{W_1} = \left(\frac{0.181Q_L - 0.0933Q_L}{0.0933Q_L}\right)(100) = 94.0\%$$

Note the large increase in energy required to reduce the temperature in a refrigerated space. And this is a minimum percentage increase, since we have assumed an ideal refrigerator.

EXAMPLE 5.6 A Carnot engine operates with air, using the cycle shown in Fig. 5-12. Determine the thermal efficiency and the work output for each cycle of operation.

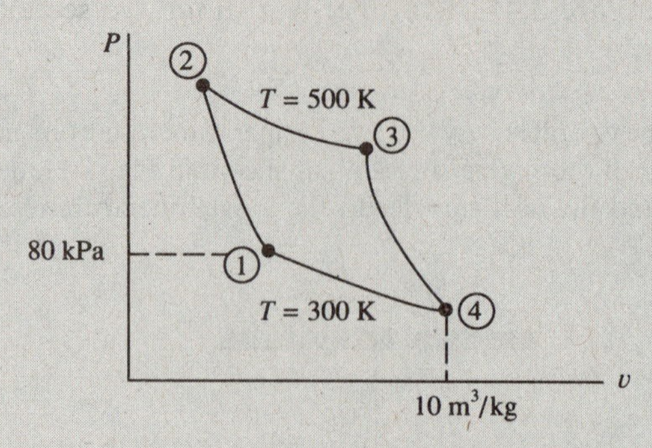

Fig. 5-12

Solution: The thermal efficiency is found to be

$$\eta = 1 - \frac{T_L}{T_H} = 1 - \frac{300}{500} = 0.4 \quad \text{or } 40\%$$

To find the work output we can determine the heat added during the constant temperature expansion and determine w from $\eta = W/Q_H = w/q_H$. We find q_H from the first law using $\Delta u = 0$:

$$q_H = w_{2\text{-}3} = \int P\,dv = RT_H \int_{v_2}^{v_3} \frac{dv}{v} = RT_H \ln \frac{v_3}{v_2}$$

To find v_2 first we must find v_1:

$$v_1 = \frac{RT_l}{P_1} = \frac{(287)(300)}{80\,000} = 1.076 \text{ m}^3/\text{kg}$$

Using (*4.49*), we have

$$v_2 = v_1 \left(\frac{T_1}{T_2}\right)^{1/(k-1)} = (1.076)(300/500)^{1/(1.4-1)} = 0.300 \text{ m}^3/\text{kg}$$

Likewise, $v_3 = v_4(T_4/T_3)^{1/(k-1)} = (10)(300/500)^{2.5} = 2.789 \text{ m}^3/\text{kg}$. Hence,

$$q_H = (287)(500) \ln \frac{2.789}{0.300} = 320.0 \text{ kJ/kg}$$

Finally, the work for each cycle is $w = \eta q_H = (0.4)(320.0) = 128$ kJ/kg.

Solved Problems

5.1 A refrigerator is rated at a COP of 4. The refrigerated space that it cools requires a peak cooling rate of 30 000 kJ/h. What size electrical motor (rated in horsepower) is required for the refrigerator?

The definition of the COP for a refrigerator is $\text{COP}_R = \dot{Q}_L/\dot{W}_{\text{net}}$. The net power required is then

$$\dot{W}_{\text{net}} = \frac{\dot{Q}_L}{\text{COP}_R} = \frac{30\,000/3600}{4} = 2.083 \text{ kW} \quad \text{or } 2.793 \text{ hp}$$

5.2 A Carnot heat engine produces 10 hp by transferring energy between two reservoirs at 40 °F and 212 °F. Calculate the rate of heat transfer from the high-temperature reservoir.

The engine efficiency is

$$\eta = 1 - \frac{T_L}{T_H} = 1 - \frac{500}{672} = 0.2560$$

The efficiency is also given by $\eta = \dot{W}/\dot{Q}_H$. Thus,

$$\dot{Q}_H = \frac{\dot{W}}{\eta} = \frac{(10 \text{ hp})(2545 \text{ Btu/hr/hp})}{0.2560} = 99{,}410 \text{ Btu/hr}$$

5.3 An inventor proposes an engine that operates between the 27 °C warm surface layer of the ocean and a 10 °C layer a few meters down. The inventor claims that the engine produces 100 kW by pumping 20 kg/s of seawater. Is this possible? Assume $(C_p)_{\text{seawater}} \cong 4.18$ kJ/kg·K.

The maximum temperature drop for the seawater is 17 °C. The maximum rate of heat transfer from the high-temperature water is then

$$\dot{Q}_H = \dot{m}C_p\Delta T = (20)(4.18)(17) = 1421 \text{ kW}$$

The efficiency of the proposed engine is then $\eta = \dot{W}/\dot{Q}_H = 100/1421 = 0.0704$ or 7.04%. The efficiency of a Carnot engine operating between the same two temperatures is

$$\eta = 1 - \frac{T_L}{T_H} = 1 - \frac{283}{300} = 0.0567 \quad \text{or } 5.67\%$$

The proposed engine's efficiency exceeds that of a Carnot engine; hence, the inventor's claim is impossible.

5.4 A power utility company desires to use the hot groundwater from a hot spring to power a heat engine. If the groundwater is at 95 °C, estimate the maximum power output if a mass flux of 0.2 kg/s is possible. The atmosphere is at 20 °C.

The maximum possible efficiency is

$$\eta = 1 - \frac{T_L}{T_H} = 1 - \frac{293}{368} = 0.2038$$

assuming the water is rejected at atmospheric temperature. The rate of heat transfer from the energy source is

$$\dot{Q}_H = \dot{m}C_p\Delta T = (0.2)(4.18)(95 - 20) = 62.7 \text{ kW}$$

The maximum power output is then

$$\dot{W} = \eta\dot{Q}_H = (0.2038)(62.7) = 12.8 \text{ kW}$$

5.5 Two Carnot engines operate in series between two reservoirs maintained at 600 °F and 100 °F, respectively. The energy rejected by the first engine is input into the second engine. If the first engine's efficiency is 20 percent greater than the second engine's efficiency, calculate the intermediate temperature.

The efficiencies of the two engines are

$$\eta_1 = 1 - \frac{T}{1060} \qquad \eta_2 = 1 - \frac{560}{T}$$

where T is the unknown intermediate temperature in °R. It is given that $\eta_1 = \eta_2 + 0.2\eta_2$. Substituting for η_1 and η_2 results in

$$1 - \frac{T}{1060} = 1.2\left(1 - \frac{560}{T}\right)$$

or

$$T^2 + 212T - 712{,}320 = 0 \qquad \therefore T = 744.6\,^\circ\text{R} \quad \text{or } 284.6\,^\circ\text{F}$$

5.6 A Carnot engine operating on air accepts 50 kJ/kg of heat and rejects 20 kJ/kg. Calculate the high and low reservoir temperatures if the maximum specific volume is 10 m³/kg and the pressure after the isothermal expansion is 200 kPa.

The thermal efficiency is

$$\eta = 1 - \frac{q_L}{q_H} = 1 - \frac{20}{50} = 0.6$$

Hence, $T_L/T_H = 0.4$. For the adiabatic processes we know that (see Fig. 5-8)

$$\frac{T_L}{T_H} = \left(\frac{v_2}{v_3}\right)^{k-1} \qquad \therefore \frac{v_2}{v_3} = 0.4^{2.5} = 0.1012$$

The maximum specific volume is v_3; thus, $v_2 = 0.1012v_3 = (0.1012)(10) = 1.012\ \text{m}^3/\text{kg}$. Now, the high temperature is

$$T_H = \frac{P_2 v_2}{R} = \frac{(200)(1.012)}{0.287} = 705.2\ \text{K} \quad \text{or } 432.2\,^\circ\text{C}.$$

The low temperature is then $T_L = 0.4T_H = (0.4)(705.2) = 282.1\ \text{K}$ or 9.1 °C.

5.7 A heat engine operates on a Carnot cycle with an efficiency of 75 percent. What COP would a refrigerator operating on the same cycle have? The low temperature is 0 °C.

The efficiency of the heat engine is given by $\eta = 1 - T_L/T_H$. Hence,

$$T_H = \frac{T_L}{1-\eta} = \frac{273}{1-0.75} = 1092\ \text{K}$$

The COP for the refrigerator is then

$$\text{COP}_R = \frac{T_L}{T_H - T_L} = \frac{273}{1092 - 273} = 0.3333$$

5.8 Two Carnot refrigerators operate in series between two reservoirs maintained at 20 °C and 200 °C, respectively. The energy output by the first refrigerator is used as the heat energy input to the second refrigerator. If the COPs of the two refrigerators are the same, what should the intermediate temperature be?

The COP for a refrigerator is given by $\text{COP}_R = T_L/(T_H - T_L)$. Requiring that the two COPs be equal gives

$$\frac{293}{T - 293} = \frac{T}{473 - T} \qquad \text{or} \qquad T^2 = 138\,589 \qquad \text{or} \qquad T = 372.3\ \text{K} = 99.3\,^\circ\text{C}$$

5.9 A heat pump is proposed in which 50 °F groundwater is used to heat a house to 70 °F. The groundwater is to experience a temperature drop of 12 °F, and the house requires 75,000 Btu/hr. Calculate the minimum mass flux of the groundwater and the minimum horsepower required.

The COP for the heat pump is

$$\text{COP}_\text{HP} = \frac{T_H}{T_H - T_L} = \frac{530}{530 - 510} = 26.5$$

This is also given by

$$\text{COP}_\text{HP} = \frac{\dot{Q}_H}{\dot{Q}_H - \dot{Q}_L} \qquad 26.5 = \frac{75{,}000}{75{,}000 - \dot{Q}_L} \qquad \dot{Q}_L = 72{,}170\ \text{Btu/hr}$$

The groundwater mass flux is then

$$\dot{Q}_L = \dot{m}C_p\Delta T \qquad 72{,}170 = (\dot{m})(1.00)(12) \qquad \dot{m} = 6014\ \text{lbm/hr}$$

The minimum horsepower required is found as follows:

$$\text{COP}_\text{HP} = \frac{\dot{Q}_H}{\dot{W}} \qquad 26.5 = \frac{75{,}000}{\dot{W}} \qquad \dot{W} = 2830\ \text{Btu/hr} \quad \text{or } 1.11\ \text{hp}$$

Supplementary Problems

5.10 A heat pump provides 75 MJ/h to a house. If the compressors require an electrical energy input of 4 kW, calculate the COP.

5.11 A power plant burns 1000 kg of coal each hour and produces 500 kW of power. Calculate the overall thermal efficiency if each kg of coal produces 6 MJ of energy.

5.12 An automobile that has a gas mileage of 13 km/L is traveling at 100 km/h. At this speed essentially all the power produced by the engine is used to overcome air drag. If the air drag force is given by $12\rho V^2 AC_D$ determine the thermal efficiency of the engine at this speed using projected area $A = 2\,\text{m}^2$, drag coefficient $C_D = 0.28$, and heating value of gasoline 9000 kJ/kg. Gasoline has a density of 740 kg/m^2.

5.13 Show that a violation of the Kelvin–Planck statement of the second law implies a violation of the Clausius statement.

5.14 A battery does work by producing an electric current while transferring heat with a constant-temperature atmosphere. Is this a violation of the second law? Explain.

5.15 Show that all reversible engines, operating between two given temperature reservoirs, have the same efficiency as a Carnot engine operating between the same two temperature reservoirs.

5.16 A Carnot cycle operates between 200 °C and 1200 °C. Calculate (*a*) its thermal efficiency if it operates as a power cycle, (*b*) its COP if it operates as a refrigerator, and (*c*) its COP if it operates as a heat pump.

5.17 A Carnot engine rejects 80 MJ of energy every hour by transferring heat to a reservoir at 10 °C. Determine the temperature of the high-temperature reservoir and the power produced if the rate of energy addition is 40 kW.

5.18 A proposed power cycle is designed to operate between temperature reservoirs, as shown in Fig. 5-13. It is supposed to produce 43 hp from the 2500 kJ of energy extracted each minute. Is the proposal feasible?

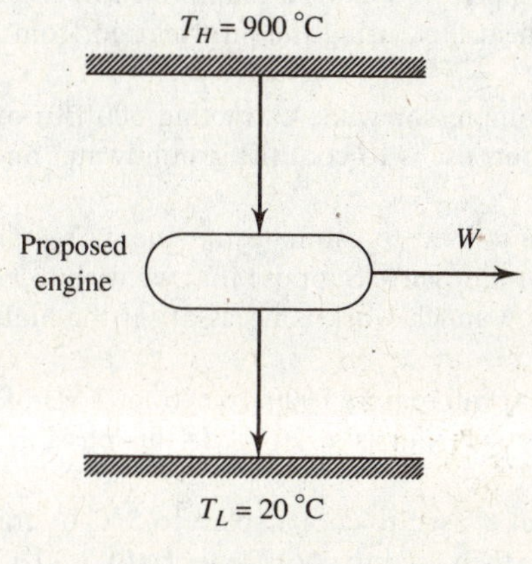

Fig. 5-13

5.19 (*a*) What is the maximum efficiency that can result from an engine that operates on the thermal gradients in the ocean? The surface waters at the proposed location are at 85 °F and those at a reasonable depth are at 50 °F. (*b*) What would be the maximum COP of a heat pump, operating between the two layers, used to heat an off-shore oil rig?

5.20 A Carnot engine operates between reservoirs at temperatures T_1 and T_2, and a second Carnot engine operates between reservoirs maintained at T_2 and T_3. Express the efficiency η_3 of the third engine operating between T_1 and T_3 in terms of the efficiencies η_1 and η_2 of the other two engines.

5.21 Two Carnot engines operate in series between two reservoirs maintained at 500 °C and 40 °C, respectively. The energy rejected by the first engine is utilized as energy input to the second engine. Determine the temperature of this intermediate reservoir between the two engines if the efficiencies of both engines are the same.

5.22 A Carnot engine operates on air with the cycle shown in Fig. 5-14. If there are 30 kJ/kg of heat added from the high-temperature reservoir maintained at 200 °C determine the work produced.

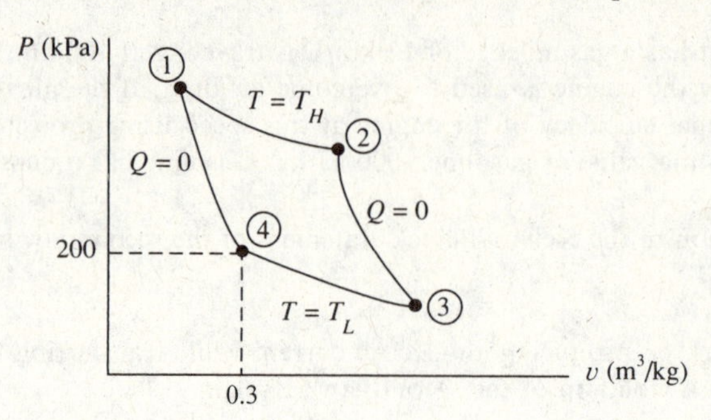

Fig. 5-14

5.23 A Carnot engine operates between a low pressure of 15 psia and a high pressure of 400 psia. The corresponding volumes are 250 and 25 in^3. If there is 0.01 lbm of air, calculate the work output.

5.24 A Carnot engine using hydrogen gas operates with the high-temperature reservoir maintained at 600 K. The pressure ratio for the adiabatic compression is 15 to 1 and the volume during the heat-addition process is tripled. If the minimum pressure is 100 kPa, determine the thermal efficiency and work produced.

5.25 A heat pump is to maintain a house at 20 °C when the outside air is at −25 °C. It is determined that 1800 kJ is required each minute to accomplish this. Calculate the minimum horsepower required.

5.26 If the heat pump of Prob. 5.25 is to be used as an air conditioner, calculate the maximum outside temperature for which the inside temperature can be maintained at 23 °C. Assume a linear relationship between temperature difference and heat flux, using the information from Prob. 5.25.

5.27 A heat pump uses a 5-hp compressor while extracting 500 Btu of energy from groundwater each minute. What is the COP (*a*) if the purpose is to cool the groundwater and (*b*) if the purpose is to heat a building?

5.28 A Carnot refrigeration cycle is used to estimate the energy requirement in an attempt to reduce the temperature of a specimen to absolute zero. Suppose that we wish to remove 0.01 J of energy from the specimen when it is at 2×10^{-6} K. How much work is necessary if the high-temperature reservoir is at 20 °C?

5.29 A refrigerator is proposed that will require 10 hp to extract 3 MJ of energy each minute from a space which is maintained at −18 °C. The outside air is at 20 °C. Is this possible?

5.30 A reversible refrigeration unit is used to cool a space to 5 °C by transferring heat to the surroundings which are at 25 °C. The same unit is then used to cool the space to −20 °C. Estimate the cooling rate for the second condition if the cooling rate for the first is 5 tons.

Review Questions for the FE Examination

5.1FE Select an acceptable paraphrase of the Kelvin–Planck statement of the second law.

(A) No process can produce more work than the heat it accepts.
(B) No engine can produce more work than the heat it intakes.
(C) An engine cannot produce work without accepting heat.
(D) An engine has to reject heat.

5.2FE Which of the following can be assumed to be reversible?
(A) A paddle wheel.
(B) A burst membrane.
(C) A resistance heater.
(D) A piston compressing gas in a race engine.

5.3FE An inventor claims that a thermal engine, operating between ocean layers at 27 °C and 10 °C, produces 10 kW of power while discharging 9900 kJ/min. This engine is:
(A) impossible
(B) reversible
(C) possible
(D) probable

5.4FE A Carnot engine, operating between reservoirs at 20 °C and 200 °C, produces 10 kW of power. The rejected heat is nearest:
(A) 26.3 kJ/s
(B) 20.2 kJ/s
(C) 16.3 kJ/s
(D) 12.0 kJ/s

5.5FE A Carnot cycle is a cycle of special interest because:
(A) It establishes a lower limit on cycle efficiency.
(B) It operates between two constant-temperature thermal reservoirs.
(C) It provides the maximum efficiency for any cycle.
(D) When it is carefully constructed in a laboratory, it provides an upper limit on cycle efficiency.

5.6FE Select an incorrect statement relating to a Carnot cycle.
(A) There are two adiabatic processes.
(B) There are two constant-pressure processes.
(C) Work occurs for all four processes.
(D) Each process is a reversible process.

5.7FE A Carnot refrigerator requires 10 kW to remove 20 kJ/s from a 20 °C reservoir. The temperature of the high-temperature reservoir is nearest:
(A) 440 K
(B) 400 K
(C) 360 K
(D) 320 K

5.8FE A heat pump is to provide 2000 kJ/hr to a house maintained at 20 °C. If it is −20 °C outside, what is the minimum power requirement?
(A) 385 kJ/hr
(B) 316 kJ/hr
(C) 273 kJ/hr
(D) 184 kJ/hr

5.9FE An engine operates on 100 °C geothermal water. It exhausts to a 20 °C stream. Its maximum efficiency is:
(A) 21%
(B) 32%
(C) 58%
(D) 80%

Answers to Supplementary Problems

5.10 5.21

5.11 30%

5.12 51.9%

5.14 No. This is not a cycle.

5.16 (*a*) 67.9% (*b*) 0.473 (*c*) 1.473

5.17 236.4 °C, 17.78 kW

5.18 No

5.19 (*a*) 6.42% (*b*) 15.57

5.20 $\eta_1 + \eta_2 - \eta_1 \eta_2$

5.21 218.9 °C

5.22 16.74 kJ/kg

5.23 178 ft-lbf

5.24 54.4%, 103 kJ/kg

5.25 6.18 hp

5.26 71.7 °C

5.27 (*a*) 2.36 (*b*) 3.36

5.28 1465 kJ

5.29 Yes

5.30 80 kW

Answers to Review Questions for the FE Examination

5.1FE (D) **5.2FE** (D) **5.3FE** (A) **5.4FE** (C) **5.5FE** (C) **5.6FE** (B) **5.7FE** (A) **5.8FE** (C) **5.9FE** (A)

Entropy

6.1 INTRODUCTION

To allow us to apply the second law of thermodynamics to a process we will identify a property called *entropy*. This will parallel our discussion of the first law; first we stated the first law for a cycle and then derived a relationship for a process.

6.2 DEFINITION

Consider the reversible Carnot engine operating on a cycle consisting of the processes described in Sec. 5.5. The quantity $\oint \delta Q/T$ is the cyclic integral of the heat transfer divided by the absolute temperature at which the heat transfer occurs. Since the temperature T_H is constant during the heat transfer Q_H, and T_L is constant during heat transfer Q_L, the integral is given by

$$\oint \frac{\delta Q}{T} = \frac{Q_H}{T_H} - \frac{Q_L}{T_L} \tag{6.1}$$

where the heat Q_L leaving the Carnot engine is considered to be positive. Using *(5.4)* and *(5.9)* we see that, for the Carnot cycle,

$$\frac{Q_L}{Q_H} = \frac{T_L}{T_H} \qquad \text{or} \qquad \frac{Q_H}{T_H} = \frac{Q_L}{T_L} \tag{6.2}$$

Substituting this into *(6.1)*, we find the interesting result

$$\oint \frac{\delta Q}{T} = 0 \tag{6.3}$$

Thus, the quantity $\delta Q/T$ is a perfect differential, since its cyclic integral is zero. We let this perfect differential be denoted by dS, where S represents a scalar function that depends only on the state of the system. This, in fact, was our definition of a property of a system. We shall call this extensive property *entropy*; its differential is given by

$$dS = \left.\frac{\delta Q}{T}\right|_{\text{rev}} \tag{6.4}$$

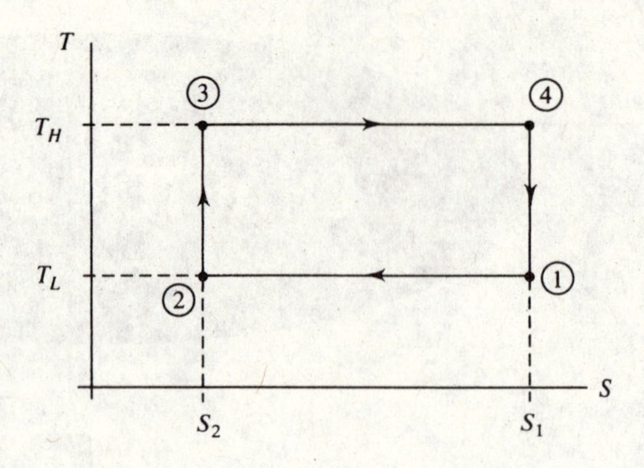

Fig. 6-1 The Carnot cycle.

where the subscript "rev" emphasizes the reversibility of the process. This can be integrated for a process to give

$$\Delta S = \int_1^2 \frac{\delta Q}{T}\bigg|_{\text{rev}} \tag{6.5}$$

From the above equation we see that the entropy change for a reversible process can be either positive or negative depending on whether energy is added to or extracted from the system during the heat transfer process. For a reversible adiabatic process the entropy change is zero.

We often sketch a temperature-entropy diagram for cycles or processes of interest. The Carnot cycle provides a simple display when plotting temperature vs. entropy. It is shown in Fig. 6-1. The change in entropy for the first process from state 1 to state 2 is

$$S_2 - S_1 = \int_1^2 \frac{\delta Q}{T} = -\frac{Q_L}{T_L} \tag{6.6}$$

The entropy change for the reversible adiabatic process from state 2 to state 3 is zero. For the process from state 3 to state 4 the entropy change is numerically equal to that of the first process; the process from state 4 to state 1 is also a reversible adiabatic process and is accompanied with a zero entropy change.

The heat transfer during a reversible process can be expressed in differential form [see (*6.4*)] as

$$\delta Q = T\, dS \tag{6.7}$$

Hence, the area under the curve in the T-S diagram represents the heat transfer during any reversible process. The rectangular area in Fig. 6-1 thus represents the net heat transfer during the Carnot cycle. Since the heat transfer is equal to the work done for a cycle, the area also represents the net work accomplished by the system during the cycle. Here, $Q_{\text{net}} = W_{\text{net}} = \Delta T \Delta S$.

The first law of thermodynamics, for a reversible infinitesimal change, becomes, using (*6.7*),

$$T\, dS - P\, dV = dU \tag{6.8}$$

This is an important relationship in our study of simple systems. We arrived at it assuming a reversible process. However, since it involves only properties of the system, it holds for an irreversible process also. If we have an irreversible process, in general, $\delta W \neq P\, dV$ and $\delta Q \neq T\, dS$ but (*6.8*) still holds as a relationship between the properties. Dividing by the mass, we have

$$T\, ds - P\, dv = du \tag{6.9}$$

where the specific entropy is defined to be

$$s = \frac{S}{m} \tag{6.10}$$

To relate the entropy change to the enthalpy change we differentiate (*4.12*) and obtain

$$dh = du + P\,dv + v\,dP \tag{6.11}$$

Substituting into (*6.9*) for *du*, we have

$$T\,ds = dh - v\,dP \tag{6.12}$$

Equations (*6.9*) and (*6.12*) will be used in subsequent sections of our study of thermodynamics for various reversible and irreversible processes.

6.3 ENTROPY FOR AN IDEAL GAS WITH CONSTANT SPECIFIC HEATS

Assuming an ideal gas, (*6.9*) becomes

$$ds = \frac{du}{T} + \frac{P\,dv}{T} = C_v\frac{dT}{T} + R\frac{dv}{v} \tag{6.13}$$

where we have used

$$du = C_v\,dT \qquad Pv = RT \tag{6.14}$$

(*6.13*) is integrated, assuming constant specific heat, to yield

$$s_2 - s_1 = C_v \ln\frac{T_2}{T_1} + R\ln\frac{v_2}{v_1} \tag{6.15}$$

Similarly, (*6.12*) is rearranged and integrated to give

$$s_2 - s_1 = C_p \ln\frac{T_2}{T_1} - R\ln\frac{P_2}{P_1} \tag{6.16}$$

Note again that the above equations were developed assuming a reversible process; however, they relate the change in entropy to other thermodynamic properties at the two end states. Since the change of a property is independent of the process used in going from one state to another, the above relationships hold for any process, reversible or irreversible, providing the working substance can be approximated by an ideal gas with constant specific heats.

If the entropy change is zero, i.e., an *isentropic process*, (*6.15*) and (*6.16*) can be used to obtain

$$\frac{T_2}{T_1} = \left(\frac{v_1}{v_2}\right)^{k-1} \qquad \frac{T_2}{T_1} = \left(\frac{P_2}{P_1}\right)^{(k-1)/k} \tag{6.17}$$

These two equations are combined to give

$$\frac{P_2}{P_1} = \left(\frac{v_1}{v_2}\right)^{k} \tag{6.18}$$

These are, of course, identical to the equations obtained in Chap. 4 when an ideal gas undergoes a quasiequilibrium adiabatic process.

EXAMPLE 6.1 Air is contained in an insulated, rigid volume at 20 °C and 200 kPa. A paddle wheel, inserted in the volume, does 720 kJ of work on the air. If the volume is 2 m^3, calculate the entropy increase assuming constant specific heats.

Solution: To determine the final state of the process we use the energy equation, assuming zero heat transfer. We have $-W = \Delta U = mC_v\Delta T$. The mass m is found from the ideal-gas equation to be

$$m = \frac{PV}{RT} = \frac{(200)(2)}{(0.287)(293)} = 4.76 \text{ kg}$$

The first law, taking the paddle-wheel work as negative, is then

$$720 = (4.76)(0.717)(T_2 - 293) \qquad \therefore T_2 = 504.0 \text{ K}$$

Using (*6.15*) for this constant-volume process there results

$$\Delta S = mC_v \ln \frac{T_2}{T_1} = (4.76)(0.717) \ln \frac{504}{293} = 1.851 \text{ kJ/K}$$

EXAMPLE 6.2 After a combustion process in a cylinder the pressure is 1200 kPa and the temperature is 350 °C. The gases are expanded to 140 kPa with a reversible adiabatic process. Calculate the work done by the gases, assuming they can be approximated by air with constant specific heats.

Solution: The first law can be used, with zero heat transfer, to give $-w = \Delta u = C_v(T_2 - T_1)$. The temperature T_2 is found from (*6.17*) to be

$$T_2 = T_1 \left(\frac{P_2}{P_1}\right)^{(k-1)/k} = (623)\left(\frac{140}{1200}\right)^{(1.4-1)/1.4} = 337 \text{ K}$$

This allows the specific work to be calculated: $w = C_v(T_1 - T_2) = (0.717)(623 - 337) = 205$ kJ/kg.

6.4 ENTROPY FOR AN IDEAL GAS WITH VARIABLE SPECIFIC HEATS

If the specific heats for an ideal gas cannot be assumed constant over a particular temperature range we return to (*6.12*) and write

$$ds = \frac{dh}{T} - \frac{v\,dP}{T} = \frac{C_p}{T}\,dT - \frac{R}{P}\,dP \tag{6.19}$$

The gas constant R can be removed from the integral, but $C_p = C_p(T)$ cannot. Hence, we integrate (*6.19*) and obtain

$$s_2 - s_1 = \int_{T_1}^{T_2} \frac{C_p}{T}\,dT - R \ln \frac{P_2}{P_1} \tag{6.20}$$

The integral in the above equation depends only on temperature, and we can evaluate its magnitude from the gas tables. It is found, using the tabulated function s^o, to be

$$s_2^\text{o} - s_1^\text{o} = \int_{T_1}^{T_2} \frac{C_p}{T}\,dT \tag{6.21}$$

Thus, the entropy change is (in some textbooks ϕ is used rather than s^o)

$$s_2 - s_1 = s_2^\text{o} - s_1^\text{o} - R \ln \frac{P_2}{P_1} \tag{6.22}$$

This more exact expression for the entropy change is used only when improved accuracy is desired.

For an isentropic process we cannot use (*6.17*) and (*6.18*) if the specific heats are not constant. However, we can use (*6.22*) and obtain, for an isentropic process,

$$\frac{P_2}{P_1} = \exp\left(\frac{s_2^\text{o} - s_1^\text{o}}{R}\right) = \frac{\exp(s_2^\text{o}/R)}{\exp(s_1^\text{o}/R)} = \frac{f(T_2)}{f(T_1)} \tag{6.23}$$

Thus, we define a *relative pressure* P_r, which depends only on the temperature, as

$$P_r = e^{s^\text{o}/R} \tag{6.24}$$

It is included as an entry in the gas table E-1. The pressure ratio for an isentropic process is then

$$\frac{P_2}{P_1} = \frac{P_{r2}}{P_{r1}} \tag{6.25}$$

The volume ratio can be found using the ideal-gas equation of state. It is

$$\frac{v_2}{v_1} = \frac{P_2}{P_1}\frac{T_2}{T_1} \tag{6.26}$$

where we would assume an isentropic process when using the relative pressure ratio. Consequently, we define a *relative specific volume* v_r, dependent solely on the temperature, as

$$v_r = \text{const.} \times \frac{T}{P_r} \tag{6.27}$$

Using its value from the gas tables we find the specific volume ratio for an isentropic process; it is

$$\frac{v_2}{v_1} = \frac{v_{r2}}{v_{r1}} \tag{6.28}$$

With the entries from the gas tables we can perform the calculations required in working problems involving an ideal gas with variable specific heats.

EXAMPLE 6.3 Repeat Example 6.1 assuming variable specific heats.

Solution: Using the gas tables, we write the first law as $-W = \Delta U = m(u_2 - u_1)$. The mass is found from the ideal-gas equation to be

$$m = \frac{PV}{RT} = \frac{(200)(2)}{(0.287)(293)} = 4.76 \text{ kg}$$

The first law is then written as

$$u_2 = -\frac{W}{m} + u_1 = -\frac{-720}{4.76} + 209.1 = 360.4 \text{ kJ/kg}$$

where u_1 is found at 293 K in the gas tables by interpolation. Now, using this value for u_2, we can interpolate to find

$$T_2 = 501.2 \text{ K} \qquad s_2^\circ = 2.222$$

The value for s_1° is interpolated to be $s_1^\circ = 1.678$. The pressure at state 2 is found using the ideal-gas equation for our constant-volume process:

$$\frac{P_2}{T_2} = \frac{P_1}{T_1} \qquad P_2 = P_1\frac{T_2}{T_1} = (200)\left(\frac{501.2}{293}\right) = 342.1 \text{ kPa}$$

Finally, the entropy change is

$$\Delta S = m\left(s_2^\circ - s_1^\circ - R\ln\frac{P_2}{P_1}\right) = 4.76\left(2.222 - 1.678 - 0.287\ \ln\frac{342.1}{200}\right) = 1.856 \text{ kJ/K}$$

The approximate result of Example 6.1 is seen to be less than 0.3% in error because T_2 is not very large.

EXAMPLE 6.4 After a combustion process in a cylinder the pressure is 1200 kPa and the temperature is 350 °C. The gases are expanded to 140 kPa in a reversible, adiabatic process. Calculate the work done by the gases, assuming they can be approximated by air with variable specific heats.

Solution: First, at 623 K the relative pressure P_{r1} is interpolated to be $P_{r1} = (\frac{3}{20})(20.64 - 18.36) + 18.36 = 18.70$. For an isentropic process,

$$P_{r2} = P_{r1}\frac{P_2}{P_1} = (18.70)\left(\frac{140}{1200}\right) = 2.182$$

With this value for the relative pressure at state 2,

$$T_2 = \left(\frac{2.182 - 2.149}{2.626 - 2.149}\right)(20) + 340 = 341 \text{ K}$$

The work is found from the first law to be

$$w = u_1 - u_2$$
$$= \left[\frac{3}{20}(465.5 - 450.1) + 450.1\right] - \left[\left(\frac{2.182 - 2.149}{2.626 - 2.149}\right)(257.2 - 242.8) + 242.8\right] = 208.6 \text{ kJ/kg}$$

The approximate result of Example 6.2 is observed to have an error less than 1.5%.

6.5 ENTROPY FOR SUBSTANCES SUCH AS STEAM, SOLIDS, AND LIQUIDS

The entropy change has been found for an ideal gas with constant specific heats and for an ideal gas with variable specific heats. For pure substances, such as steam, entropy is included as an entry in the tables. In the quality region, it is found using the relation

$$s = s_f + xs_{fg} \tag{6.29}$$

Note that the entropy of saturated liquid water at 0 °C is arbitrarily set equal to zero. It is only the change in entropy that is of interest; hence, this arbitrary datum for entropy is of no consequence. In the superheated region it is tabulated as a function of temperature and pressure along with the other properties.

For a compressed liquid it is included as an entry in Table C-4, the compressed liquid table, or it can be approximated by the saturated liquid values s_f at the given temperature. From the compressed liquid table at 10 MPa and 100 °C, $s = 1.30$ kJ/kg·K, and from the saturated steam table at 100 °C, $s = 1.31$ kJ/kg·K; this is an insignificant difference.

The temperature-entropy diagram is of particular interest and is often sketched during the problem solution. A *T-s* diagram is sketched in Fig. 6-2*a*; for steam it is essentially symmetric about the critical point. Note that the high-pressure lines in the compressed liquid region are indistinguishable from the saturated liquid line. It is often helpful to visualize a process on a *T-s* diagram, since such a diagram illustrates assumptions regarding irreversibilites.

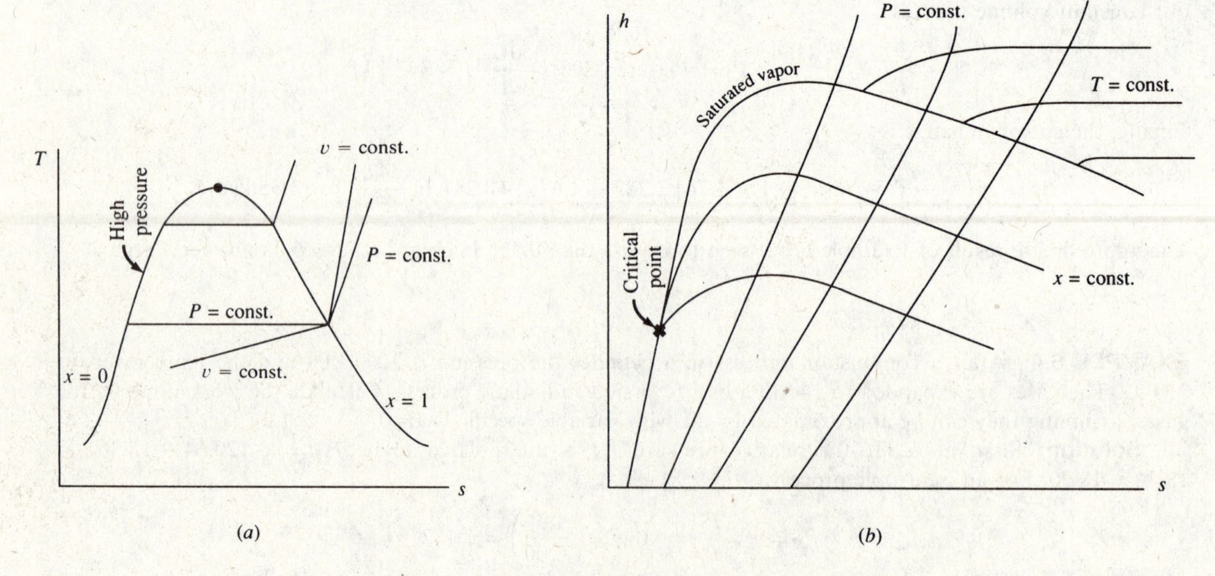

Fig. 6-2 The *T-s* and *h-s* diagrams for steam.

In addition to a *T-s* diagram, an *h-s* diagram, which is also called a *Mollier diagram*, is often useful in solving particular types of problems. The general shape of an *h-s* diagram is sketched in Fig. 6-2*b*.

For a solid or a liquid, the entropy change can be found quite easily if we can assume the specific heat to be constant. Returning to (*6.9*), we can write, assuming the solid or liquid to be incompressible so that $dv = 0$,

$$T\,ds = du = C\,dT \tag{6.30}$$

where we have dropped the subscript on the specific heat since for solids and liquids $C_p \cong C_v$. Tables usually list values for C_p; these are assumed to be equal to C. Assuming a constant specific heat, we find that

$$\Delta s = \int C\frac{dT}{T} = C\ln\frac{T_2}{T_1} \tag{6.31}$$

If the specific heat is a known function of temperature, the integration can be performed. Specific heats for solids and liquids are listed in Table B-4.

EXAMPLE 6.5 Steam is contained in a rigid container at an initial pressure of 100 psia and 600°F. The pressure is reduced to 10 psia by removing energy via heat transfer. Calculate the entropy change and the heat transfer and sketch a *T-s* diagram.

Solution: From the steam tables, $v_1 = v_2 = 6.216\ \text{ft}^3/\text{lbm}$. State 2 is in the quality region. Using the above value for v_2, the quality is found as follows:

$$6.216 = 0.0166 + x(38.42 - 0.0166) \qquad x = 0.1614$$

The entropy at state 2 is $s_2 = 0.2836 + (0.1614)(1.5041) = 0.5264\ \text{Btu/lbm-°R}$; the entropy change is then

$$\Delta s = s_2 - s_1 = 0.5264 - 1.7582 = -1.232\ \text{Btu/lbm-°R}$$

The heat transfer is found from the first law using $w = 0$:

$$q = u_2 - u_1 = [161.2 + (0.1614)(911.01)] - 1214.2 = -906\ \text{Btu/lbm}$$

The process is displayed in the *T-s* diagram shown in Fig. 6-3.

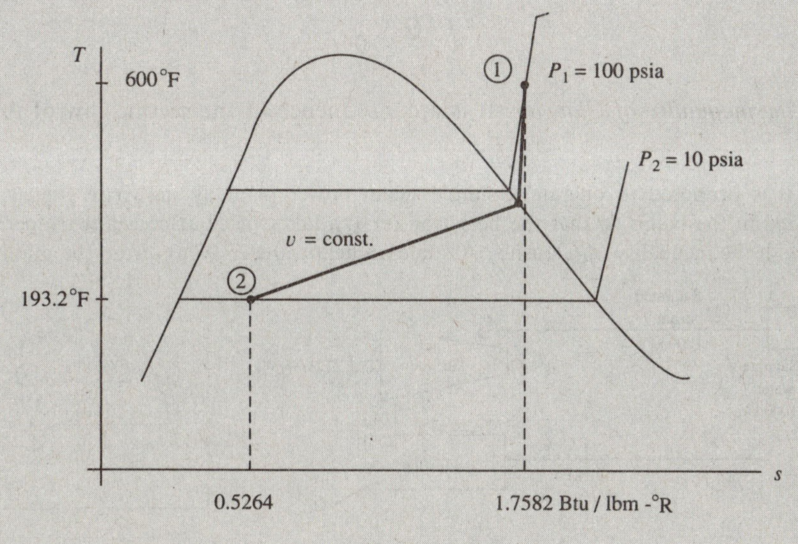

Fig. 6-3

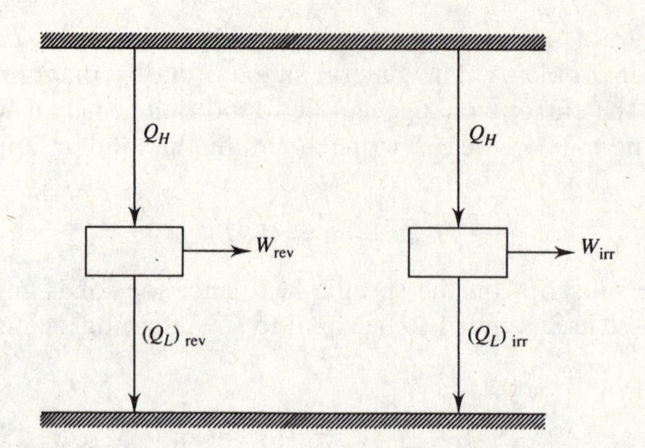

Fig. 6-4 A reversible and an irreversible engine operating between two reservoirs.

6.6 THE INEQUALITY OF CLAUSIUS

The Carnot cycle is a reversible cycle and produces work which we will refer to as W_{rev}. Consider an irreversible cycle operating between the same two reservoirs, shown in Fig. 6-4. Obviously, since the Carnot cycle possesses the maximum possible efficiency, the efficiency of the irreversible cycle must be less than that of the Carnot cycle. In other words, for the same amount of heat addition Q_H, we must have

$$W_{irr} < W_{rev} \tag{6.32}$$

From the first law applied to a cycle ($W = Q_H - Q_L$) we see that, assuming that $(Q_H)_{irr}$ and $(Q_H)_{rev}$ are the same,

$$(Q_L)_{rev} < (Q_L)_{irr} \tag{6.33}$$

This requires, referring to (*6.1*) and (*6.3*),

$$\oint \left(\frac{\delta Q}{T}\right)_{irr} < 0 \tag{6.34}$$

since the above integral for a reversible cycle is zero.

If we were considering an irreversible refrigerator rather than an engine, we would require more work for the same amount of refrigeration Q_L. By applying the first law to refrigerators, we would arrive at the same inequality as in (*6.34*). Hence, for all cycles, reversible or irreversible, we can write

$$\oint \frac{\delta Q}{T} \leq 0 \tag{6.35}$$

This is known as *the inequality of Clausius*. It is a consequence of the second law of thermodynamics.

EXAMPLE 6.6 It is proposed to operate a simple steam power plant as shown in Fig. 6-5. The water is completely vaporized in the boiler so that the heat transfer Q_B takes place at constant temperature. Does this proposal comply with the inequality of Clausius? Assume no heat transfer occurs from the pump or the turbine.

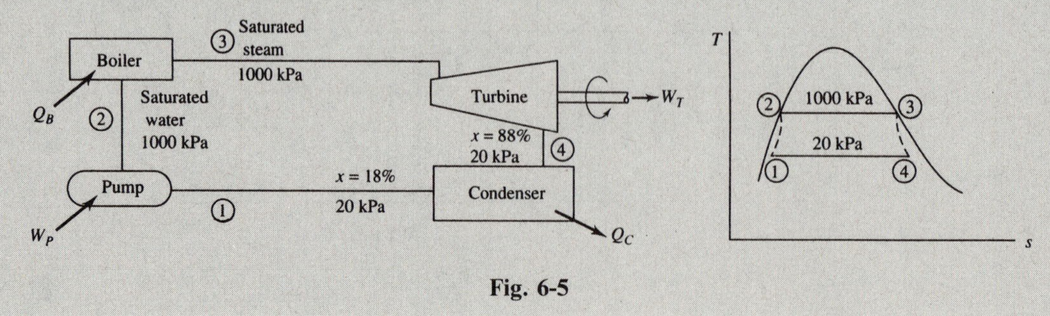

Fig. 6-5

Solution: The quantity that we seek is $\oint \delta Q/T$. Since the proposed heat transfer occurs at constant temperature, this takes the form

$$\oint \frac{\delta Q}{T} = \frac{Q_B}{T_B} - \frac{Q_C}{T_C}$$

From the steam tables we can find the following for each kilogram of water ($m = 1$ kg):

$$T_B = 179.9\,^\circ\text{C} \qquad T_C = 60.1\,^\circ\text{C} \qquad Q_B = m(h_3 - h_2) = 2778 - 763 = 2015 \text{ kJ}$$
$$Q_C = m(h_4 - h_1) = [251 + (0.88)(2358)] - [251 + (0.18)(2358)] = 1651 \text{ kJ}$$

Thus, we have

$$\oint \frac{\delta Q}{T} = \frac{2015}{452.9} - \frac{1651}{333.1} = -\,0.507 \text{ kJ/K}$$

This is negative, as it must be if the proposed power plant is to satisfy the inequality of Clausius.

6.7 ENTROPY CHANGE FOR AN IRREVERSIBLE PROCESS

Consider a cycle to be composed of two reversible processes, shown in Fig. 6-6. Suppose that we can also return from state 2 to state 1 along the irreversible process marked by path C. For the reversible cycle we have

$$\underset{\text{along } A}{\int_1^2 \frac{\delta Q}{T}} + \underset{\text{along } B}{\int_2^1 \frac{\delta Q}{T}} = 0 \tag{6.36}$$

For the cycle involving the irreversible process, the Clausius inequality demands that

$$\underset{\text{along } A}{\int_1^2 \frac{\delta Q}{T}} + \underset{\text{along } C}{\int_2^1 \frac{\delta Q}{T}} < 0 \tag{6.37}$$

Subtracting (*6.36*) from (*6.37*),

$$\underset{\text{along } B}{\int_1^2 \frac{\delta Q}{T}} > \underset{\text{along } C}{\int_1^2 \frac{\delta Q}{T}} \tag{6.38}$$

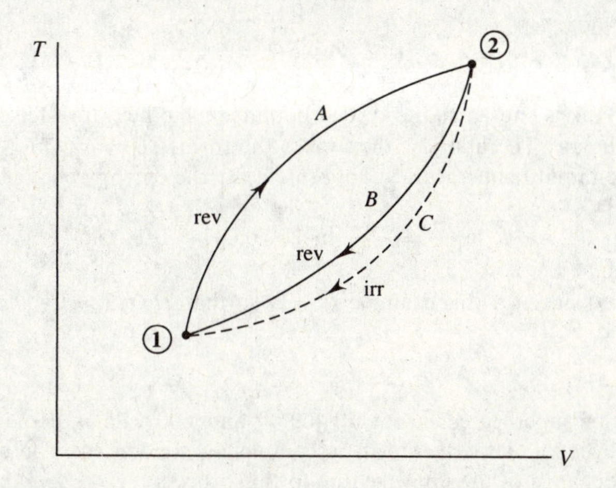

Fig. 6-6 A cycle with an irreversible process.

But, along the reversible path B, $\delta Q/T = dS$. Thus, for any path representing any process,

$$\Delta S \geq \int \frac{\delta Q}{T} \quad \text{or} \quad dS \geq \frac{\delta Q}{T} \tag{6.39}$$

The equality holds for a reversible process and the inequality for an irreversible process.

Relationship (*6.39*) leads to an important conclusion in thermodynamics. Consider an infinitesimal heat transfer δQ to a system at absolute temperature T. If the process is reversible, the differential change in entropy is $\delta Q/T$; if the process is irreversible, the change in entropy is greater than $\delta Q/T$. We thus conclude that the effect of irreversibility (e.g., friction) is to increase the entropy of a system.

Finally, in our application of the second law to a process, (*6.39*) can summarize our results. If we wish to investigate whether a proposed process satisfies the second law, we simply check using (*6.39*). We see that entropy and the second law are synonymous in the same way that energy and the first law are synonymous.

Finally, consider an *isolated* system, a system which exchanges no work or heat with its surroundings. For such a system the first law demands that $U_2 = U_1$ for any process. Equation (*6.39*) takes the form

$$\Delta S \geq 0 \tag{6.40}$$

demanding that the entropy of an isolated system either remains constant or increases, depending on whether the process is reversible or irreversible. Hence, for any real process the entropy of an isolated system increases.

We can generalize the above by considering a larger system to include both the system under consideration and its surroundings, often referred to as the *universe*. For the universe we can write

$$\Delta S_{\text{univ}} = \Delta S_{\text{sys}} + \Delta S_{\text{surr}} \geq 0 \tag{6.41}$$

where the equality applies to a (ideal) reversible process and the inequality to a (real) irreversible process. Relation (*6.41*), the *principle of entropy increase*, is often used as the mathematical statement of the second law. In (*6.41*) ΔS_{univ} is also referred to as ΔS_{gen}, the entropy generated, or ΔS_{net}, the net increase in entropy.

EXAMPLE 6.7 Air is contained in one half of the insulated tank shown in Fig. 6-7. The other side is completely evacuated. The membrane is punctured and the air quickly fills the entire volume. Calculate the specific entropy change of this isolated system.

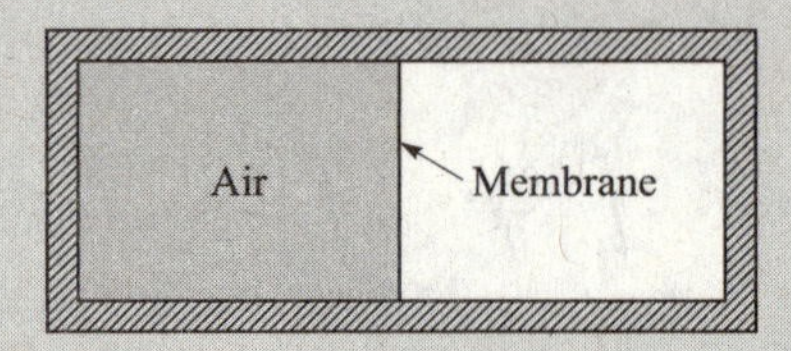

Fig. 6-7

Solution: The entire tank is chosen as the system boundary. No heat transfer occurs across the boundary and no work is done by the air. The first law then takes the form $\Delta U = mC_v(T_2 - T_1) = 0$. Hence, the final temperature is equal to the initial temperature. Using (*6.15*) for the entropy change, we have, with $T_1 = T_2$,

$$\Delta s = R \ln \frac{v_2}{v_1} = \frac{53.3}{778} \ln 2 = 0.04749 \text{ Btu/lbm-°R}$$

Note that this satisfies (*6.39*) since for this example $Q = 0$, so that $\int \delta Q/T = 0 < m\Delta s$.

EXAMPLE 6.8 Two kg of superheated steam at 400 °C and 600 kPa is cooled at constant pressure by transferring heat from a cylinder until the steam is completely condensed. The surroundings are at 25 °C. Determine the net entropy change of the universe due to this process.

Solution: The entropy of the steam which defines our system decreases since heat is transferred from the system to the surroundings. From the steam tables this change is found to be

$$\Delta S_{\text{sys}} = m(s_2 - s_1) = (2)(1.9316 - 7.7086) = -11.55 \text{ kJ/K}$$

The heat transfer to the surroundings occurs at constant temperature. Hence, the entropy change of the surroundings is

$$\Delta S_{\text{surr}} = \int \frac{\delta Q}{T} = \frac{Q}{T}$$

The heat transfer for the constant-pressure process is

$$Q = m\Delta h = 2(3270.2 - 670.6) = 5199 \text{ kJ}$$

giving $\Delta S_{\text{surr}} = 5199/298 = 17.45$ kJ/K and

$$\Delta S_{\text{univ}} = \Delta S_{\text{surr}} + \Delta S_{\text{sys}} = 17.45 - 11.55 = 5.90 \text{ kJ/K} > 0$$

6.8 THE SECOND LAW APPLIED TO A CONTROL VOLUME

The second law has been applied thus far in this chapter to a system, a particular collection of mass particles. We now wish to apply the second law to a control volume, following the same strategy used in our study of the first law. In Fig. 6-8 a control volume is enclosed by the control surface shown with the dashed lines surrounding some device or volume of interest. The second law can then be expressed over a time increment Δt as

$$\begin{pmatrix}\text{Entropy change}\\ \text{of control volume}\end{pmatrix} + \begin{pmatrix}\text{Entropy}\\ \text{exiting}\end{pmatrix} - \begin{pmatrix}\text{Entropy}\\ \text{entering}\end{pmatrix} + \begin{pmatrix}\text{Entropy change}\\ \text{of surroundings}\end{pmatrix} \geq 0 \tag{6.42}$$

This is expressed as

$$\Delta S_{\text{c.v.}} + m_2 s_2 - m_1 s_1 + \frac{Q_{\text{surr}}}{T_{\text{surr}}} \geq 0 \tag{6.43}$$

If we divide the above equation by dt and use dots to denote rates, we arrive at the rate equation

$$\dot{S}_{\text{c.v.}} + \dot{m}_2 s_2 - \dot{m}_1 s_1 + \frac{\dot{Q}_{\text{surr}}}{T_{\text{surr}}} \geq 0 \tag{6.44}$$

The equality is associated with a reversible process. The inequality is associated with irreversibilities such as viscous effects, which are always present in a material flow; separations of the flow from boundaries where abrupt changes in geometry occur; and shock waves in high-speed compressible flow.

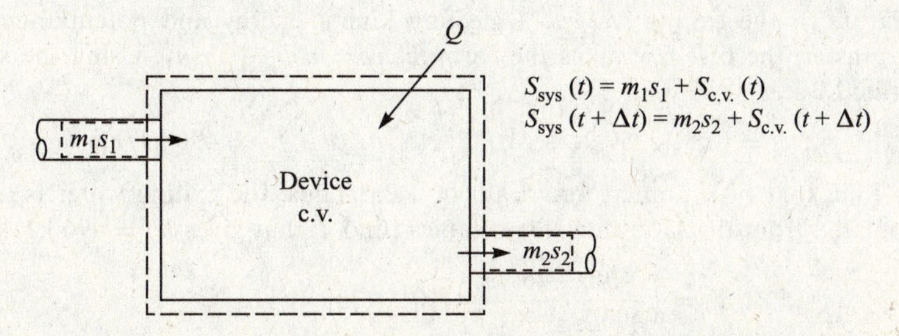

Fig. 6-8 The control volume used in the second-law analysis.

For a steady-flow process the entropy of the control volume remains constant with time. We can then write, recognizing that $\dot{m}_2 = \dot{m}_1 = \dot{m}$,

$$\dot{m}(s_2 - s_1) + \frac{\dot{Q}_{\text{surr}}}{T_{\text{surr}}} \geq 0 \tag{6.45}$$

By transferring energy to the body via heat transfer, we can obviously increase the entropy of the fluid flowing from the control volume. However, we also note that for an adiabatic steady-flow process the entropy also increases from inlet to exit due to irreversibilities since, for that case, (*6.45*) reduces to

$$s_2 \geq s_1 \tag{6.46}$$

For the reversible adiabatic process the inlet entropy and exit entropy are equal, an isentropic process. We use this fact when solving reversible adiabatic processes involving steam, such as flow through an ideal turbine.

We may be particularly interested in the *entropy production*; we define the rate of entropy production to be the left side of (*6.44*):

$$\dot{S}_{\text{prod}} \equiv \dot{S}_{\text{c.v.}} + \dot{m}_2 s_2 - \dot{m}_1 s_1 + \frac{\dot{Q}_{\text{surr}}}{T_{\text{surr}}} \tag{6.47}$$

This production rate is zero for reversible processes and positive for irreversible processes.

One last comment is in order regarding irreversible steady-flow processes, such as that in an actual turbine. We desire a quantity that can easily be used as a measure of the irreversibilities that exist in a particular device. The *efficiency* of a device is one such measure; it is defined as the ratio of the actual performance of a device to the ideal performance. The ideal performance is often that associated with an isentropic process. For example, the efficiency of a turbine would be

$$\eta_T = \frac{w_a}{w_s} \tag{6.48}$$

where w_a is the actual (specific) work output and w_s is the (specific) work output associated with an isentropic process. In general, the efficiency is defined using the desired output as the measure; for a diffuser we would use the pressure increase and for a nozzle the kinetic energy increase. For a compressor the actual work required is greater than the ideal work requirement of an isentropic process. For a compressor or pump the efficiency is defined to be

$$\eta_C = \frac{w_s}{w_a} \tag{6.49}$$

The efficiencies above are also called the *adiabatic efficiencies* since each efficiency is based on an adiabatic process.

EXAMPLE 6.9 A preheater is used to preheat water in a power plant cycle, as shown in Fig. 6-9. The superheated steam is at a temperature of 250 °C and the entering water is subcooled at 45 °C. All pressures are 600 kPa. Calculate the rate of entropy production.

Solution: From conservation of mass, $\dot{m}_3 = \dot{m}_2 + \dot{m}_1 = 0.5 + 4 = 4.5$ kg/s. The first law allows us to calculate the temperature of the exiting water. Neglecting kinetic-energy and potential-energy changes and assuming zero heat transfer, the first law takes the form $\dot{m}_3 h_3 = \dot{m}_2 h_2 + \dot{m}_1 h_1$. Using the steam tables (h_1 is the enthalpy of saturated water at 45 °C),

$$4.5h_3 = (0.5)(2957.2) + (4)(188.4) \qquad \therefore h_3 = 496 \text{ kJ/kg}$$

This enthalpy is less than that of saturated liquid at 600 kPa. Thus, the exiting water is also subcooled. Its temperature is interpolated from the saturated steam tables (find T that gives $h_f = 496$ kJ/kg) to be

$$T_3 = \left(\frac{496 - 461.3}{503.7 - 461.3}\right)(10) + 110 = 118\,°\text{C}$$

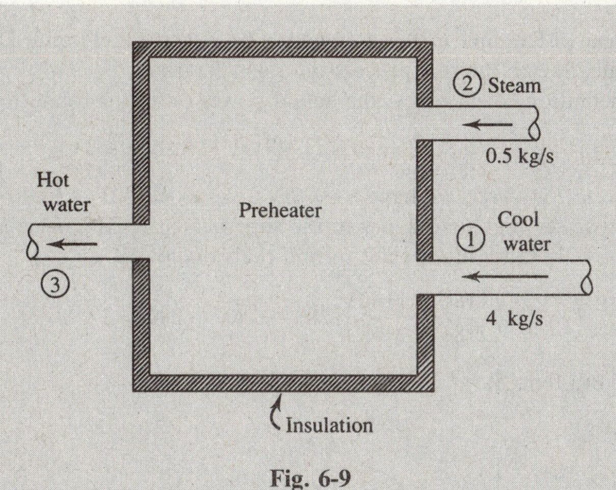

Fig. 6-9

The entropy at this temperature is then interpolated (using s_f) to be $s_3 = 1.508\,\text{kJ/kg·K}$. The entropy of the entering superheated steam is found to be $s_2 = 7.182\,\text{kJ/kg·K}$. The entering entropy of the subcooled water is s_f at $T_1 = 45\,°\text{C}$, or $s_1 = 0.639\,\text{kJ/kg·K}$. Finally, modifying (*6.47*), to account for two inlets, we have, with $\dot{Q} = 0$,

$$\dot{S}_{\text{prod}} = \dot{m}_3 s_3 - \dot{m}_2 s_2 - m_1 s_1 = (4.5)(1.508) - (0.5)(7.182) - (4)(0.639) = 0.639\ \text{kW/K}$$

This is positive, indicating that entropy is produced, a consequence of the second law. The mixing process between the superheated steam and the subcooled water is indeed an irreversible process.

EXAMPLE 6.10 Superheated steam enters a turbine, as shown in Fig. 6-10*a*, and exits at 2 psia. If the mass flux is 4 lbm/sec, determine the power output if the process is assumed to be reversible and adiabatic. Sketch the process on a *T-s* diagram.

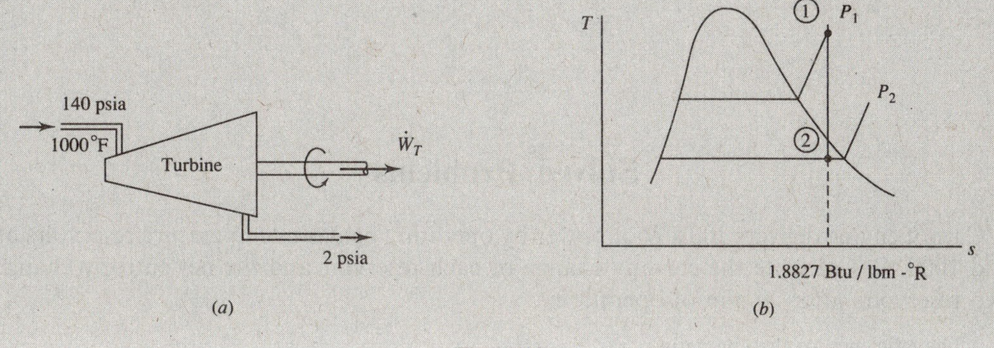

Fig. 6-10

Solution: If we neglect kinetic-energy and potential-energy changes, the first law, for an adiabatic process, is $-W_T = \dot{m}(h_2 - h_1)$. Since the process is also assumed to be reversible, the entropy exiting is the same as the entropy entering, as shown in Fig. 6-10*b* (such a sketch is quite useful in visualizing the process). From the steam tables,

$$h_1 = 1531\ \text{Btu/lbm} \qquad s_1 = s_2 = 1.8827\ \text{Btu/lbm-°R}$$

With the above value for s_2, we see that state 2 is in the quality region. The quality is determined as follows:

$$s_2 = s_f + x_2 s_{fg} \qquad 1.8827 = 0.1750 + 1.7448 x_2 \qquad x_2 = 0.9787$$

Then $h_2 = h_f + x_2 h_{fg} = 94.02 + (0.9787)(1022.1) = 1094\ \text{Btu/lbm}$ and

$$\dot{W}_T = (4)(1531 - 1094) = 1748\ \text{Btu/sec} \quad \text{or } 2473\ \text{hp}$$

EXAMPLE 6.11 The turbine of Example 6.10 is assumed to be 80 percent efficient. Determine the entropy and temperature of the final state. Sketch the real process on a T-s diagram.

Solution: Using the definition of efficiency, the actual power output is found to be

$$\dot{W}_a = (0.8)\dot{W}_s = (0.8)(1748) = 1398 \text{ Btu/sec}$$

From the first law, $-\dot{W}_a = \dot{m}(h_{2'} - h_1)$, we have $h_{2'} = h_1 - \dot{W}_a/\dot{m} = 1521 - 1398/4 = 1182$ Btu/lbm. Using this value and $P_{2'} = 2$ psia, we see that state 2′ lies in the superheated region, since $h_{2'} > h_g$. This is shown in Fig. 6-11. At $P_2 = 2$ and $h_{2'} = 1182$ we interpolate to find the value of $T_{2'}$:

$$T_{2'} = -\left(\frac{1186 - 1182}{1186 - 1168}\right)(280 - 240) + 280 = 271\,^\circ\text{F}$$

The entropy is $s_{2'} = 2.0526$ Btu/lbm-°R.

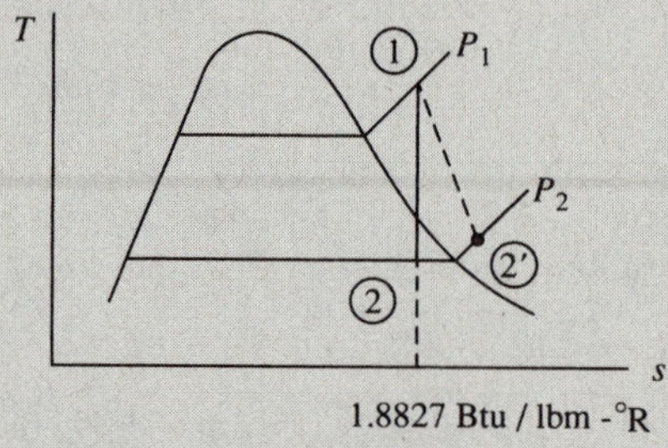

Fig. 6-11

Note that the irreversibility has the desired effect of moving state 2 into the superheated region, thereby eliminating the formation of droplets due to the condensation of moisture. In an actual turbine, moisture formation cannot be tolerated because of damage to the turbine blades.

Solved Problems

6.1 A Carnot engine delivers 100 kW of power by operating between temperature reservoirs at 100 °C and 1000 °C. Calculate the entropy change of each reservoir and the net entropy change of the two reservoirs after 20 min of operation.

The efficiency of the engine is

$$\eta = 1 - \frac{T_L}{T_H} = 1 - \frac{373}{1273} = 0.7070$$

The high-temperature heat transfer is then $\dot{Q}_H = \dot{W}/\eta = 100/0.7070 = 141.4$ kW. The low-temperature heat transfer is

$$\dot{Q}_L = \dot{Q}_H - \dot{W} = 141.4 - 100 = 41.4 \text{ kW}$$

The entropy changes of the reservoirs are then

$$\Delta S_H = \frac{Q_H}{T_H} = -\frac{\dot{Q}_H \Delta t}{T_H} = -\frac{(141.4)[(20)(60)]}{1273} = -133.3 \text{ kJ/K}$$

$$\Delta S_L = \frac{Q_L}{T_L} = \frac{\dot{Q}_L \Delta t}{T_L} = \frac{(41.4)[(20)(60)]}{373} = 133.2 \text{ kJ/K}$$

The net entropy change of the two reservoirs is $\Delta S_{net} = \Delta S_H + \Delta S_L = -133.3 + 133.2 = -0.1$ kJ/K. This is zero, except for round-off error, in compliance with (*6.2*).

6.2 Two kilograms of air is heated at constant pressure of 200 kPa to 500 °C. Calculate the entropy change if the initial volume is 0.8 m^3.

The initial temperature is found to be

$$T_1 = \frac{P_1 V_1}{mR} = \frac{(200)(0.8)}{(2)(0.287)} = 278.7 \text{ K}$$

The entropy change is then found, using (*6.16*) to be

$$\Delta S = m\left[C_p \ln \frac{T_2}{T_1} - R \ln 1\right] = (2)(1.00) \ln \frac{773}{278.7} = 2.040 \text{ kJ/K}$$

6.3 Air is compressed in an automobile cylinder from 14.7 to 2000 psia. If the initial temperature is 60 °F, estimate the final temperature.

Compression occurs very rapidly in an automobile cylinder ($Q \cong 0$); hence, we approximate the process with an adiabatic reversible process. Using (*6.17*), we find the final temperature to be

$$T_2 = T_1\left(\frac{P_2}{P_1}\right)^{(k-1)/k} = (520)\left(\frac{2000}{14.7}\right)^{0.4/1.4} = 2117\,°\text{R} \quad \text{or } 1657\,°\text{F}$$

6.4 A piston allows air to expand from 6 MPa to 200 kPa. The initial volume and temperature are 500 cm^3 and 800 °C. If the temperature is held constant, calculate the heat transfer and the entropy change.

The first law, using the work for an isothermal process, provides us with

$$Q = W = mRT \ln \frac{P_1}{P_2} = \left(\frac{P_1 V_1}{RT_1}\right) RT_1 \ln \frac{P_1}{P_2} = (6000)(500 \times 10^{-6}) \ln \frac{6000}{200} = 10.20 \text{ kJ}$$

The entropy change is then

$$\Delta S = mC_p \cancelto{0}{\ln}\, 1 - mR \ln \frac{P_2}{P_1} = -\frac{P_1 V_1}{T_1} \ln \frac{P_2}{P_1} = -\frac{(6000)(500 \times 10^{-6})}{1073} \ln \frac{200}{6000} = 9.51 \text{ J/K}$$

6.5 A paddle wheel provides 200 kJ of work to the air contained in a 0.2-m^3 rigid volume, initially at 400 kPa and 40 °C. Determine the entropy change if the volume is insulated.

The first law, with zero heat transfer because of the insulation, provides

$$-W = m\Delta u = mC_v \Delta T \qquad -(-200) = \frac{(400)(0.2)}{(0.287)(313)}(0.717)(T_2 - 313) \qquad T_2 = 626.2 \text{ K}$$

The entropy change is then found to be

$$\Delta S = mC_v \ln \frac{T_2}{T_1} + mR \cancelto{0}{\ln}\, 1 = \frac{(400)(0.2)}{(0.287)(313)}(0.717) \ln \frac{626.2}{313} = 0.4428 \text{ kJ/K}$$

6.6 Air is compressed in an automobile cylinder from 14.7 to 2000 psia. Predict the final temperature if the initial temperature is 60 °F. Do not assume constant specific heat.

Since the process is quite rapid, with little chance for heat transfer, we will assume an adiabatic reversible process. For such a process we may use (*6.25*) and find

$$P_{r2} = P_{r1} \frac{P_2}{P_1} = (1.2147)\left(\frac{2000}{14.7}\right) = 165.3$$

where P_{r1} is found in Table E-1E. The temperature is now interpolated, using P_{r2}, to be

$$T_2 = \left(\frac{165.3 - 141.5}{174.0 - 141.5}\right)(2000 - 1900) + 1900 = 1973\,^\circ\text{R}$$

This compares with 2117°R of Prob. 6.3, in which the specific heat was assumed constant. Note the significant error (over 7 percent) in T_2 of Prob. 6.3. This occurs for large ΔT.

6.7 Air expands from 200 to 1000 cm^3 in a cylinder while the pressure is held constant at 600 kPa. If the initial temperature is 20 °C, calculate the heat transfer assuming (*a*) constant specific heat and (*b*) variable specific heat.

(*a*) The air mass is

$$m = \frac{PV}{RT} = \frac{(600)(200 \times 10^{-6})}{(0.287)(293)} = 0.001427 \text{ kg}$$

The final temperature is found using the ideal-gas law:

$$T_2 = T_1 \frac{V_2}{V_1} = (293)\left(\frac{1000}{200}\right) = 1465 \text{ K}$$

The heat transfer is then (constant-pressure process)

$$Q = mC_p(T_2 - T_1) = (0.001427)(1.00)(1465 - 293) = 1.672 \text{ kJ}$$

(*b*) The mass and T_2 are as computed in part (*a*). The first law again provides, using h_2 and h_1 from Table E-1,

$$Q = m(h_2 - h_1) = (0.001427)(1593.7 - 293.2) = 1.856 \text{ kJ}$$

This shows that a 9.9 percent error results from assuming constant specific heat. This is due to the large temperature difference between the end states of the process.

6.8 Water is maintained at a constant pressure of 400 kPa while the temperature changes from 20 °C to 400 °C. Calculate the heat transfer and the entropy change.

Using $v_1 = v_f$ at 20 °C [state 1 is compressed liquid],

$$w = P(v_2 - v_1) = (400)(0.7726 - 0.001002) = 308.6 \text{ kJ/kg}$$

The first law gives $q = u_2 - u_1 + w = 2964.4 - 83.9 + 308.6 = 3189$ kJ/kg and the entropy change is

$$\Delta s = s_2 - s_1 = 7.8992 - 0.2965 = 7.603 \text{ kJ/kg·K}$$

6.9 Two kilograms of steam is contained in a 6-liter tank at 60 °C. If 1 MJ of heat is added, calculate the final entropy.

The initial quality is found as follows:

$$v_1 = \frac{V_1}{m} = \frac{6 \times 10^{-3}}{2} = 0.001017 + x_1(7.671 - 0.001) \qquad \therefore x_1 = 0.0002585$$

The initial specific internal energy is then

$$u_1 = u_f + x_1(u_g - u_f) = 251.1 + (0.0002585)(2456.6 - 251.1) = 251.7 \text{ kJ/kg}$$

The first law, with $W = 0$, gives

$$Q = m(u_2 - u_1) \qquad \text{or} \qquad u_2 = u_1 + \frac{\dot{Q}}{m} = 251.7 + \frac{1000}{2} = 751.7 \text{ kJ/kg}$$

Using $v_2 = v_1 = 0.003\ \text{m}^3/\text{kg}$ and $u_2 = 751.7$ kJ/kg, we locate state 2 by trial and error. The quality must be the same for the temperature selected:

$$
\begin{aligned}
T_2 = 170\,^\circ\text{C}: \quad & 0.003 = 0.0011 + x_2(0.2428 - 0.0011) && \therefore x_2 = 0.00786\\
& 751.7 = 718.3 + x_2(2576.5 - 718.3) && \therefore x_2 = 0.01797\\
T_2 = 177\,^\circ\text{C}: \quad & 0.003 = 0.0011 + x_2(0.2087 - 0.0011) && \therefore x_2 = 0.00915\\
& 751.7 = 750.0 + x_2(2581.5 - 750.0) && \therefore x_2 = 0.00093
\end{aligned}
$$

A temperature of 176 °C is chosen. The quality from v_2 is used since it is less sensitive to temperature change. At 176 °C, we interpolate to find

$$0.003 = 0.0011 + x_2(0.2136 - 0.0011) \qquad \therefore x_2 = 0.00894$$

Hence $S_2 = m(s_f + x_2 s_{fg}) = (2)[2.101 + (0.00894)(4.518)] = 4.28$ kJ/K.

6.10 Five ice cubes (each 1.2 in^3) at 0 °F are placed in a 16-oz glass of water at 60 °F. Calculate the final equilibrium temperature and the net entropy change, assuming an insulated glass.

The first law allows us to determine the final temperature. We will assume that not all of the ice melts so that $T_2 = 32\,^\circ$F. The ice warms up and some of it then melts. The original water cools. First, we calculate the mass of the ice (see Table C-5E) and the water:

$$m_i = \frac{(5)(1.2/1728)}{0.01745} = 0.199\ \text{lbm}, \qquad m_w = 1\ \text{lbm} \quad (\textit{a pint's a pound})$$

The first law is expressed as $m_i(C_p)_i \Delta T + m_I \Delta h_I = m_w (C_p)_w\, \Delta T$, where m_I is the amount of ice that melts. This becomes

$$(0.199)(0.49)(32 - 0) + (m_I)(140) = (1)(1.0)(60 - 32) \qquad \therefore m_I = 0.1777\ \text{lbm}$$

The net entropy change of the ice and water is then

$$
\begin{aligned}
\Delta S_{\text{net}} &= m_i C_p \ln \frac{T_2}{T_{1i}} + m_I(s_w - s_i) + m_w C_p \ln \frac{T_2}{T_{1w}}\\
&= (0.199)(0.49) \ln \frac{492}{460} + (0.1777)[0.0 - (-0.292)] + (1)(1.0) \ln \frac{492}{520} = 0.00311\ \text{Btu}/^\circ\text{R}
\end{aligned}
$$

6.11 The steam in a Carnot engine is compressed adiabatically from 10 kPa to 6 MPa with saturated liquid occurring at the end of the process. If the work output is 500 kJ/kg, calculate the quality at the end of the isothermal expansion.

For a cycle, the work output equals the net heat input, so that

$$W = \Delta T \Delta s \qquad 500 = (275.6 - 45.8)(s_2 - 3.0273) \qquad s_2 = 5.203\ \text{kJ/kg·K}$$

This s_2 is the entropy at the end of the isothermal expansion. Using the values of s_f and s_{fg} at 6 MPa, we have

$$5.203 = 3.0273 + 2.8627 x_2 \qquad \therefore x_2 = 0.760$$

6.12 The R134a in a Carnot refrigerator operates between saturated liquid and saturated vapor during the heat rejection process. If the cycle has a high temperature of 52 °C and a low temperature of −20 °C, calculate the heat transfer from the refrigerated space and the quality at the beginning of the heat addition process.

The cycle COP is given as

$$\text{COP} = \frac{T_L}{T_H - T_L} = \frac{253}{325 - 253} = 3.51$$

The COP is also given by COP $= q_L/w$, where

$$w = \Delta T \Delta s = [52 - (-20)](0.9004 - 0.4432) = 32.92\ \text{kJ/kg}$$

Hence, the heat transfer that cools is $q_L = (\text{COP})(w) = (3.51)(32.92) = 115.5$ kJ/kg.

The quality at the beginning of the heat addition process is found by equating the entropy at the end of the heat rejection process to the entropy at the beginning of the heat addition process:

$$0.4432 = 0.0996 + (0.9332 - 0.0996)x \qquad \therefore x = 0.412$$

6.13 Show that the inequality of Clausius is satisfied by a Carnot engine operating with steam between pressures of 40 kPa and 4 MPa. The work output is 350 kJ/kg, and saturated vapor enters the adiabatic expansion process.

Referring to Table C-2, the high and low temperatures are 250.4 °C and 75.9 °C. The work output allows us to calculate the entropy at the beginning of the heat-addition process as follows:

$$w = \Delta T \Delta s \qquad 350 = (250.4 - 75.9)\Delta s \qquad \therefore \Delta s = 2.006 \text{ kJ/kg·K}$$

The heat addition is then $q_H = T_H \Delta s = (250.4 + 273)(2.006) = 1049.9$ kJ/kg, and the heat extraction is

$$q_L = T_L \Delta s = (75.9 + 273)(2.006) = 699.9 \text{ kJ/kg}$$

For the (reversible) Carnot cycle the inequality of Clausius should become an equality:

$$\oint \frac{\delta Q}{T} = \frac{Q_H}{T_H} - \frac{Q_L}{T_L} = \frac{1049.9}{523.4} - \frac{699.9}{348.9} = 2.006 - 2.006 = 0 \quad \text{(O.K.)}$$

6.14 A 5-lb block of copper at 200 °F is submerged in 10 lbm of water at 50 °F, and after a period of time, equilibrium is established. If the container is insulated, calculate the entropy change of the universe.

First, we find the final equilibrium temperature. Since no energy leaves the container, we have, using specific heat values from Table B-4E,

$$m_c(C_p)_c(\Delta T)_c = m_w(C_p)_w(\Delta T)_w \qquad 5 \times 0.093(200 - T_2) = (10)(1.00)(T_2 - 50) \qquad T_2 = 56.66\,°\text{F}$$

The entropy changes are found to be

$$(\Delta S)_c = m_c(C_p)_c \ln \frac{T_2}{(T_1)_c} = (5)(0.093) \ln \frac{516.7}{660} = -0.1138 \text{ Btu/°R}$$

$$(\Delta S)_w = m_w(C_p)_w \ln \frac{T_2}{(T_1)_w} = (10)(1.00) \ln \frac{516.7}{510} = 0.1305 \text{ Btu/°R}$$

Since no heat leaves the container, there is no entropy change of the surroundings. Hence

$$\Delta S_{\text{universe}} = (\Delta S)_c + (\Delta S)_w = -0.1138 + 0.1305 = 0.0167 \text{ Btu/°R}$$

6.15 Two kilograms of saturated steam is contained in 0.2-m^3 rigid volume. Heat is transferred to the surroundings at 30 °C until the quality reaches 20 percent. Calculate the entropy change of the universe.

The initial specific volume is $v_1 = 0.2/2 = 0.1$ m^3/kg. By studying Tables C-1 and C-2 for the nearest v_g we see that this occurs at $P_1 = 2$ MPa. We also observe that $T_1 = 212.4$ °C, $s_1 = 6.3417$ kJ/kg·K, and $u_1 = 2600.3$ kJ/kg. Since the volume is rigid, we can locate state 2 by trial and error as follows.

$$\text{Try P}_2 = 0.4 \text{ MPa:} \quad v_2 = 0.0011 + 0.2(0.4625 - 0.0011) = 0.0934 \text{ m}^3/\text{kg}$$

$$\text{Try } P_2 = 0.3 \text{ MPa:} \quad v_2 = 0.0011 + 0.2(0.6058 - 0.0011) = 0.122 \text{ m}^2/\text{kg}$$

Obviously, $v_2 = 0.1$, so that state 2 is between 0.4 and 0.3 MPa. We interpolate to find

$$P_2 = \left(\frac{0.122 - 0.1}{0.122 - 0.0934}\right)(0.1) + 0.3 = 0.377 \text{ MPa}$$

The entropy and internal energy are also interpolated as follows:

$$s_2 = 1.753 + (0.2)(5.166) = 2.786 \text{ kJ/kg·K} \qquad u_2 = 594.3 + (0.2)(2551.3 - 594.3) = 986 \text{ kJ/kg}$$

The heat transfer is then, with $W = 0$ for the rigid volume,

$$Q = m(u_2 - u_1) = (2)(986 - 2600) = -3230 \text{ kJ} \quad \text{[heat to surroundings]}$$

The entropy change for the universe is calculated as

$$\Delta S_{\text{universe}} = m\Delta S_{\text{sys}} + \Delta S_{\text{surr}} = (2)(2.786 - 6.3417) + \frac{3230}{273 + 30} = 3.55 \text{ kJ/K}$$

6.16 A steam turbine accepts 2 kg/s of steam at 6 MPa and 600 °C and exhausts saturated steam at 20 kPa while producing 2000 kW of work. If the surroundings are at 30 °C and the flow is steady, calculate the rate of entropy production.

The first law for a control volume allows us to calculate the heat transfer from the turbine to the surroundings:

$$\dot{Q}_T = \dot{m}(h_2 - h_1) + \dot{W}_T = (2)(2609.7 - 3658.4) + 2000 = -97.4 \text{ kW}$$

Hence, $\dot{Q}_{\text{surr}} = -\dot{Q}_T = +97.4$ kW. The rate of entropy production is then found from (*6.47*) to be

$$\dot{S}_{\text{prod}} = \dot{S}_{\text{c.v.}} + \dot{m}(s_2 - s_1) + \frac{\dot{Q}_{\text{surr}}}{T_{\text{surr}}} = 0 + (2)(7.9093 - 7.1685) + \frac{97.4}{303} = 1.80 \text{ kW/K}$$

6.17 A rigid tank is sealed when the temperature is 0 °C. On a hot day the temperature in the tank reaches 50 °C. If a small hole is drilled in the tank, estimate the velocity of the escaping air.

As the tank heats up, the volume remains constant. Assuming atmospheric pressure at the initial state, the ideal-gas law yields

$$P_2 = P_1\frac{T_2}{T_1} = (100)\left(\frac{323}{273}\right) = 118.3 \text{ kPa}$$

The temperature at the exit, as the air expands from P_2 to P_3 as it escapes out of the hole, is found by assuming an isentropic process:

$$T_3 = T_2\left(\frac{P_3}{P_2}\right)^{(k-1)/k} = (323)\left(\frac{100}{118.3}\right)^{(1.4-1)/1.4} = 307.9 \text{ K}$$

where we have assumed pressure P_3 outside the tank to be atmospheric. The control-volume energy equation is now used to find the exit velocity $\mathcal{V}_3$:

$$0 = \frac{\mathcal{V}_3^2 - \cancel{\mathcal{V}_2^2}^{\,0}}{2} + C_p(T_3 - T_2) \qquad \mathcal{V}_3 = \sqrt{2C_p(T_2 - T_3)} = \sqrt{(2)(1000)(323 - 307.9)} = 173.8 \text{ m/s}$$

Note that we have used $C_p = 1000$ J/kg·K, not $C_p = 1.00$ kJ/kg·K. This provides the correct units; that is, J/kg·K = N·m/kg·K = m^2/s^2·K.

6.18 Steam expands isentropically through a turbine from 6 MPa and 600 °C to 10 kPa. Calculate the power output if the mass flux is 2 kg/s.

The exit state is at the same entropy as the inlet. This allows us to determine the exit quality as follows (use entries at 10 kPa):

$$s_2 = s_1 = 7.1685 = 0.6491 + 7.5019x_2 \qquad \therefore x_2 = 0.8690$$

The exit enthalpy is $h_2 = h_f + x_2 h_{fg} = 191.8 + (0.8690)(2392.8) = 2271$ kJ/kg. The control-volume energy equation then allows us to calculate

$$\dot{W}_T = -\dot{m}(h_2 - h_1) = -(2)(2271 - 3658.4) = 2774 \text{ kW}$$

This is the maximum possible power ouput for this turbine operating between the temperature and pressure limits imposed.

6.19 A steam turbine produces 3000 hp from a mass flux of 20,000 lbm/hr. The steam enters at 1000 °F and 800 psia and exits at 2 psia. Calculate the efficiency of the turbine.

The maximum possible work output is calculated first. For an isentropic process, state 2 is located as follows:

$$s_2 = s_1 = 1.6807 = 0.1750 + 1.7448x_2 \qquad \therefore x_2 = 0.8630$$

The exit enthalpy is then $h_2 = h_f + x_2 h_{fg} = 94.02 + (0.8630)(1022.1) = 976.1$ Btu/lbm. The work output w_s associated with the isentropic process is

$$w_s = -(h_2 - h_1) = -(976.1 - 1511.9) = 535.8 \text{ Btu/lbm}$$

The actual work output w_a is calculated from the given information:

$$w_a = \frac{\dot{W}_T}{\dot{m}} = \frac{(3000)(550)/778}{20\,000/3600} = 381.7 \text{ Btu/lbm}$$

The efficiency is found, using (*6.48*), to be

$$\eta_T = \frac{w_a}{w_s} = \frac{381.7}{535.8} = 0.712 \quad \text{or} \quad 71.2\%$$

6.20 Calculate the efficiency of the Rankine cycle operating on steam shown in Fig. 6-12 if the maximum temperature is 700 °C. The pressure is constant in the boiler and condenser.

The isentropic process from 2 to 3 allows us to locate state 3. Since $P_2 = 10$ MPa and $T_2 = 700\,°\text{C}$, we find

$$s_3 = s_2 = 7.1696 = 0.6491 + 7.5019 x_3 \qquad \therefore x_3 = 0.8692$$

The enthalpy of state 3 is then $h_3 = h_f + x_3 h_{fg} = 191.8 + (0.8692)(2392.8) = 2272$ kJ/kg. The turbine output is

$$w_T = -(h_3 - h_2) = -(2272 - 3870.5) = 1598 \text{ kJ/kg}$$

The energy input to the pump is

$$w_p = \frac{p_1 - p_4}{\rho} = -\frac{10\,000 - 10}{1000} = -9.99 \text{ kJ/kg}$$

and, since $-W_p = h_1 - h_4$,

$$h_1 = h_4 - w_p = 191.8 - (-9.99) = 201.8 \text{ kJ/kg}$$

The energy input to the boiler is $q_B = h_2 - h_1 = 3870.9 - 201.8 = 3669$ kJ/kg, from which

$$\eta_{\text{cycle}} = \frac{w_T + w_P}{q_B} = \frac{1598 - 9.99}{3669} = 0.433 \quad \text{or} \quad 43.3\%$$

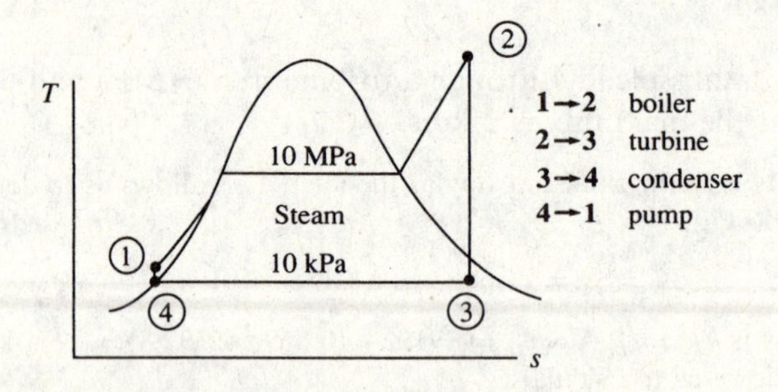

Fig. 6-12

Supplementary Problems

6.21 A Carnot engine extracts 100 kJ of heat from an 800 °C reservoir and rejects to the surroundings at 20 °C. Calculate the entropy change (*a*) of the reservoir and (*b*) of the surroundings.

6.22 A Carnot refrigerator removes 200 kJ of heat from a refrigerated space maintained at $-10\,°\text{C}$. Its COP is 10. Calculate the entropy change (*a*) of the refrigerated space and (*b*) of the high-temperature reservoir.

6.23 A reversible heat pump requires 4 hp while providing 50,000 Btu/hr to heat a space maintained at $70\,°\text{F}$. Calculate the entropy change of the space and the low-temperature reservoir after 10 min of operation.

6.24 Compare the entropy increase of the high-temperature reservoir and the entropy decrease of the specimen of Prob. 5.28.

6.25 Verify that (*6.17*) results from (*6.15*) and (*6.16*).

6.26 A gas of mass 0.2 kg is compressed slowly from 150 kPa and $40\,°\text{C}$ to 600 kPa, in an adiabatic process. Determine the final volume if the gas is (*a*) air, (*b*) carbon dioxide, (*c*) nitrogen, and (*d*) hydrogen.

6.27 Two kilograms of gas changes state from 120 kPa and $27\,°\text{C}$ to 600 kPa in a rigid container. Calculate the entropy change if the gas is (*a*) air, (*b*) carbon dioxide, (*c*) nitrogen, and (*d*) hydrogen.

6.28 Determine the entropy change of a gas in a rigid container that is heated from the conditions shown in Fig. 6-13 to 100 psia, if the gas is (*a*) air, (*b*) carbon dioxide, (*c*) nitrogen, and (*d*) hydrogen. Atmospheric pressure is 13 psia. The initial pressure is 0 psi gage

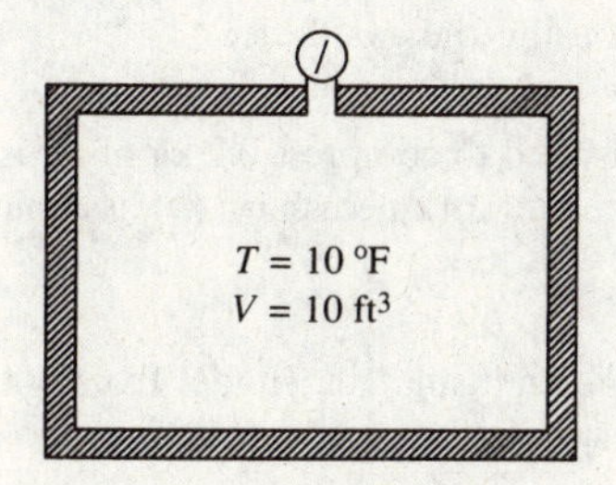

Fig. 6-13

6.29 The entropy change in a certain expansion process is 5.2 kJ/K. The gas, initially at 80 kPa, $27\,°\text{C}$, and $4\ \text{m}^3$, achieves a final temperature of $127\,°\text{C}$. Calculate the final volume if the gas is (*a*) air, (*b*) carbon dioxide, (*c*) nitrogen, and (*d*) hydrogen.

6.30 Nine kJ of heat is added to the cylinder shown in Fig. 6-14. If the initial conditions are 200 kPa and $47\,°\text{C}$, compute the work done and the entropy change for (*a*) air, (*b*) carbon dioxide, (*c*) nitrogen, and (*d*) hydrogen.

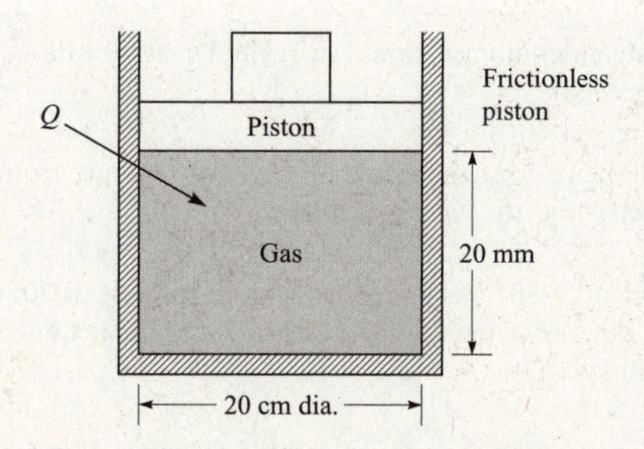

Fig. 6-14

6.31 A piston is inserted into a cylinder, causing the pressure to change from 50 to 4000 kPa while the temperature remains constant at 27 °C. To accomplish this, heat transfer must occur. Determine the heat transfer and the entropy change, if the working substance is (*a*) air, (*b*) carbon dioxide, (*c*) nitrogen, and (*d*) hydrogen.

6.32 The temperature of a gas changes from 60 °F to 900 °F while the pressure remains constant at 16 psia. Compute the heat transfer and the entropy change if the gas is (*a*) air, (*b*) carbon dioxide, (*c*) nitrogen, and (*d*) hydrogen.

6.33 A rigid, insulated 4-m^3 volume is divided in half by a membrane. One chamber is pressurized with 20 °C air to 100 kPa and the other is completely evacuated. The membrane is ruptured and after a period of time equilibrium is restored. What is the entropy change?

6.34 Four hundred kJ of paddle-wheel work is transferred to air in a rigid, insulated 2-m^3 volume, initially at 100 kPa and 57 °C. Calculate the entropy change if the working substance is (*a*) air, (*b*) carbon dioxide, (*c*) nitrogen, and (*d*) hydrogen.

6.35 A torque of 40 N·m is needed to rotate a shaft at 40 rad/s. It is attached to a paddle wheel located in a rigid 2-m^3 volume. Initially the temperature is 47 °C and the pressure is 200 kPa; if the paddle wheel rotates for 10 min and 500 kJ of heat is transferred to the air in the volume, determine the entropy increase (*a*) assuming constant specific heats and (*b*) using the gas table.

6.36 Two pounds of air is contained in an insulated piston-cylinder arrangement. The air is compressed from 16 psia and 60 °F by applying 2×10^5 ft-lbf of work. Compute the final pressure and temperature, (*a*) assuming constant specific heats and (*b*) using the gas table.

6.37 A piston-cylinder arrangement is used to compress 0.2 kg of air isentropically from initial conditions of 120 kPa and 27 °C to 2000 kPa. Calculate the work necessary, (*a*) assuming constant specific heats and (*b*) using the gas table.

6.38 Four kilograms of air expands in an insulated cylinder from 500 kPa and 227 °C to 20 kPa. What is the work output (*a*) assuming constant specific heats and (*b*) using the gas table?

6.39 Steam, at a quality of 85 percent, is expanded in a cylinder at a constant pressure of 800 kPa by adding 2000 kJ/kg of heat. Compute the entropy increase and the final temperature.

6.40 Two pounds of steam, initially at a quality of 40 percent and a pressure of 600 psia, is expanded in a cylinder at constant temperature until the pressure is halved. Determine the entropy change and the heat transfer.

6.41 0.1 kg water is expanded in a cylinder at a constant pressure of 4 MPa from saturated liquid until the temperature is 600 °C. Calculate the work necessary and the entropy change.

6.42 Two kilograms of steam at 100 °C is contained in a 3.4-m^3 cylinder. If the steam undergoes an isentropic expansion to 20 kPa, determine the work output.

6.43 Five kilograms of steam contained in a 2-m^3 cylinder at 40 kPa is compressed isentropically to 5000 kPa. What is the work needed?

6.44 Ten pounds of water at 14.7 psia is heated at constant pressure from 40 °F to saturated vapor. Compute the heat transfer necessary and the entropy change.

6.45 Five kilograms of ice at −20 °C is mixed with water initially at 20 °C. If there is no significant heat transfer from the container, determine the final temperature and the net entropy change if the initial mass of water is (*a*) 10 kg and (*b*) 40 kg.

6.46 A Carnot engine operates with steam on the cycle shown in Fig. 6-15. What is the thermal efficiency? If the work output is 300 kJ/kg, what is the quality of state 1?

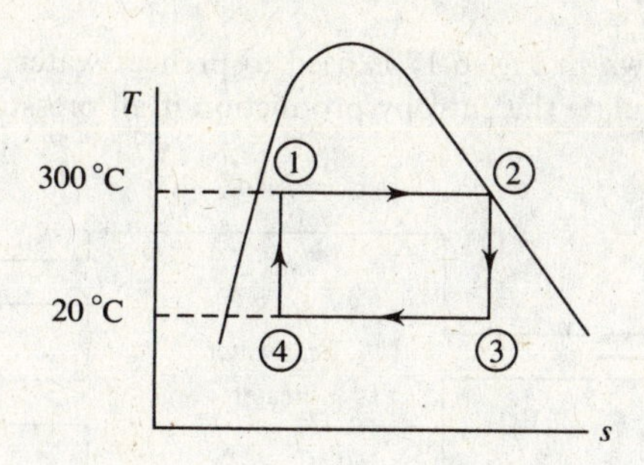

Fig. 6-15

6.47 The steam in a Carnot engine is compressed adiabatically from 20 kPa to 800 kPa. The heat addition results in saturated vapor. If the final quality is 15 percent, calculate the net work per cycle and the thermal efficiency.

6.48 A Carnot engine which operates with steam has a pressure of 8 psia and a quality of 20 percent at the beginning of the adiabatic compression process. If the thermal efficiency is 40 percent and the adiabatic expansion process begins with a saturated vapor, determine the heat added.

6.49 A Carnot engine operates at 4000 cycles per minute with 0.02 kg of steam, as shown in Fig. 6-16. If the quality of state 4 is 15 percent, (*a*) What is the power output? (*b*) what is the quality of state 3?

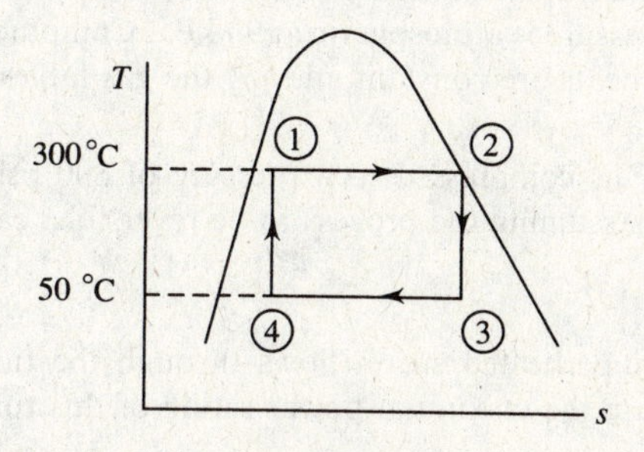

Fig. 6-16

6.50 For a Carnot engine operating under the conditions of Prob. 5.17, show that the inequality of Clausius is satisfied.

6.51 Using the information given in Prob. 5.22, verify that the inequality of Clausius is satisfied.

6.52 For the steam cycle of Prob. 6.46 show that the inequality of Clausius is satisfied.

6.53 One pound of air is contained in a 6 ft^3 volume at a pressure of 30 psia. Heat is transferred to the air from a high-temperature reservoir until the temperature is tripled in value while the pressure is held constant. Determine the entropy change of (*a*) the air, (*b*) the high-temperature reservoir which is at 1000 °F, and (*c*) the universe.

6.54 Two kilograms of air is stored in a rigid volume of 2 m^3 with the temperature initially at 300 °C. Heat is transferred from the air until the pressure reaches 120 kPa. Calculate the entropy change of (*a*) the air and (*b*) the universe if the surroundings are at 27 °C.

6.55 Three kilograms of saturated steam at 200 °C is cooled at constant pressure until the steam is completely condensed. What is the net entropy change of the universe if the surroundings are at 20 °C?

6.56 Steam at a quality of 80 percent is contained in a rigid vessel of a volume 400 cm^3. The initial pressure is 200 kPa. Energy is added to the steam by heat transfer from a source maintained at 700 °C until the pressure is 600 kPa. What is the entropy change of the universe?

6.57 The feedwater heater shown in Fig. 6-17 is used to preheat water in a power plant cycle. Saturated water leaves the preheater. Calculate the entropy production if all pressures are 60 psia.

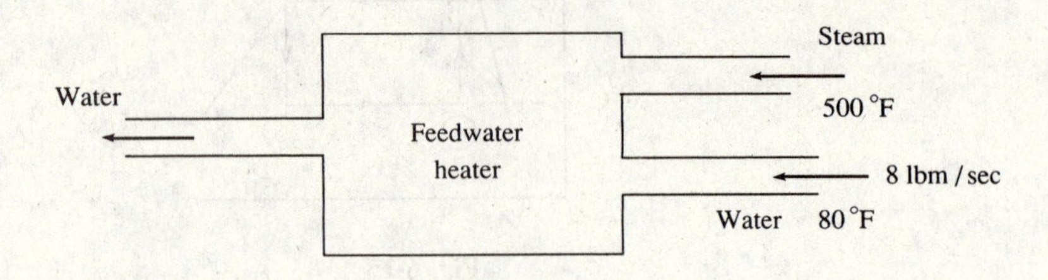

Fig. 6-17

6.58 Air flows from a tank maintained at 140 kPa and 27 °C from a 25-mm-diameter hole. Estimate the mass flux from the hole assuming an isentropic process.

6.59 Air flows from a nozzle in which the diameter is reduced from 100 to 40 mm. The inlet conditions are 130 kPa and 150 °C with a velocity of 40 m/s. Assuming an isentropic process, calculate the exit velocity if the exit pressure is 85 kPa.

6.60 The gases flowing through a turbine have essentially the same properties as air. The inlet gases are at 800 kPa and 900 °C and the exit pressure is atmospheric at 90 kPa. Compute the work output assuming an isentropic process if (*a*) the specific heats are constant and (*b*) the gas tables are used.

6.61 Saturated steam at 300 °F is compressed to a pressure of 800 psia. The device used for the compression process is well-insulated. Assuming the process to be reversible, calculate the power needed if 6 lbm/sec of steam is flowing.

6.62 Every second 3.5 kg of superheated steam flows through the turbine shown in Fig. 6-18. Assuming an isentropic process, calculate the maximum power rating of this turbine.

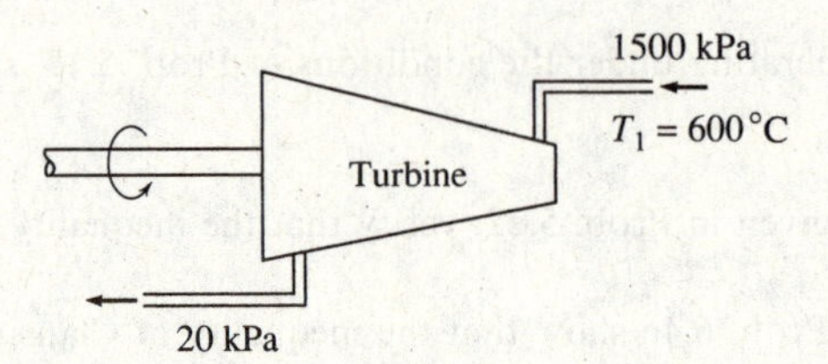

Fig. 6-18

6.63 Two hundred kW is to be produced by a steam turbine. The outlet steam is to be saturated at 80 kPa and the steam entering will be at 600 °C. For an isentropic process determine the mass flux of steam.

6.64 A turbine produces 3 MW by extracting energy from 4 kg of steam which flows through the turbine every second. The steam enters at 250 °C and 1500 kPa and exits as saturated steam at 2 kPa. Calculate the turbine efficiency.

6.65 A steam turbine is 85% efficient. Steam enters at 900 °F and 300 psia and leaves at 4 psia: (*a*) How much energy can be produced? (*b*) If 3000 hp must be produced, what must the mass flux be?

6.66 Determine the efficiency of an ideal piston engine operating on the Otto cycle shown in Fig. 6-19, if $T_1 = 60\,°C$ and $T_3 = 1600\,°C$.

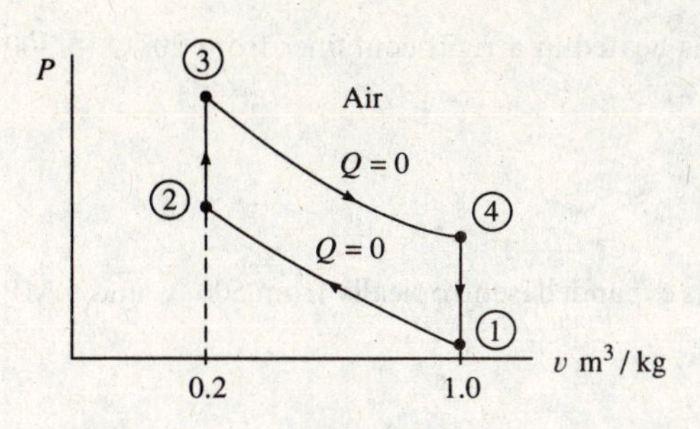

Fig. 6-19

6.67 Calculate the efficiency of the Rankine cycle shown in Fig. 6-20, if $P_4 = 20$ kPa, $P_1 = P_2 = 4$ MPa, and $T_2 = 600\,°C$.

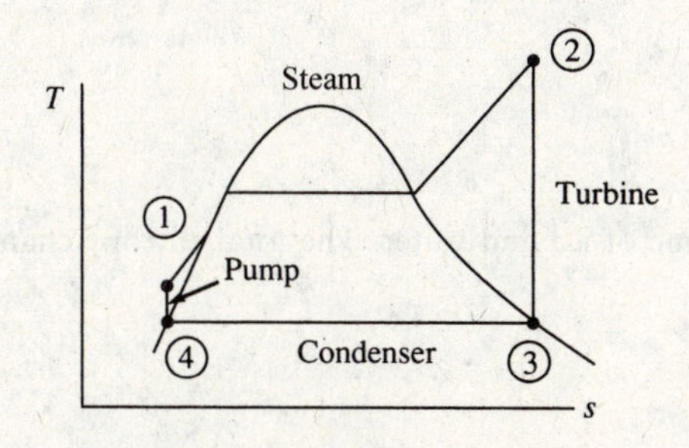

Fig. 6-20

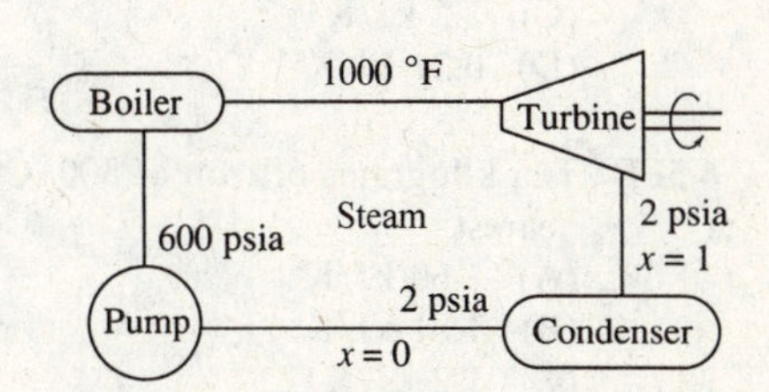

Fig. 6-21

6.68 Determine the efficiency of the Rankine cycle shown schematically in Fig. 6-21.

6.69 For the diesel cycle shown in Fig. 6-22 the compression ratio v_1/v_2 is 15 and the added heat is 1800 kJ per kilogram of air. If $T_1 = 20\,°C$, calculate the thermal efficiency.

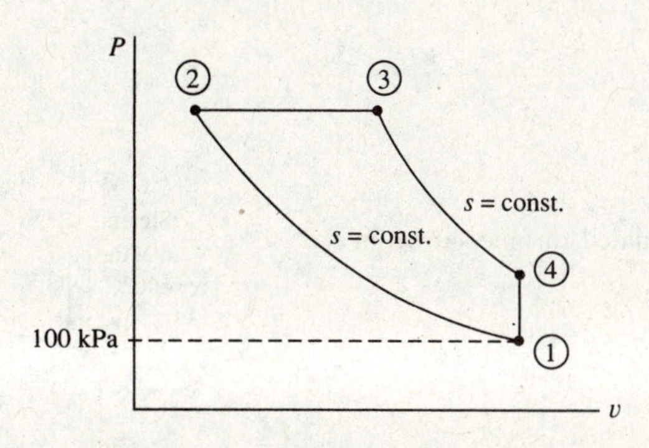

Fig. 6-22

Review Questions for the FE Examination

6.1FE Which of the following entropy relationships for a process is incorrect?

(A) Air, $V =$ const.: $\Delta s = C_v \ln T_2/T_1$

(B) Water: $\Delta s = C_p \ln T_2/T_1$

(C) Reservoir: $\Delta s = C_p \ln T_2/T_1$

(D) Copper: $\Delta s = C_p \ln T_2/T_1$

6.2FE One kilogram of air is heated in a rigid container from 20 °C to 300 °C. The entropy change is nearest:
(A) 0.64 kJ/K
(B) 0.54 kJ/K
(C) 0.48 kJ/K
(D) 0.34 kJ/K

6.3FE Ten kilograms of air is expanded isentropically from 500 °C and 6 MPa to 400 kPa. The work accomplished is nearest:
(A) 7400 kJ
(B) 6200 kJ
(C) 4300 kJ
(D) 2990 kJ

6.4FE Calculate the total entropy change if 10 kg of ice at 0 °C is mixed in an insulated container with 20 kg of water at 20 °C. The heat of melting for ice is 340 kJ/K.
(A) 6.1 kJ/K
(B) 3.9 kJ/K
(C) 1.2 kJ/K
(D) 0.21 kJ/K

6.5FE Ten kilograms of iron at 300 °C is chilled in a large volume of ice and water. The total entropy change is nearest:
(A) 1.60 kJ/K
(B) 1.01 kJ/K
(C) 1.2 kJ/K
(D) 0.21 kJ/K

6.6FE Which of the following second-law statements is incorrect?
(A) The entropy of an isolated system must remain constant or increase.
(B) The entropy of a hot copper block decreases as it cools.
(C) If ice is melted in water in an insulated container, the net entropy decreases.
(D) Work must be input if energy is transferred from a cold body to a hot body.

6.7FE The work needed to isentropically compress 2 kg of air in a cylinder at 400 kPa and 400 °C to 2 MPa is nearest:
(A) 1020 kJ
(B) 940 kJ
(C) 780 kJ
(D) 560 kJ

6.8FE Find w_T of the insulated turbine shown in Fig. 6-23.
(A) 1410 kJ/kg
(B) 1360 kJ/kg
(C) 1200 kJ/kg
(D) 1020 kJ/kg

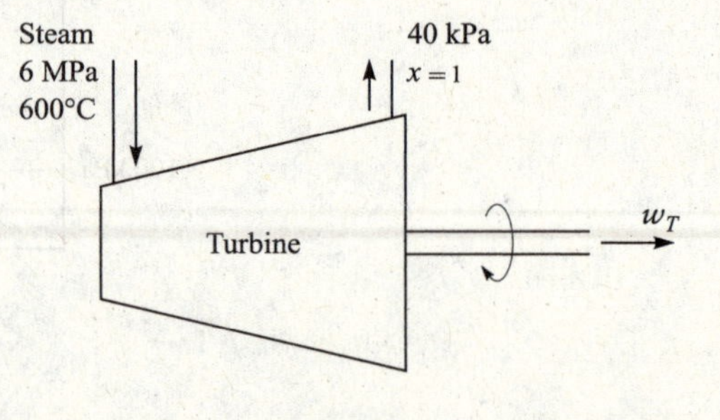

Fig. 6-23

6.9FE For the turbine of Fig. 6-23, find $(w_T)_{max}$.
(A) 1410 kJ/kg
(B) 1360 kJ/kg
(C) 1200 kJ/kg
(D) 1020 kJ/kg

6.10FE For the turbine of Fig. 6-23, find T_2.
(A) 64 °C
(B) 76 °C
(C) 88 °C
(D) 104 °C

6.11FE The efficiency of the turbine of Fig. 6-23 is nearest:

(A) 85%
(B) 89%
(C) 91%
(D) 93%

Answers to Supplementary Problems

6.21 (*a*) −0.0932 kJ/K (*b*) 0.0932 kJ/K

6.22 (*a*) −0.76 kJ/K (*b*) 0.76 kJ/s

6.23 15.72 Btu/°R, −15.72 Btu/°R

6.24 5 kJ/K, −5 kJ/K

6.26 (*a*) 0.0445 m^3 (*b*) 0.0269 m^3 (*c*) 0.046 m^3 (*d*) 0.0246 m^3

6.27 (*a*) 2.31 kJ/K (*b*) 2.1 kJ/K (*c*) 2.4 kJ/K (*d*) 32.4 kJ/K

6.28 (*a*) 0.349 Btu/°R (*b*) 0.485 Btu/°R (*c*) 0.352 Btu/°R (*d*) 0.342 Btu/°R

6.29 (*a*) 254 m^3 (*b*) 195 m^3 (*c*) 255 m^3 (*d*) 259 m^3

6.30 (*a*) 35.4 J, 15.4 J/K; (*b*) 42 J, 16.9 J/K; (*c*) 34 J, 15.3 J/K; (*d*) 2.48 J, 15.2 J/K.

6.31 (*a*) −377 kJ/kg, −1.26 kJ/kg·K; (*b*) −248 kJ/kg, −0.828 kJ/kg·K; (*c*) −390 kJ/kg, −1.30 kJ/kg·K; (*d*) −5420 kJ/kg, −18.1 kJ/kg·K

6.32 (*a*) 202 Btu/lbm, 0.24 Btu/lbm-°R; (*b*) 170 Btu/lbm, 0.194 Btu/lbm-°R (*c*) 208 Btu/lbm, 0.238 Btu/lbm-°R; (*d*) 2870 Btu/lbm, 3.29 Btu/lbm-°R

6.33 0.473 kJ/K

6.34 (*a*) 0.889 kJ/K (*b*) 0.914 kJ/K (*c*) 0.891 kJ/K (*d*) 0.886 kJ/K

6.35 (*a*) 2.81 kJ/K (*b*) 2.83 kJ/K

6.36 (*a*) 366 psia, 812°F; (*b*) 362 psia, 785°F

6.37 (*a*) −53.1 kJ (*b*) −53.4 kJ

6.38 (*a*) 863 kJ (*b*) 864 kJ

6.39 2.95 kJ/kg·K, 930°C

6.40 1.158 Btu/°R, 1108 Btu

6.41 39 kJ, 0.457 kJ/K

6.42 446 kJ

6.43 185 kJ

6.44 11,420 Btu, 17.4 Btu/°R

6.45 (*a*) 0°C, 0.135 kJ/K; (*b*) 8.0°C, 0.378 kJ/K

6.46 48.9%, 0.563

6.47 433 kJ/kg, 24.9%

6.48 796 Btu/lbm

6.49 (*a*) 19.5 kW (*b*) 0.678

6.53 (*a*) 0.264 Btu/°R (*b*) −0.156 Btu/°R (*c*) 0.108 Btu/°R

6.54 (*a*) −0.452 kJ/K (*b*) 0.289 kJ/K

6.55 7.56 kJ/K

6.56 0.611 J/K

6.57 0.432 Btu/sec-°R

6.58 0.147 kg/s

6.59 309 m/s

6.60 (*a*) 545 kJ/kg (*b*) 564 kJ/kg

6.61 2280 hp

6.62 3.88 MW

6.63 0.198 kg/s

6.64 39.9%

6.65 (*a*) 348 Btu/lbm (*b*) 6.096 lbm/sec

6.66 47.5%

6.67 31.0%

6.68 28%

6.69 55.3

Answers to Review Questions for the FE Examination

6.1FE (C) **6.2FE** (C) **6.3FE** (D) **6.4FE** (D) **6.5FE** (A) **6.6FE** (C) **6.7FE** (D) **6.8FE** (D) **6.9FE** (C) **6.10FE** (B) **6.11FE** (A)

Reversible Work, Irreversibility, and Availability

7.1 BASIC CONCEPTS

Reversible work for a process is defined as the work associated by taking a reversible-process path from state A to state B. As stated previously, a *reversible process* is a process that, having taken place, can be reversed and, having been reversed, leaves no change in either the system or the surroundings. A reversible process must be a quasiequilibrium process and is subject to the following restrictions:

- No friction exists.
- Heat transfer is due only to an infinitesimal temperature difference.
- Unrestrained expansion does not occur.
- There is no mixing.
- There is no turbulence.
- There is no combustion.

It can be easily shown that the reversible work or the work output from a reversible process going from state A to state B is the maximum work that can be achieved for the state change from A to B.

It is of interest to compare the actual work for a process to the reversible work for a process. This comparison is done in two ways. First, a *second-law efficiency* for a process or a device can be defined as

$$\eta_{\mathrm{II}} = \frac{W_a}{W_{\mathrm{rev}}} \quad \text{(turbine or engine)} \tag{7.1}$$

$$\eta_{\mathrm{II}} = \frac{W_{\mathrm{rev}}}{W_a} \quad \text{(pump or compressor)} \tag{7.2}$$

where W_a is the actual work and W_{rev} is the reversible work for the fictitious reversible process. Second-law efficiency is different from the adiabatic efficiency of a device introduced in Chap. 6. It is generally higher and provides a better comparison to the ideal.

Second, *irreversibility* is defined as the difference between the reversible work and the actual work for a process, or

$$I = W_{rev} - W_a \tag{7.3}$$

On a per-unit-mass basis,

$$i = w_{rev} - w_a \tag{7.4}$$

Both irreversibility and second-law efficiency will allow us to consider how close an actual process or device is to the ideal. Once the irreversibilities for devices in an actual engineering system, such as a steam power cycle, have been calculated, attempts to improve the performance of the system can be guided by attacking the largest irreversibilities. Similarly, since the maximum possible work will be reversible work, irreversibility can be used to evaluate the feasibility of a device. If the irreversibility of a proposed device is less than zero, the device is not feasible. [Section 7.2 develops the concepts of reversible work and irreversibility.]

Availability is defined as the maximum amount of reversible work that can be extracted from a system:

$$\Psi = (W_{rev})_{max} \tag{7.5}$$

or, on a per-unit-mass basis,

$$\psi = (w_{rev})_{max} \tag{7.6}$$

The maximization in (*7.5*) and (*7.6*) is over the reversible path joining the prescribed initial state to a final *dead state* in which system and surroundings are in equilibrium. [Section 7.3 develops the notion of availability.]

7.2 REVERSIBLE WORK AND IRREVERSIBILITY

To obtain expressions for reversible work and irreversibility, we will consider a transient process with specified work output and heat input and a uniform through-flow. We begin by allowing this to be an irreversible process. Consider the control volume shown in Fig. 7-1. The first law for this control volume can be written as

$$\dot{Q} - \dot{W}_S = \left(h_2 + \frac{V_2^2}{2} + gz_2\right)\dot{m}_2 - \left(h_1 + \frac{V_1^2}{2} + gz_1\right)\dot{m}_1 + \dot{E}_{c.v.} \tag{7.7}$$

Using (*6.47*), with $T_{surr} = T_0$ and $\dot{Q}_{surr} = -\dot{Q}$, we may write the second law as

$$\dot{S}_{c.v.} + s_2\dot{m}_2 - s_1\dot{m}_1 - \frac{\dot{Q}}{T_0} - \dot{S}_{prod} = 0 \tag{7.8}$$

Eliminate $\dot{Q}$ between (*7.7*) and (*7.8*) to obtain

$$\begin{aligned}\dot{W}_S = & -\dot{E}_{c.v.} + T_0\dot{S}_{c.v.} - \left(h_2 + \frac{V_2^2}{2} + gz_2 - T_0s_2\right)\dot{m}_2 \\ & + \left(h_1 + \frac{V_1^2}{2} + gz_1 - T_0s_1\right)\dot{m}_1 - T_0\dot{S}_{prod}\end{aligned} \tag{7.9}$$

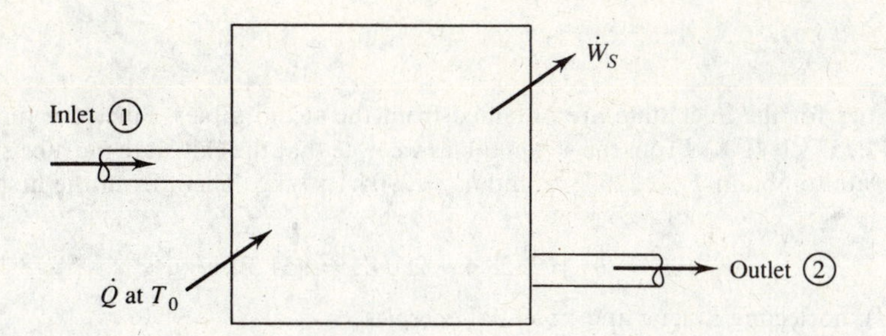

Fig. 7-1 The control volume used in the second-law analysis.

Since $\dot{S}_{\text{prod}}$ is due to the irreversibilities, the reversible work rate is given by (*7.9*) when $\dot{S}_{\text{prod}}$ is set equal to zero:

$$\dot{W}_{\text{rev}} = -\dot{E}_{\text{c.v.}} + T_0\dot{S}_{\text{c.v.}} - \left(h_2 + \frac{V_2^2}{2} + gz_2 - T_0 s_2\right)\dot{m}_2 + \left(h_1 + \frac{V_1^2}{2} + gz_1 - T_0 s_1\right)\dot{m}_1 \tag{7.10}$$

Then a time integration yields

$$\begin{aligned} W_{\text{rev}} = &\left[m_i\left(u_i + \frac{V_i^2}{2} + gz_i - T_0 s_i\right) - m_f\left(u_f + \frac{V_f^2}{2} + gz_f - T_0 s_f\right)\right]_{\text{c.v.}} \\ &+ m_1\left(h_1 + \frac{V_1^2}{2} + gz_1 - T_0 s_1\right) - m_2\left(h_2 + \frac{V_2^2}{2} + gz_2 - T_0 s_2\right) \end{aligned} \tag{7.11}$$

where the subscripts i and f pertain to the initial and final states of the control volume.

The actual work, if not given, can be determined from a first-law analysis:

$$\begin{aligned} W_a = &\left[m_i\left(u_i + \frac{V_i^2}{2} + gz_i\right) - m_f\left(u_f + \frac{V_f^2}{2} + gz_f\right)\right]_{\text{c.v.}} \\ &+ m_1\left(h_1 + \frac{V_1^2}{2} + gz_1\right) - m_2\left(h_2 + \frac{V_2^2}{2} + gz_2\right) + Q \end{aligned} \tag{7.12}$$

From (*7.3*), (*7.11*), and (*7.12*),

$$I = (m_f T_0 s_f - m_i T_0 s_i)_{\text{c.v.}} + T_0 m_2 s_2 - T_0 m_1 s_1 - Q \tag{7.13}$$

For a steady flow with negligible changes in kinetic and potential energies, we have

$$\dot{W}_{\text{rev}} = \dot{m}[h_1 - h_2 + T_0(s_2 - s_1)] \tag{7.14}$$

$$\dot{I} = \dot{m}T_0(s_2 - s_1) - \dot{Q} \tag{7.15}$$

It is important to realize that the basic results of this Section—(*7.11*), (*7.12*), and (*7.13*)—also hold for a system, which is nothing other than a control volume for which $m_1 = m_2 = 0$ (and thus $m_i = m_f = m$). Because time plays no part in the thermodynamics of a system, we generally replace the indices i and f by 1 and 2.

EXAMPLE 7.1 An ideal steam turbine is supplied with steam at 12 MPa and 700 °C, and exhausts at 0.6 MPa.

(*a*) Determine the reversible work and irreversibility.

(*b*) If the turbine has an adiabatic efficiency of 0.88, what is the reversible work, irreversibility, and second-law efficiency?

Solution:

(*a*) The properties for the inlet state are obtained from the steam tables. Since the turbine is isentropic, $s_2 = s_1 = 7.0757$ kJ/kg·K. From the steam tables we note that the exit state must be superheated vapor. We interpolate to obtain $T_2 = 225.2\,°\text{C}$ and $h_2 = 2904.1$ kJ/kg. Then, from the first law for a control volume,

$$w_a = h_1 - h_2 = 3858.4 - 2904.1 = 954.3 \text{ kJ/kg}$$

From (*7.11*), neglecting kinetic and potential energies,

$$w_{\text{rev}} = h_1 - h_2 - T_0(s_1 - s_2)^{0} = 3858.4 - 2904.1 = 954.3 \text{ kJ/kg}$$

The irreversibility for an ideal turbine is $i = w_{\text{rev}} - w_a = 954.3 - 954.3 = 0$ kJ/kg.

(*b*) Now let the adiabatic turbine have $\eta_T = 0.88$. The isentropic or ideal work was calculated in (*a*), so that the actual work is $w_a = \eta_T w_{\text{ideal}} = (0.88)(954.3) = 839.8$ kJ/kg. For this adiabatic process,

$$h_2 = h_1 - w_a = 3858.4 - 839.8 = 3018.6 \text{ kJ/kg}$$

From the steam tables we find that the exit state with $P_2 = 0.6$ MPa is superheated vapor, with $T_2 = 279.4\,°\text{C}$ and $s_2 = 7.2946$ kJ/kg. Then, assuming $T_0 = 298$ K,

$$w_{\text{rev}} = h_1 - h_2 - T_0(s_1 - s_2) = 3858.4 - 3018.6 - (298)(7.0757 - 7.2946) = 905 \text{ kJ/kg}$$

The second-law efficiency is $\eta_{\text{II}} = w_a/w_{\text{rev}} = 0.928$, which is greater than the adiabatic efficiency. The irreversibility is

$$i = w_{\text{rev}} - w_a = 905.0 - 839.8 = 65.2 \text{ kJ/kg}$$

EXAMPLE 7.2 Measurements are made on an adiabatic compressor with supply air at 15 psia and 80 °F. The exhaust air is measured at 75 psia and 440 °F. Can these measurements be correct?

Solution: For steady flow in the control volume, with $Q = 0$, (*7.15*) becomes

$$i = T_0(s_2 - s_1)$$

The entropy change is found, using values from the air tables, to be

$$s_2 - s_1 = s_2^{\circ} - s_1^{\circ} - R \ln \frac{P_2}{P_1} = 0.72438 - 0.60078 - \frac{53.3}{778} \ln \frac{75}{15}$$
$$= 0.01334 \text{ Btu/lbm-°R}$$

The irreversibility is then $i = (537)(0.01334) = 7.16$ Btu/lbm. As this is positive, the measurements can be correct. We assumed T_0 to be 537 °R.

7.3 AVAILABILITY AND EXERGY

According to the discussion in Section 7.1, Ψ is given by (*7.11*) when the final state (*f*) is identified with the state of the surroundings (0):

$$\Psi = \left[m_i\left(u_i + \frac{V_i^2}{2} + gz_i - T_0 s_i\right) - m_f\left(u_0 + \frac{V^2}{2} + gz_0 - T_0 s_0\right)\right]_{\text{c.v.}} + m_1\left(h_1 + \frac{V_1^2}{2} + gz_1 - T_0 s_1\right) - m_2\left(h_0 + \frac{V_0^2}{2} + gz_0 - T_0 s_0\right) \quad (7.16)$$

For a steady-flow process (*7.16*) becomes

$$\psi = h_1 - h_0 + \frac{V_1^2 - V_0^2}{2} + g(z_1 - z_0) - T_0(s_1 - s_0) \quad (7.17)$$

In carrying out a second-law analysis, it is often useful to define a new thermodynamic function (analogous to enthalpy), called *exergy*:

$$E \equiv h + \frac{V^2}{2} + gz - T_0 s \tag{7.18}$$

Comparing (*7.18*) to (*7.17*), we see that $E_1 - E_0 = \psi$. We interpret this equation as a work-energy relation: the extractable specific work ψ exactly equals the decrease in useful exergy E between the entrance and dead states of the system. More generally, when the system passes from one state to another, specific work in the amount $-\Delta E$ is made available.

Certain engineering devices have useful outputs or inputs that are not in the form of work; a nozzle is an example. Consequently, we generalize the notion of second-law efficiency to that of *second-law effectiveness*:

$$\varepsilon_{\text{II}} = \frac{(\text{availability produced}) + (\text{work produced}) + (\text{adjusted heat produced})}{(\text{availability supplied}) + (\text{work used}) + (\text{adjusted heat used})} \tag{7.19}$$

Heat to or from a device is "adjusted" in (*7.19*) on the basis of the temperature $-T_{\text{h.r.}}$ of the heat reservoir which is interacting with the device:

$$\text{adjusted heat} = \left(1 - \frac{T_0}{T_{\text{h.r.}}}\right) Q \tag{7.20}$$

EXAMPLE 7.3 Which system can do more useful work, 0.1 lbm of CO_2 at 440 °F and 30 psia or 0.1 lbm of N_2 at 440 °F and 30 psia?

Solution: Assuming a dead state at 77 °F (537 °R) and 14.7 psia, we use Table E-4E to calculate the availability of the CO_2:

$$\begin{aligned}\Psi &= m\left[h - h_0 - T_0\left(s_1^\circ - s_0^\circ - R \ln \frac{P}{P_0}\right)\right] \\ &= \left(\frac{0.1}{44}\right)\left[7597.6 - 4030.2 - 537\left(56.070 - 51.032 - 1.986 \ln \frac{30}{14.7}\right)\right] = 3.77 \text{ Btu}\end{aligned}$$

Similarly, for the N_2,

$$\begin{aligned}\Psi &= m\left[h - h_0 - T_0\left(s_1^\circ - s_0^\circ - R \ln \frac{P}{P_0}\right)\right] \\ &= \left(\frac{0.1}{28}\right)\left[6268.1 - 3279.5 - (537)\left(49.352 - 45.743 - 1.986 \ln \frac{30}{14.7}\right)\right] = 6.47 \text{ Btu}\end{aligned}$$

Hence, the N_2 can do more useful work.

EXAMPLE 7.4 How much useful work is wasted in the condenser of a power plant which takes in steam of quality 0.85 and 5 kPa and delivers saturated liquid at the same pressure?

Solution: The maximum specific work available at the condenser inlet is $\psi_1 = h_1 - h_0 - T_0(s_1 - s_0)$; at the outlet it is $\psi_2 = h_2 - h_0 - T_0(s_2 - s_0)$. The useful work wasted is $\psi_1 - \psi_2 = h_1 - h_2 - T_0(s_1 - s_2)$.

From the steam tables, assuming $T_0 = 298$ K and using the quality to find h_1 and s_1, we find

$$\psi_1 - \psi_2 = h_1 - h_2 - T_0(s_1 - s_2) = 2197.2 - 136.5 - (298)(7.2136 - 0.4717) = 51.6 \text{ kJ/kg}$$

EXAMPLE 7.5 Calculate the exergy of steam at 500°F and 300 psia. The surroundings are at 76°F.

Solution: From the superheated steam tables,

$$E = h - T_0 s = 1257.5 - (536)(1.5701) = 415.9 \text{ Btu/lbm}$$

EXAMPLE 7.6 Determine the second-law effectiveness for an ideal isentropic nozzle. Air enters the nozzle at 1000 K and 0.5 MPa with negligible kinetic energy and exits to a pressure of 0.1 MPa.

Solution: Since the process is isentropic, we use the air tables to find

$$s_2^\circ = s_1^\circ - R \ln \frac{P_1}{P_2} = 2.968 - 0.286 \ln 5 = 2.506 \text{ kJ/kg·K}$$

Thus

$$T_2 = 657.5 \text{ K} \qquad h_2 = 667.8 \text{ kJ/kg} \qquad h_1 = 1046.1 \text{ kJ/kg} \qquad h_0 = 298.2 \text{ kJ/kg}$$

By the first law,

$$h_1 = h_2 + \frac{V_2^2}{2} \qquad \text{or} \qquad V_2 = \sqrt{2}(h_1 - h_2)^{0.5} = \sqrt{2}[(1046.1 - 667.8)(10^3)]^{0.5} = 1230 \text{ m/s}$$

To evaluate the second-law effectiveness we need the availability produced:

$$\psi_2 = h_2 - h_0 + \frac{V_2^2}{2} - T_0\left(s_2^\circ - s_1^\circ - R \ln \frac{P_2}{P_0}\right)$$

$$= 667.8 - 298.2 + \frac{1230^2}{(2)(1000)} - (298)[2.506 - 1.695 - (0.287)(0)] = 884 \text{ kJ/kg}$$

where $P_2 = P_0 = 0.1$ MPA. The availability supplied is

$$\psi_1 = h_1 - h_0 - T_0\left(s_1^\circ - s_0^\circ - R \ln \frac{P_1}{P_0}\right) = 1046.1 - 298.2 - (298)(2.968 - 1.695 - 0.287 \ln 5) = 506 \text{ kJ/kg}$$

Since there is no work or heat transfer, (*7.19*) gives

$$\varepsilon_{II} = \frac{\psi_2}{\psi_1} = \frac{884}{506} = 1.75$$

Note that second-law effectiveness is not bounded by 1 (much like the COP for a refrigeration cycle).

7.4 SECOND-LAW ANALYSIS OF A CYCLE

You may choose to study this section after Chapters 8 and 9.

In applying second-law concepts to a cycle two approaches may be employed. The first is simply to evaluate the irreversibilities associated with each device or process in the cycle; this will identify sources of large irreversibilities which will adversely affect the efficiency of the cycle. The second is to evaluate ε_{II} for the whole cycle.

EXAMPLE 7.7 Consider the simple cycle with steam extraction shown in Fig. 7-2. Calculate the second-law effectiveness for the cycle if the boiler produces steam at 1 MPa and 300°C and the turbine exhausts to the condenser at 0.01 MPa. The steam extraction occurs at 0.1 MPa, where 10 percent of the steam is removed. Make-up water is supplied as saturated liquid at the condenser pressure, and saturated liquid leaves the condenser.

Solution: We begin by traversing the cycle starting at state 1:

$$1 \rightarrow 2 \quad \text{Ideal turbine:} \quad s_2 = s_1 = 7.1237 \text{ kJ/kg·K}$$

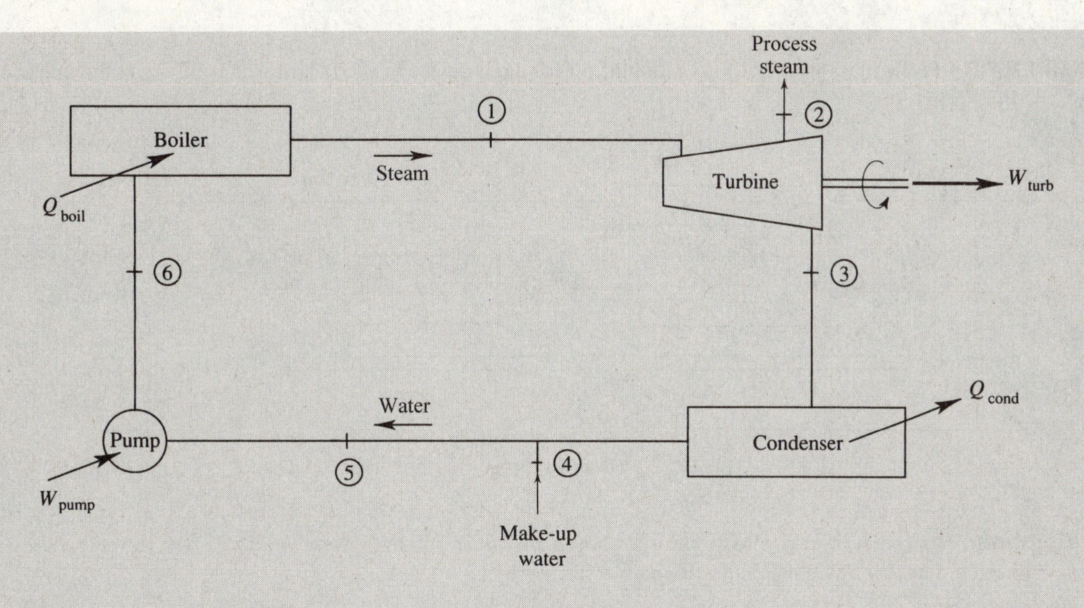

Fig. 7-2

Comparing to s_f and s_g at 0.1 MPa, we have a two-phase mixture at state 2 with

$$x_2 = \frac{s_2 - s_f}{s_{fg}} = 0.96$$

so that $h_2 = h_f + 0.96h_{fg} = 2587.3\,\text{kJ/kg}$.

$$2 \rightarrow 3 \quad \text{Ideal turbine:} \quad s_3 = s_2 = 7.1237\ \text{kJ/kg·K}$$

Comparing to s_f and s_g at 0.01 MPa, we have a two-phase mixture at state 3 with

$$x_3 = \frac{s_3 - s_f}{s_{fg}} = 0.86$$

so that $h_3 = h_f + 0.86h_{fg} = 2256.9\,\text{kJ/kg}$. The second-law effectiveness is given by

$$\varepsilon_{\text{II}} = \frac{\Psi_2 + W_{\text{turb}}}{\Psi_4 + W_{\text{pump}} + [1 - (T_0/T_1)]Q_{\text{boil}}}$$

The dead state for water is liquid at 100 kPa and 25 °C:

$$h_0 = h_f = 104.9\ \text{kJ/kg} \qquad s_0 = s_f = 0.3672\ \text{kJ/kg·K}$$

Now the various quantities of interest may be calculated, assuming $m_1 = 1\,\text{kg}$:

$$\Psi_2 = m_2[h_2 - h_0 - T_0(s_2 - s_0)] = (0.1)[2587.3 - 104.9 - (298)(7.1237 - 0.3672)] = 46.89\ \text{kJ}$$

$$W_{\text{turb}} = m_1(h_1 - h_2) + m_3(h_2 - h_3) = (1.0)(3051.2 - 2587.3) + (0.9)(2587.3 - 2256.9) = 761.3\ \text{kJ}$$

$$\Psi_4 = m_4[h_4 - h_0 - T_0(s_4 - s_0)] = (0.1)[191.8 - 104.9 - (298)(0.6491 - 0.3671)] = 0.28\ \text{kJ}$$

$$W_{\text{pump}} = m_1\frac{\Delta P}{\rho} = (1.0)\left(\frac{1000 - 10}{1000}\right) = 0.99\ \text{kJ} \qquad Q_{\text{boil}} = m_1(h_1 - h_6) = (1.0)(3051.2 - 192.8) = 2858\ \text{kJ}$$

whence

$$\varepsilon_{\text{II}} = \frac{46.89 + 761.3}{0.28 + 0.99 + (1 - 298/573)(2858)} = 0.59$$

EXAMPLE 7.8 Perform an irreversibility calculation for each device in the ideal regenerative gas turbine cycle shown in Fig. 7-3.

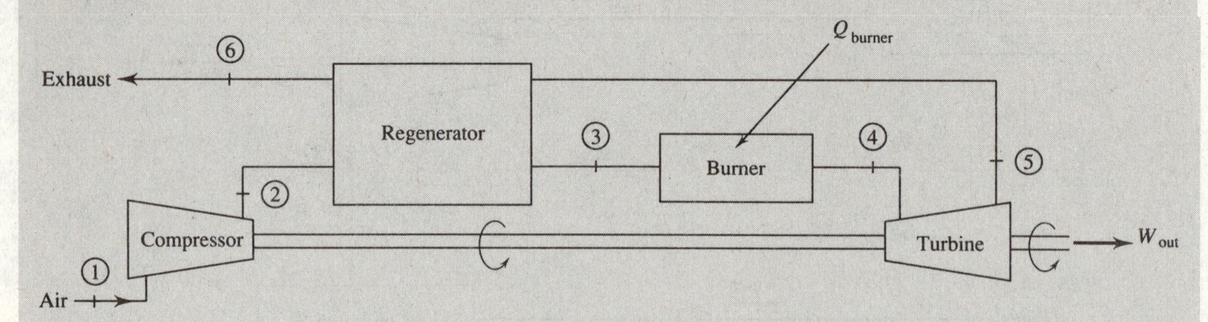

Fig. 7-3

Solution: The temperatures and pressures shown in Table 7-1 are given; h and s° are found in the air tables. For each device we will calculate the irreversibility by

$$i = T_0\left(s_1^\circ - s_2^\circ - R\ln\frac{P_1}{P_2}\right) - q$$

except for the burner, where we assume the heat transfer to occur at T_4. The irreversibilities are:

Compressor: 0
Regenerator: 0
Burner: 206.3 kJ/kg
Turbine: 0

Table 7-1

State	T (K)	P (MPa)	h (kJ/kg)	s° (kJ/kg·K)
1	294	0.1	294.2	1.682
2	439	0.41	440.7	2.086
3	759	0.41	777.5	2.661
4	1089	0.41	1148.3	2.764
5	759	0.1	777.5	2.661
6	439	0.1	440.7	2.086

The only irreversibility is associated with the burner. This suggests that large savings are possible by improving the performance of the burner. However, in attempting such improvement we must bear in mind that much of the irreversibility in the burner arises out of the combustion process, an irreversible process, which is essential for the operation of the turbine.

Solved Problems

7.1 The intake stroke for the cylinder of an internal combustion engine may be considered as a transient polytropic process with exponent -0.04. The initial pressure, temperature, and volume are 13.5 psia, 560 °R, and 0.0035 ft^3. Air is supplied at 14.7 psia and 520 °R, and the final volume

Table 7-2

Inlet State	Initial State of C.V.	Final State of C.V.
$T_1 = 520\,°\text{R}$	$T_i = 560\,°\text{R}$	$T_f = 520\,°\text{R}$
$P_1 = 14.7$ psia	$P_i = 13.5$ psia	$u_f = 88.62$ Btu/lbm
$h_1 = 124.27$ Btu/lbm	$u_i = 95.47$ Btu/lbm	$s_f^\circ = 0.5917$ Btu/lbm-°R
$s_1^\circ = 0.5917$ Btu/lbm-°R	$s_i^\circ = 0.6095$ Btu/lbm-°R	$V_f = 0.025\ \text{ft}^3$
	$V_i = 0.0035\ \text{ft}^3$	

and temperature are 0.025 ft^3 and 520 °R. Determine the reversible work and the irreversibility associated with the intake process.

At the various states either we are given, or the air tables provide, the values shown in Table 7-2. In the initial state,

$$m_i = \frac{P_i V_i}{RT_i} = \frac{(13.5)(144)(0.0035)}{(53.3)(560)} = 2.28 \times 10^{-4}\ \text{lbm}$$

The final state is produced by a polytropic process, so that

$$P_f = P_i\left(\frac{V_i}{V_f}\right)^n = (13.5)\left(\frac{0.0035}{0.025}\right)^{-0.04} = 14.6\ \text{psia}$$

$$m_f = \frac{P_f V_f}{RT_f} = \frac{(14.6)(144)(0.025)}{(53.3)(520)} = 1.90 \times 10^{-3}\ \text{lbm}$$

From conservation of mass, $m_1 = m_f - m_i = (1.90 \times 10^3) - (2.28 \times 10^{-4}) = 1.67 \times 10^{-3}$ lbm. Only boundary work is actually performed; for the polytropic process we have

$$W_a = \frac{P_f V_f - P_i V_i}{1-n} = \frac{[(14.6)(0.025) - (13.5)(0.0035)](144)}{(1 + 0.04)(778)} = 0.057\ \text{Btu}$$

The reversible work is given by (*7.11*) (neglect *KE* and *PE*, as usual):

$$W_{\text{rev}} = m_i(u_i - T_0 s_i) - m_f(u_f - T_0 s_f) + m_1(h_1 - T_0 s_1)$$

The needed values of s_i and s_f are obtained from the ideal-gas relation

$$s = s^\circ - R \ln \frac{P}{P_0}$$

where P_0 is some reference pressure. Normally, we do not have to worry about P_0, since when we consider an entropy change, P_0 cancels. It can be shown that even for this problem it will cancel, so that

$$W_{\text{rev}} = m_i(u_i - T_0 s_i^\circ + T_0 R \ln P_i) - m_f(u_f - T_0 s_f^\circ + T_0 R \ln P_f)$$
$$+ m_1(h_1 - T_0 s_1^\circ + T_0 R \ln P_1) = 0.058\ \text{Btu}$$

and, finally, $I = W_{\text{rev}} - W_a = 0.058 - 0.057 = 0.001$ Btu.

7.2 A supply pump for a power plant takes in saturated water at 0.01 MPa and boosts its pressure to 10 MPa. The pump has an adiabatic efficiency of 0.90. Calculate the irreversibility and second-law efficiency.

At the inlet and exit states either we are given, or the steam tables provide, the values given in Table 7-3. The actual work is

$$w_a = \frac{w_{\text{ideal}}}{\eta} = -\frac{\Delta P}{\eta \rho} = -\frac{10\,000 - 10}{(0.9)(1000)} = -11.1\ \text{kJ/kg}$$

Table 7-3

Inlet state 1: saturated liquid phase	Exit state 2: compressed liquid phase
$T = 45.8\,°\text{C}$	$P = 10$ MPa
$P = 0.01$ MPa	
$h = 191.8$ kJ/kg	
$s = 0.6491$ kJ/kg·K	

Then, by the first law, $h_2 = -w_a + h_1 = -(-11.1) + 191.8 = 202.9\,\text{kJ/kg}$. Using this enthalpy, we can interpolate for the entropy from the compressed liquid table and find $s_2 = 0.651\,\text{kJ/kg·K}$. As in Example 7.2, the irreversibility is given by

$$i = T_0(s_2 - s_1) = (298)(0.651 - 0.6491) = 0.57\,\text{kJ/kg}$$

whence

$$w_{\text{rev}} = i + w_a = 0.57 + (-11.1) = -10.5\,\text{kJ/kg} \qquad \eta_{\text{II}} = \frac{w_{\text{rev}}}{w_a} = \frac{-10.5}{-11.1} = 0.95$$

7.3 A power plant utilizes groundwater in a secondary coolant loop. Water enters the loop at 40 °F and 16 psia and exits at 80 °F and 15 psia. If the heat transfer in the loop occurs at 100 °F, what is the irreversibility?

Data are presented in Table 7-4. The heat transfer is $q = h_2 - h_1 = 48.1 - 8.02 = 40.1\,\text{Btu/lbm}$. The irreversibility is given by

$$i = T_0(s_2 - s_1) - q = (560)(0.09332 - 0.01617) - 40.1 = 3.1\,\text{Btu/lbm}$$

Table 7-4

Inlet state 1: compressed liquid phase	Exit state 2: compressed liquid phase
$T = 40\,°\text{F}$	$T = 80\,°\text{F}$
$P = 16$ psia	$P = 15$ psia
$h = 8.02$ Btu/lbm	$h = 48.1$ Btu/lbm
$s = 0.01617$ Btu/lbm-°R	$s = 0.09332$ Btu/lbm-°R

7.4 A reservoir of water is perched in the hills overlooking a valley. The water is at 25 °C and 100 kPa. If the reservoir is 1 km above the valley floor, calculate the availability of the water from the perspective of a farmer living in the valley.

The inlet and exit states are identified as follows:

Inlet state 1:	$T = 25\,°\text{C}$	$P = 0.1$ MPa	$z = 1$ km
Dead state 2:	$T = 25\,°\text{C}$	$P = 0.1$ MPa	$z = 0$ km

We have assumed that the availability of the water in the reservoir is due entirely to the elevation. Then

$$\psi = g(z_1 - z_0) = (9.8)(1 - 0) = 9.8\,\text{kJ/kg}$$

Table 7-5

Inlet state 1: superheated vapor	Inlet state 2: compressed liquid	Exit state 3: saturated liquid
$T = 250\ °\mathrm{C}$	$T = 150\ °\mathrm{C}$	$P = 0.6$ MPa
$P = 0.6$ MPa	$P = 0.6$ MPa	$T = 158.9\ °\mathrm{C}$
$h = 2957.2$	$h = 632.2$ kJ/kg	$h = 670.6$ kJ/kg
$s = 7.1824$ kJ/kg·K	$s = 1.8422$ kJ/kg·K	$s = 1.9316$ kJ/kg·K

7.5 A feedwater heater extracts steam from a turbine at 600 kPa and 250 °C which it combines with 0.3 kg/s of liquid at 600 kPa and 150 °C. The exhaust is saturated liquid at 600 kPa. Determine the second-law effectiveness of the heater.

For data, see Table 7-5. By conservation of mass, $\dot{m}_3 = \dot{m}_1 + \dot{m}_2$. Then, the first law demands $\dot{m}_3 h_3 = \dot{m}_1 h_1 + \dot{m}_2 h_2$. Solving simultaneously for $\dot{m}_1$ and $\dot{m}_3$:

$$\dot{m}_1 = 0.00504\ \mathrm{kg/s} \qquad \dot{m}_3 = 0.305\ \mathrm{kg/s}$$

The second-law effectiveness is $\varepsilon_{\mathrm{II}} = \dot{\Psi}_3/(\dot{\Psi}_1 + \dot{\Psi}_2)$. Taking the dead state as liquid water at 25 °C and 100 kPa, we have

$$h_0 = 105\ \mathrm{kJ/kg} \qquad s_0 = 0.3672\ \mathrm{kJ/kg{\cdot}K}$$

Then

$$\dot{\Psi}_3 = \dot{m}[h_3 - h_0 - T_0(s_3 - s_0)] = (0.305)[670.6 - 105 - 298(1.9316 - 0.3672)] = 30.33\ \mathrm{kW}$$
$$\dot{\Psi}_1 = \dot{m}_1[h_1 - h_0 - T_0(s_1 - s_0)] = (0.00504)[2957.2 - 105 - 298(7.1824 - 0.3672)] = 4.14\ \mathrm{kW}$$
$$\dot{\Psi}_2 = \dot{m}_2[h_2 - h_0 - T_0(s_2 - s_0)] = (0.30)[632.2 - 105 - 298(1.8422 - 0.3672)] = 23.63\ \mathrm{kW}$$

and

$$\varepsilon_{\mathrm{II}} = \frac{30.33}{4.14 + 23.63} = 1.09$$

7.6 Consider the ideal refrigeration cycle shown in Fig. 7-4 which utilizes R134a. The condenser operates at 800 kPa while the evaporator operates at 120 kPa. Calculate the second-law effectiveness for the cycle.

The given values and the R134a tables in Appendix D allow us to set up Table 7-6.

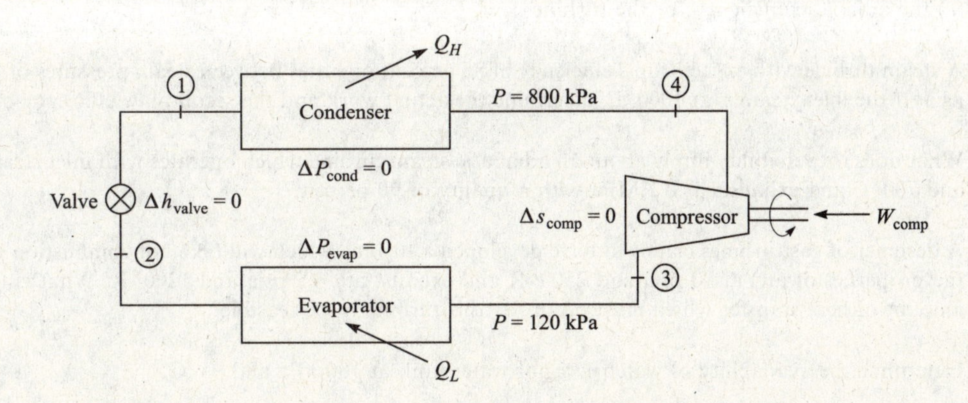

Fig. 7-4

Table 7-6

State	T (°C)	P (kPa)	h (kJ/kg)	s (kJ/kg·K)
1 (Saturated liquid phase)	31.3	800	93.42	0.3459
2 (Two-phase)	–22.4	120		
3 (Saturated vapor phase)	–22.4	120	233.9	0.9354
4 (Superheated phase)		800		

Now, traversing the cycle, the enthalpy remains constant across a valve, so that $h_2 = h_1 = 30.84$ Btu/lbm. State 2 is two-phase, so that

$$x = \frac{h_2 - h_f}{h_g - h_f} = \frac{93.42 - 21.32}{233.9 - 21.32} = 0.339$$

and

$$s_2 = s_f + x(s_g - s_f) = 0.0879 + (0.339)(0.9354 - 0.0879) = 0.375 \text{ kJ/kg·K}$$
$$h_2 = h_f + xh_{fg} = 21.32 + 0.339(212.54) = 93.4$$

State 4 results from an isentropic compression across the ideal compressor. At $P_4 = 800$ kPa and $s_4 = 0.9354$ kJ/kg·K, we interpolate to find $h_4 = 273$ kJ/kg. We now calculate the second-law effectiveness for the cycle assuming $T_0 = 298$ K:

$$\text{availability produced} = \left(1 - \frac{T_0}{T_3}\right)Q_L = \left(1 - \frac{298}{251}\right)(93.4 - 233.9) = 26.3 \text{ kJ/kg}$$
$$\text{work used} = W_{\text{comp}} = h_4 - h_3 = 273.0 - 233.9 = 39.1 \text{ kJ/kg}$$
$$\varepsilon_{\pi} = \frac{26.3}{39.1} = 0.673$$

Supplementary Problems

7.7 Steam enters a turbine at 6 MPa and 500 °C and exits at 100 kPa and 150 °C. Determine (*a*) the reversible work and (*b*) the irreversibility of the process.

7.8 The inlet conditions to an adiabatic steam turbine are 800 psia and 700 °F. At the exit the pressure is 30 psia and the steam has a quality of 93 percent. Determine (*a*) the irreversibility, (*b*) the reversible work, and (*c*) the adiabatic efficiency for the turbine.

7.9 A steam turbine with an isentropic efficiency of 85 percent operates between steam pressures of 1500 and 100 psia. If the inlet steam is at 1000 °F, determine the actual work and the second-law efficiency of the turbine.

7.10 What does irreversibility imply about an adiabatic steam turbine which operates with inlet steam at 10 MPa and 700 °C and exhausts at 0.2 MPa with a quality of 90 percent?

7.11 A designer of gas turbines claims to have developed a turbine which will take hot combustion gases (having the properties of air) at 80 psia and 2500 °R and exhaust at 14.7 psia and 1200 °R. What is the minimum amount of heat transfer which must occur for this turbine to be feasible?

7.12 Determine the availability of water in a hot water tank at 100 kPa and 95 °C.

7.13 What is the availability of a 2-in^3 ice cube at 10 °F and 14.7 psia?

7.14 Ideally, which fluid can do more work: air at 600 psia and 600 °F or steam at 600 psia and 600 °F?

7.15 A piston-cylinder system with air undergoes a polytropic compression with $n = 1.1$ from 75 °F, 15 psia, and 0.2 liter to 0.04 liter. Determine (*a*) actual work, (*b*) heat transfer, (*c*) reversible work, and (*d*) irreversibility.

7.16 Methane gas at 800 K and 3 MPa is contained in a piston-cylinder system. The system is allowed to expand to 0.1 MPa in a polytropic process with $n = 2.3$. What is the second-law efficiency of the process?

7.17 Argon is contained in a sealed tank of 10 liters at 400 psia and 50 °F. What is the maximum work the argon can do on earth at 536 °R?

7.18 A rigid tank initially contains 0.5 lbm of R134a as saturated liquid at 30 psia. It is then allowed to come to equilibrium with its surroundings at 70 °F. Determine (*a*) the final state of the refrigerant and (*b*) the irreversibility.

7.19 Air enters a compressor at 100 kPa and 295 K and exits at 700 kPa and 530 K with 40 kJ/kg of heat transfer to the surroundings. Determine (*a*) reversible work, (*b*) irreversibility, and (*c*) second-law efficiency for the compressor.

7.20 A compressor with an adiabatic efficiency of 90 percent intakes air at 500 °R and 15 psia and exhausts at 120 psia. Determine (*a*) the actual work and (*b*) the reversible work associated with this compressor.

7.21 The evaporator for an air-conditioning system is a heat exchanger. R134a enters at 0.05 kg/s and −20 °C as saturated liquid and leaves as saturated vapor. Air enters at 34 °C and leaves at 18 °C. (*a*) What is the mass flow rate of air? (*b*) What is the irreversibility rate of the evaporator?

7.22 A direct contact heat exchanger serves as the condenser for a steam power plant. Steam with quality of 50 percent at 100 kPa flows into the mixing tank at 2 kg/s. Groundwater at 10 °C and 100 kPa is available to produce saturated liquid flowing out of the mixing tank. The mixing tank is well-insulated. Determine (*a*) the mass flow rate of groundwater required and (*b*) the irreversibility rate.

7.23 Steam is throttled across an adiabatic valve from 250 psia and 450 °F to 60 psia. Determine (*a*) the reversible work and (*b*) the irreversibility.

7.24 It has been proposed to utilize a nozzle in conjunction with a wind turbine system. Air enters the adiabatic nozzle at 9 m/s, 300 K, and 120 kPa and exits at 100 m/s and 100 kPa. Determine (*a*) the irreversibility and (*b*) the reversible work.

7.25 In the burner for a gas turbine system 0.2 lbm/sec of air at 20 psia and 900 °R is heated to 2150 °R in a constant-pressure process while hot combustion gases (assumed to be air) are cooled from 3000 °R to 2400 °R. What is the irreversibility rate of this process?

7.26 Saturated water enters an adiabatic pump at 10 kPa and exits at 1 MPa. If the pump has an adiabatic efficiency of 95 percent, determine (*a*) the reversible work and (*b*) the second-law efficiency.

7.27 The pressure of water is increased, by the use of a pump, from 14 to 40 psia. A rise in the water temperature from 60 °F to 60.1 °F is observed. Determine (*a*) the irreversibility, (*b*) the reversible work, and (*c*) the adiabatic efficiency of the pump.

7.28 Air at 2200 °R and 40 psia enters a gas turbine with an adiabatic efficiency of 75 percent and exhausts at 14.7 psia. Determine (*a*) the availability of the exhaust air and (*b*) the reversible work.

Answers to Supplementary Problems

7.7 (*a*) 864.2 kJ/kg (*b*) 218.5 kJ/kg

7.8 (*a*) 31.8 Btu/lbm (*b*) 272 Btu/lbm (*c*) 85.1%

7.9 259 Btu/lbm, 90.2%

7.10 $i = -179$ kJ/kg (*impossible*)

7.11 −25.2 Btu/lbm

7.12 29.8 kJ/kg

7.13 1.06 Btu

7.14 Steam (471 Btu/lbm vs. 173 Btu/lbm)

7.15 (*a*) −26.64 ft-lbf (*b*) −0.0257 Btu (*c*) −25.09 ft-lbf (*d*) 1.55 ft-lbf

7.16 65.0%

7.17 92.2 Btu

7.18 (*a*) compressed liquid (*b*) 0.463 Btu

7.19 (*a*) −228 kJ/kg (*b*) 48.2 kJ/kg (*c*) 82.5%

7.20 (*a*) −107.7 Btu/lbm (*b*) −101.4 Btu/lbm

7.21 (*a*) 0.659 kg/s (*b*) 22.9 kW

7.22 (*a*) 6.00 kg/s (*b*) 261 kW

7.23 (*a*) 61,800 ft-lbf/lbm (*b*) 61,800 ft-lbf/lbm

7.24 (*a*) 10.6 kJ/kg (*b*) 10.6 kJ/kg

7.25 11.3 Btu/sec

7.26 (*a*) −1.053 kJ/kg (*b*) 95.2%

7.27 (*a*) 0.432 Btu/lbm (*b*) −0.0634 Btu/lbm (*c*) 15.6%

7.28 (*a*) 156 Btu/lbm (*b*) 110 Btu/lbm

Gas Power Cycles

8.1 INTRODUCTION

Several cycles utilize a gas as the working substance, the most common being the Otto cycle and the diesel cycle used in internal combustion engines. The word "cycle" used in reference to an internal combustion engine is technically incorrect since the working fluid does not undergo a thermodynamic cycle; air enters the engine, mixes with a fuel, undergoes combustion, and exits the engine as exhaust gases. This is often referred to as an *open cycle*, but we should keep in mind that a thermodynamic cycle does not really occur; the engine itself operates in what we could call a *mechanical cycle*. We do, however, analyze an internal combustion engine as though the working fluid operated on a cycle; it is an approximation that allows us to predict influences of engine design on such quantities as efficiency and fuel consumption.

8.2 GAS COMPRESSORS

We have already utilized the gas compressor in the refrigeration cycles discussed earlier and have noted that the control volume energy equation relates the power input to the enthalpy change as follows:

$$\dot{W}_{\text{comp}} = \dot{m}(h_e - h_i) \tag{8.1}$$

where h_e and h_i are the exit and inlet enthalpies, respectively. In this form we model the compressor as a fixed volume into which and from which a gas flows; we assume that negligible heat transfer occurs from the compressor and ignore the difference between inlet and outlet kinetic and potential energy changes.

There are three general types of compressors: reciprocating, centrifugal, and axial-flow. Reciprocating compressors are especially useful for producing high pressures, but are limited to relatively low flow rates; upper limits of about 200 MPa with inlet flow rates of 160 m^3/min are achievable with a two-stage unit. For high flow rates with relatively low pressure rise, a centrifugal or axial-flow compressor would be selected; a pressure rise of several MPa for an inlet flow rate of over 10 000 m^3/min is possible.

The Reciprocating Compressor

A sketch of the cylinder of a reciprocating compressor is shown in Fig. 8-1. The intake and exhaust valves are closed when state 1 is reached, as shown on the *P-v* diagram of Fig. 8-2*a*. An isentropic

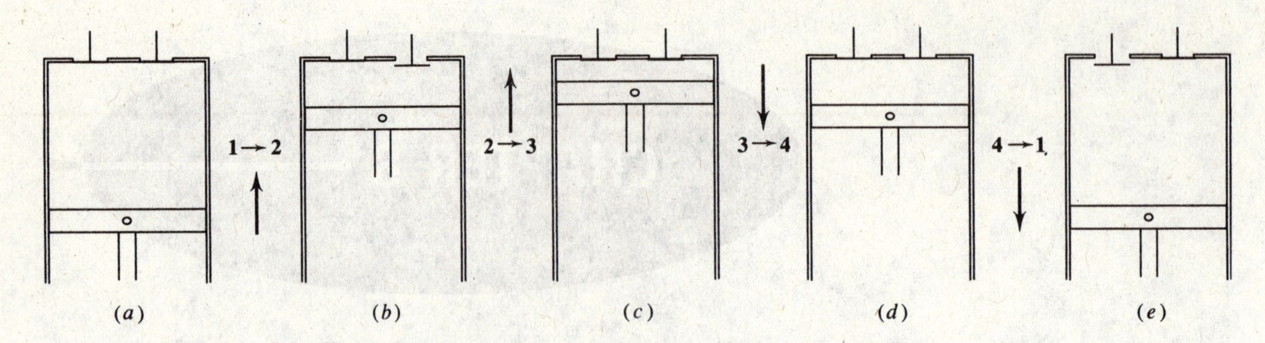

Fig. 8-1 A reciprocating compressor.

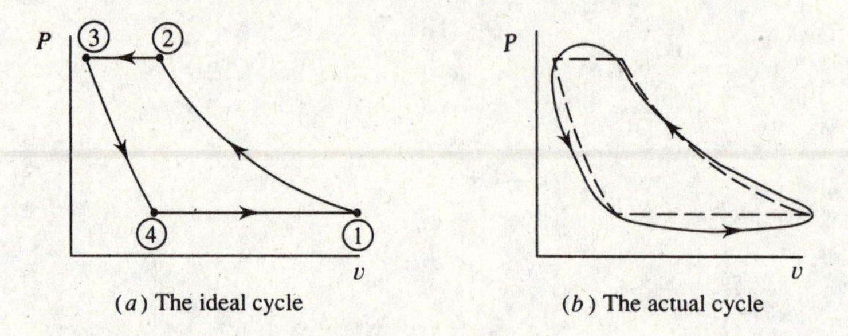

Fig. 8-2 The *P-v* diagram.

compression follows as the piston travels inward until the maximum pressure at state 2 is reached. The exhaust valve then opens and the piston continues its inward motion while the air is exhausted until state 3 is reached at top dead center. The exhaust valve then closes and the piston begins its outward motion with an isentropic expansion process until state 4 is reached. At this point the intake valve opens and the piston moves outward during the intake process until the cycle is completed.

During actual operation the *P-v* diagram would more likely resemble that of Fig. 8-2*b*. Intake and exhaust valves do not open and close instantaneously, the airflow around the valves results in pressure gradients during the intake and exhaust strokes, losses occur due to the valves, and some heat transfer may take place. The ideal cycle does, however, allow us to predict the influence of proposed design changes on work requirements, maximum pressure, flow rate, and other quantities of interest.

The effectiveness of a compressor is partially measured by the *volumetric efficiency*, which is defined as the volume of gas drawn into the cylinder divided by the displacement volume. That is, referring to Fig. 8-2,

$$\eta_{\text{vol}} = \frac{V_1 - V_4}{V_1 - V_3} \tag{8.2}$$

The higher the volumetric efficiency the greater the volume of air drawn in as a percentage of the displacement volume. This can be increased if the clearance volume V_3 is decreased.

To improve the performance of the reciprocating compressor, we can remove heat from the compressor during the compression process $1 \to 2$. The effect of this is displayed in Fig. 8-3, where a polytropic process is shown. The temperature of state $2'$ would be significantly lower than that of state 2 and the work requirement for the complete cycle would be less since the area under the *P-v* diagram would decrease. To analyze this situation let us return to the control volume inlet-outlet description, as used with (*8.1*): The required work is, for an adiabatic compressor,

$$w_{\text{comp}} = h_2 - h_1 = C_p(T_2 - T_1) \tag{8.3}$$

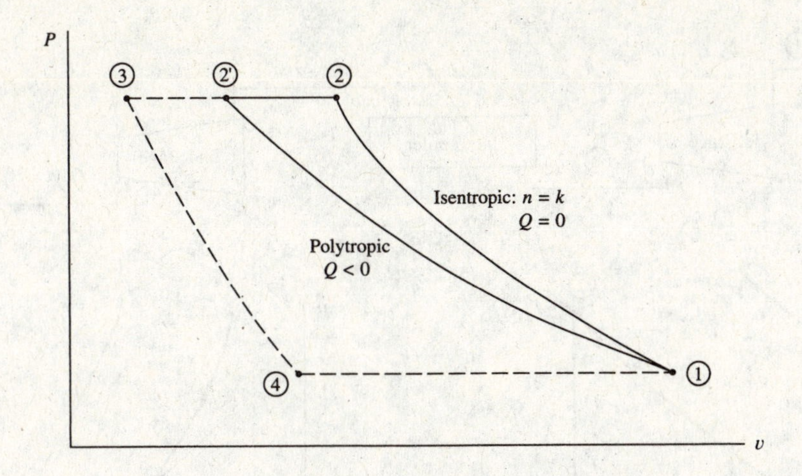

Fig. 8-3 Removal of heat during compression.

assuming an ideal gas with constant specific heat. For an isentropic compression between inlet and outlet we know that

$$T_2 = T_1\left(\frac{P_2}{P_1}\right)^{(k-1)/k} \tag{8.4}$$

This allows the work to be expressed as, using C_p given in (*4.30*),

$$w_{\text{comp}} = \frac{kR}{k-1}T_1\left[\left(\frac{P_2}{P_1}\right)^{(k-1)/k} - 1\right] \tag{8.5}$$

For a polytropic process we simply replace k with n and obtain

$$w_{\text{comp}} = \frac{nR}{n-1}T_1\left[\left(\frac{P_2}{P_1}\right)^{(n-1)/n} - 1\right] \tag{8.6}$$

The heat transfer is then found from the first law.

By external cooling, with a water jacket surrounding the compressor, the value of n when compressing air can be reduced to about 1.35. This reduction from 1.4 is difficult since heat transfer must occur from the rapidly moving air through the compressor casing to the cooling water, or from fins. This is an ineffective process, and multistage compressors with interstage cooling are often a desirable alternative. With a single stage and with a high P_2 the outlet temperature T_2 would be too high even if n could be reduced to, say, 1.3.

Consider a two-stage compressor with a single intercooler, as shown in Fig. 8-4*a*. The compression processes are assumed to be isentropic and are shown in the T-s and P-v diagrams of Fig. 8-4*b*.

Referring to (*8.5*), the work is written as

$$\begin{aligned} w_{\text{comp}} &= C_pT_1\left[\left(\frac{P_2}{P_1}\right)^{(k-1)/k} - 1\right] + C_pT_3\left[\left(\frac{P_4}{P_3}\right)^{(k-1)/k} - 1\right] \\ &= C_pT_1\left[\left(\frac{P_2}{P_1}\right)^{(k-1)/k} + \frac{P_4}{P_2}^{(k-1)/k} - 2\right] \end{aligned} \tag{8.7}$$

where we have used $P_2 = P_3$ and $T_1 = T_3$, for an ideal intercooler. To determine the intercooler pressure P_2 that minimizes the work, we let $\partial w_{\text{comp}}/\partial P_2 = 0$. This gives

$$P_2 = (P_1P_4)^{1/2} \qquad \text{or} \qquad \frac{P_2}{P_1} = \frac{P_4}{P_3} \tag{8.8}$$

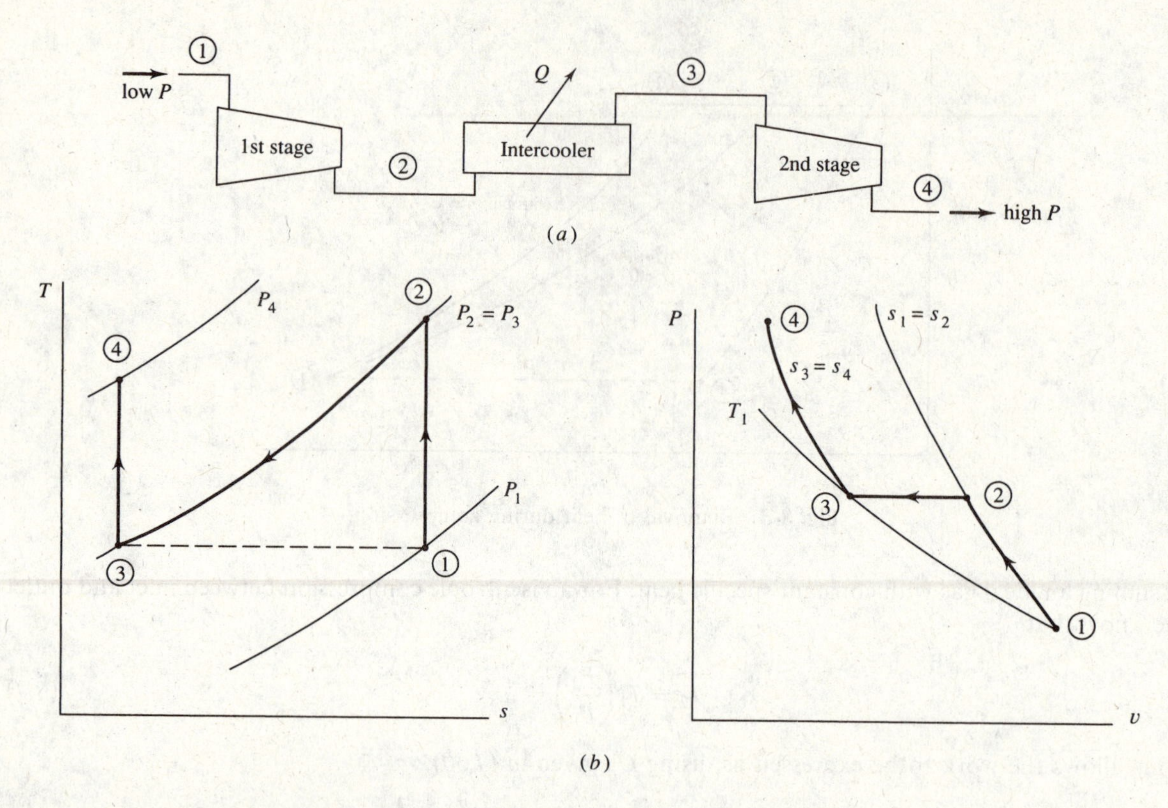

Fig. 8-4 A two-stage compressor with an intercooler.

That is, the pressure ratio is the same across each stage. If three stages were used, the same analysis would lead to a low-pressure intercooler pressure of

$$P_2 = (P_1^2 P_6)^{1/3} \tag{8.9}$$

and a high-pressure intercooler pressure of

$$P_4 = (P_1 P_6^2)^{1/3} \tag{8.10}$$

where P_6 is the highest pressure. This is also equivalent to equal pressure ratios across each stage. Additional stages may be necessary for extremely high outlet pressures; an equal pressure ratio across each stage would yield the minimum work for the ideal compressor.

Centrifugal and Axial-Flow Compressors

A centrifugal compressor is sketched in Fig. 8-5. Air enters along the axis of the compressor and is forced to move outward along the rotating impeller vanes due to the effects of centrifugal forces. This results in an increased pressure from the axis to the edge of the rotating impeller. The diffuser section results in a further increase in the pressure as the velocity is reduced due to the increasing area in each subsection of the diffuser. Depending on the desired pressure-speed characteristics, the rotating impeller can be fitted with radial impeller vanes, as shown; with backward-curved vanes; or with forward-curved vanes.

An axial-flow compressor is illustrated in Fig. 8-6. It is similar in appearance to the steam turbine used in the Rankine power cycle. Several stages of blades are needed to provide the desired pressure rise, with a relatively small rise occurring over each stage. Each stage has a *stator*, a series of blades that are attached to the stationary housing, and a *rotor*. All the rotors are attached to a common rotating shaft which utilizes the power input to the compressor. The specially designed airfoil-type

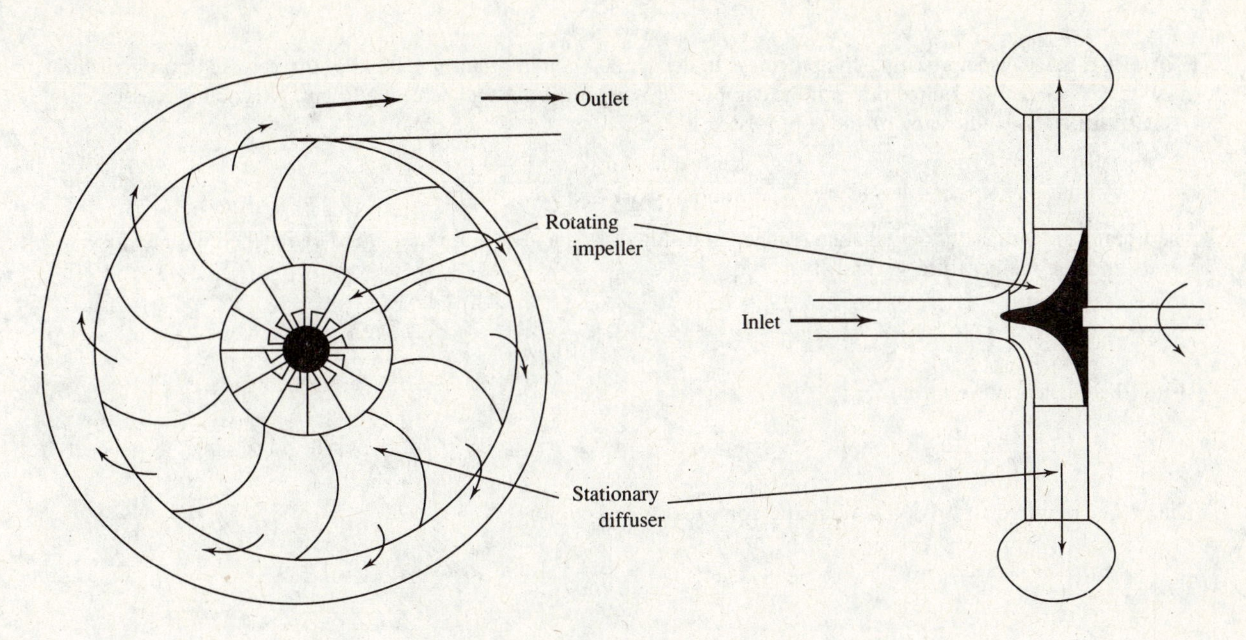

Fig. 8-5 A centrifugal compressor.

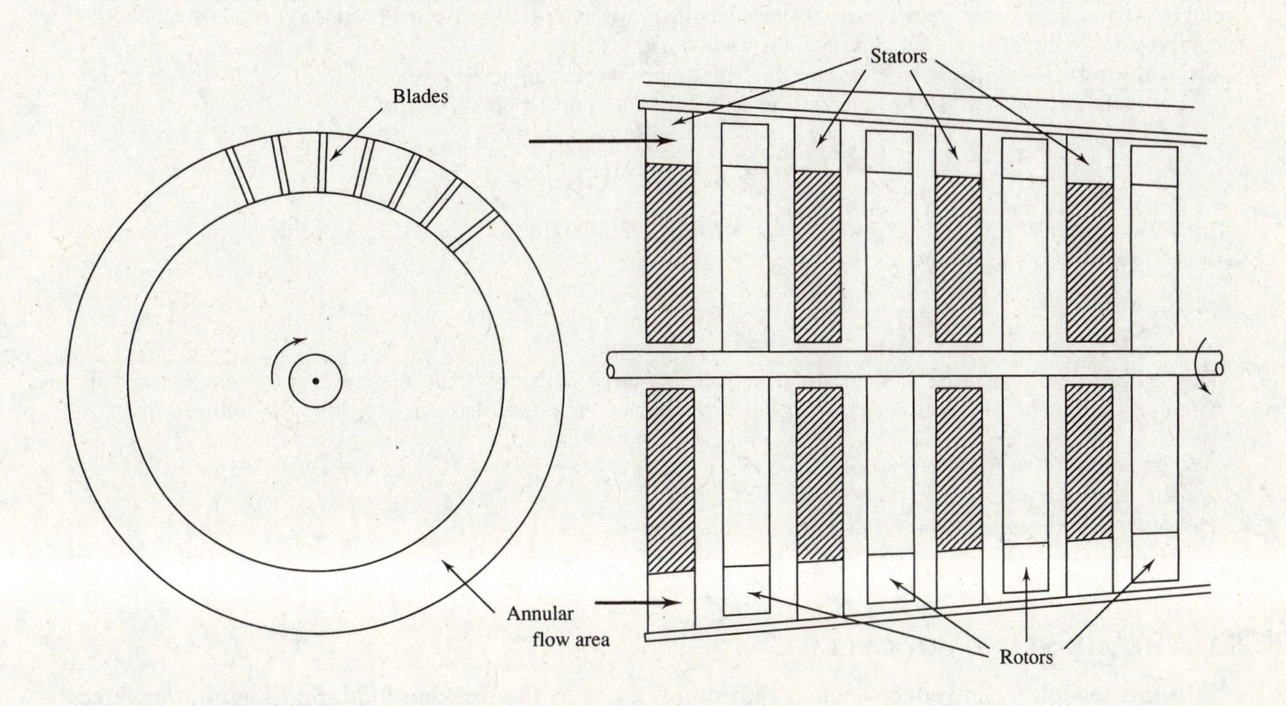

Fig. 8-6 An axial-flow compressor.

blades require extreme precision in manufacturing and installation to yield the maximum possible pressure rise while avoiding flow separation. The area through which the air passes decreases slightly as the pressure rises due to the increased density in the higher-pressure air. In fluid mechanics the velocity and pressure at each stage can be analyzed; in thermodynamics we are concerned only with inlet and outlet conditions.

EXAMPLE 8.1 A reciprocating compressor is to deliver 20 kg/min of air at 1600 kPa. It receives atmospheric air at 20 °C. Calculate the required power if the compressor is assumed to be 90 percent efficient. No cooling is assumed.

Solution: The efficiency of the compressor is defined as

$$\eta = \frac{\text{isentropic work}}{\text{actual work}} = \frac{h_{2'} - h_1}{h_2 - h_1}$$

where state 2 identifies the actual state reached and state $2'$ is the ideal state that could be reached with no losses. Let us find the temperature $T_{2'}$ first. It is

$$T_{2'} = T_1 = \left(\frac{P_2}{P_1}\right)^{(k-1)/k} = (293)\left(\frac{1600}{100}\right)^{(1.4-1)/1.4} = 647 \text{ K}$$

Using the efficiency, we have

$$\eta = \frac{C_p(T_{2'} - T_1)}{C_p(T_2 - T_1)} \qquad \text{or} \qquad T_2 = T_1 + \frac{1}{\eta}(T_{2'} - T_1) = 293 + \left(\frac{1}{0.9}\right)(647 - 293) = 686 \text{ K}$$

The power required to drive the adiabatic compressor (no cooling) is then

$$\dot{W}_{\text{comp}} = \dot{m}(h_2 - h_1) = \dot{m}C_p(T_2 - T_1) = \left(\frac{20}{60}\right)(1.006)(686 - 293) = 131.9 \text{ kW}$$

EXAMPLE 8.2 Suppose that, for the compressor of Example 8.1, it is decided that because T_2 is too high, two stages with an intercooler are necessary. Determine the power requirement for the proposed two-stage adiabatic compressor. Assume 90 percent efficiency for each stage.

Solution: The intercooler pressure for minimum power input is given by (*8.8*) as $P_2 = \sqrt{P_1 P_4} = \sqrt{(100)(1600)} = 400$ kPa. This results in a temperature entering the intercooler of

$$T_{2'} = T_1\left(\frac{P_2}{P_1}\right)^{(1.4-1)/1.4} = 293\left(\frac{400}{100}\right)^{0.2857} = 435 \text{ K}$$

Since $T_3 = T_1$ and $P_4/P_3 = P_2/P_1$, we also have $T_{4'} = (293)(400/100)^{0.2857} = 435$ K. Considering the efficiency of each stage allows us to find

$$T_2 = T_1 + \frac{1}{\eta}(T_{2'} - T_1) = 293 + \left(\frac{1}{0.9}\right)(435 - 293) = 451 \text{ K}$$

This will also be the exiting temperature $T_{4'}$. Note the large reduction from the single-stage temperature of 686 K. Assuming no heat transfer in the compressor stages, the power necessary to drive the compressor is

$$\dot{W}_{\text{comp}} = \dot{m}C_p(T_2 - T_1) + \dot{m}C_p(T_4 - T_3) = \left(\frac{20}{60}\right) \times (1.00)(451 - 293) + \left(\frac{20}{60}\right)(1.00)(451 - 293) = 105 \text{ kW}$$

This is a 20 percent reduction in the power requirement.

8.3 THE AIR-STANDARD CYCLE

In this section we introduce engines that utilize a gas as the working fluid. Spark-ignition engines that burn gasoline and compression-ignition (diesel) engines that burn fuel oil are the two most common engines of this type.

The operation of a gas engine can be analyzed by assuming that the working fluid does indeed go through a complete thermodynamic cycle. The cycle is often called an *air-standard cycle*. All the air-standard cycles we will consider have certain features in common:

- Air is the working fluid throughout the entire cycle. The mass of the small quantity of injected fuel is negligible.
- There is no inlet process or exhaust process.

- The combustion process is replaced by a heat transfer process with energy transferred from an external source. It may be a constant-volume process or a constant-pressure process, or a combination.
- The exhaust process, used to restore the air to its original state, is replaced with heat transfer to the surroundings.
- All processes are assumed to be in quasiequilibrium.
- The air is assumed to be an ideal gas with constant specific heats.

A number of the engines we will consider make use of a closed system with a piston-cylinder arrangement, as shown in Fig. 8-7. The cycle shown on the *P-v* and *T-s* diagrams in the figure is representative only. The diameter of the piston is called the *bore*, and the distance the piston travels in one direction is the *stroke*. When the piston is at top dead center (TDC), the volume occupied by the air in the cylinder is at a minimum; this volume is the *clearance volume*. When the piston moves to bottom dead center (BDC), the air occupies the *maximum volume*. The difference between the maximum volume and the clearance volume is the *displacement volume*. The clearance volume is often implicitly presented as the *percent clearance c*, the ratio of the clearance volume to the displacement volume. The *compression ratio r* is defined to be the ratio of the volume occupied by the air at BDC to the volume occupied by the air at TDC, that is, referring to Fig. 8-7,

$$r = \frac{V_1}{V_2} \tag{8.11}$$

The *mean effective pressure* (MEP) is another quantity that is often used when rating piston-cylinder engines; it is the pressure that, if acting on the piston during the power stroke, would produce an amount of work equal to that actually done during the entire cycle. Thus,

$$W_{\text{cycle}} = (\text{MEP})(V_{\text{BDC}} - V_{\text{TDC}}) \tag{8.12}$$

In Fig. 8-7 this means that the enclosed area of the actual cycle is equal to the area under the MEP dotted line.

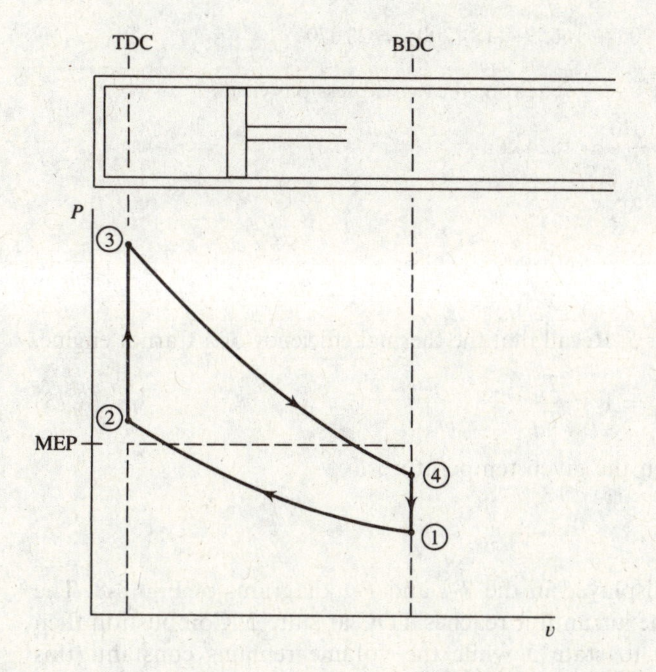

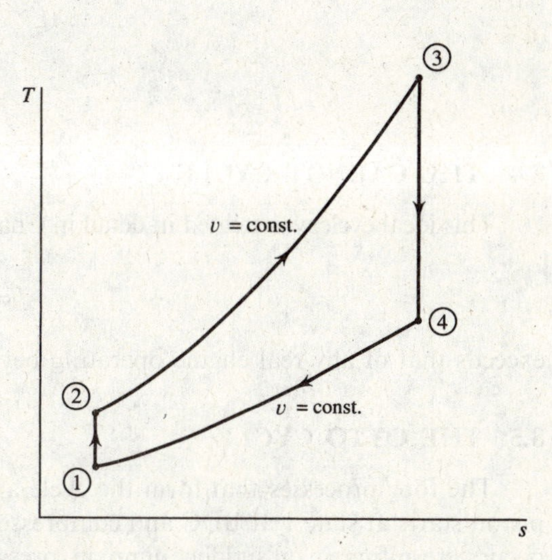

Fig. 8-7 A piston-cylinder engine.

EXAMPLE 8.3 An engine operates with air on the cycle shown in Fig. 8-7 with isentropic processes $1 \to 2$ and $3 \to 4$. If the compression ratio is 12, the minimum pressure is 200 kPa, and the maximum pressure is 10 MPa determine (*a*) the percent clearance and (*b*) the MEP.

Solution:

(*a*) The percent clearance is given by

$$c = \frac{V_2}{V_1 - V_2}(100)$$

But the compression ratio is $r = V_1/V_2 = 12$. Thus,

$$c = \frac{V_2}{12V_2 - V_2}(100) = \frac{100}{11} = 9.09\%$$

(*b*) To determine the MEP we must calculate the area under the *P-V* diagram; this is equivalent to calculating the work. The work from $3 \to 4$ is, using $PV^k = C$,

$$W_{3-4} = \int P\,dV = C\int \frac{dV}{V^k} = \frac{C}{1-k}(V_4^{1-k} - V_3^{1-k}) = \frac{P_4V_4 - P_3V_3}{1-k}$$

where $C = P_4V_4^k = P_3V_3^k$. But we know that $V_4/V_3 = 12$, so

$$W_{3-4} = \frac{V_3}{1-k}(12P_4 - P_3)$$

Likewise, the work from $1 \to 2$ is

$$W_{1-2} = \frac{V_2}{1-k}(P_2 - 12P_1)$$

Since no work occurs in the two constant-volume processes, we find, using $V_2 = V_3$,

$$W_{\text{cycle}} = \frac{V_2}{1-k}(12P_4 - P_3 + P_2 - 12P_1)$$

The pressures P_2 and P_4 are found as follows:

$$P_2 = P_1\left(\frac{V_1}{V_2}\right)^k = (200)(12)^{1.4} = 1665\text{ kPa} \qquad P_4 = P_3\left(\frac{V_3}{V_4}\right)^k = (10\,000)\left(\frac{1}{12}\right)^{1.4} = 308\text{ kPa}$$

whence

$$W_{\text{cycle}} = \frac{V_2}{-0.4}[(12)(308) - 10\,000 + 1665 - (12)(200)] = 20\,070V_2$$

But $W_{\text{cycle}} = (\text{MEP})(V_1 - V_2) = (\text{MEP})(12V_2 - V_2)$; equating the two expressions yields

$$\text{MEP} = \frac{20\,070}{11} = 1824\text{ kPa}$$

8.4 THE CARNOT CYCLE

This ideal cycle was treated in detail in Chapter 5. Recall that the thermal efficiency of a Carnot engine,

$$\eta_{\text{carnot}} = 1 - \frac{T_L}{T_H} \tag{8.13}$$

exceeds that of any real engine operating between the given temperatures.

8.5 THE OTTO CYCLE

The four processes that form the cycle are displayed in the *T-s* and *P-v* diagrams of Fig. 8-8. The piston starts at state 1 at BDC and compresses the air until it reaches TDC at state 2. Combustion then occurs, resulting in a sudden jump in pressure to state 3 while the volume remains constant (this combustion process is simulated with a quasiequilibrium heat addition process). The process that follows

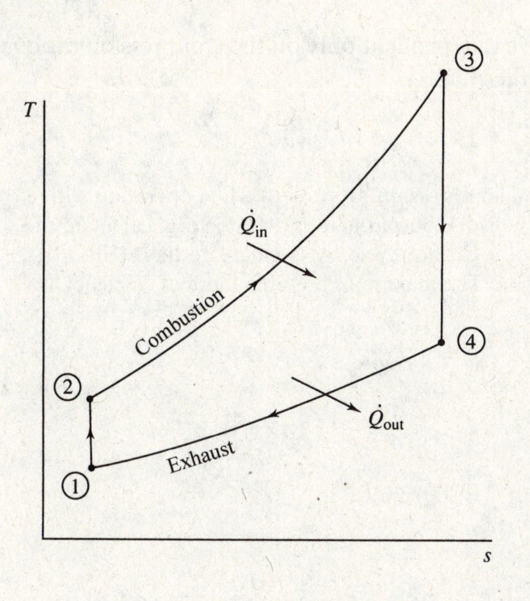

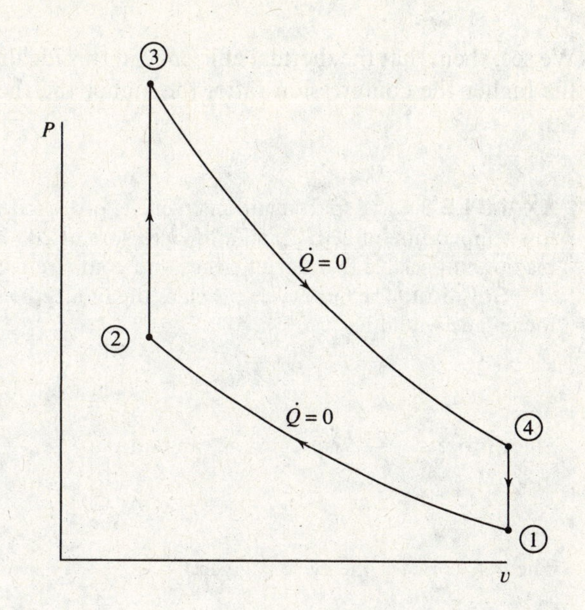

Fig. 8-8 The Otto cycle.

is the power stroke as the air (simulating the combustion products) expands isentropically to state 4. In the final process heat transfer to the surroundings occurs and the cycle is completed. The *spark-ignition engine* is modeled with this Otto cycle.

The thermal efficiency of the Otto cycle is found from

$$\eta = \frac{\dot{W}_{\text{net}}}{\dot{Q}_{\text{in}}} = \frac{\dot{Q}_{\text{in}} - \dot{Q}_{\text{out}}}{\dot{Q}_{\text{in}}} = 1 - \frac{\dot{Q}_{\text{out}}}{\dot{Q}_{\text{in}}} \tag{8.14}$$

Noting that the two heat transfer processes occur during constant-volume processes, for which the work is zero, there results

$$\dot{Q}_{\text{in}} = \dot{m}C_v(T_3 - T_2) \qquad \dot{Q}_{\text{out}} = \dot{m}C_v(T_4 - T_1) \tag{8.15}$$

where we have assumed each quantity to be positive. Then

$$\eta = 1 - \frac{T_4 - T_1}{T_3 - T_2} \tag{8.16}$$

This can be written as

$$\eta = 1 - \frac{T_1}{T_2}\frac{T_4/T_1 - 1}{T_3/T_2 - 1} \tag{8.17}$$

For the isentropic processes we have

$$\frac{T_2}{T_1} = \left(\frac{V_1}{V_2}\right)^{k-1} \qquad \text{and} \qquad \frac{T_3}{T_4} = \left(\frac{V_4}{V_3}\right)^{k-1} \tag{8.18}$$

But, using $V_1 = V_4$ and $V_3 = V_2$, we see that

$$\frac{T_2}{T_1} = \frac{T_3}{T_4} \tag{8.19}$$

Thus, (*8.17*) gives the thermal efficiency as

$$\eta = 1 - \frac{T_1}{T_2} = 1 - \left(\frac{V_2}{V_1}\right)^{k-1} = 1 - \frac{1}{r^{k-1}} \tag{8.20}$$

We see, then, that the thermal efficiency in this idealized cycle is dependent only on the compression ratio r; the higher the compression ratio, the higher the thermal efficiency.

EXAMPLE 8.4 A spark-ignition engine is proposed to have a compression ratio of 10 while operating with a low temperature of 200 °C and a low pressure of 200 kPa. If the work output is to be 1000 kJ/kg, calculate the maximum possible thermal efficiency and compare with that of a Carnot cycle. Also calculate the MEP.

Solution: The Otto cycle provides the model for this engine. The maximum possible thermal efficiency for the engine would be

$$\eta = 1 - \frac{1}{r^{k-1}} = 1 - \frac{1}{(10)^{0.4}} = 0.602 \quad \text{or } 60.2\%$$

Since process 1 → 2 is isentropic, we find that

$$T_2 = T_1\left(\frac{v_1}{v_2}\right)^{k-1} = (473)(10)^{0.4} = 1188\ \text{K}$$

The net work for the cycle is given by

$$w_{\text{net}} = w_{1-2} + \cancel{w}^{0}_{2-3} + w_{3-4} + \cancel{w}^{0}_{4-1} = c_v(T_1 - T_2) + c_v(T_3 - T_4) \quad \text{or}$$
$$1000 = (0.717)(473 - 1188 + T_3 - T_4)$$

But, for the isentropic process 3 → 4,

$$T_3 = T_4\left(\frac{v_4}{v_3}\right)^{k-1} = (T_4)(10)^{0.4} = 2.512\,T_4$$

Solving the last two equations simultaneously, we find $T_3 = 3508$ K and $T_4 = 1397$ K, so that

$$\eta_{\text{carnot}} = 1 - \frac{T_L}{T_H} = 1 - \frac{473}{3508} = 0.865 \quad \text{or } 86.5\%$$

The Otto cycle efficiency is less than that of a Carnot cycle operating between the limiting temperatures because the heat transfer processes in the Otto cycle are not reversible.

The MEP is found by using the equation

$$w_{\text{net}} = (\text{MEP})(v_1 - v_2)$$

We have

$$v_1 = \frac{RT_1}{P_1} = \frac{(0.287)(473)}{200} = 0.6788\ \text{m}^3/\text{kg} \qquad \text{and} \qquad v_2 = \frac{v_1}{10}$$

Thus

$$\text{MEP} = \frac{w_{\text{net}}}{v_1 - v_2} = \frac{1000}{(0.9)(0.6788)} = 1640\ \text{kPa}$$

8.6 THE DIESEL CYCLE

If the compression ratio is large enough, the temperature of the air in the cylinder when the piston approaches TDC will exceed the ignition temperature of diesel fuel. This will occur if the compression ratio is about 14 or greater. No external spark is needed; the diesel fuel is simply injected into the cylinder and combustion occurs because of the high temperature of the compressed air. This type of engine is referred to as a *compression-ignition engine*. The ideal cycle used to model the compression-ignition engine is the diesel cycle, shown in Fig. 8-9. The difference between this cycle and the Otto cycle is that, in the diesel cycle, the heat is added during a constant-pressure process.

The cycle begins with the piston at BDC, state 1; compression of the air occurs isentropically to state 2 at TDC; heat addition takes place (this models the injection and combustion of fuel) at constant

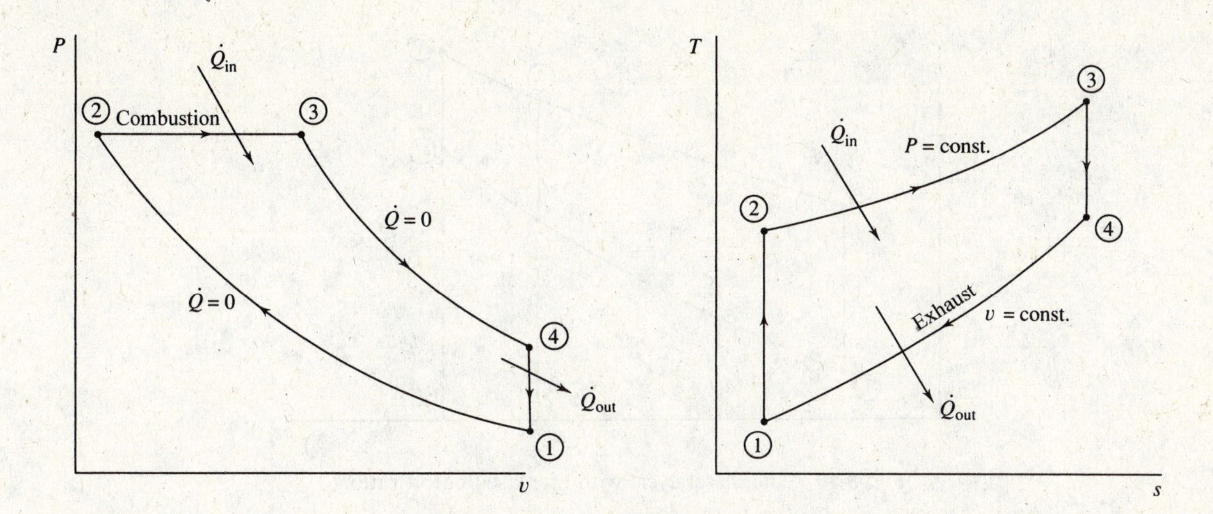

Fig. 8-9 The diesel cycle.

pressure until state 3 is reached; expansion occurs isentropically to state 4 at BDC; constant volume heat rejection completes the cycle and returns the air to the original state. Note that the power stroke includes the heat addition process and the expansion process.

The thermal efficiency of the diesel cycle is expressed as

$$\eta = \frac{\dot{W}_{net}}{\dot{Q}_{in}} = 1 - \frac{\dot{Q}_{out}}{\dot{Q}_{in}} \tag{8.21}$$

For the constant-volume process and the constant-pressure process

$$\dot{Q}_{out} = \dot{m}C_v(T_4 - T_1) \qquad \dot{Q}_{in} = \dot{m}C_p(T_3 - T_2) \tag{8.22}$$

The efficiency is then

$$\eta = 1 - \frac{C_v(T_4 - T_1)}{C_p(T_3 - T_2)} = 1 - \frac{T_4 - T_1}{k(T_3 - T_2)} \tag{8.23}$$

This can be put in the form

$$\eta = 1 - \frac{T_1}{kT_2}\frac{T_4/T_1 - 1}{T_3/T_2 - 1} \tag{8.24}$$

This expression for the thermal efficiency is often written in terms of the compression ratio r and the *cutoff ratio* r_c which is defined as V_3/V_2; there results

$$\eta = 1 - \frac{1}{r^{k-1}}\frac{r_c^k - 1}{k(r_c - 1)} \tag{8.25}$$

From this expression we see that, for a given compression ratio r, the efficiency of the diesel cycle is less than that of an Otto cycle. For example, if $r = 10$ and $r_c = 2$, the Otto cycle efficiency is 60.2 percent and the diesel cycle efficiency is 53.4 percent. As r_c increases, the diesel cycle efficiency decreases. In practice, however, a compression ratio of 20 or so can be achieved in a diesel engine; using $r = 20$ and $r_c = 2$, we would find $\eta = 64.7$ percent. Thus, because of the higher compression ratios, a diesel engine typically operates at a higher efficiency than a gasoline engine.

The decrease in diesel cycle efficiency with an increase in r_c can also be observed by considering the T-s diagram shown in Fig. 8-10. If we increase r_c, the end of the heat input process moves to state $3'$. The increased work output is then represented by area $3-3'-4'-4-3$. The heat input increases considerably, as represented by area $3-3'-a-b-3$. The net effect is a decrease in cycle efficiency, caused

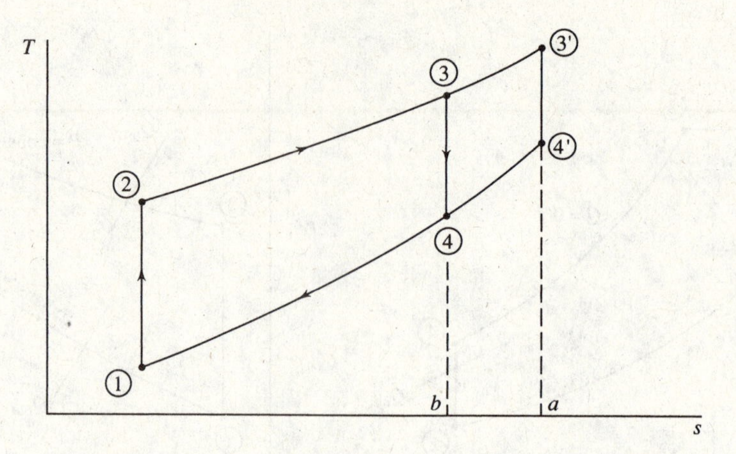

Fig. 8-10 The diesel cycle with increased cutoff ratio.

obviously by the convergence of the constant-pressure and constant-volume lines on the T-s diagram. For the Otto cycle note that two constant-volume lines diverge, thereby giving an increase in cycle efficiency with increasing T_3.

EXAMPLE 8.5 A diesel cycle, with a compression ratio of 18, operates on air with a low pressure of 200 kPa and a low temperature of 200 °C. If the work output is 1000 kJ/kg, determine the thermal efficiency and the MEP. Also, compare with the efficiency of an Otto cycle operating with the same maximum pressure.

Solution: The cutoff ratio r_c is found first. We have

$$v_1 = \frac{RT_1}{P_1} = \frac{(0.287)(473)}{200} = 0.6788\ \text{m}^3/\text{kg} \qquad \text{and} \qquad v_2 = v_1/18 = 0.03771\ \text{m}^3/\text{kg}$$

Since process 1 → 2 is isentropic, we find

$$T_2 = T_1\left(\frac{v_1}{v_2}\right)^{k-1} = (473)(18)^{0.4} = 1503\ \text{K} \qquad \text{and} \qquad P_2 = P_1\left(\frac{v_1}{v_2}\right)^k = (200)(18)^{1.4} = 11.44\ \text{MPa}$$

The work for the cycle is given by

$$w_{\text{net}} = q_{\text{net}} = q_{2-3} + q_{4-1} = C_p(T_3 - T_2) + C_v(T_1 - T_4)$$
$$1000 = (1.00)(T_3 - 1503) + (0.717)(473 - T_4)$$

For the isentropic process 3 → 4 and the constant-pressure process 2 → 3, we have

$$T_4 = T_3\left(\frac{v_3}{v_4}\right)^{k-1} = T_3\left(\frac{v_3}{0.6788}\right)^{0.4} \qquad \frac{T_3}{v_3} = \frac{T_2}{v_2} = \frac{1503}{0.03771} = 39\,860$$

The last three equations can be combined to yield

$$1000 = (1.00)(39\,860 v_3 - 1503) + (0.717)(473 - 46\,540 v_3^{1.4})$$

This equation is solved by trial and error to give

$$v_3 = 0.0773\ \text{m}^3/\text{kg} \qquad \therefore T_3 = 3080\ \text{K} \qquad T_4 = 1290\ \text{K}$$

This gives the cutoff ratio as $r_c = v_3/v_2 = 2.05$. The thermal efficiency is now calculated as

$$\eta = 1 - \frac{1}{r^{k-1}}\frac{r_c^k - 1}{k(r_c - 1)} = 1 - \frac{1}{(18)^{0.4}}\frac{(2.05)^{1.4} - 1}{(1.4)(2.05 - 1)} = 0.629 \quad \text{or } 62.9\%$$

Also, MEP $= w_{\text{net}}/(v_1 - v_2) = 1000/(0.6788 - 0.0377) = 641$ kPa.

For the comparison Otto cycle,

$$r_{\text{otto}} = v_1/v_3 = \frac{0.6788}{0.0773} = 8.78 \qquad \eta_{\text{otto}} = 1 - \frac{1}{r^{k-1}} = 0.581 \quad \text{or } 58.1\%$$

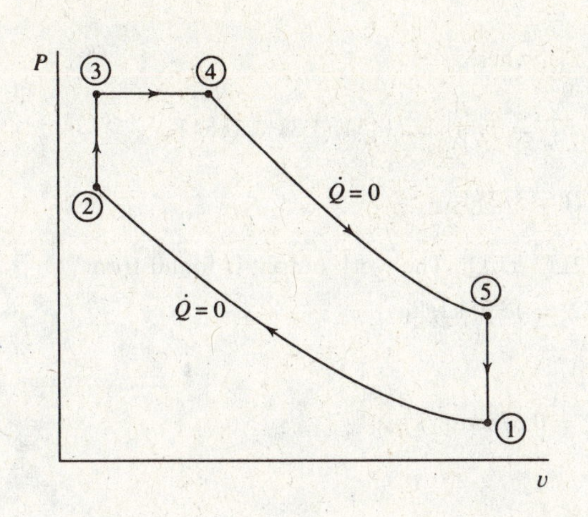

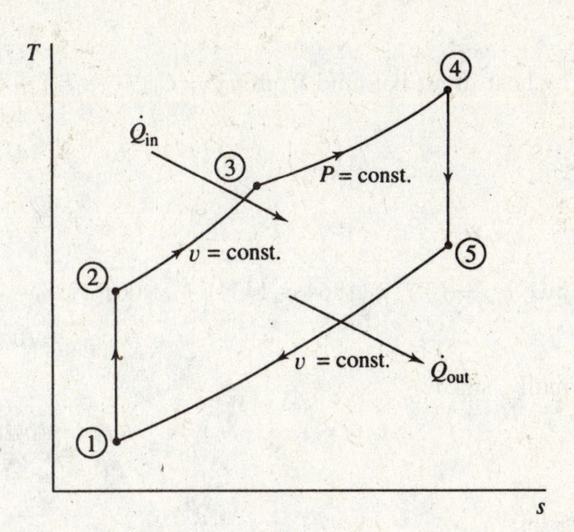

Fig. 8-11 The dual cycle.

8.7 THE DUAL CYCLE

An ideal cycle that better approximates the actual performance of a compression-ignition engine is the *dual cycle*, in which the combustion process is modeled by two heat-addition processes: a constant-volume process and a constant-pressure process, as shown in Fig. 8-11. The thermal efficiency is found from

$$\eta = 1 - \frac{\dot{Q}_{\text{out}}}{\dot{Q}_{\text{in}}} \tag{8.26}$$

where

$$\dot{Q}_{\text{out}} = \dot{m}C_v(T_5 - T_1) \qquad \dot{Q}_{in} = \dot{m}C_v(T_3 - T_2) + \dot{m}C_p(T_4 - T_3) \tag{8.27}$$

Hence, we have

$$\eta = 1 - \frac{T_5 - T_1}{T_3 - T_2 + k(T_4 - T_3)} \tag{8.28}$$

If we define the *pressure ratio* $r_p = P_3/P_2$, the thermal efficiency can be expressed as

$$\eta = 1 - \frac{1}{r^{k-1}}\frac{r_p r_c^k - 1}{kr_p(r_c - 1) + r_p - 1} \tag{8.29}$$

If we let $r_p = 1$, the diesel cycle efficiency results; if we let $r_c = r_p = 1$, the Otto cycle efficiency results. If $r_p > 1$, the thermal efficiency will be less than the Otto cycle efficiency but greater than the diesel cycle efficiency.

EXAMPLE 8.6 A dual cycle, which operates on air with a compression ratio of 16, has a low pressure of 200 kPa and a low temperature of 200 °C. If the cutoff ratio is 2 and the pressure ratio is 1.3, calculate the thermal efficiency, the heat input, the work output, and the MEP.

Solution: By (*8.29*),

$$\eta = 1 - \frac{1}{(16)^{0.4}}\frac{(1.3)(2)^{1.4} - 1}{(1.4)(1.3)(2-1) + 1.3 - 1} = 0.622 \quad \text{or } 62.2\%$$

The heat input is found from $q_{in} = C_v(T_3 - T_2) + C_p(T_4 - T_3)$, where

$$T_2 = T_1\left(\frac{v_1}{v_2}\right)^{k-1} = (473)(16)^{0.4} = 1434\,\text{K} \qquad T_3 = T_2\frac{P_3}{P_2} = (1434)(1.3) = 1864\,\text{K}$$

$$T_4 = T_3\frac{v_4}{v_3} = (1864)(2) = 3728\,\text{K}$$

Thus, $q_{in} = (0.717)(1864 - 1434) + (1.00)(3728 - 1864) = 2172$ kJ/kg. The work output is found from

$$w_{out} = \eta q_{in} = (0.622)(2172) = 1350\ \text{kJ/kg}$$

Finally, since

$$v_1 = \frac{RT_1}{P_1} = \frac{(0.287)(473)}{200} = 0.6788\ \text{m}^3/\text{kg}$$

we have

$$\text{MEP} = \frac{w_{out}}{v_1(1 - v_2/v_1)} = \frac{1350}{(0.6788)(15/16)} = 2120\ \text{kPa}$$

8.8 THE STIRLING AND ERICSSON CYCLES

The Stirling and Ericsson cycles, although not extensively used to model actual engines, are presented to illustrate the effective use of a *regenerator*, a heat exchanger which utilizes waste heat. A schematic diagram is shown in Fig. 8-12. Note that for both the constant-volume processes of the Stirling cycle (Fig. 8-13) and the constant-pressure processes of the Ericsson cycle (Fig. 8-14) the heat transfer q_{2-3} required by the gas is equal in magnitude to the heat transfer q_{4-1} discharged by the gas.

This suggests the use of a regenerator that will, internally to the cycle, transfer the otherwise wasted heat from the air during the process $4 \rightarrow 1$ to the air during the process $2 \rightarrow 3$. The net result of this is that the thermal efficiency of each of the two ideal cycles shown equals that of a Carnot cycle operating between the same two temperatures. This is obvious because the heat transfer in and out of each cycle occurs at constant temperature. Thus, the thermal efficiency is

$$\eta = 1 - \frac{T_L}{T_H} \tag{8.30}$$

Note that the heat transfer (the purchased energy) needed for the turbine can be supplied from outside an actual engine, that is, external combustion. Such external combustion engines have lower emissions but have not proved to be competitive with the Otto and diesel cycle engines because of problems inherent in the regenerator design and the isothermal compressor and turbine.

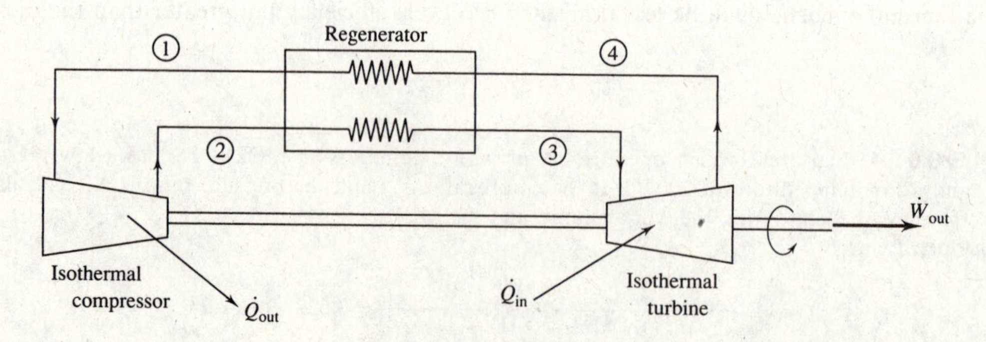

Fig. 8-12 The components of the Stirling and Ericsson cycles.

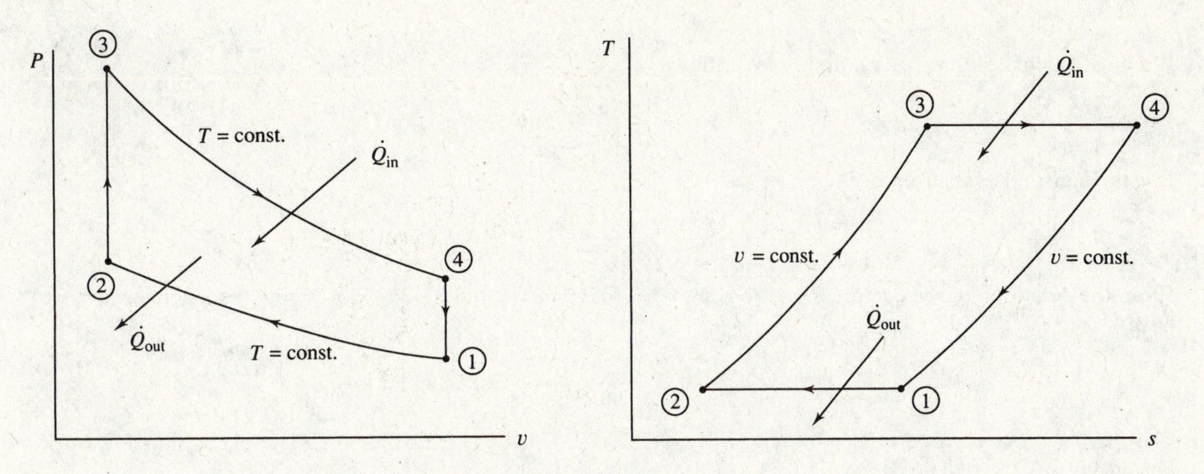

Fig. 8-13 The Stirling cycle.

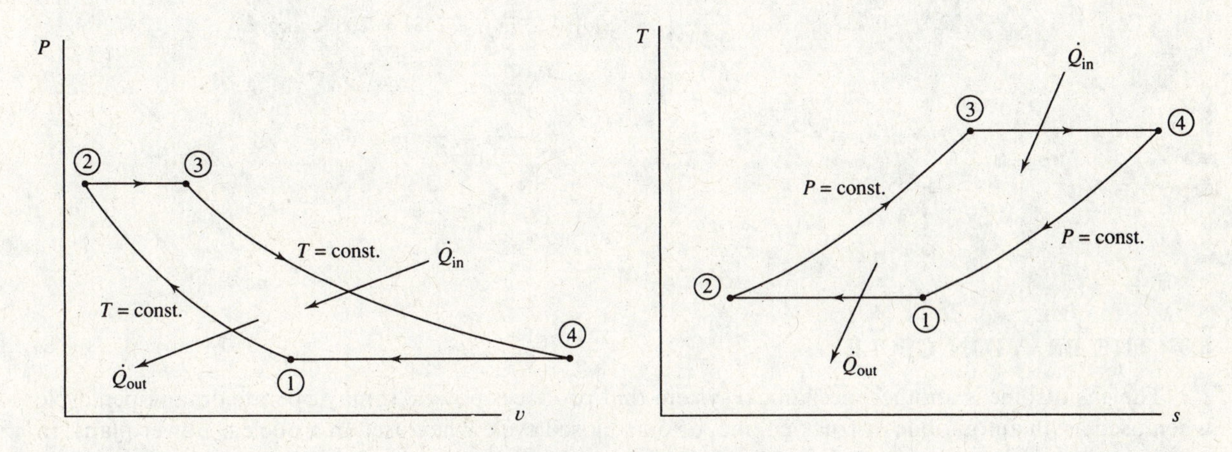

Fig. 8-14 The Ericsson cycle.

EXAMPLE 8.7 A Stirling cycle operates on air with a compression ratio of 10. If the low pressure is 30 psia, the low temperature is 200 °F, and the high temperature is 1000 °F, calculate the work output and the heat input.

Solution: For the Stirling cycle the work output is

$$w_{out} = w_{3-4} + w_{1-2} = RT_3 \ln \frac{v_4}{v_3} + RT_1 \ln \frac{v_2}{v_1} = (53.3)(1460 \ln 10 + 660 \ln 0.1) = 98{,}180 \text{ ft-lbf/lbm}$$

where we have used (*4.36*) for the isothermal process. Consequently,

$$\eta = 1 - \frac{T_L}{T_H} = 1 - \frac{660}{1460} = 0.548 \qquad q_{in} = \frac{w_{out}}{\eta} = \frac{98{,}180/778}{0.548} = 230 \text{ Btu/lbm}$$

EXAMPLE 8.8 An Ericsson cycle operates on air with a compression ratio of 10. For a low pressure of 200 kPa, a low temperature of 100 °C, and a high temperature of 600 °C, calculate the work output and the heat input.

Solution: For the Ericsson cycle the work output is

$$w_{out} = w_{1-2} + w_{2-3} + w_{3-4} + w_{4-1} = RT_1 \ln \frac{v_2}{v_1} + P_2(v_3 - v_2) + RT_3 \ln \frac{v_4}{v_3} + P_1(v_1 - v_4)$$

We must calculate P_2, v_1, v_2, v_3, and v_4. We know

$$v_1 = \frac{RT_1}{P_1} = \frac{(0.287)(373)}{200} = 0.5353\,\text{m}^3/\text{kg}$$

For the constant-pressure process $4 \to 1$,

$$\frac{T_4}{v_4} = \frac{T_1}{v_1} \qquad \frac{873}{v_4} = \frac{373}{0.5353} \qquad v_4 = 1.253\,\text{m}^3/\text{kg}$$

From the definition of the compression ratio, $v_4/v_2 = 10$, giving $v_2 = 0.1253\,\text{m}^3/\text{kg}$. Using the ideal-gas law, we have

$$P_3 = P_2 = \frac{RT_2}{v_2} = \frac{(0.287)(373)}{0.1253} = 854.4\,\text{kPa}$$

The final necessary property is $v_3 = RT_3/P_3 = (0.287)(873)/854.4 = 0.2932\,\text{m}^3/\text{kg}$. The expression for work output gives

$$w_{\text{out}} = (0.287)(373)\ln\frac{0.1253}{0.5353} + (854.4)(0.2932 - 0.1253)$$
$$+0.287 \times 873 \ln\frac{1.253}{0.2932} + (200)(0.5353 - 1.253) = 208\,\text{kJ/kg}$$

Finally,

$$\eta = 1 - \frac{T_L}{T_H} = 1 - \frac{378}{873} = 0.573 \qquad q_{\text{in}} = \frac{w_{\text{out}}}{\eta} = \frac{208}{0.573} = 364\,\text{kJ/kg}$$

8.9 THE BRAYTON CYCLE

The gas turbine is another mechanical system that produces power. It may operate on an open cycle when used as an automobile or truck engine, or on a closed cycle when used in a nuclear power plant. In open cycle operation, air enters the compressor, passes through a constant-pressure combustion chamber, passes through a turbine, and then exits as products of combustion to the atmosphere, as shown in Fig. 8-15*a*. In closed cycle operation the combustion chamber is replaced with a heat exchanger in which energy enters the cycle from some exterior source; an additional heat exchanger transfers heat from the cycle so that the air is returned to its initial state, as shown in Fig. 8-15*b*.

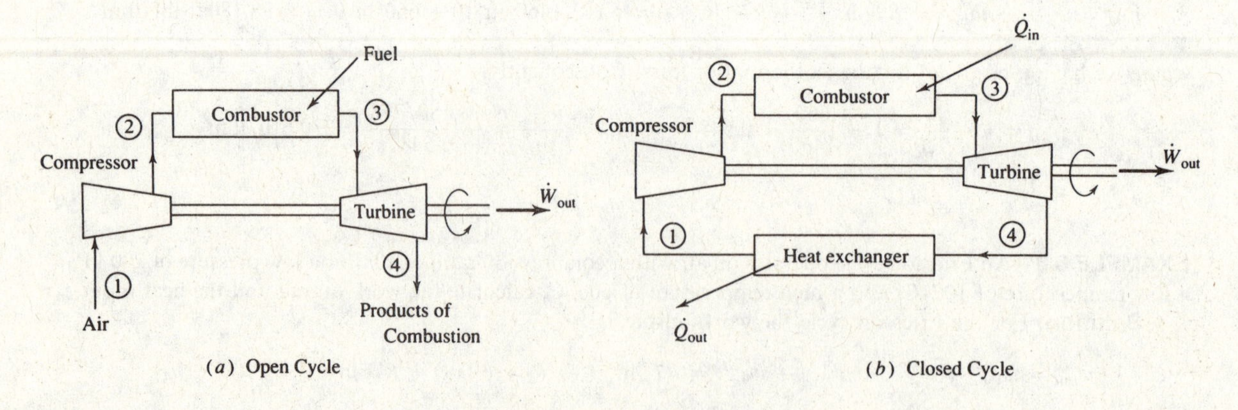

Fig. 8-15 The Brayton cycle components.

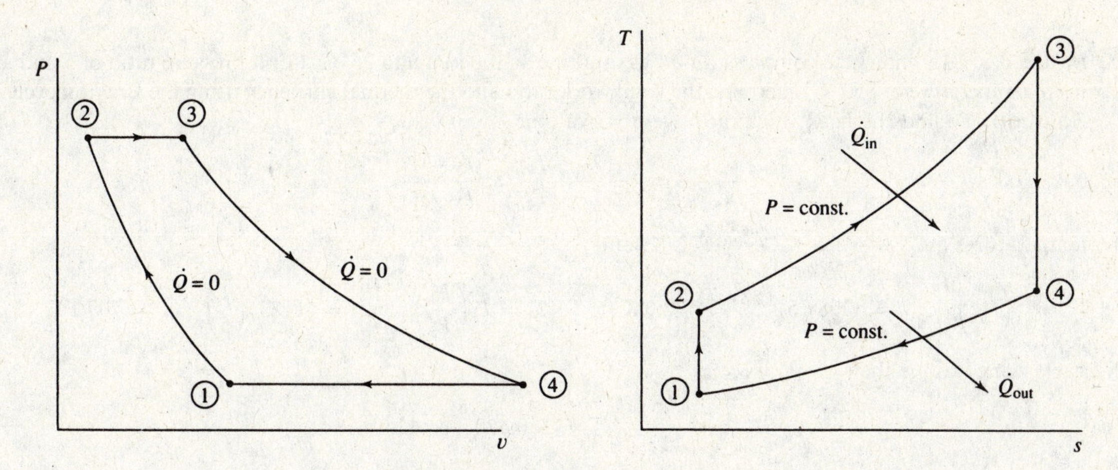

Fig. 8-16 The Brayton cycle.

The ideal cycle used to model the gas turbine is the Brayton cycle. It utilizes isentropic compression and expansion, as indicated in Fig. 8-16. The efficiency of such a cycle is given by

$$\eta = 1 - \frac{\dot{Q}_{out}}{\dot{Q}_{in}} = 1 - \frac{C_p(T_4 - T_1)}{C_p(T_3 - T_2)} = 1 - \frac{T_1}{T_2}\frac{T_4/T_1 - 1}{T_3/T_2 - 1} \tag{8.31}$$

Using the isentropic relations

$$\frac{P_2}{P_1} = \left(\frac{T_2}{T_1}\right)^{k/(k-1)} \qquad \frac{P_3}{P_4} = \left(\frac{T_3}{T_4}\right)^{k/(k-1)} \tag{8.32}$$

and observing that $P_2 = P_3$ and $P_1 = P_4$, we see that

$$\frac{T_2}{T_1} = \frac{T_3}{T_4} \qquad \text{or} \qquad \frac{T_4}{T_1} = \frac{T_3}{T_2} \tag{8.33}$$

Hence, the thermal efficiency can be written as

$$\eta = 1 - \frac{T_1}{T_2} = 1 - \left(\frac{P_1}{P_2}\right)^{(k-1)/k} \tag{8.34}$$

In terms of the pressure ratio $r_p = P_2/P_1$ the thermal efficiency is

$$\eta = 1 - r_p^{(1-k)/k} \tag{8.35}$$

Of course, this expression for thermal efficiency was obtained using constant specific heats. For more accurate calculations the gas tables should be used.

In an actual gas turbine the compressor and the turbine are not isentropic; some losses do occur. These losses, usually in the neighborhood of 85 percent, significantly reduce the efficiency of the gas turbine engine.

Another important feature of the gas turbine that seriously limits thermal efficiency is the high work requirement of the compressor, measured by the *back work ratio* $\dot{W}_{comp}/\dot{W}_{turb}$. The compressor may require up to 80 percent of the turbine's output (a back work ratio of 0.8), leaving only 20 percent for net work output. This relatively high limit is experienced when the efficiencies of the compressor and turbine are too low. Solved problems illustrate this point.

EXAMPLE 8.9 Air enters the compressor of a gas turbine at 100 kPa and 25 °C. For a pressure ratio of 5 and a maximum temperature of 850 °C determine the back work ratio and the thermal efficiency using the Brayton cycle.

Solution: To find the back work ratio we observe that

$$\frac{w_{\text{comp}}}{w_{\text{turb}}} = \frac{C_p(T_2 - T_1)}{C_p(T_3 - T_4)} = \frac{T_2 - T_1}{T_3 - T_4}$$

The temperatures are $T_1 = 298$ K, $T_3 = 1123$ K, and

$$T_2 = T_1\left(\frac{P_2}{P_1}\right)^{(k-1)/k} = (298)(5)^{0.2857} = 472.0\ \text{K} \qquad T_4 = T_3\left(\frac{P_4}{P_5}\right)^{(k-1)/k} = (1123)\left(\frac{1}{5}\right)^{0.2857} = 709.1\ \text{K}$$

The back work ratio is then

$$\frac{w_{\text{comp}}}{w_{\text{turb}}} = \frac{472.0 - 298}{1123 - 709} = 0.420 \quad \text{or } 42.0\%$$

The thermal efficiency is $\eta = 1 - r^{(1-k)/k} = 1 - (5)^{-0.2857} = 0.369$ (36.9%).

EXAMPLE 8.10 Assume the compressor and the gas turbine in Example 8.9 both have an efficiency of 80 percent. Using the Brayton cycle determine the back work ratio and the thermal efficiency.

Solution: We can calculate the quantities asked for if we determine W_{comp}, w_{turb}, and q_{in}. The compressor work is

$$w_{\text{comp}} = \frac{w_{\text{comp},s}}{\eta_{\text{comp}}} = \frac{C_p}{\eta_{\text{comp}}}(T_{2'} - T_1)$$

where $w_{\text{comp},s}$ is the isentropic work. $T_{2'}$ is the temperature of state 2′ assuming an isentropic process; state 2 is the actual state. We then have, using $T_{2'} = T_2$ from Example 8.9,

$$w_{\text{comp}} = \left(\frac{1.00}{0.8}\right)(472 - 298) = 217.5\ \text{kJ/kg}$$

Likewise, there results $w_{\text{turb}} = \eta_{\text{turb}} w_{\text{turb},s} = \eta_{\text{turb}}\, C_p(T_3 - T_4) = (0.8)(1.00)(1123 - 709.1) = 331.1$ kJ/kg, where $T_{4'} = T_4$ is calculated in Example 8.9. State 4′ is the isentropic state and state 4 is the actual state. The back work ratio is then

$$\frac{w_{\text{comp}}}{w_{\text{turb}}} = \frac{217.5}{331.1} = 0.657 \quad \text{or } 65.7\%$$

The heat transfer input necessary in this cycle is $q_{\text{in}} = h_3 - h_2 = C_p(T_3 - T_2)$, where T_2 is the actual temperature of the air leaving the compressor. It is found by returning to the compressor:

$$w_{\text{comp}} = C_p(T_2 - T_1) \qquad 217.5 = (1.00)(T_2 - 298) \qquad \therefore T_2 = 515.5\ \text{K}$$

Thus, $q_{\text{in}} = (1.00)(1123 - 515.5) = 607.5$ kJ/kg. The thermal efficiency of the cycle can then be written as

$$\eta = \frac{w_{\text{net}}}{q_{\text{in}}} = \frac{w_{\text{turb}} - w_{\text{comp}}}{q_{\text{in}}} = \frac{331.1 - 217.5}{607.5} = 0.187 \quad \text{or } 18.7\%$$

8.10 THE REGENERATIVE GAS-TURBINE CYCLE

The heat transfer from the simple gas-turbine cycle of the previous section is simply lost to the surroundings—either directly, with the products of combustion, or from a heat exchanger. Some of this exit energy can be utilized since the temperature of the flow exiting the turbine is greater than the temperature of the flow entering the compressor. A counterflow heat exchanger, a regenerator, is used to transfer some of this energy to the air leaving the compressor, as shown in Fig. 8-17. For an ideal regenerator the exit temperature T_3 would equal the entering temperature T_5; and, similarly, T_2

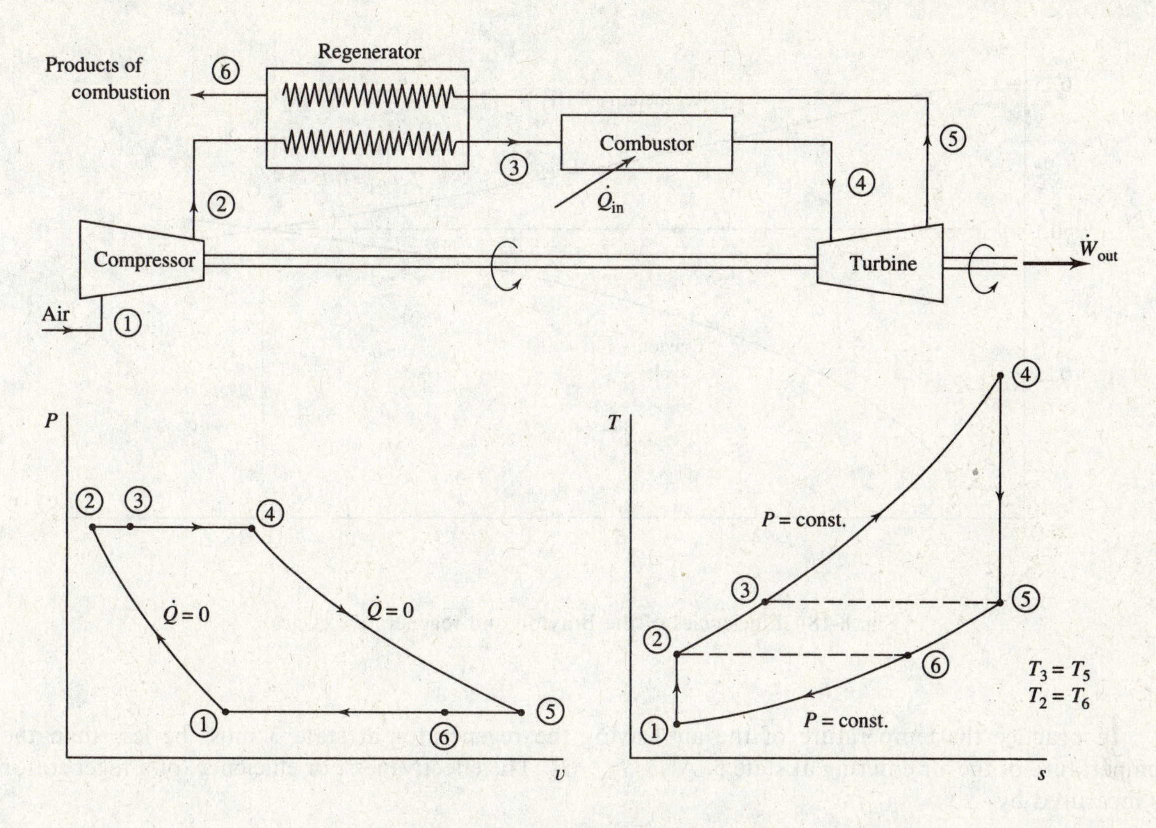

Fig. 8-17 The regenerative Brayton cycle.

would equal T_6. Since less energy is rejected from the cycle, the thermal efficiency is expected to increase. It is given by

$$\eta = \frac{w_{\text{turb}} - w_{\text{comp}}}{q_{\text{in}}} \tag{8.36}$$

Using the first law, expressions for q_{in} and w_{turb} are found to be

$$q_{\text{in}} = C_p(T_4 - T_3) \qquad w_{\text{turb}} = C_p(T_4 - T_5) \tag{8.37}$$

Hence, for the ideal regenerator in which $T_3 = T_5$, $q_{\text{in}} = w_{\text{turb}}$ and the thermal efficiency can be written as

$$\eta = 1 - \frac{w_{\text{comp}}}{w_{\text{turb}}} = 1 - \frac{T_2 - T_1}{T_4 - T_5} = 1 - \frac{T_1}{T_4}\frac{T_2/T_1 - 1}{1 - T_5/T_4} \tag{8.38}$$

Using the appropriate isentropic relation, this can be written in the form

$$\eta = 1 - \frac{T_1}{T_4}\frac{(P_2/P_1)^{(k-1)/k} - 1}{1 - (P_1/P_2)^{(k-1)/k}} = 1 - \frac{T_1}{T_4}r_p^{(k-1)/k} \tag{8.39}$$

Note that this expression for thermal efficiency is quite different from that for the Brayton cycle. For a given pressure ratio, the efficiency increases as the ratio of minimum to maximum temperature decreases. But, perhaps more surprisingly, as the pressure ratio increases the efficiency decreases, an effect opposite to that of the Brayton cycle. Hence it is not surprising that for a given regenerative cycle temperature ratio, there is a particular pressure ratio for which the efficiency of the Brayton cycle will equal the efficiency of the regenerative cycle. This is shown for a temperature ratio of 0.25 in Fig. 8-18.

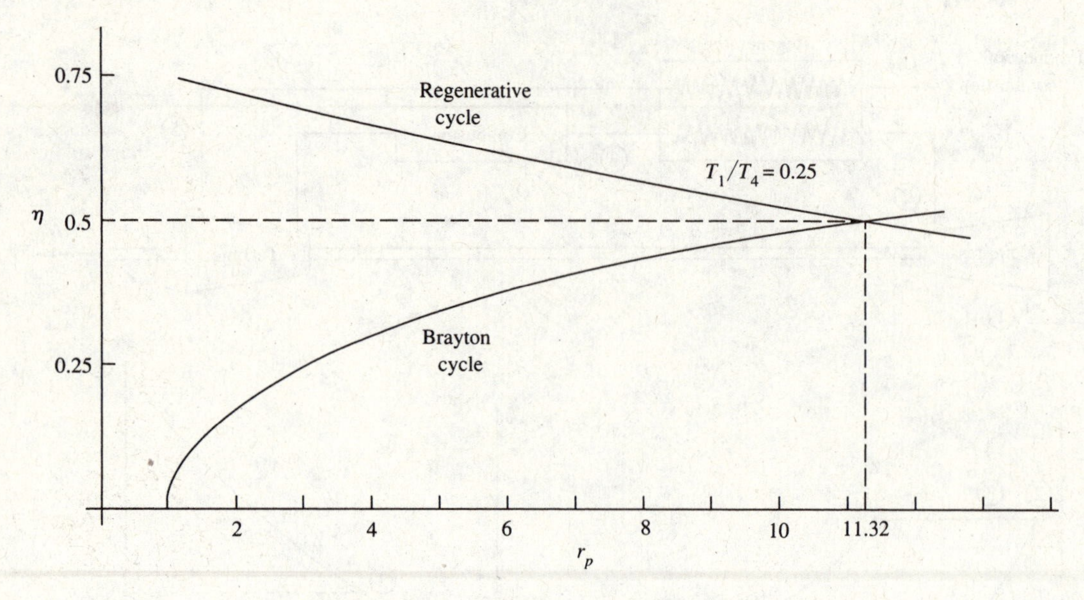

Fig. 8-18 Efficiencies of the Brayton and regenerative cycles.

In practice the temperature of the air leaving the regenerator at state 3 must be less than the temperature of the air entering at state 5. Also, $T_6 > T_2$. The effectiveness, or efficiency, of a regenerator is measured by

$$\eta_{\text{reg}} = \frac{h_3 - h_2}{h_5 - h_2} \tag{8.40}$$

This is equivalent to

$$\eta_{\text{reg}} = \frac{T_3 - T_2}{T_5 - T_2} \tag{8.41}$$

if we assume an ideal gas with constant specific heats. Obviously, for the ideal regenerator $T_3 = T_5$ and $\eta_{\text{reg}} - 1$. Regenerator efficiencies exceeding 80 percent are common.

EXAMPLE 8.11 Add an ideal regenerator to the gas-turbine cycle of Example 8.9 and calculate the thermal efficiency and the back work ratio.

Solution: The thermal efficiency is found using (*8.39*):

$$\eta = 1 - \frac{T_1}{T_4}\left(\frac{P_2}{P_1}\right)^{(k-1)/k} = 1 - \left(\frac{298}{1123}\right)(5)^{0.2857} = 0.580 \quad \text{or } 58.0\%$$

This represents a 57 percent increase in efficiency, a rather large effect. Note that, for the information given, the back work ratio does not change; hence, $w_{\text{comp}}/w_{\text{turb}} = 0.420$.

8.11 THE INTERCOOLING, REHEAT, REGENERATIVE GAS-TURBINE CYCLE

In addition to the regenerator of the previous section there are two other common techniques for increasing the thermal efficiency of the gas-turbine cycle. First, an intercooler can be inserted into the compression process; air is compressed to an intermediate pressure, cooled in an intercooler, and then compressed to the final pressure. This reduces the work required for the compressor, as was discussed in Sec. 8.2, and it reduces the maximum temperature reached in the cycle. The intermediate pressure is

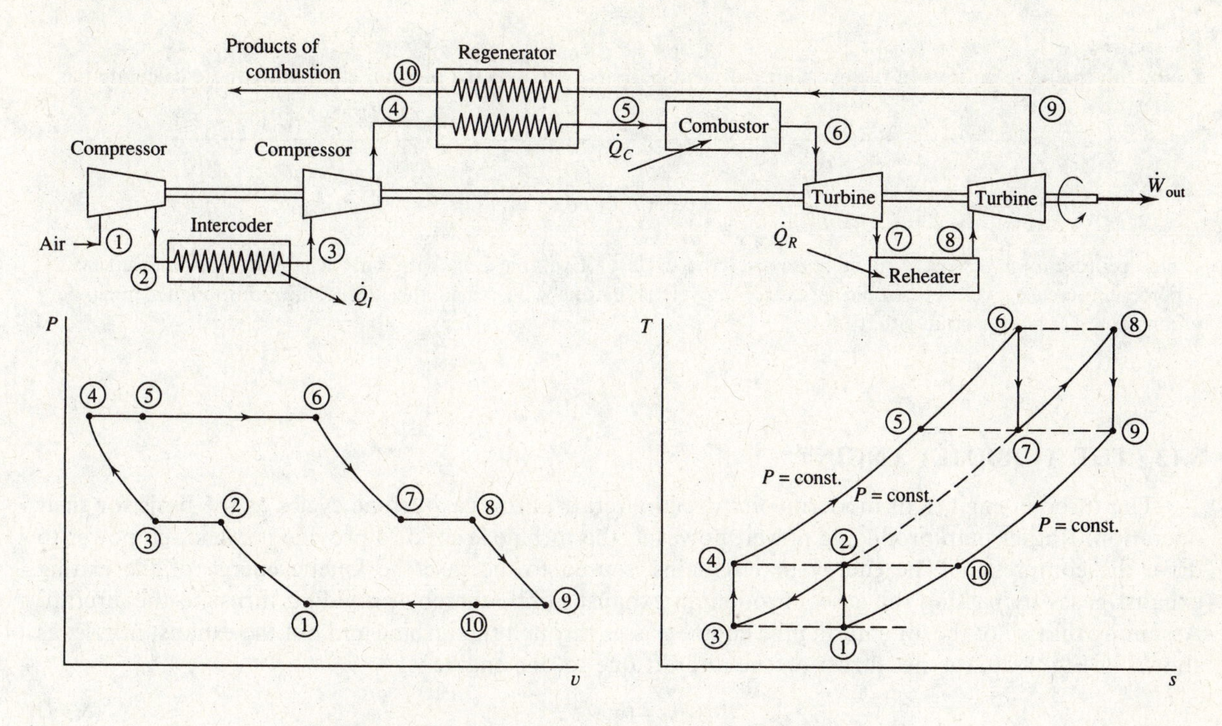

Fig. 8-19 The Brayton cycle with regeneration and reheat.

determined by equating the pressure ratio for each stage of compression; that is, referring to Fig. 8.19 [see (*8.8*)],

$$\frac{P_2}{P_1} = \frac{P_4}{P_3} \tag{8.42}$$

The second technique for increasing thermal efficiency is to use a second combustor, called a *reheater*. The intermediate pressure is determined as in the compressor; we again require that the ratios be equal; that is,

$$\frac{P_6}{P_7} = \frac{P_8}{P_9} \tag{8.43}$$

Since $P_9 = P_1$ and $P_6 = P_4$, we see that the intermediate turbine pressure is equal to the intermediate compressor pressure for our ideal-gas turbine.

Finally, we should note that intercooling and reheating are never used without regeneration. In fact, if regeneration is not employed, intercooling and reheating reduce the efficiency of a gas-turbine cycle.

EXAMPLE 8.12 Add an ideal intercooler, reheater, and regenerator to the gas-turbine cycle of Example 8.9 and calculate the thermal efficiency. Keep all given quantities the same.

Solution: The intermediate pressure is found to be $P_2 = \sqrt{P_1 P_4} = \sqrt{(100)(500)} = 223.6$ kPa. Hence, for the ideal isentropic process,

$$T_2 = T_1 \left(\frac{P_2}{P_1}\right)^{(k-1)/k} = (298)\left(\frac{223.6}{100}\right)^{0.2857} = 375.0\,\text{K}$$

The maximum temperature $T_6 = T_8 = 1123$ K. Using $P_7 = P_2$ and $P_6 = P_4$, we have

$$T_7 = T_6 \left(\frac{P_7}{P_6}\right)^{(k-1)/k} = (1123)\left(\frac{223.6}{500}\right)^{0.2857} = 892.3\,\text{K}$$

Now all the temperatures in the cycle are known (see Fig. 8-19) and the thermal efficiency can be calculated as

$$\eta = \frac{w_{\text{out}}}{q_{\text{in}}} = \frac{w_{\text{turb}} - w_{\text{comp}}}{q_C + q_R} = \frac{C_p(T_6 - T_7) + C_p(T_8 - T_9) - C_p(T_2 - T_1) - C_p(T_4 - T_3)}{C_p(T_6 - T_5) + C_p(T_8 - T_7)}$$

$$= \frac{230.7 + 230.7 - 77.0 - 77.0}{230.7 + 230.7} = 0.666 \quad \text{or } 66.6\%$$

This represents a 14.9 percent increase over the cycle of Example 8.11 with only a regenerator, and an 80.5 percent increase over the simple gas-turbine cycle. Obviously, losses in the additional components must be considered for any actual situation.

8.12 THE TURBOJET ENGINE

The turbojet engines of modern commercial aircraft utilize gas-turbine cycles as the basis for their operation. Rather than producing power, however, the turbine is sized to provide just enough power to drive the compressor. The energy that remains is used to increase the kinetic energy of the exiting exhaust gases by passing the gases through an exhaust nozzle thereby providing thrust to the aircraft. Assuming that all of the air entering the engine passes through the turbine and out the exhaust nozzle, as shown in Fig. 8-20, the net thrust on the aircraft due to one engine is

$$\text{thrust} = \dot{m}(V_5 - V_1) \tag{8.44}$$

where $\dot{m}$ is the mass flux of air passing through the engine. The mass flux of fuel is assumed to be negligibly small. In our ideal engine we assume that the pressures at section 1 and section 5 are equal to atmospheric pressure and that the velocity at section 1 is equal to the aircraft speed. A solved problem will illustrate the calculations for this application.

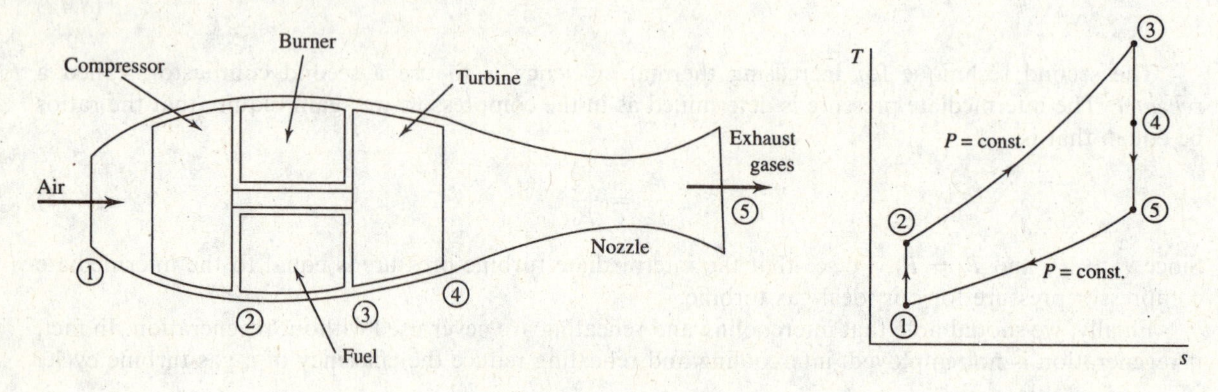

Fig. 8-20 The turbojet engine.

EXAMPLE 8.13 A turbojet engine inlets 100 lbm/sec of air at 5 psia and − 50 °F with a velocity of 600 ft/sec. The compressor discharge pressure is 50 psia and the turbine inlet temperature is 2000 °F. Calculate the thrust and the horsepower developed by the engine.

Solution: To calculate the thrust we must first calculate the exit velocity. To do this we must know the temperatures T_4 and T_5 exiting the turbine and the nozzle, respectively. Then the energy equation can be applied across the nozzle as

$$\frac{\overset{\text{neglect}}{V_4^2}}{2} + h_4 = \frac{V_5^2}{2} + h_5 \qquad \text{or} \qquad V_5^2 = 2C_p(T_4 - T_5)$$

Let us find the temperatures T_4 and T_5. The temperature T_2 is found to be (using $T_1 = 410\,^\circ\text{R}$)

$$T_2 = T_1\left(\frac{P_2}{P_1}\right)^{(k-1)/k} = (410)\left(\frac{50}{5}\right)^{0.2857} = 791.6\,^\circ\text{R}$$

Since the work from the turbine equals the work required by the compressor, we have $h_2 - h_1 = h_3 - h_4$ or $T_3 - T_4 = T_2 - T_1$. Thus, $T_4 = 2460 - (791.6 - 410) = 2078\,^\circ\text{R}$. The isentropic expansion through the turbine yields

$$P_4 = P_3\left(\frac{T_4}{T_3}\right)^{k/(k-1)} = (50)\left(\frac{2078}{2460}\right)^{3.5} = 27.70 \text{ psia}$$

The temperature T_5 at the nozzle exit where $P_5 = 5$ psia is found, assuming isentropic nozzle expansion, to be

$$T_5 = T_4\left(\frac{P_5}{P_4}\right)^{(k-1)/k} = (2078)\left(\frac{5}{27.7}\right)^{0.2857} = 1274\,^\circ\text{R}$$

The energy equation then gives

$$V_5 = [2C_p(T_4 - T_5)]^{1/2} = [(2)(0.24)(778)(32.2)(2078 - 1274)]^{1/2} = 3109 \text{ ft/sec}$$

[*Note:* We use $C_p = (0.24 \text{ Btu/lbm-}^\circ\text{R}) \times (778 \text{ ft-lbf/Btu}) \times (32.2 \text{ lbm-ft/lbf-sec}^2)$. This provides the appropriate units for C_p.]

The thrust is: thrust $= \dot{m}(V_5 - V_1) = (100/32.2)(3109 - 600) = 7790$ lbf. The horsepower is

$$\text{hp} = \frac{(\text{thrust})(\text{velocity})}{550} = \frac{(7790)(600)}{550} = 8500 \text{ hp}$$

where we have used the conversion 550 ft-lbf/sec = 1 hp.

Solved Problems

8.1 An adiabatic compressor receives 20 m^3/min of air from the atmosphere at 20 °C and compresses it to 10 MPa. Calculate the minimum power requirement.

An isentropic compression requires the minimum power input for an adiabatic compressor. The outlet temperature for such a process is

$$T_2 = T_1\left(\frac{P_2}{P_1}\right)^{(k-1)/k} = (293)\left(\frac{10\,000}{100}\right)^{0.2857} = 1092 \text{ K}$$

To find the mass flux, we must know the density. It is $\rho = P/RT = 100/(0.287)(293) = 1.189 \text{ kg/m}^3$. The mass flux is then (the flow rate is given) $\dot{m} = \rho(AV) = (1.189)(20/60) = 0.3963$ kg/s. The minimum power requirement is now calculated to be

$$\dot{W}_{\text{comp}} = \dot{m}(h_2 - h_1) = \dot{m}C_p(T_2 - T_1) = (0.3963)(1.00)(1092 - 293) = 317 \text{ kW}$$

8.2 A compressor receives 4 kg/s of 20 °C air from the atmosphere and delivers it at a pressure of 18 MPa. If the compression process can be approximated by a polytropic process with $n = 1.3$, calculate the power requirement and the rate of heat transfer.

The power requirement is [see (*8.6*)]

$$\dot{W}_{\text{comp}} = \dot{m}\frac{nR}{n-1}T_1\left[\left(\frac{P_2}{P_1}\right)^{(n-1)/n} - 1\right] = (4)\frac{(1.3)(0.287)}{1.3-1}(293)\left[\left(\frac{18\,000}{100}\right)^{0.3/1.3} - 1\right] = 3374 \text{ kW}$$

The first law for the control volume [see (*4.66*)] surrounding the compressor provides us with

$$\dot{Q} = \dot{m}\Delta h + \dot{W}_{\text{comp}} = \dot{m}C_p(T_2 - T_1) + \dot{W}_{\text{comp}} = \dot{m}C_pT_1\left[\left(\frac{P_2}{P_1}\right)^{(n-1)/n} - 1\right] + \dot{W}_{\text{comp}}$$

$$= (4)(1.00)(293)\left[\left(\frac{18\,000}{100}\right)^{0.3/1.3} - 1\right] - 3374 = -661 \text{ kW}$$

In the above, we have used the compressor power as negative since it is a power input. The expression of (*8.6*) is the magnitude of the power with the minus sign suppressed, but when the first law is used we must be careful with the signs. The negative sign on the heat transfer means that heat is leaving the control volume.

8.3 An adiabatic compressor is supplied with 2 kg/s of atmospheric air at 15 °C and delivers it at 5 MPa. Calculate the efficiency and power input if the exiting temperature is 700 °C.

Assuming an isentropic process and an inlet temperature of 15 °C, the exit temperature would be

$$T_{2'} = T_1\left(\frac{P_2}{P_1}\right)^{(k-1)/k} = (288)\left(\frac{5000}{100}\right)^{0.2857} = 880.6 \text{ K}$$

The efficiency is then

$$\eta = \frac{w_s}{w_a} = \frac{C_p(T_{2'} - T_1)}{C_p(T_2 - T_1)} = \frac{880.6 - 288}{973 - 288} = 0.865 \quad \text{or } 86.5\%$$

The power input is $\dot{W}_{\text{comp}} = \dot{m}C_p(T_2 - T_1) = (2)(1.00)(973 - 288) = 1370$ kW.

8.4 An ideal compressor is to compress 20 lbm/min of atmospheric air at 70 °F at 1500 psia. Calculate the power requirement for (*a*) one stage, (*b*) two stages, and (*c*) three stages.

(*a*) For a single stage, the exit temperature is

$$T_2 = T_1\left(\frac{P_2}{P_1}\right)^{(k-1)/k} = (530)\left(\frac{1500}{14.7}\right)^{0.2857} = 1987\,^\circ\text{R}$$

The required power is

$$\dot{W}_{\text{comp}} = \dot{m}C_p(T_2 - T_1) = \left(\frac{20}{60}\right)[(0.24)(778)](1987 - 530)$$

$$= 90{,}680 \text{ ft-lbf/sec} \quad \text{or } 164.9 \text{ hp}$$

(*b*) With two stages, the intercooler pressure is $P_2 = (P_1P_4)^{1/2} = [(14.7)(1500)]^{1/2} = 148.5$ psia. The intercooler inlet and exit temperatures are (see Fig. 8-4)

$$T_2 = T_1\left(\frac{P_2}{P_1}\right)^{(k-1)/k} = 530\left(\frac{148.5}{14.7}\right)^{0.2857} = 1026\,^\circ\text{R}$$

$$T_4 = T_3\left(\frac{P_4}{P_3}\right)^{(k-1)/k} = 530\left(\frac{1500}{148.5}\right)^{0.2857} = 1026\,^\circ\text{R}$$

The power required for this two-stage compressor is

$$\dot{W}_{\text{comp}} = \dot{m}C_p(T_2 - T_1) + \dot{m}C_p(T_4 - T_3)$$

$$= \left(\frac{20}{60}\right)[(0.24)(778)](1026 - 530 + 1026 - 530) = 61{,}740 \text{ ft-lbf/sec}$$

or 112.3 hp. This represents a 31.9 percent reduction compared to the single-stage compressor.

(c) For three stages, we have, using (8.9) and (8.10),

$$P_2 = (P_1^2 P_6)^{1/3} = \left[(14.7)^2(1500)\right]^{1/3} = 68.69 \text{ psia}$$

$$P_4 = (P_1 P_6^2)^{1/3} = \left[(14.7)(1500)^2\right]^{1/3} = 321.0 \text{ psia}$$

The high temperature and power requirement are then

$$T_2 = T_4 = T_6 = T_1\left(\frac{P_2}{P_1}\right)^{(k-1)/k} = (530)\left(\frac{68.69}{14.7}\right)^{0.2857} = 823.3\,^\circ\text{R}$$

$$\dot{W}_{\text{comp}} = 3\dot{m}C_p(T_2 - T_1) = (3)\left(\frac{20}{60}\right)[(0.24)(778)](823.3 - 530) = 54{,}770 \text{ ft-lbf/sec}$$

or 99.6 hp. This represents a 39.6 percent reduction compared to the single-stage compressor.

8.5 **The calculations in Prob. 8.4 were made assuming constant specific heats. Recalculate the power requirements for (a) and (b) using the more accurate air tables (Appendix E).**

(a) For one stage, the exit temperature is found using P_r. At stage $T_1 = 530\,^\circ\text{R}$: $h_1 = 126.7$ Btu/lbm, $(P_r)_1 = 1.300$. Then,

$$(P_r)_2 = (P_r)_1 \frac{P_2}{P_1} = (1.300)\left(\frac{1500}{14.7}\right) = 132.7$$

This provides us with $T_2 = 1870\,^\circ\text{R}$ and $h_2 = 469.0$ Btu/lbm. The power requirement is

$$\dot{W}_{\text{comp}} = \dot{m}(h_2 - h_1) = \left(\frac{20}{60}\right)(469 - 126.7)(778) = 88{,}760 \text{ ft-lbf/sec} \quad \text{or } 161.4 \text{ hp}$$

(b) With two stages, the intercooler pressure remains at 148.5 psia. The intercooler inlet condition is found as follows:

$$(P_r)_2 = (P_r)_1 \frac{P_2}{P_1} = (1.300)\left(\frac{148.5}{14.7}\right) = 13.13$$

where $T_2 = 1018\,^\circ\text{R}$ and $h_2 = 245.5$ Btu/lbm. These also represent the compressor exit (see Fig. 8-4), so that

$$\dot{W}_{\text{comp}} = \dot{m}(h_2 - h_1) + \dot{m}(h_4 - h_3)$$
$$= \left(\frac{20}{60}\right)(245.5 - 126.7 + 245.5 - 126.7)(778) = 61{,}620 \text{ ft-lbf/sec}$$

or 112.0 hp. Obviously, the assumption of constant specific heats is quite acceptable. The single-stage calculation represents an error of only 2 percent.

8.6 **A Carnot engine operates on air between high and low pressures of 3 MPa and 100 kPa with a low temperature of 20 °C. For a compression ratio of 15, calculate the thermal efficiency, the MEP, and the work output.**

The specific volume at TDC (see Fig. 6-1) is $v_1 = RT_1/P_1 = (0.287)(293)/100 = 0.8409\,\text{m}^3/\text{kg}$. For a compression ratio of 15 (we imagine the Carnot engine to have a piston-cylinder arrangement), the specific volume at BDC is

$$v_3 = \frac{v_1}{15} = \frac{0.8409}{15} = 0.05606\,\text{m}^3/\text{kg}$$

The high temperature is then $T_3 = P_3 v_3/R = (3000)(0.05606)/0.287 = 586.0$ K.

The cycle efficiency is calculated to be $\eta = 1 - T_L/T_H = 1 - 293/586 = 0.500$. To find the work output, we must calculate the specific volume of state 2 as follows:

$$P_2v_2 = P_1v_1 = (100)(0.8409) = 84.09 \qquad P_2v_2^{1.4} = P_3v_3^{1.4} = (3000)(0.05606)^{1.4} = 53.12$$

$$\therefore v_2 = 0.3171 \text{ m}^3/\text{kg}$$

The entropy change $(s_2 - s_1)$ is then

$$\Delta s = C_v \ln 1 + R \ln \frac{v_2}{v_1} = 0 + 0.287 \ln \frac{0.3171}{0.8409} = -0.2799 \text{ kJ/kg·K}$$

The work output is now found to be $w_{\text{net}} = \Delta T|\Delta s| = (586 - 293)(0.2799) = 82.0$ kJ/kg. Finally,

$$w_{\text{net}} = (\text{MEP})(v_1 - v_2) \qquad 82.0 = (\text{MEP})(0.8409 - 0.3171) \qquad \text{MEP} = 156.5 \text{ kPa}$$

8.7 An inventor proposes a reciprocating engine with a compression ratio of 10, operating on 1.6 kg/s of atmospheric air at 20 °C, that produces 50 hp. After combustion the temperature is 400 °C. Is the proposed engine feasible?

We will consider a Carnot engine operating between the same pressure and temperature limits; this will establish the ideal situation without reference to the details of the proposed engine. The specific volume at state 1 (see Fig. 6-1) is

$$v_1 = \frac{RT_1}{P_1} = \frac{(0.287)(293)}{100} = 0.8409 \text{ m}^3/\text{kg}$$

For a compression ratio of 10, the minimum specific volume must be $v_3 = v_1/10 = 0.8409/10 = 0.08409$. The specific volume at state 2 is now found by considering the isothermal process from 1 to 2 and the isentropic process from 2 to 3:

$$P_2v_2 = P_1v_1 = 100 \times 0.8409 = 84.09 \qquad P_2v_2^k = \frac{0.287(673)}{0.08409}(0.08409)^{1.4} = 71.75$$

$$\therefore v_2 = 0.6725 \text{ m}^3/\text{kg}$$

The change in entropy is

$$\Delta s = R \ln \frac{v_2}{v_1} = 0.287 \ln \frac{0.6725}{0.8409} = -0.0641 \text{ kJ/kg·K}$$

The work output is then $w_{\text{net}} = \Delta T|\Delta s| = (400 - 20)(0.0641) = 24.4$ kJ/kg. The power output is

$$\dot{W} = \dot{m}w_{\text{net}} = (1.6)(24.4) = 39.0 \text{ kW} \quad \text{or } 52.2 \text{ hp}$$

The maximum possible power output is 52.2 hp; the inventor's claims of 50 hp is highly unlikely, though not impossible.

8.8 A six-cylinder engine with a compression ratio of 8 and a total volume at TDC of 600 mL intakes atmospheric air at 20 °C. The maximum temperature during a cycle is 1500 °C. Assuming an Otto cycle, calculate (*a*) the heat supplied per cycle, (*b*) the thermal efficiency, and (*c*) the power output for 400 rpm.

(*a*) The compression ratio of 8 allows us to calculate T_2 (see Fig. 8-8):

$$T_2 = T_1\left(\frac{V_1}{V_2}\right)^{k-1} = (293)(8)^{0.4} = 673.1 \text{ K}$$

The heat supplied is then $q_{\text{in}} = C_v(T_3 - T_2) = (0.717)(1773 - 673.1) = 788.6$ kJ/kg. The mass of air in the six cylinders is

$$m = \frac{P_1V_1}{RT_1} = \frac{(100)(600 \times 10^{-6})}{(0.287)(293)} = 0.004281 \text{ kg}$$

The heat supplied per cycle is $Q_{\text{in}} = mq_{\text{in}} = (0.004281)(788.6) = 3.376$ kJ.

(*b*) $\eta = 1 - r^{1-k} = 1 - 8^{-0.4} = 0.5647$ or 56.5%.

(*c*) $W_{out} = \eta Q_{in} = (0.5647)(3.376) = 1.906$ kJ.

For the idealized Otto cycle, we assume that one cycle occurs each revolution. Consequently,

$$\dot{W}_{out} = (W_{out})(\text{cycles per second}) = (1.906)(4000/60) = 127 \text{ kW} \quad \text{or } 170 \text{ hp}$$

8.9 A diesel engine intakes atmospheric air at 60 °F and adds 800 Btu/lbm of energy. If the maximum pressure is 1200 psia calculate (*a*) the cutoff ratio, (*b*) the thermal efficiency, and (*c*) the power output for an airflow of 0.2 lbm/sec.

(*a*) The compression process is isentropic. The temperature at state 2 (see Fig. 8-9) is calculated to be

$$T_2 = T_1\left(\frac{P_2}{P_1}\right)^{(k-1)/k} = (520)\left(\frac{1200}{14.7}\right)^{0.2857} = 1829\,°\text{R}$$

The temperature at state 3 is found from the first law as follows:

$$q_{in} = C_p(T_3 - T_2) \qquad 800 = (0.24)(T_3 - 1829) \qquad \therefore T_3 = 5162\,°\text{R}$$

The specific volumes of the three states are

$$v_1 = \frac{RT_1}{P_1} = \frac{(53.3)(520)}{(14.7)(144)} = 13.09 \text{ ft}^3/\text{lbm} \qquad v_2 = \frac{RT_2}{P_2} = \frac{(53.3)(1829)}{(1200)(144)} = 0.5642 \text{ ft}^3/\text{lbm}$$

$$v_3 = \frac{RT_3}{P_3} = \frac{(53.3)(5162)}{(1200)(144)} = 1.592 \text{ ft}^3/\text{lbm}$$

The cutoff ratio is then $r_c = v_3/v_2 = 1.592/0.5642 = 2.822$.

(*b*) The compression ratio is $r = v_1/v_2 = 13.09/0.5642 = 23.20$. The thermal efficiency can now be calculated, using (*8.25*):

$$\eta = 1 - \frac{1}{r^{k-1}}\frac{r_c^k - 1}{k(r_c - 1)} = 1 - \frac{1}{(23.2)^{0.4}}\frac{(2.822)^{1.4} - 1}{(1.4)(2.822 - 1)} = 0.6351 \quad \text{or } 63.51\%$$

(*c*) $\dot{W}_{out} = \eta\dot{Q}_{in} = \eta\dot{m}q_{in} = [(0.6351)(0.2)(800)](778) = 79{,}060$ ft-lbf/sec or 143.7 hp.

8.10 A dual cycle is used to model a piston engine. The engine intakes atmospheric air at 20 °C, compresses it to 10 MPa, and then combustion increases the pressure to 20 MPa. For a cutoff ratio of 2, calculate the cycle efficiency and the power output for an airflow of 0.1kg/s.

The pressure ratio (refer to Fig. 8-11) is $r_p = P_3/P_2 = 20/10 = 2$. The temperature after the isentropic compression is

$$T_2 = T_1\left(\frac{P_2}{P_1}\right)^{(k-1)/k} = (293)\left(\frac{10\,000}{100}\right)^{0.2857} = 1092 \text{ K}$$

The specific volumes are

$$v_1 = \frac{RT_1}{P_1} = \frac{(0.287)(293)}{100} = 0.8409 \text{ m}^3/\text{kg} \qquad v_2 = \frac{RT_2}{P_2} = \frac{(0.287)(1092)}{10\,000} = 0.03134 \text{ m}^3/\text{kg}$$

The compression ratio is then $r = v_1/v_2 = 0.8409/0.03134 = 26.83$. This allows us to calculate the thermal efficiency:

$$\eta = 1 - \frac{1}{r^{k-1}}\frac{r_p r_c^{k-1} - 1}{kr_p(r_c - 1) + r_p - 1} = 1 - \frac{1}{(26.83)^{0.4}}\frac{(2)(2)^{0.4} - 1}{(1.4)(2)(2 - 1) + 2 - 1} = 0.8843$$

To find the heat input, the temperatures of states 3 and 4 must be known. For the constant-volume heat addition,

$$\frac{T_3}{P_3} = \frac{T_2}{P_2} \qquad \therefore T_3 = T_2\frac{P_3}{P_2} = (1092)(2) = 2184 \text{ K}$$

For the constant-pressure heat addition,

$$\frac{T_3}{v_3} = \frac{T_4}{v_4} \qquad \therefore T_4 = T_3 \frac{v_4}{v_3} = (2184)(2) = 4368 \text{ K}$$

The heat input is then

$$q_{in} = C_v(T_3 - T_2) + C_p(T_4 - T_3) = (0.717)(2184 - 1092) + (1.00)(4368 - 2184) = 2967 \text{ kJ/kg}$$

so that

$$w_{out} = \eta\, q_{in} = (0.8843)(2967) = 2624 \text{ kJ/kg}$$

The power output is $\dot{W}_{out} = \dot{m} w_{out} = (0.1)(2624) = 262.4$ kW.

8.11 Air at 90 kPa and 15 °C is supplied to an ideal cycle at intake. If the compression ratio is 10 and the heat supplied is 300 kJ/kg, calculate the efficiency and the maximum temperature for (*a*) a Stirling cycle, and (*b*) an Ericsson cycle.

(*a*) For the constant-temperature process, the heat transfer equals the work. Referring to Fig. 8-13, the first law gives

$$q_{out} = w_{1-2} = RT_1 \ln \frac{v_1}{v_2} = (0.287)(288) \ln 10 = 190.3 \text{ kJ/kg}$$

The work output for the cycle is then $w_{out} = q_{in} - q_{out} = 300 - 190.3 = 109.7$ kJ/kg. The efficiency is

$$\eta = \frac{w_{out}}{q_{in}} = \frac{109.7}{300} = 0.366$$

The high temperature is found from

$$\eta = 1 - \frac{T_L}{T_H} \qquad \therefore T_H = \frac{T_L}{1-\eta} = \frac{288}{1 - 0.366} = 454 \text{ K}$$

(*b*) For the Ericsson cycle of Fig. 8-14, the compression ratio is v_4/v_2. The constant-temperature heat addition $3 \rightarrow 4$ provides

$$q_{in} = w_{3-4} = RT_4 \ln \frac{v_4}{v_3} \qquad \therefore 300 = (0.287)T_4 \ln \frac{v_4}{v_3}$$

The constant-pressure process $2 \rightarrow 3$ allows

$$\frac{T_3}{v_3} = \frac{T_2}{v_2} = \frac{288}{v_4/10}$$

The constant-pressure process $4 \rightarrow 1$ demands

$$\frac{T_4}{v_4} = \frac{T_1}{v_1} = \frac{P_1}{R} = \frac{90}{0.287} = 313.6$$

Recognizing that $T_3 = T_4$, the above can be combined to give

$$300 = (0.287)(313.6 v_4) \ln \frac{v_4}{v_3} \qquad v_3 = 0.1089 v_4^2$$

The above two equations are solved simultaneously by trial and error to give

$$v_4 = 3.94 \text{ m}^3/\text{kg} \qquad v_3 = 1.69 \text{ m}^3/\text{kg}$$

Thus, from the compression ratio, $v_2 = v_4/10 = 0.394 \text{ m}^3/\text{kg}$. The specific volume of state 1 is

$$v_1 = \frac{RT}{P_1} = \frac{(0.287)(288)}{90} = 0.9184 \text{ m}^3/\text{kg}$$

The heat rejected is then

$$q_{out} = RT_1 \ln \frac{v_1}{v_2} = (0.287)(288) \ln \frac{0.9184}{0.394} = 70.0 \text{ kJ/kg}$$

The net work for the cycle is $w_{out} = q_{in} - q_{out} = 300 - 70.0 = 230$ kJ/kg. The efficiency is then $\eta = w_{out}/q_{in} = 230/300 = 0.767$. This allows us to calculate the high temperature:

$$\eta = 1 - \frac{T_L}{T_H} \qquad 0.767 = 1 - \frac{288}{T_H} \qquad \therefore T_H = 1240 \text{ K}$$

8.12 A gas-turbine power plant is to produce 800 kW of power by compressing atmospheric air at 20 °C to kPa. If the maximum temperature is 800 °C, calculate the minimum mass flux of the air.

The cycle is modeled as an ideal Brayton cycle. The cycle efficiency is given by (*8.35*):

$$\eta = 1 - r_p^{(1-k)/k} = 1 - \left(\frac{800}{100}\right)^{-0.4/1.4} = 0.4479$$

The energy added in the combustor is (see Fig. 8-15) $\dot{Q}_{in} = \dot{W}_{out}/\eta = 800/0.4479 = 1786$ kW. The temperature into the combustor is

$$T_2 = T_1\left(\frac{P_2}{P_1}\right)^{(k-1)/k} = (293)\left(\frac{800}{100}\right)^{0.2857} = 530.7 \text{ K}$$

With a combustor outlet temperature of 1073 K, the mass flux follows from a combustor energy balance:

$$\dot{Q}_{in} = \dot{m}C_p(T_3 - T_2) \qquad 1786 = (\dot{m})(1.00)(1073 - 530.7) \qquad \therefore \dot{m} = 3.293 \text{ kg/s}$$

This represents a minimum, since losses have not been included.

8.13 If the efficiency of the turbine of Prob. 8.12 is 85 percent and that of the compressor is 80 percent, calculate the mass flux of air needed, keeping the other quantities unchanged. Also calculate the cycle efficiency.

The compressor work, using $T_{2'} = 530.7$ from Prob. 8.12, is

$$w_{comp} = \frac{w_{comp,s}}{\eta_{comp}} = \frac{1}{\eta_{comp}}C_p(T_{2'} - T_1) = \left(\frac{1}{0.8}\right)(1.00)(530.7 - 293) = 297.1 \text{ kJ/kg}$$

The temperature of state 4′, assuming an isentropic process, is

$$T_{4'} = T_3\left(\frac{P_4}{P_3}\right)^{(k-1)/k} = (1073)\left(\frac{100}{800}\right)^{0.2857} = 592.4 \text{ K}$$

The turbine work is then

$$w_{turb} = \eta_{turb}w_{turb,s} = \eta_{turb}C_p(T_{4'} - T_3) = (0.85)(1.00)(592.4 - 1073) = 408.5 \text{ kJ/kg}$$

The work output is then $w_{out} = w_{turb} - w_{comp} = 408.5 - 297.1 = 111.4$ kJ/kg. This allows us to determine the mass flux:

$$\dot{W}_{out} = \dot{m}w_{out} \qquad 800 = (\dot{m})(111.4) \qquad \therefore \dot{m} = 7.18 \text{ kg/s}$$

To calculate the cycle efficiency, we find the actual temperature T_2. It follows from an energy balance on the actual compressor:

$$w_{comp} = C_p(T_2 - T_1) \qquad 297.1 = (1.00)(T_2 - 293) \qquad \therefore T_2 = 590.1 \text{ K}$$

The combustor rate of heat input is thus $\dot{Q}_{in} = \dot{m}(T_3 - T_2) = (7.18)(1073 - 590.1) = 3467$ kW. The efficiency follows as

$$\eta = \frac{\dot{W}_{out}}{\dot{Q}_{in}} = \frac{800}{3467} = 0.2307$$

Note the sensitivity of the mass flux and the cycle efficiency to the compressor and turbine efficiency.

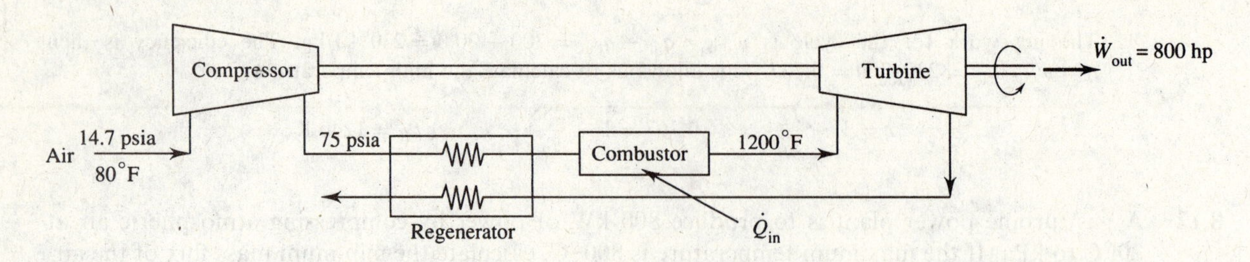

Fig. 8-21

8.14 Assuming the ideal-gas turbine and regenerator shown in Fig. 8.21, find $\dot{Q}_{in}$ and the back work ratio.

The cycle efficiency is (see Fig. 8-17)

$$\eta = 1 - \frac{T_1}{T_4} r_p^{(k-1)/k} = 1 - \left(\frac{540}{1660}\right)\left(\frac{75}{14.7}\right)^{0.2857} = 0.4818$$

The rate of energy input to the combustor is

$$\dot{Q}_{in} = \frac{\dot{W}_{out}}{\eta} = \frac{(800)(550/778)}{0.4818} = 1174 \text{ Btu/sec}$$

The compressor outlet temperature is

$$T_2 = T_1\left(\frac{P_2}{P_1}\right)^{(k-1)/k} = (540)\left(\frac{75}{14.7}\right)^{0.2857} = 860.2\,°\text{R}$$

The turbine outlet temperature is

$$T_4 = T_3\left(\frac{P_4}{P_3}\right)^{(k-1)/k} = (1660)\left(\frac{14.7}{75}\right)^{0.2857} = 1042\,°\text{R}$$

The turbine and compressor work are then

$$w_{comp} = C_p(T_2 - T_1) = (1.00)(860.2 - 540) = 320.2 \text{ Btu/lbm}$$
$$w_{turb} = C_p(T_3 - T_4) = (1.00)(1660 - 1042) = 618 \text{ Btu/lbm}$$

The back work ratio is then $w_{comp}/w_{turb} = 320.2/618 = 0.518$.

8.15 To Prob. 8.14 add an intercooler and a reheater. Calculate the ideal cycle efficiency and the back work ratio.

The intercooler pressure is (see Fig. 8-19), $P_2 = \sqrt{P_1P_4} = \sqrt{(14.7)(75)} = 33.2$ psia. The temperatures T_2 and T_4 are

$$T_4 = T_2 = T_1\left(\frac{P_2}{P_1}\right)^{(k-1)/k} = (540)\left(\frac{33.2}{14.7}\right)^{0.2857} = 681.5\,°\text{R}$$

Using $P_7 = P_2$ and $P_6 = P_4$, there results

$$T_9 = T_7 = T_6\left(\frac{P_7}{P_6}\right)^{(k-1)/k} = (1660)\left(\frac{33.2}{75}\right)^{0.2857} = 1315\,°\text{R}$$

The work output of the turbine and input to the compressor are

$$w_{turb} = C_p(T_8 - T_9) + C_p(T_6 - T_7) = (0.24)(778)(1660 - 1315)(2) = 128{,}800 \text{ ft-lbf/lbm}$$
$$w_{comp} = C_p(T_4 - T_3) + C_p(T_2 - T_1) = (0.24)(778)(681.5 - 540)(2) = 52{,}840 \text{ ft-lbf/lbm}$$

The heat inputs to the combustor and the reheater are

$$q_{comb} = C_p(T_6 - T_5) = (0.24)(1660 - 1315) = 82.8 \text{ Btu/lbm}$$
$$q_{reheater} = C_p(T_8 - T_7) = (0.24)(1660 - 1315) = 82.8 \text{ Btu/lbm}$$

The cycle efficiency is now calculated to be

$$\eta = \frac{w_{out}}{q_{in}} = \frac{w_{turb} - w_{comp}}{q_{comb} + q_{reheater}} = \frac{(128{,}800 - 52{,}840)/778}{82.8 + 82.8} = 0.590$$

The back work ratio is $w_{comp}/w_{turb} = 52{,}840/128{,}800 = 0.410$.

8.16 A turbojet aircraft flies at a speed of 300 m/s at an elevation of 10 000 m. If the compression ratio is 10, the turbine inlet temperature is 1000 °C, and the mass flux of air is 30 kg/s, calculate the maximum thrust possible from this engine. Also, calculate the rate of fuel consumption if the heating value of the fuel is 8400 kJ/kg.

The inlet temperature and pressure are found from Table B-1 to be (see Fig. 8-20)

$$T_1 = 223.3 \text{ K} \qquad P_1 = 0.2615 \qquad P_0 = 26.15 \text{ kPa}$$

The temperature exiting the compressor is

$$T_2 = T_1\left(\frac{P_2}{P_1}\right)^{(k-1)/k} = (223.3)(10)^{0.2857} = 431.1 \text{ K}$$

Since the turbine drives the compressor, the two works are equal so that

$$C_p(T_2 - T_1) = C_p(T_3 - T_4) \qquad \therefore T_3 - T_4 = T_2 - T_1$$

Since $T_3 = 1273$, we can find T_4 as $T_4 = T_3 + T_1 - T_2 = 1273 + 223.3 - 431.1 = 1065.2$ K. We can now calculate the pressure at the turbine exit to be, using $P_3 = P_2 = 261.5$ kPa,

$$P_4 = P_3\left(\frac{T_4}{T_3}\right)^{k/(k-1)} = (261.5)\left(\frac{1065.2}{1273}\right)^{3.5} = 140.1 \text{ kPa}$$

The temperature at the nozzle exit, assuming an isentropic expansion, is

$$T_5 = T_4\left(\frac{P_5}{P_4}\right)^{(k-1)/k} = (1065.2)\left(\frac{26.15}{140.1}\right)^{0.2857} = 659.4 \text{ K}$$

The energy equation provides us with the exit velocity $V_5 = [2C_p(T_4 - T_5)]^{1/2} = [(2)(1000)\ (1065.2 - 659.4)]^{1/2} = 901$ m/s, where $C_p = 1000$ J/kg·K must be used in the expression. The thrust can now be calculated as

$$\text{thrust} = \dot{m}(V_5 - V_1) = (30)(901 - 300) = 18\,030 \text{ N}$$

This represents a maximum since a cycle composed of ideal processes was used.

The heat transfer rate in the burner is $\dot{Q} = \dot{m}C_p(T_3 - T_2) = (30)(1.00)(1273 - 431.1) = 25.26$ MW. This requires that the mass flux of fuel $\dot{m}_f$ be

$$8400\,\dot{m}_f = 25\,260 \qquad \therefore \dot{m}_f = 3.01 \text{ kg/s}$$

Supplementary Problems

8.17 An ideal compressor receives 100 m^3/min of atmospheric air at 10 °C and delivers it at 20 MPa. Determine the mass flux and the power required.

8.18 An adiabatic compressor receives 1.5 kg/s of atmospheric air at 25 °C and delivers it at 4 MPa. Calculate the required power and the exiting temperature if the efficiency is assumed to be (*a*) 100 percent, and (*b*) 80 percent.

8.19 An adiabatic compressor receives atmospheric air at 60 °F at a flow rate of 4000 ft^3/min and delivers it at 10,000 psia. Calculate the power requirement assuming a compressor efficiency of (*a*) 100 percent and (b) 82 percent.

8.20 A compressor delivers 2 kg/s of air at 2 MPa having received it from the atmosphere at 20 °C. Determine the required power input and the rate of heat removed if the compression process is polytropic with (*a*) $n = 1.4$, (*b*) $n = 1.3$, (*c*) $n = 1.2$, and (*d*) $n = 1.0$.

8.21 The heat transfer from a compressor is one-fifth the work input. If the compressor receives atmospheric air at 20 °C and delivers it at 4 MPa, determine the polytropic exponent assuming an ideal compressor.

8.22 The maximum temperature in the compressor of Prob. 8.19(*a*) is too high. To reduce it, several stages are suggested. Calculate the maximum temperature and the isentropic power requirement assuming (*a*) two stages and (*b*) three stages.

8.23 A compressor receives 0.4 lbm/sec of air at 12 psia and 50 °F and delivers it at 500 psia. For an 85 percent efficient compressor calculate the power requirement assuming (*a*) one stage, and (*b*) two stages.

8.24 Rather than assuming constant specific heats, use the air tables (Appendix E) and rework (*a*) Prob. 8.17 and (*b*) Prob. 8.19(*a*). Compute the percentage error for the constant specific heat assumption.

8.25 A three-stage compressor receives 2 kg/s of air at 95 kPa and 22 °C and delivers it at 4 MPa. For an ideal compressor calculate (*a*) the intercooler pressures, (*b*) the temperatures at each state, (*c*) the power required, and (*d*) the intercooler heat transfer rates.

8.26 An engine with a bore and a stroke of 0.2 × 0.2 m and a clearance of 5 percent experiences a minimum pressure of 120 kPa and a maximum pressure of 12 MPa. If it operates with air on the cycle of Fig. 8-7, determine (*a*) the displacement volume, (*b*) the compression ratio, and (*c*) the MEP.

8.27 An air-standard cycle operates in a piston-cylinder arrangement with the following four processes: 1 → 2—isentropic compression from 100 kPa and 15 °C to 2 MPa; 2 → 3—constant-pressure heat addition to 1200 °C; 3 → 4—isentropic expansion; and 4 → 1—constant-volume heat rejection. (*a*) Show the cycle on *P-v* and *T-s* diagrams, (*b*) calculate the heat addition and (*c*) calculate the cycle efficiency.

8.28 An air-standard cycle operates in a piston-cylinder arrangement with the following four processes: 1 → 2—constant-temperature compression from 12 psia and 70 °F to 400 psia; 2 → 3—constant-pressure expansion to 1400 °F; 3 → 4—isentropic expansion; and 4 → 1—constant-volume process. (*a*) Show the cycle on *P-v* and *T-s* diagrams, (*b*) calculate the work output, and (*c*) calculate the cycle efficiency.

8.29 A Carnot piston engine operates with air between 20 °C and 600 °C with a low pressure of 100 kPa. If it is to deliver 800 kJ/kg of work calculate (*a*) the thermal efficiency, (*b*) the compression ratio, and (*c*) the MEP. See Fig. 6-1.

8.30 A Carnot engine operates on air as shown in Fig. 8-22. Find (*a*) the power output, (*b*) the thermal efficiency, and (*c*) the MEP. See Fig. 6-1.

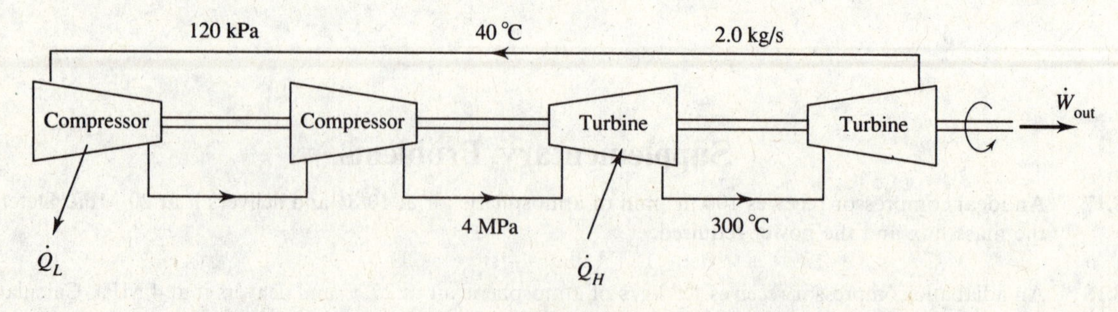

Fig. 8-22

8.31 A Carnot engine has heat addition during the combustion process of 4000 Btu/sec. If the temperature limits are 1200 °F and 30 °F, with high and low pressures of 1500 psia and 10 psia, determine the mass flux of air and the MEP. See Fig. 6-1.

8.32 A Carnot engine operates between the temperatures of 100 °C and 600 °C with pressure limits of 150 kPa and 10 MPa. Calculate the mass flux of air if the rejected heat flux is (*a*) 100 kW, (*b*) 400 kW, and (*c*) 2 MW. See Fig. 6-1.

8.33 A piston engine with a 0.2 × 0.2 m bore and stroke is modeled as a Carnot engine. It operates on 0.5 kg/s of air between temperatures of 20 °C and 500 °C with a low pressure of 85 kPa and a clearance of 2 percent. Find (*a*) the power delivered, (*b*) the compression ratio, (*c*) the MEP, and (*d*) the volume at top dead center. See Fig. 6-1.

8.34 A spark-ignition engine operates on an Otto cycle with a compression ratio of 9 and temperature limits of 30 °C and 1000 °C. If the power output is 500 kW, calculate the thermal efficiency and the mass flux of air.

8.35 An Otto cycle operates with air entering the compression process at 15 psia and 90 °F. If 600 Btu/lbm of energy is added during combustion and the compression ratio is 10, determine the work output and the MEP.

8.36 The maximum allowable pressure in an Otto cycle is 8 MPa. Conditions at the beginning of the air compression are 85 kPa and 22 °C. Calculate the required heat addition and the MEP, if the compression ratio is 8.

8.37 A maximum temperature of 1600 °C is possible in an Otto cycle in which air enters the compression process at 85 kPa and 30 °C. Find the heat addition and the MEP, if the compression ratio is 6.

8.38 If the Otto cycle shown in Fig. 8-23 operates on air, calculate the thermal efficiency and the MEP.

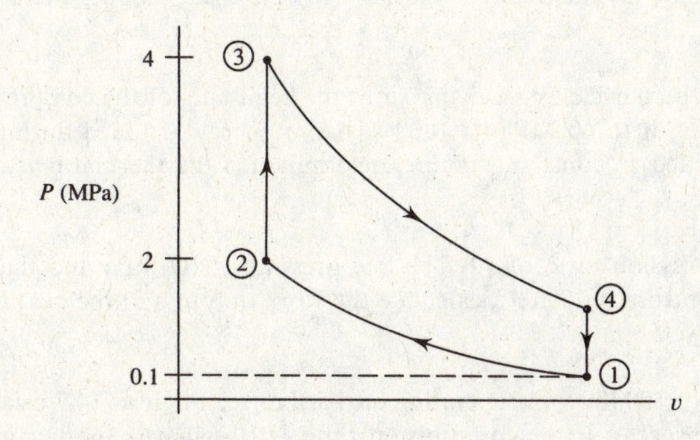

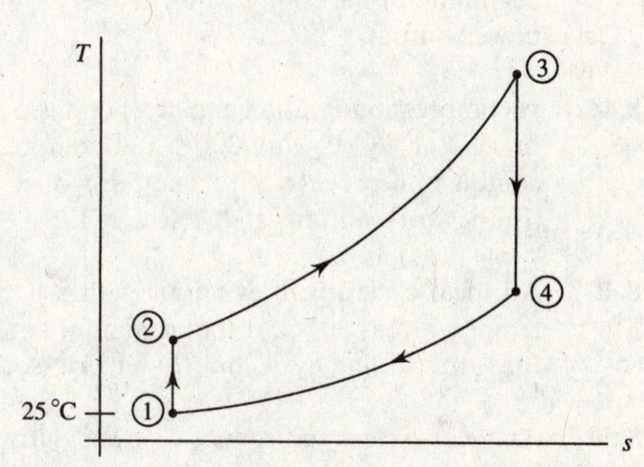

Fig. 8-23

8.39 A spark-ignition engine with a compression ratio of 8 operates on an Otto cycle using air with a low temperature of 60 °F and a low pressure of 14.7 psia. If the energy addition during combustion is 800 Btu/lbm, determine (*a*) the work output and (*b*) the maximum pressure.

8.40 Use the air tables (Appendix E) to solve (*a*) Prob. 8.35 and (*b*) Prob. 8.38. Do not assume constant specific heats.

8.41 A diesel engine is designed to operate with a compression ratio of 16 and air entering the compression stroke at 110 kPa and 20 °C. If the energy added during combustion is 1800 kJ/kg, calculate (*a*) the cutoff ratio, (*b*) the thermal efficiency, and (*c*) the MEP.

8.42 A diesel cycle operates on air which enters the compression process at 85 kPa and 30 °C. If the compression ratio is 16, the power output is 500 hp, and the maximum temperature is 2000 °C, calculate (*a*) the cutoff ratio, (*b*) the thermal efficiency, and (*c*) the mass flux of air.

8.43 Air enters the compression process of a diesel cycle at 120 kPa and 15 °C. The pressure after compression is 8 MPa and 1500 kJ/kg is added during combustion. What are (*a*) the cutoff ratio, (*b*) the thermal efficiency, and (*c*) the MEP?

8.44 For the cycle shown in Fig. 8-24 find the thermal efficiency and the work output.

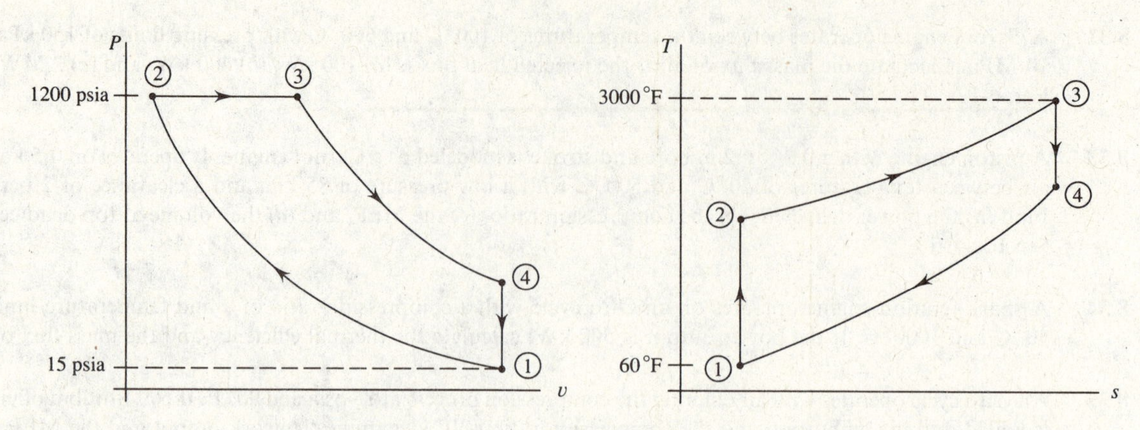

Fig. 8-24

8.45 A diesel engine has a 0.6 × 1.2 m bore and stroke and operates with 5 percent clearance. For a power output of 5000 hp calculate the compression ratio and the rate of heat input if the cutoff ratio is 2.5.

8.46 Use the air tables (Appendix E) to solve (*a*) Prob. 8.41 and (*b*) Prob. 8.44. Do not assume constant specific heats.

8.47 A dual cycle with $r = 18$, $r_c = 2$, and $r_p = 1.2$ operates on 0.5 kg/s of air at 100 kPa and 20 °C at the beginning of the compression process. Calculate (*a*) the thermal efficiency, (*b*) the energy input, and (*c*) the power output.

8.48 A compression-ignition engine operates on a dual cycle by receiving air at the beginning of the compression process at 80 kPa and 20 °C and compressing it to 60 MPa. If 1800 kJ/kg of energy is added during the combustion process, with one-third of it added at constant volume, determine (*a*) the thermal efficiency, (*b*) the work output, and (*c*) the MEP.

8.49 An ideal cycle operates on air with a compression ratio of 12. The low pressure is 100 kPa and the low temperature is 30 °C. If the maximum temperature is 1500 °C, calculate the work output and the heat input for (*a*) a Stirling cycle and (*b*) an Ericsson cycle.

8.50 An ideal cycle is to produce a power output of 100 hp while operating on 1.2 lbm/sec of air at 14.7 psia and 70 °F at the beginning of the compression process. If the compression ratio is 10, what is the maximum temperature and the energy input for (*a*) a Stirling cycle and (*b*) an Ericsson cycle?

8.51 Calculate the work output and thermal efficiency for the cycles shown in Fig. 8-25*a* and *b*. Air is the operating fluid.

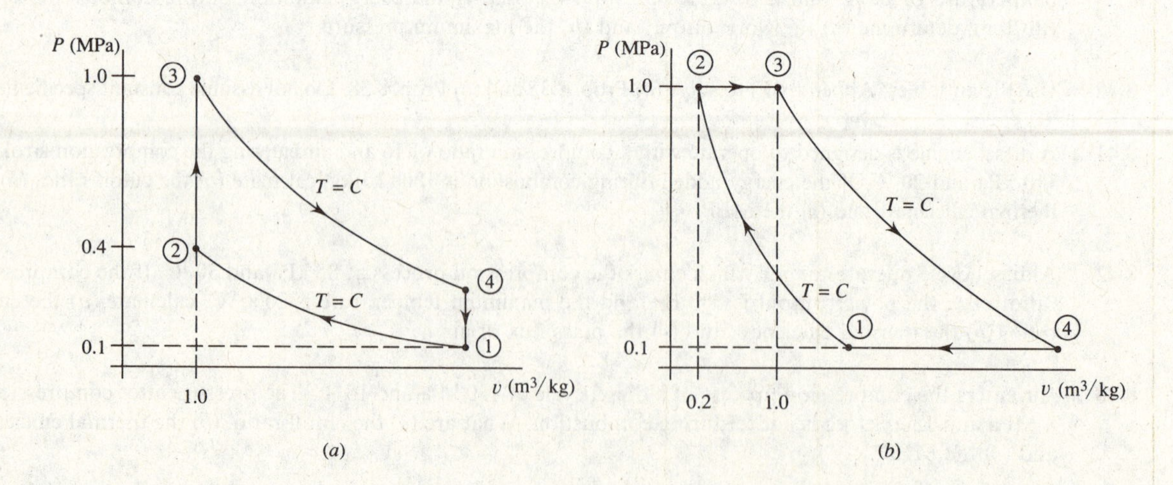

Fig. 8-25

8.52 Air enters the compressor of a gas turbine at 85 kPa and 0 °C. If the pressure ratio is 6 and the maximum temperature is 1000 °C, find (*a*) the thermal efficiency and (*b*) the back work ratio for the associated Brayton cycle.

8.53 Three kilograms of air enters the compressor of a gas turbine each second at 100 kPa and 10 °C. If the pressure ratio is 5 and the maximum temperature is 800 °C, determine (*a*) the horsepower output, (*b*) the back work ratio, and (*c*) the thermal efficiency for the associated Brayton cycle.

8.54 Determine the compressor outlet pressure that will result in maximum work output for a Brayton cycle in which the compressor inlet air conditions are 14.7 psia and 65 °F and the maximum temperature is 1500 °F.

8.55 Air enters the compressor of a Brayton cycle at 80 kPa and 30 °C and compresses it to 500 kPa. If 1800 kJ/kg of energy is added in the combustor, calculate (*a*) the compressor work requirement, (*b*) the net turbine output, and (*c*) the back work ratio.

8.56 Find the back work ratio and the horsepower output of the cycle shown in Fig. 8-26.

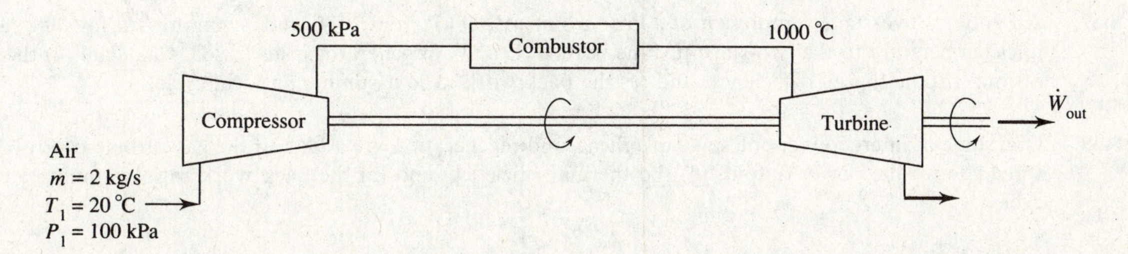

Fig. 8-26

8.57 Calculate the thermal efficiency and the back work ratio of the gas turbine of Prob. 8.52 if the respective compressor and turbine efficiencies are (*a*) 80%, 80% and (*b*) 83%, 86%.

8.58 Determine the efficiency of the compressor and turbine (the efficiencies are equal) that would result in a zero thermal efficiency for the gas turbine of Prob. 8.52.

8.59 Calculate the thermal efficiency and the back work ratio of the Brayton cycle of Prob. 8.55 if the compressor and turbine efficiencies are (*a*) 83%, 83% and (*b*) 81%, 88%.

8.60 Determine the efficiency of the compressor and turbine (the efficiencies are equal) of the Brayton cycle of Prob. 8.55 that would result in no net work output.

8.61 The efficiency of the turbine of Prob. 8.56 is 83 percent. What compressor efficiency would reduce the Brayton cycle thermal efficiency to zero?

8.62 Use the air tables to find the thermal efficiency and the back work ratio for (*a*) Prob. 8.52, (*b*) Prob. 8.55, and (*c*) Prob. 8.56. Do not assume constant specific heats.

8.63 A regenerator is installed in the gas turbine of Prob. 8.55. Determine the cycle efficiency if its effectiveness is (*a*) 100 percent and (*b*) 80 percent.

8.64 For the ideal-gas turbine with regenerator shown in Fig. 8-27 find $\dot{W}_{out}$ and the back work ratio.

8.65 Assume that the efficiencies of the compressor and turbine of Prob. 8.64 are 83 percent and 86 percent, respectively, and that the effectiveness of the regenerator is 90 percent. Determine the power output and the back work ratio.

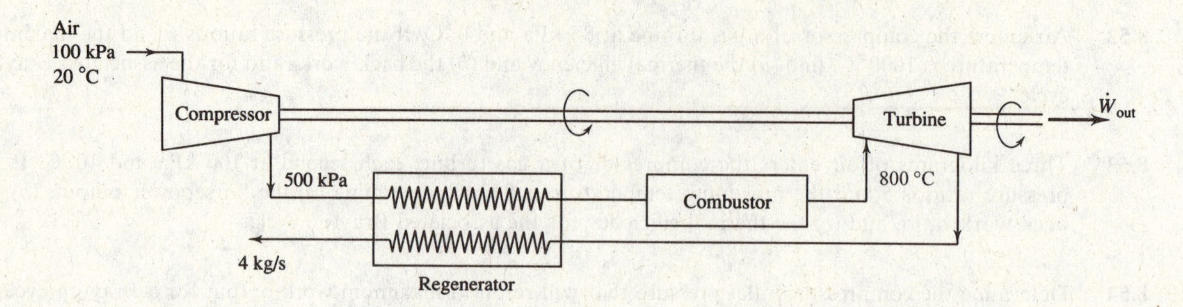

Fig. 8-27

8.66 Temperatures for the ideal regenerative gas-turbine cycle of Fig. 8-17 are $T_1 = 60\,°F$, $T_2 = 500\,°F$, $T_3 = 700\,°F$, and $T_4 = 1600\,°F$. Calculate the thermal efficiency and the back work ratio if air is the working fluid.

8.67 Air enters a two-stage compressor of a gas turbine at 100 kPa and 20 °C and is compressed to 600 kPa. The inlet temperature to the two-stage turbine is 1000 °C and a regenerator is also used. Calculate (*a*) the work output, (*b*) the thermal efficiency, and (*c*) the back work ratio assuming an ideal cycle.

8.68 One stage of intercooling, one stage of reheat, and regeneration are added to the gas turbine of Prob. 8.56. Calculate (*a*) the power output, (*b*) the thermal efficiency, and (*c*) the back work ratio assuming an ideal cycle.

8.69 (*a*) For the ideal components shown in Fig. 8-28 calculate the thermal efficiency. (*b*) For the same components, with an air mass flux of 2 kg/s, determine $\dot{W}_{out}$, $\dot{Q}_C$, $\dot{Q}_R$, and $\dot{Q}_{out}$.

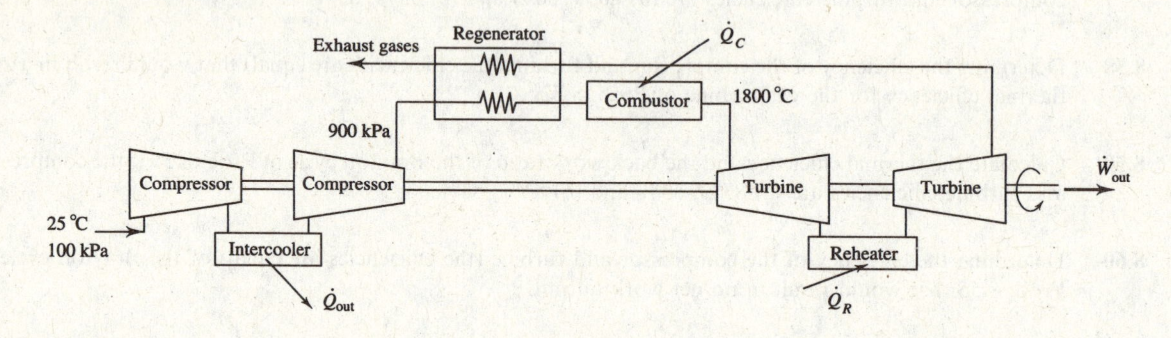

Fig. 8-28

8.70 A turbojet engine inlets 70 kg/s of air at an altitude of 10 km while traveling at 300 m/s. The compressor provides a pressure ratio of 9 and the turbine inlet temperature is 1000 °C. What is the maximum thrust and horsepower that can be expected from this engine?

8.71 Rework Prob. 8.70 with realistic efficiencies of 85 percent and 89 percent in the compressor and turbine, respectively. Assume the nozzle to be 97 percent efficient.

8.72 An aircraft with two turbojet engines requires a thrust of 4300 lbf for cruise conditions of 800 ft/sec. If each engine has a mass flux of 30 lbm/sec of air, calculate the pressure ratio if the maximum temperature is 2000 °F. The aircraft flies at an altitude of 30,000 ft.

Review Questions for the FE Examination

8.1FE Select the primary reason why an Otto cycle is less efficient than a Carnot cycle.
(A) The temperature after compression is too high.
(B) Heat transfer occurs across a large temperature difference.
(C) Friction exists between the piston and the cylinder.
(D) The sudden expansion process results in very large losses.

8.2FE Which of the following statements is not true of the diesel cycle?
(A) The expansion process is an isentropic process.
(B) The combustion process is a constant-volume process.
(C) The exhaust process is a constant-volume process.
(D) The compression process is an adiabatic process.

8.3FE The engines on a commercial jet aircraft operate on which of the basic cycles?
(A) Otto
(B) Diesel
(C) Carnot
(D) Brayton

8.4FE The exhaust process in the Otto and diesel cycles is replaced with a constant-volume process for what primary reason?
(A) To simulate zero work of the actual exhaust process.
(B) To simulate zero heat transfer of the actual exhaust process.
(C) To restore the air to its original state.
(D) To ensure that the first law is satisfied.

8.5FE The heat rejected in the Otto cycle shown in Fig. 8-29 is:
(A) $C_p(T_4 - T_1)$
(B) $C_p(T_3 - T_4)$
(C) $C_p(T_4 - T_1)$
(D) $C_v(T_3 - T_4)$

8.6FE The cycle efficiency of Fig. 8-29 is:
(A) $1 - T_1/T_3$
(B) $(T_3 - T_2)/(T_4/T_1)$
(C) $1 - \dfrac{T_4 - T_1}{T_3 - T_2}$
(D) $\dfrac{T_3 - T_2}{T_4 - T_1} - 1$

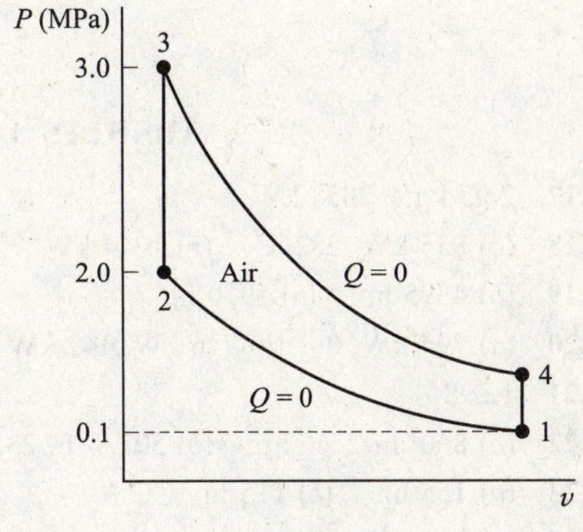

8.7FE If $T_1 = 27\,^\circ\text{C}$ in Fig. 8-29, what mass of air would exist in a 1000 cm³ cylinder?
(A) 0.00187 kg
(B) 0.00116 kg
(C) 0.00086 kg
(D) 0.00062 kg

Fig. 8-29

8.8FE If $T_1 = 27\,^\circ\text{C}$ in Fig. 8-29, Estimate T_2 assuming constant specific heats.
(A) 700 °C
(B) 510 °C
(C) 480 °C
(D) 430 °C

8.9FE Determine T_2 in the Brayton cycle of Fig. 8-30.
(A) 531 °C
(B) 446 °C
(C) 327 °C
(D) 258 °C

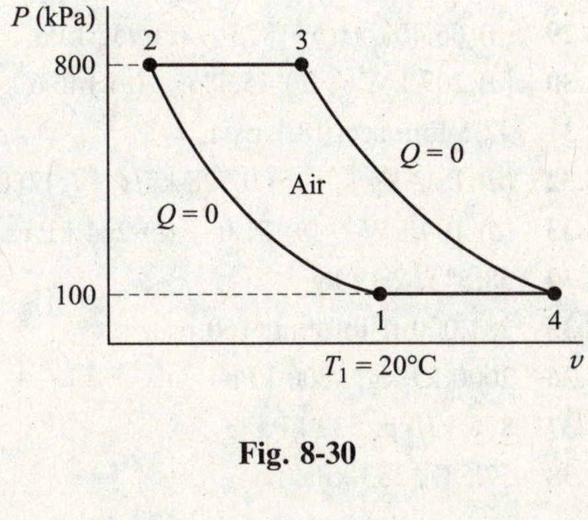

Fig. 8-30

8.10FE If $T_{high} = 1200\,^{\circ}\mathrm{C}$, find w_T (not w_{net}) of the turbine of Fig. 8-30.
(A) 720 kJ/kg
(B) 660 kJ/kg
(C) 590 kJ/kg
(D) 540 kJ/kg

8.11FE Estimate the back work ratio for the cycle of Fig. 8-30.
(A) 0.40
(B) 0.38
(C) 0.36
(D) 0.34

8.12FE If an ideal regenerator is added to the cycle of Fig. 8-30, the temperature entering the burner is:
(A) T_1
(B) T_2
(C) T_3
(D) T_4

8.13FE If $\dot{m} = 0.2$ kg/s for the cycle of Fig. 8-30, what should be the inlet diameter of the compressor if a maximum velocity of 80 m/s is allowed?
(A) 5.2 cm
(B) 6.4 cm
(C) 8.6 cm
(D) 12.4 cm

Answers to Supplementary Problems

8.17 2.05 kg/s, 2058 kW
8.18 (*a*) 835 kW, 582 °C (*b*) 1044 kW, 721 °C
8.19 (*a*) 4895 hp (*b*) 5970 hp
8.20 (*a*) 794 kW, 0 (*b*) 726 kW, 142 kW (*c*) 653 kW, 274 kW (*d*) 504 kW, 504 kW
8.21 1.298
8.22 (*a*) 860 °F, 2766 hp (*b*) 507.8 °F, 2322 hp
8.23 (*a*) 155 hp (*b*) 115 hp
8.24 (*a*) 2003 kW, 2.6% (*b*) 4610 hp, 6.5%
8.25 (*a*) 330 kPa, 1150 kPa (*b*) 148 °C, 22 °C (*c*) 756 kW (*d*) 252 kW
8.26 (*a*) 6.28 liters (*b*) 21 (*c*) 306 kPa
8.27 (*b*) 522 kJ/kg (*c*) 22.3%
8.28 (*b*) 118,700 ft-lbf/lbm (*c*) 47.8%
8.29 (*a*) 66.4% (*b*) 1873 (*c*) 952 kPa
8.30 (*a*) 207 kW (*b*) 45.4% (*c*) 146.6 kPa
8.31 47.5 lbm/sec, 18.1 psia
8.32 (*a*) 1.23 kg/s (*b*) 0.328 kg/s (*c*) 0.0655 kg/s
8.33 (*a*) 104 kW (*b*) 51.0 (*c*) 214 kPa (*d*) 0.1257 liter
8.34 58.5%, 2.19 kg/s
8.35 281,000 ft-lbf/lbm, 160 psia
8.36 2000 kJ/kg, 1300 kPa
8.37 898 kJ/kg, 539 kPa
8.38 57.5%, 383 kPa

8.39 (*a*) 352,000 ft-lbf/lbm (*b*) 1328 psia

8.40 (*a*) 254,000 ft-lbf/lbm, 144 psia (*b*) 54.3%, 423 kPa

8.41 (*a*) 3.03 (*b*) 56.8% (*c*) 1430 kPa

8.42 (*a*) 2.47 (*b*) 59.2% (*c*) 0.465 kg/s

8.43 (*a*) 2.57 (*b*) 62.3% (*c*) 1430 kPa

8.44 67%, 205,000 ft-lbf/lbm

8.45 21, 5890 kW

8.46 (*a*) 2.76, 50.6%, 1270 kPa (*b*) 62.2%, 240,000 ft-lbf/lbm

8.47 (*a*) 63.7% (*b*) 625 kJ/kg (*c*) 534 hp

8.48 (*a*) 84.1% (*b*) 1514 kJ/kg (*c*) 1450 kPa

8.49 (*a*) 1048 kJ/kg, 1264 kJ/kg (*b*) 1048 kJ/kg, 1264 kJ/kg

8.50 (*a*) 443 °F, 142.5 Btu/lbm (*b*) 443 °F, 142.5 Btu/lbm

8.51 (*a*) 831 kJ/kg, 60% (*b*) 1839 kJ/kg, 80%

8.52 (*a*) 40.1% (*b*) 0.358

8.53 (*a*) 927 hp (*b*) 0.418 (*c*) 36.9%

8.54 147 psia

8.55 (*a*) 208 kJ/kg (*b*) 734 kJ/kg (*c*) 0.221

8.56 0.365, 799 hp

8.57 (*a*) 0.559, 23.3% (*b*) 0.502, 28.1%

8.58 59.8%

8.59 (*a*) 30.3%, 0.315 (*b*) 32.8%, 0.304

8.60 43.7%

8.61 44%

8.62 (*a*) 38.1%, 0.346 (*b*) 37.1%, 0.240 (*c*) 34.8%, 0.355

8.63 (*a*) 88.4% (*b*) 70.3%

8.64 899 kW, 0.432

8.65 540 kW, 0.604

8.66 60.0%, 0.489

8.67 (*a*) 404 kJ/kg (*b*) 70.3% (*c*) 0.297

8.68 (*a*) 992 hp (*b*) 71% (*c*) 0.29

8.69 (*a*) 80.1% (*b*) 1788 kW, 1116 kW, 1116 kW, 222 kW

8.70 55.1 kN, 22 200 hp

8.71 54.6 kN, 21 900 hp

8.72 10

Answers to Review Questions for the FE Examination

8.1FE (B) **8.2FE** (B) **8.3FE** (D) **8.4FE** (A) **8.5FE** (C) **8.6FE** (C) **8.7FE** (B) **8.8FE**(D) **8.9FE** (D)
8.10FE (B) **8.11FE** (C) **8.12FE** (D) **8.13FE** (A)

Vapor Power Cycles

9.1 INTRODUCTION

The ideal Carnot cycle is used as a model to compare all real and all other ideal cycles against. The efficiency of a Carnot power cycle is the maximum possible for any power cycle; it is given by

$$\eta = 1 - \frac{T_L}{T_H} \tag{9.1}$$

Note that the efficiency is increased by raising the temperature T_H at which heat is added or by lowering the temperature T_L at which heat is rejected. We will observe that this carries over to real cycles: the cycle efficiency can be maximized by using the highest maximum temperature and the lowest minimum temperature. We will discuss vapor cycles that are used to generate power in this chapter.

9.2 THE RANKINE CYCLE

The first class of power cycles that we consider in this chapter are those utilized by the electric power generating industry, namely, power cycles that operate in such a way that the working fluid changes phase from a liquid to a vapor. The simplest vapor power cycle is called the *Rankine cycle*, shown schematically in Fig. 9-1*a*. A major feature of such a cycle is that the pump requires very little work to deliver high-pressure water to the boiler. A possible disadvantage is that the expansion process in the turbine usually enters the quality region, resulting in the formation of liquid droplets that may damage the turbine blades.

The Rankine cycle is an idealized cycle in which friction losses in each of the four components are neglected. Such losses usually are quite small and will be neglected completely in our initial analysis. The Rankine cycle is composed of the four ideal processes shown on the *T-s* diagram in Fig. 9-1*b*:

1 → 2: Isentropic compression in a pump
2 → 3: Constant-pressure heat addition in a boiler
3 → 4: Isentropic expansion in a turbine
4 → 1: Constant-pressure heat extraction in a condenser

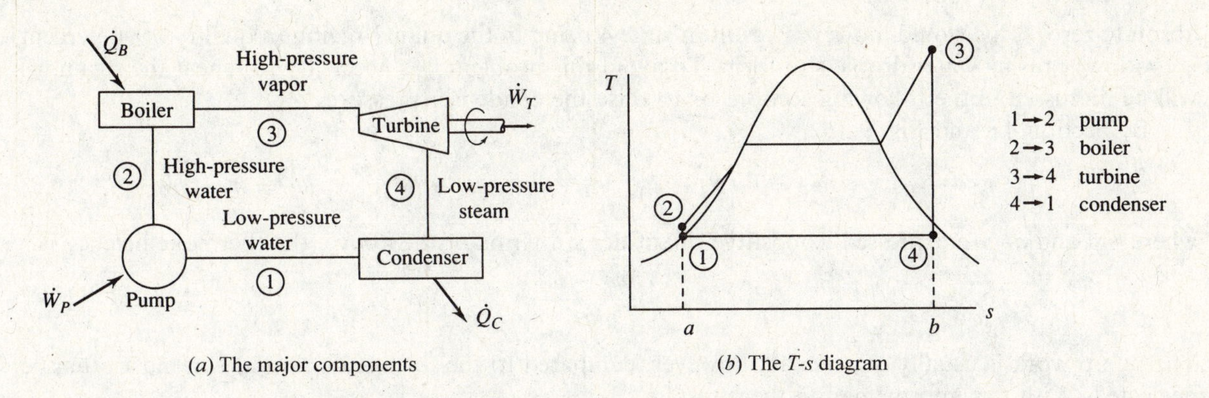

(*a*) The major components (*b*) The *T-s* diagram

Fig. 9-1 The Rankine cycle.

If we neglect kinetic energy and potential energy changes, the net work output is the area under the *T-s* diagram, represented by area 1-2-3-4-1; this is true since the first law requires that $W_{net} = Q_{net}$. The heat transfer to the working substance is represented by area *a*-2-3-*b*-*a*. Thus, the thermal efficiency η of the Rankine cycle is

$$\eta = \frac{\text{area 1-2-3-4-1}}{\text{area } a\text{-2-3-}b\text{-}a} \tag{9.2}$$

that is, the desired output divided by the energy input (the purchased energy). Obviously, the thermal efficiency can be improved by increasing the numerator or by decreasing the denominator. This can be done by increasing the pump outlet pressure P_2, increasing the boiler outlet temperature T_3, or decreasing the turbine outlet pressure P_4.

Note that the efficiency of the Rankine cycle is less than that of a Carnot cycle operating between the high temperature T_3 and the low temperature T_1 since most of the heat transfer from a high-temperature reservoir occurs across large temperature differences.

It is possible for the efficiency of a Rankine cycle to be equal to that of a Carnot cycle if the cycle is designed to operate as shown in Fig. 9-2*a*. However, the pump would be required to pump a mixture of liquid and vapor, a rather difficult and work-consuming task compared to pumping all liquid. In addition, the condensation of liquid droplets in the turbine would result in severe damage. To avoid the damage from droplets, one could propose superheating the steam at constant temperature, as shown in Fig. 9-2*b*. This, however, requires that the pressure for the constant-temperature, super-heated portion of the process decrease from the saturated vapor point to state 3. To achieve such a decrease, the flow in the boiler pipes would have to be accelerated, a task that would require pipes of decreasing diameter. This would be expensive, should it even be attempted. Thus it is proposed that P_2 and T_3 be quite large (T_3 being limited by the temperature-resistance characteristics of the pipe metal, typically about 600 °C). (See Fig. 9-2*c*.) It is also proposed that the condenser outlet pressure be very low (it can be quite close to

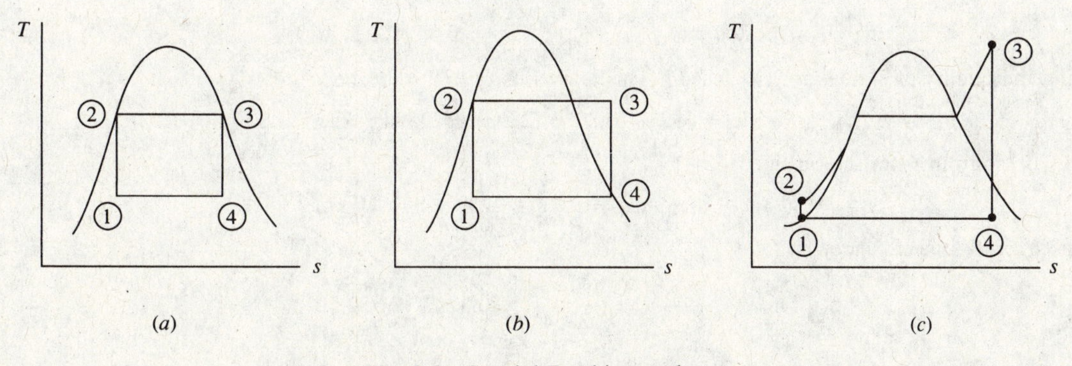

Fig. 9-2 Special Rankine cycles.

absolute zero). This would, however, result in state 4 being in the quality region (a quality of 90 percent is too low) causing water droplets to form. To avoid this problem it is necessary to reheat the steam, as will be discussed in the following section, or to raise the condenser pressure.

By Section 4.8 and Fig. 9-1*b*,

$$q_B = h_3 - h_2 \qquad w_P = v_1(P_2 - P_1) \qquad q_C = h_4 - h_1 \qquad w_T = h_3 - h_4 \tag{9.3}$$

where w_P and q_C are expressed as positive quantities. In terms of the above, the thermal efficiency is

$$\eta = \frac{w_T - w_P}{q_B} \tag{9.4}$$

The pump work is usually quite small, however, compared to the turbine work and can most often be neglected. With this approximation there results

$$\eta = \frac{w_T}{q_B} \tag{9.5}$$

This is the relation used for the thermal efficiency of the Rankine cycle.

EXAMPLE 9.1 A steam power plant is proposed to operate between the pressures of 10 kPa and 2 MPa with a maximum temperature of 400 °C, as shown in Fig. 9-3. What is the maximum efficiency possible from the power cycle?

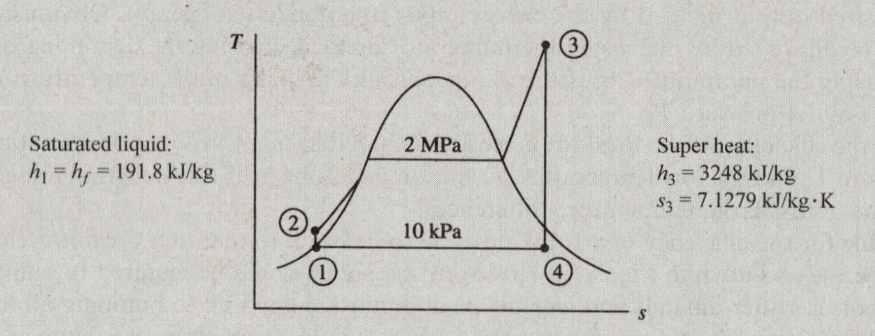

Fig. 9-3

Solution: Let us include the pump work in the calculation and show that it is negligible. Also, we will assume a unit mass of working fluid since we are only interested in the efficiency. The pump work is [see (*4.71*) with $v = 1/\rho$]

$$w_P = v_1(P_2 - P_1) = (0.001)(2000 - 10) = 1.99 \text{ kJ/kg}$$

Using (*4.67*) we find that $h_2 = h_1 + w_{\text{in}} = 191.8 + 1.99 = 194$ kJ/kg. The heat input is found using $q_B = h_3 - h_2 = 3248 - 194 = 3054$ kJ/kg. To locate state 4 we recognize that $s_4 = s_3 = 7.1279$. Hence,

$$s_4 = s_f + x_4 s_{fg} \qquad \therefore 7.1279 = 0.6491 + 7.5019x_4$$

giving the quality of state 4 as $x_4 = 0.8636$. This allows us to find h_4 to be

$$h_4 = 192 + (0.8636)(2393) = 2259 \text{ kJ/kg}$$

The work output from the turbine is

$$w_T = h_3 - h_4 = 3248 - 2259 = 989 \text{ kJ/kg}$$

Consequently, the efficiency is

$$\eta = \frac{w_T - w_P}{q_B} = \frac{989 - 2}{3054} = 0.3232 \quad \text{or } 32.32\%$$

Obviously, the work required in the pumping process is negligible, being only 0.2 percent of the turbine work. In engineering applications we often neglect quantities that have an influence of less than 3 percent, since invariably there is some quantity in the calculations that is known to only ± 3 percent; for example, the mass flux, the dimensions of a pipe, or the density of the fluid.

9.3 RANKINE CYCLE EFFICIENCY

The efficiency of the Rankine cycle can be improved by increasing the boiler pressure while maintaining the maximum temperature and the minimum pressure. The net increase in work output is the crosshatched area minus the shaded area of Fig. 9-4*a*, a relatively small change; the added heat, however, decreases by the shaded area minus the crosshatched area of Fig. 9-4*b*. This is obviously a significant decrease, and it leads to a significant increase in efficiency. Example 9.2 illustrates this effect. The disadvantage of raising the boiler pressure is that the quality of the steam exiting the turbine may become too low (less than 90 percent), resulting in severe water droplet damage to the turbine blades and impaired turbine efficiency.

Increasing the maximum temperature also results in an improvement in thermal efficiency of the Rankine cycle. In Fig. 9-5*a* the net work is increased by the crosshatched area and the heat input is increased by the sum of the crosshatched area and the shaded area, a smaller percentage increase than the work increase. Since the numerator of (*9.5*) realizes a larger percentage increase than the denominator, there will be a resulting increase in efficiency. This will be illustrated in Example 9.3. Of course, metallurgical considerations limit the maximum temperature which can be attained in the boiler. Temperatures up to about 600 °C are allowable. Another advantage of raising the boiler temperature is that the quality of state 4 is obviously increased; this reduces water droplet formation in the turbine.

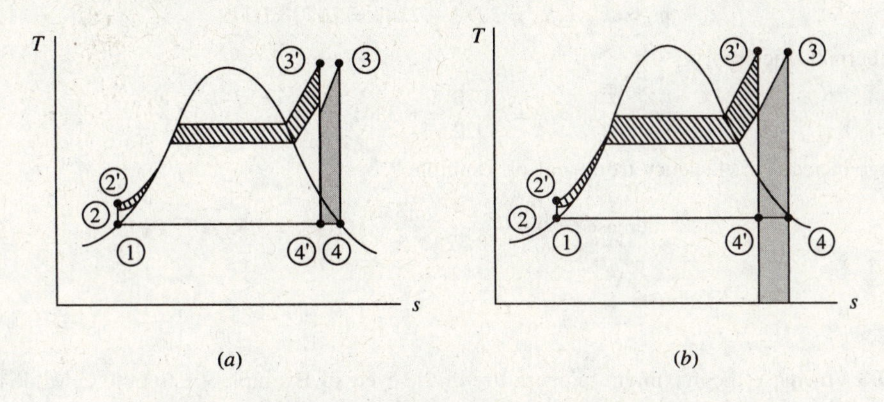

Fig. 9-4 Effect of increased pressure on the Rankine cycle.

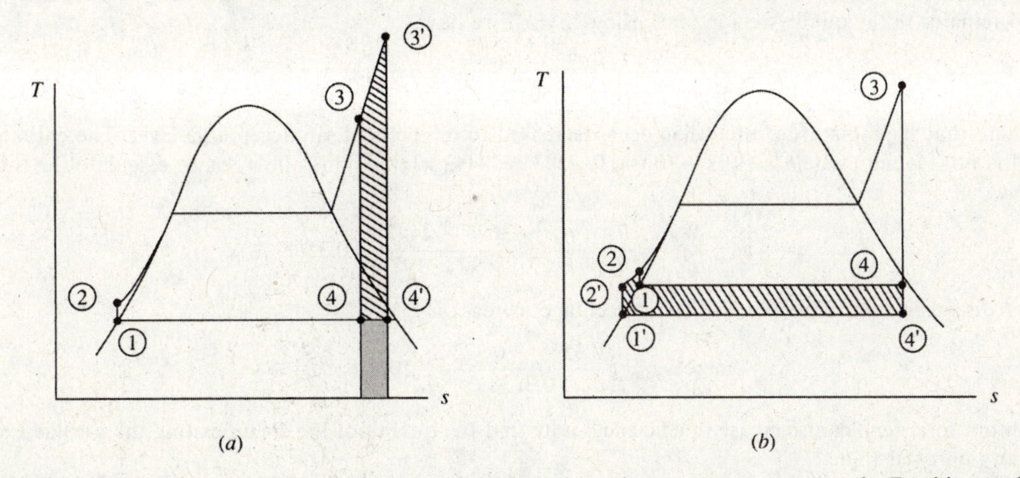

Fig. 9-5 Effect of increased temperature and decreased condenser temperature on the Rankine cycle.

A decrease in condenser pressure, illustrated in Fig. 9-5*b*, will also result in increased Rankine cycle efficiency. The net work will increase a significant amount, represented by the crosshatched area, and the heat input will increase a slight amount because state 1′ will move to a slightly lower entropy than that of state 1; this will result in an increase in the Rankine cycle efficiency. The low pressure is limited by the heat transfer process that occurs in the condenser. The heat is rejected by transferring heat to cooling water or to air which enters the condenser at about 20 °C; the heat transfer process requires a temperature differential between the cooling water and the steam of at least 10 °C. Hence, a temperature of at least 30 °C is required in the condenser; this corresponds to a minimum condenser pressure (see the saturated steam tables) of approximately 4 kPa abs. This is, of course, dependent on the temperature of the cooling water and the temperature differential required in the heat exchanger.

EXAMPLE 9.2 Increase the boiler pressure of Example 9.1 to 4 MPa while maintaining the maximum temperature and the minimum pressure. Calculate the percentage increase in the thermal efficiency.

Solution: Neglecting the work of the pump, the enthalpy h_2 remains unchanged: $h_2 = 192$ kJ/kg. At 400 °C and 4 MPa the enthalpy and entropy are $s_3 = 6.7698$ kJ/kg·K and $h_3 = 3214$ kJ/kg. State 4 is in the quality region. Using $s_4 = s_3$, the quality is found to be

$$x_4 = \frac{s_4 - s_f}{s_{fg}} = \frac{6.7698 - 0.6491}{7.5019} = 0.8159$$

Observe that the moisture content has increased to 18.4 percent, an undesirable result. The enthalpy of state 4 is then

$$h_4 = h_f + x_4 h_{fg} = 192 + (0.8159)(2393) = 2144 \text{ kJ/kg}$$

The heat addition is $q_B = h_3 - h_2 = 3214 - 192 = 3022$ kJ/kg and the turbine work output is

$$w_T = h_3 - h_4 = 3214 - 2144 = 1070 \text{ kJ/kg}$$

Finally, the thermal efficiency is

$$\eta = \frac{1070}{3022} = 0.3541$$

The percentage increase in efficiency from that of Example 9.1 is

$$\% \text{ increase} = \left(\frac{0.3541 - 0.3232}{0.3232}\right)(100) = 9.55\%$$

EXAMPLE 9.3 Increase the maximum temperature in the cycle of Example 9.1 to 600 °C, while maintaining the boiler pressure and condenser pressure, and determine the percentage increase in thermal efficiency.

Solution: At 600 °C and 2 MPa the enthalpy and entropy are $h_3 = 3690$ kJ/kg and $s_3 = 7.7032$ kJ/kg·K. State 4 remains in the quality region and, using $s_4 = s_3$, we have

$$x_4 = \frac{7.7032 - 0.6491}{7.5019} = 0.9403$$

Note here that the moisture content has been decreased to 6.0 percent, an acceptable level. The enthalpy of state 4 is now found to be $h_4 = 192 + (0.9403)(2393) = 2442$ kJ/kg. This allows us to calculate the thermal efficiency as

$$\eta = \frac{W_T}{q_B} = \frac{h_3 - h_4}{h_3 - h_2} = \frac{3690 - 2442}{3690 - 192} = 0.3568$$

where h_2 is taken from Example 9.1. The percentage increase is

$$\% \text{ increase} = \left(\frac{0.3568 - 0.3232}{0.3232}\right)(100) = 10.4\%$$

In addition to a significant increase in efficiency, note that the quality of the steam exiting the turbine exceeds 90%, an improved value.

EXAMPLE 9.4 Decrease the condenser pressure of Example 9.1 to 4 kPa while maintaining the boiler pressure and maximum temperature, and determine the percentage increase in thermal efficiency.

Solution: The enthalpies $h_2 = 192$ kJ/kg and $h_3 = 3248$ kJ/kg remain as stated in Example 9.1. Using $s_3 = s_4 = 7.1279$, with $P_4 = 4$ kPa, we find the quality to be

$$x_4 = \frac{s_4 - s_f}{s_{fg}} = \frac{7.1279 - 0.4225}{8.0529} = 0.8327$$

Note that the moisture content of 16.7 percent is quite high. The enthalpy of state 4 is $h_4 = 121 + (0.8327)(2433) = 2147$ kJ/kg. The thermal efficiency is then

$$\eta = \frac{h_3 - h_4}{h_3 - h_2} = \frac{3248 - 2147}{3248 - 192} = 0.3603$$

The percentage increase is found to be

$$\% \text{ increase} = \left(\frac{0.3603 - 0.3232}{0.3232}\right)(100) = 11.5\%$$

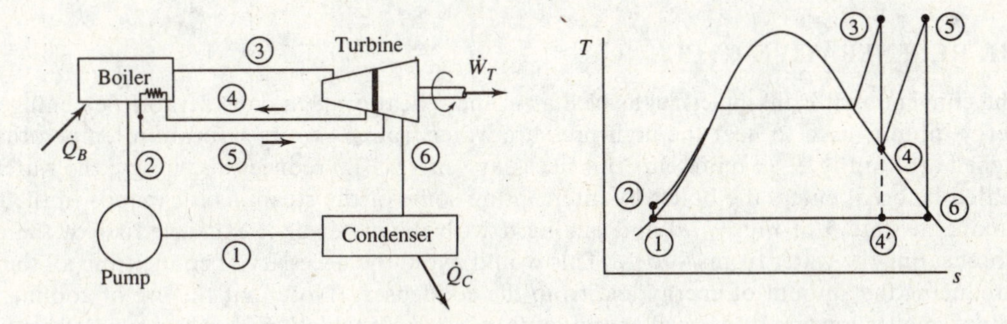

Fig. 9-6 The reheat cycle.

9.4 THE REHEAT CYCLE

It is apparent from the previous section that when operating a Rankine cycle with a high boiler pressure or a low condenser pressure it is difficult to prevent liquid droplets from forming in the low-pressure portion of the turbine. Since most metals cannot withstand temperatures above about 600 °C, the reheat cycle is often used to prevent liquid droplet formation: the steam passing through the turbine is reheated at some intermediate pressure, thereby raising the temperature to state 5 in the *T-s* diagram of Fig. 9-6. The steam then passes through the low-pressure section of the turbine and enters the condenser at state 6. This controls or completely eliminates the moisture problem in the turbine. Often the turbine is separated into a high-pressure turbine and a low-pressure turbine. The reheat cycle does not significantly influence the thermal efficiency of the cycle, but it does result in a significant additional work output, represented in the figure by area 4-5-6-4′-4. The reheat cycle demands a significant investment in additional equipment, and the use of such equipment must be economically justified by the increased work output.

EXAMPLE 9.5 High-pressure steam enters a turbine at 600 psia and 1000 °F. It is reheated at a pressure of 40 psia to 600 °F and then expanded to 2 psia. Determine the cycle efficiency. See Fig. 9-6.

Solution: At 2 psia saturated water has an enthalpy of (refer to Table C-2E) $h_1 \simeq h_2 = 94$ Btu/lbm. From Table C-3E we find $h_3 = 1518$ Btu/lbm and $s_3 = 1.716$ Btu/lbm-°R. Setting $s_4 = s_3$, we interpolate, obtaining

$$h_4 = \left(\frac{1.716 - 1.712}{1.737 - 1.712}\right)(1217 - 1197) + 1197 = 1200 \text{ Btu/lbm}$$

At 40 psia and 600 °F we have

$$h_5 = 1333 \text{ Btu/lbm} \qquad \text{and} \qquad s_5 = 1.862 \text{ Btu/lbm-°R}$$

In the quality region use $s_6 = s_5$ and find

$$x_6 = \frac{1.862 - 0.175}{1.745} = 0.9668$$

Thus, $h_6 = 94 + (0.9668)(1022) = 1082$ Btu/lbm. The energy input and output are

$$q_B = (h_5 - h_4) + (h_3 - h_2) = 1333 - 1200 + 1518 - 94 = 1557 \text{ Btu/lbm}$$
$$w_T = (h_5 - h_6) + (h_3 - h_4) = 1333 - 1082 + 1518 - 1200 = 569 \text{ Btu/lbm}$$

The thermal efficiency is then calculated to be

$$\eta = \frac{w_T}{q_B} = \frac{569}{1557} = 0.365 \qquad \text{or } 36.5\%$$

9.5 THE REGENERATIVE CYCLE

In the conventional Rankine cycle, as well as in the reheat cycle, a considerable percentage of the total energy input is used to heat the high-pressure water from T_2 to its saturation temperature. The crosshatched area in Fig. 9-7*a* represents this necessary energy. To reduce this energy, the water could be preheated before it enters the boiler by intercepting some of the steam as it expands in the turbine (for example, at state 5 of Fig. 9-7*b*) and mixing it with the water as it exits the first of the pumps, thereby preheating the water from T_2 to T_6. This would avoid the necessity of condensing all the steam, thereby reducing the amount of energy lost from the condenser. (Note that the use of cooling towers would allow smaller towers for a given energy output.) A cycle which utilizes this type of heating is a *regenerative cycle*, and the process is referred to as *regeneration*. A schematic representation of the major elements of such a cycle is shown in Fig. 9-8. The water entering the boiler is often referred to as *feedwater*, and the device used to mix the extracted steam and the condenser water is called a *feedwater heater*. When the condensate is mixed directly with the steam, it is done so in an *open* feedwater heater, as sketched in Fig. 9-8.

In analyzing a regenerative cycle we must consider a control volume surrounding the feedwater heater (see Fig. 9-9). A mass balance would result in

$$\dot{m}_6 = \dot{m}_5 + \dot{m}_2 \tag{9.6}$$

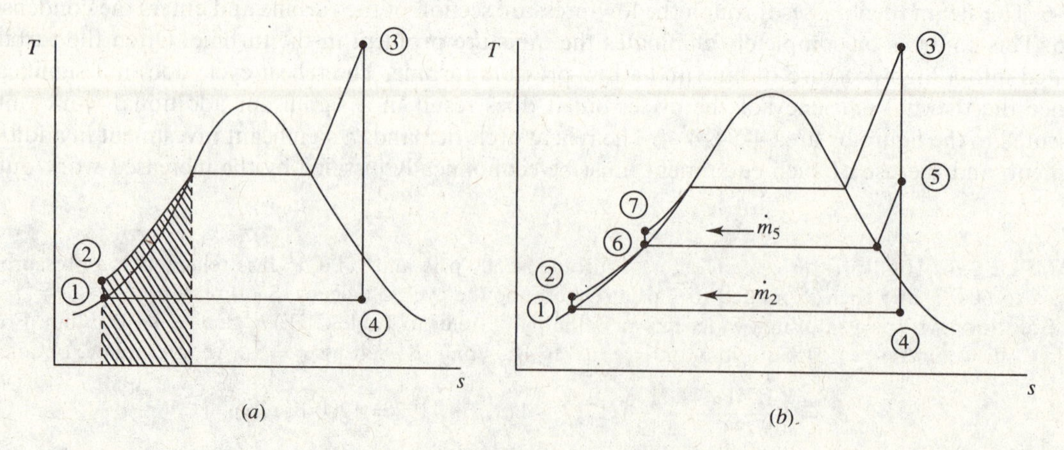

Fig. 9-7 The regenerative cycle.

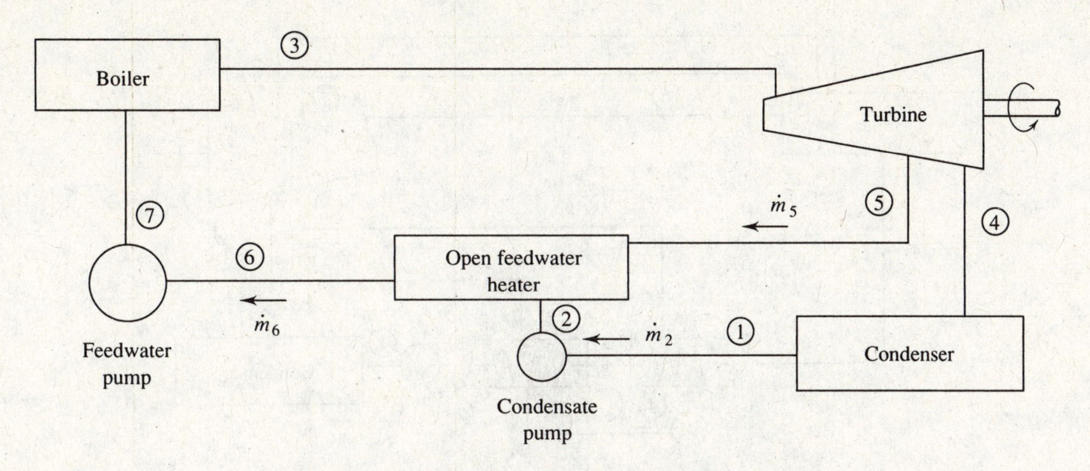

Fig. 9-8 The major elements of the regenerative cycle.

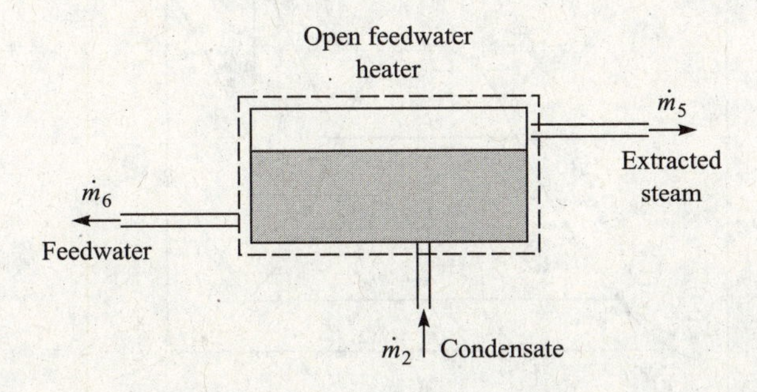

Fig. 9-9 An open feedwater heater.

An energy balance, assuming an insulated heater, neglecting kinetic and potential energy changes, gives

$$\dot{m}_6 h_6 = \dot{m}_5 h_5 + \dot{m}_2 h_2 \tag{9.7}$$

Combining the above two equations gives

$$\dot{m}_5 = \frac{h_6 - h_2}{h_5 - h_2}\dot{m}_6 \tag{9.8}$$

A *closed* feedwater heater, which can be designed into a system using only one main pump, is also a possibility. Figure 9-10 is a schematic diagram of a system using a closed feedwater heater. The

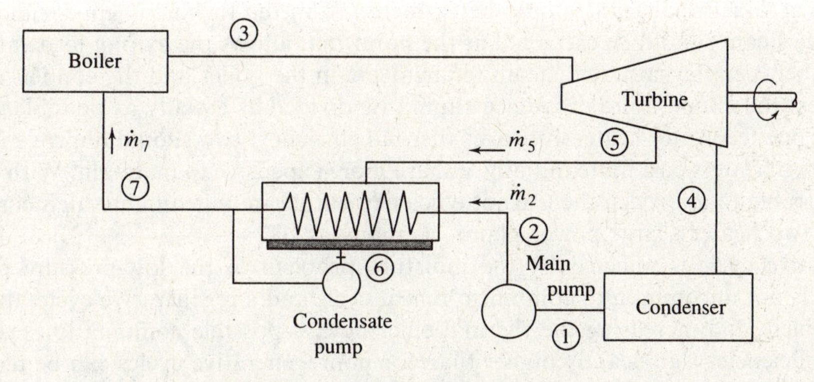

Fig. 9-10 A closed feedwater heater.

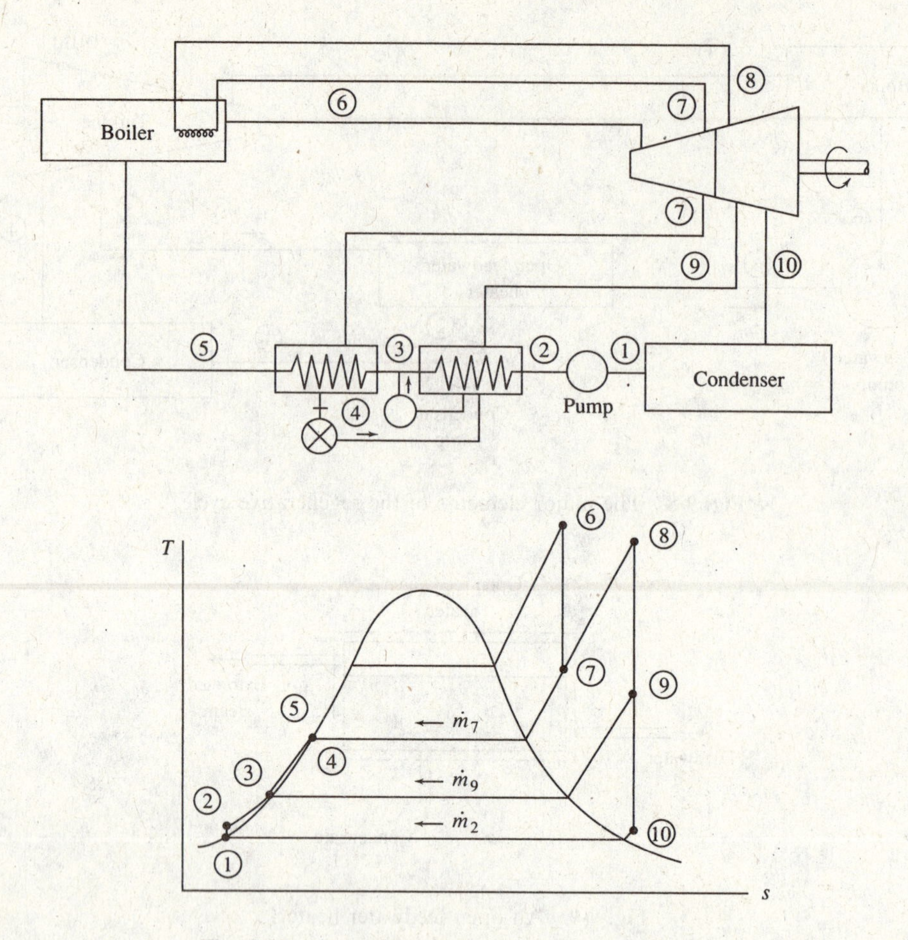

Fig. 9-11 A combined reheat/regenerative cycle.

disadvantages of such a system are that it is more expensive and its heat transfer characteristics are not as desirable as heat transfer in which the steam and water are simply mixed, as in the open heater.

The closed feedwater heater is a heat exchanger in which the water passes through in tubes and the steam surrounds the tubes, condensing on the outer surfaces. The condensate thus formed, at temperature T_6, is pumped with a small condensate pump into the main feedwater line, as shown, or it passes through a trap (a device that permits only liquid to pass through) and is fed back to the condenser or back to a lower-pressure feedwater heater. A mass and energy balance are also required when analyzing a closed feedwater heater; if pump energy requirement is neglected in the analysis, the same relationship [see (*9.8*)] results.

The pressure at which the steam should be extracted from the turbine is approximated as follows. For one heater the steam should be extracted at the point that allows the exiting feedwater temperature T_6 to be midway between the saturated steam temperature in the boiler and the condenser temperature. For several heaters the temperature difference should be divided as equally as possible.

Obviously, if one feedwater heater improves thermal efficiency, two should improve it more. This is, in fact, true, but two heaters cost more initially and are more expensive to maintain. With a large number of heaters it is possible to approach the Carnot efficiency but at an unjustifiably high cost. Small power plants may have two heaters; large power plants, as many as six.

The regenerative cycle is afflicted by the moisture problem in the low-pressure portions of the turbine; hence, it is not uncommon to combine a reheat cycle and a regenerative cycle, thereby avoiding the moisture problem and increasing the thermal efficiency. A possible combination cycle is shown in Fig. 9-11. Ideal efficiencies significantly higher than for nonregenerative cycles can be realized with this combination cycle.

A final word about efficiency. We calculate the efficiency of a cycle using the turbine work output as the desired output and consider the rejected heat from the condenser as lost energy. There are special situations where a power plant can be located strategically so that the rejected steam can be utilized to heat or cool buildings or the steam can be used in various industrial processes. This is often referred to as *cogeneration*. Often one-half of the rejected heat can be effectively used, almost doubling the "efficiency" of a power plant. Steam or hot water cannot be transported very far; thus, the power plant must be located close to an industrial area or a densely populated area. A university campus is an obvious candidate for cogeneration, as are most large industrial concerns.

EXAMPLE 9.6 The high-temperature situation of Example 9.3 is to be modified by inserting an open feedwater heater such that the extraction pressure is 200 kPa. Determine the percentage increase in thermal efficiency.

Solution: Refer to the *T-s* diagram of Fig. 9-7*b* and to Fig. 9-8. We have from Example 9.3 and the steam tables

$$h_1 \simeq h_2 = 192 \text{ kJ/kg} \qquad h_6 \simeq h_7 = 505 \text{ kJ/kg} \qquad h_3 = 3690 \text{ kJ/kg} \qquad h_4 = 2442 \text{ kJ/kg}$$

Now, locate state 5. Using $s_5 = s_3 = 7.7032$ kJ/kg·K, we interpolate and find, at 200 kPa,

$$h_5 = \left(\frac{7.7032 - 7.5074}{7.7094 - 7.5074}\right)(2971 - 2870) + 2870 = 2968 \text{ kJ/kg}$$

We now apply conservation of mass and the first law to a control volume surrounding the feedwater heater. We have, using $m_6 = 1$ kg, since we are only interested in efficiency [see (*9.8*)],

$$m_5 = \frac{505 - 192}{2968 - 192} = 0.1128 \text{ kg} \qquad \text{and} \qquad m_2 = 0.8872 \text{ kg}$$

The work output from the turbine is

$$w_T = h_3 - h_5 + (h_5 - h_4)m_2 = 3690 - 2968 + (2968 - 2442)(0.8872) = 1189 \text{ kJ/kg}$$

The energy input to the boiler is $q_B = h_3 - h_7 = 3690 - 505 = 3185$ kJ/kg. The thermal efficiency is calculated to be

$$\eta = \frac{1189}{3185} = 0.3733$$

The increase in efficiency is

$$\% \text{ increase} = \left(\frac{0.3733 - 0.3568}{0.3568}\right)(100) = 4.62\%$$

EXAMPLE 9.7 An open feedwater heater is added to the reheat cycle of Example 9.5. Steam is extracted where the reheater interrupts the turbine flow. Determine the efficiency of this reheat/regeneration cycle.

Solution: A *T-s* diagram (Fig. 9-12*a*) is sketched to aid in the calculations. From the steam tables or from Example 9.5,

$$h_1 \simeq h_2 = 94 \text{ Btu/lbm} \qquad {}_7 \simeq h_8 = 236 \text{ Btu/lbm} \qquad h_3 = 1518 \text{ Btu/lbm}$$
$$h_5 = 1333 \text{ Btu/lbm} \qquad h_6 = 1082 \text{ Btu/lbm} \qquad h_4 = 1200 \text{ Btu/lbm}$$

Continuity and the first law applied to the heater give [see (*9.8*)]

$$m_4 = \frac{h_8 - h_2}{h_4 - h_2} = \frac{236 - 94}{1200 - 94} = 0.128 \text{ lbm} \qquad \text{and} \qquad m_2 = 0.872 \text{ lbm}$$

The turbine work output is then

$$w_T = h_3 - h_4 + (h_5 - h_6)m_2 = 1518 - 1200 + (1333 - 1082)(0.872) = 537 \text{ Btu/lbm}$$

The energy input is $q_B = h_3 - h_8 = 1518 - 236 = 1282$ Btu/lbm. The efficiency is calculated to be

$$\eta = \frac{537}{1282} = 0.419 \quad \text{or } 41.9\%$$

Note the substantial improvement in cycle efficiency.

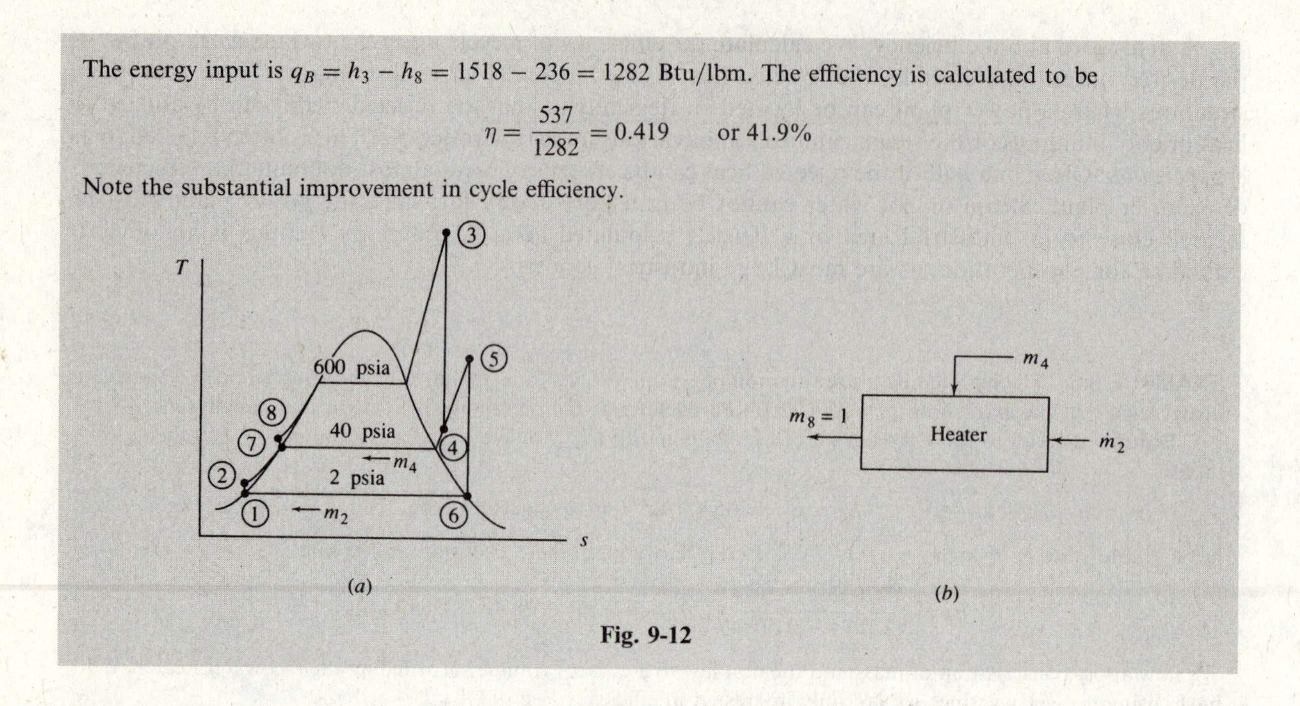

Fig. 9-12

9.6 THE SUPERCRITICAL RANKINE CYCLE

The Rankine cycle and variations of the Rankine cycle presented thus far have involved heat addition during the vaporization process; this heat transfer process occurs at a relatively low temperature, say 250 °C, at a pressure of 4 MPa, yet the hot gases surrounding the boiler after combustion are around 2500 °C. This large temperature difference makes the heat transfer process quite irreversible; recall that to approach reversibility the heat transfer process must occur over a small temperature difference. Hence, to improve the plant efficiency it is desirable to increase the temperature at which the heat transfer takes place. This will, of course, also improve the cycle efficiency since the area representing work will be increased. To get closer to the Carnot cycle efficiency, the temperature of the working fluid should be as near the temperature of the hot gases as possible. The *supercritical Rankine cycle* accomplishes this, as sketched on the *T-s* diagram in Fig. 9.13*a*. Note that the quality

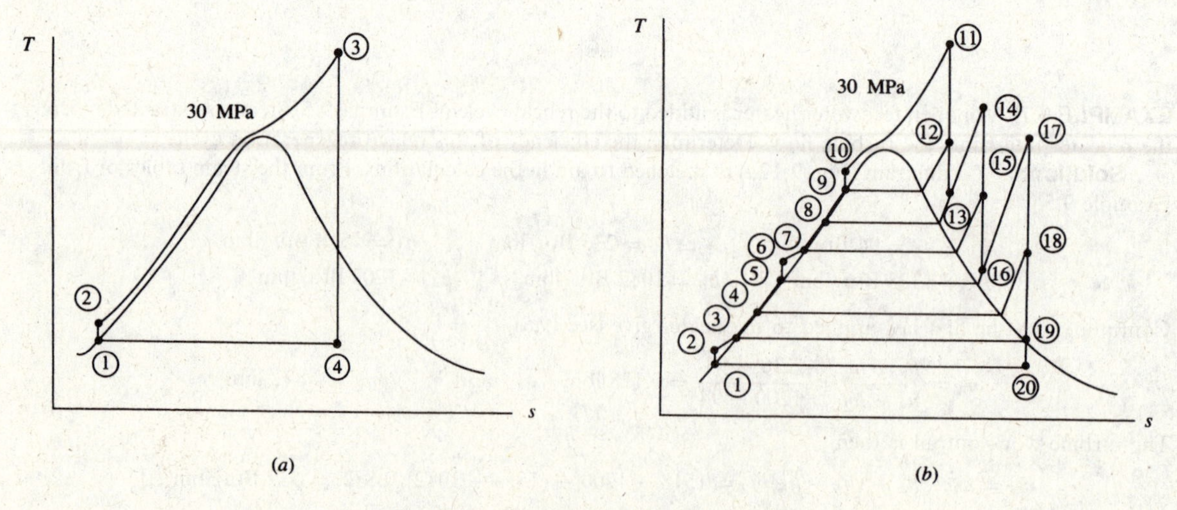

Fig. 9-13 Supercritical Rankine cycles.

region is never entered during the heat-addition process. At these high pressures the pipes and associated fluid handling equipment must be quite massive, capable of resisting the large pressure forces. The added cost of this more massive structure must be justified by the increase in efficiency and power output.

If the high-pressure superheated steam is expanded isentropically (insulated and without losses) through the turbine to a relatively low condenser pressure, it is obvious that a Rankine cycle will result in too high a moisture content in the low-pressure portion of the turbine. To eliminate this problem two reheat stages may be employed, and to maximize the cycle efficiency several regenerative stages may be utilized. Figure 9-13*b* shows six regenerative stages and two reheat stages. Example 9.8 illustrates a cycle with two reheat and two regenerative stages.

EXAMPLE 9.8 A supercritical reheat/regeneration cycle is proposed to operate as shown in the *T-s* diagram in Fig. 9-14, with two reheat stages and two open feedwater heaters. Determine the maximum possible cycle efficiency.

Solution: The enthalpies are found from the steam tables to be

$$h_1 \simeq h_2 = 192\ \text{kJ/kg} \qquad h_4 \cong h_5 = 1087\ \text{kJ/kg} \qquad h_8 = 3674\ \text{kJ/kg}$$

$$h_3 = 505\ \text{kJ/kg} \qquad h_6 = 3444\ \text{kJ/kg} \qquad h_{10} = 3174\ \text{kJ/kg}$$

$$s_6 = s_7 = 6.2339 \qquad \therefore h_7 = \left(\frac{6.2339 - 6.0709}{6.3622 - 6.0709}\right)(2961 - 2801) + 2801 = 2891\ \text{kJ/kg}$$

$$s_8 = s_9 = 7.3696$$

$$\therefore h_9 = \left(\frac{7.3696 - 7.2803}{7.5074 - 7.2803}\right)(2870 - 2769) + 2769 = 2809\ \text{kJ/kg}$$

$$s_{10} = s_{11} = 8.0636\ \text{kJ/kg·K}$$

$$\therefore x_{11} = \frac{8.0636 - 0.6491}{7.5019} = 0.9883$$

$$\therefore h_{11} = 192 + (0.9883)(2393) = 2557\ \text{kJ/kg}$$

Next, we apply the first law to each of the two heaters. Assume that $\dot{m} = 1$ kg/s. The other mass fluxes are shown on the *T-s* diagram in Fig. 9-15. We find, from the first law applied to the high-pressure heater,

$$h_5 = h_7\dot{m}_7 + (1 - \dot{m}_7)h_3 \qquad \therefore \dot{m}_7 = \frac{h_5 - h_3}{h_7 - h_3} = \frac{1087 - 505}{2891 - 505} = 0.2439\ \text{kg/s}$$

From the first law applied to the low-pressure heater, we find

$$(1 - \dot{m}_7)h_3 = \dot{m}_9 h_9 + (1 - \dot{m}_7 - \dot{m}_9)h_2$$

$$\therefore \dot{m}_9 = \frac{(1 - \dot{m}_7)h_3 - h_2 + \dot{m}_7 h_2}{h_9 - h_2}$$

$$= \frac{(1 - 0.2439)(505) - 192 + (0.2434)(192)}{2809 - 192} = 0.0904\ \text{kg/s}$$

The power from the turbine is calculated to be

$$\dot{W}_T = (1)(h_6 - h_7) + (1 - \dot{m}_7)(h_8 - h_9) + (1 - \dot{m}_7 - \dot{m}_9)(h_{10} - h_{11})$$

$$= 3444 - 2891 + (0.7561)(3674 - 2809) + (0.6657)(3174 - 2557) = 1609\ \text{kW}$$

The boiler energy input is

$$\dot{Q}_B = (1)(h_6 - h_5) + (1 - \dot{m}_7)(h_8 - h_7) + (1 - \dot{m}_7 - \dot{m}_9)(h_{10} - h_9)$$

$$= 3444 - 1087 + (0.7561)(3674 - 2891) + (0.6657)(3174 - 2809) = 3192\ \text{kW}$$

The cycle efficiency is fairly high at

$$\eta = \frac{1609}{3192} = 0.504 \qquad \text{or } 50.4\%$$

This higher efficiency results from the extremely high pressure of 30 MPa during the heat addition process. The associated savings must justify the increased costs of the massive equipment needed in a high-pressure system.

Note: The fact that state 11 is in the quality region is not of concern since x_{11} is quite close to unity. As the next section demonstrates, losses will increase the entropy of state 11, with the result that state 11 will actually be in the superheated region.

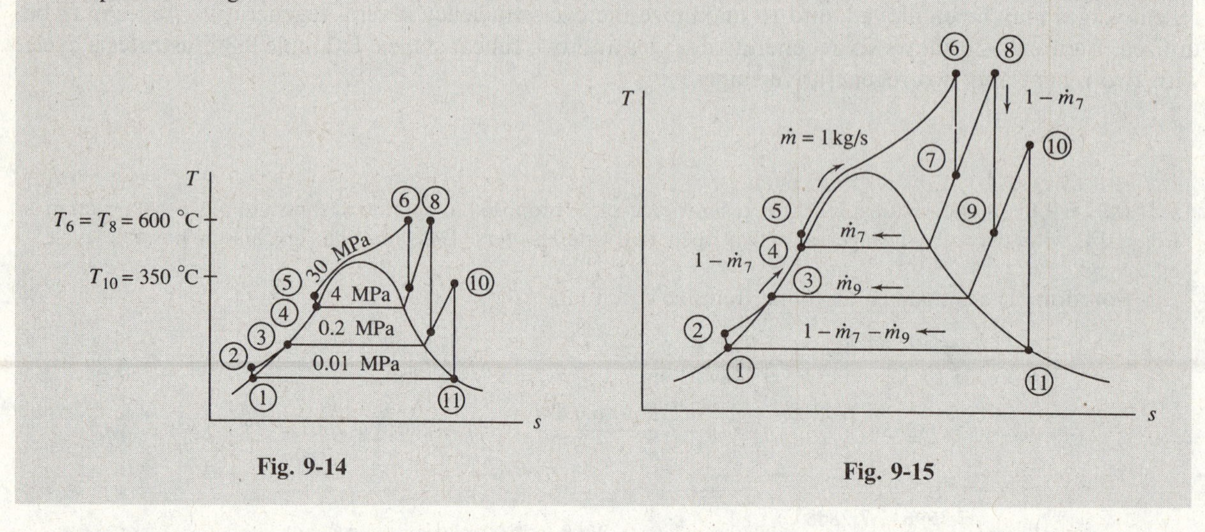

Fig. 9-14

Fig. 9-15

9.7 EFFECT OF LOSSES ON POWER CYCLE EFFICIENCY

The preceding sections dealt with ideal cycles assuming no pressure drop through the pipes in the boiler, no losses as the superheated steam passes over the blades in the turbine, no subcooling of the water leaving the condenser, and no pump losses during the compression process. The losses in the combustion process and the inefficiencies in the subsequent heat transfer to the fluid in the pipes of the boiler are not included here; those losses, which are in the neighborhood of 15 percent of the input energy in the coal or oil, would be included in the overall plant efficiency.

There is actually only one substantial loss that must be accounted for when we calculate the actual cycle efficiency: the loss that occurs when the steam is expanded through the rows of turbine blades in the turbine. As the steam passes over a turbine blade, there is friction on the blade and the steam may separate from the rear portion of the blade. In addition, heat transfer from the turbine may occur, although this is usually quite small. These losses result in a turbine efficiency of 80 to 89 percent. Turbine efficiency is defined as

$$\eta_T = \frac{w_a}{w_s} \tag{9.9}$$

where w_a is the actual work and w_s is the isentropic work.

The definition of pump efficiency, with pump work taken into account, is

$$\eta_P = \frac{w_s}{w_a} \tag{9.10}$$

where the isentropic work input is obviously less than the actual input.

There is a substantial loss in pressure, probably 10 to 20 percent, as the fluid flows from the pump exit through the boiler to the turbine inlet. The loss can be overcome by simply increasing the exit pressure from the pump. This does require more pump work, but the pump work is still less than 1 percent of the turbine output and is thus negligible. Consequently, we ignore the boiler pipe losses.

The condenser can be designed to operate such that the exiting water is very close to the saturated liquid condition. This will minimize the condenser losses so that they can also be neglected. The resulting

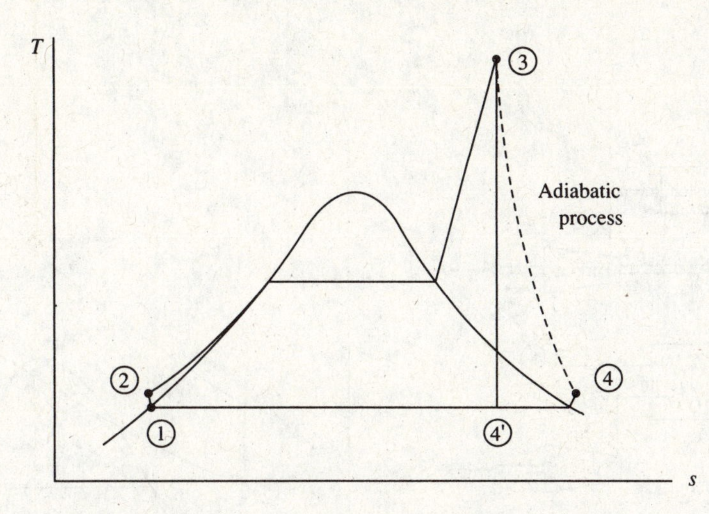

Fig. 9-16 The Rankine cycle with turbine losses.

actual Rankine cycle is shown on the T-s diagram in Fig. 9-16; the only significant loss is the turbine loss. Note the increase in entropy of state 4 as compared to state 3. Also, note the desirable effect of the decreased moisture content of state 4; in fact, state 4 may even move into the superheated region, as shown.

EXAMPLE 9.9 A Rankine cycle operates between pressures of 2 MPa and 10 kPa with a maximum temperature of 600 °C. If the insulated turbine has an efficiency of 80 percent, calculate the cycle efficiency and the temperature of steam at the turbine outlet.

Solution: From the steam tables we find $h_1 \simeq h_2 = 192$ kJ/kg, $h_3 = 3690$ kJ/kg, and $s_3 = 7.7032$ kJ/kg·K. Setting $s_{4'} = s_3$ we find the quality and enthalpy of state 4′ (see Fig. 9-16) to be

$$x_{4'} = \frac{7.7032 - 0.6491}{7.5019} = 0.9403 \qquad \therefore h_{4'} = 192 + (0.9403)(2393) = 2442 \text{ kJ/kg}$$

From the definition of turbine efficiency,

$$0.8 = \frac{w_a}{3690 - 2442} \qquad w_a = 998 \text{ kJ/kg}$$

The cycle efficiency is then

$$\eta = \frac{w_a}{q_B} = \frac{998}{3690 - 192} = 0.285 \qquad \text{or } 28.5\%$$

Note the substantial reduction from the ideal cycle efficiency of 35.7 percent as calculated in Example 9.3.

If we neglect kinetic and potential energy changes, the adiabatic process from state 3 to state 4 allows us to write

$$w_a = h_3 - h_4 \qquad 998 = 3690 - h_4 \qquad h_4 = 2692 \text{ kJ/kg}$$

At 10 kPa we find that state 4 is in the superheated region. The temperature is interpolated to be

$$T_4 = \left(\frac{2692 - 2688}{2783 - 2688}\right)(150 - 100) + 100 = 102\,°\text{C}$$

Obviously, the moisture problem has been eliminated by the losses in the turbine; the losses tend to act as a small reheater.

9.8 THE COMBINED BRAYTON-RANKINE CYCLE

The Brayton cycle efficiency is quite low primarily because a substantial amount of the energy input is exhausted to the surroundings. This exhausted energy is usually at a relatively high temperature and

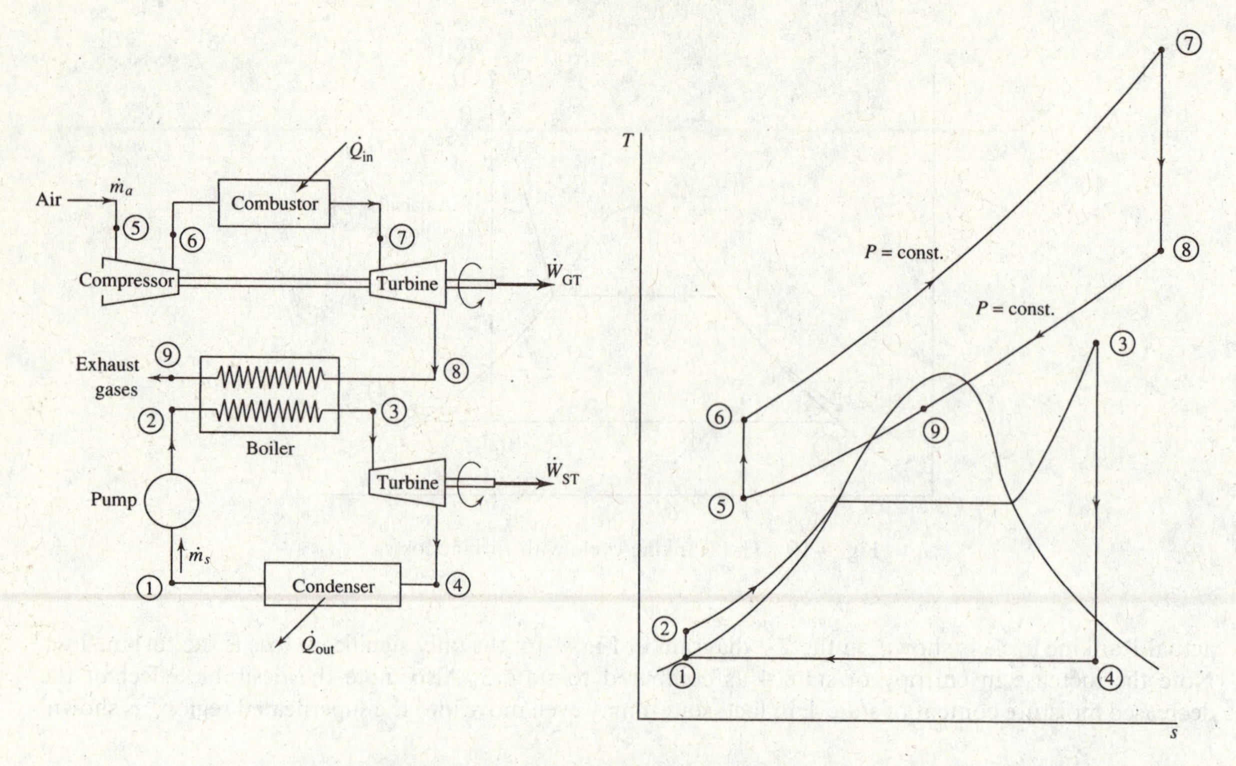

Fig. 9-17 The combined Brayton-Rankine cycle.

thus it can be used effectively to produce power. One possible application is the combined Brayton-Rankine cycle in which the high-temperature exhaust gases exiting the gas turbine are used to supply energy to the boiler of the Rankine cycle, as illustrated in Fig. 9-17. Note that the temperature T_9 of the Brayton cycle gases exiting the boiler is less than the temperature T_3 of the Rankine cycle steam exiting the boiler; this is possible in the counterflow heat exchanger, the boiler.

To relate the air mass flux $\dot{m}_a$ of the Brayton cycle to the steam mass flux $\dot{m}_s$ of the Rankine cycle, we use an energy balance in the boiler; it gives (see Fig. 9-17),

$$\dot{m}_a(h_8 - h_9) = \dot{m}_s(h_3 - h_2) \tag{9.11}$$

assuming no additional energy addition in the boiler, which would be possible with an oil burner, for example.

The cycle efficiency would be found by considering the purchased energy as $\dot{Q}_{in}$, the energy input in the combustor. The output is the sum of the net output $\dot{W}_{GT}$ from the gas turbine and the ouput $\dot{W}_{ST}$ from the steam turbine. The combined cycle efficiency is thus given by

$$\eta = \frac{\dot{W}_{GT} + \dot{W}_{ST}}{\dot{Q}_{in}} \tag{9.12}$$

An example will illustrate the increase in efficiency of such a combined cycle.

EXAMPLE 9.10 A simple steam power plant operates between pressures of 10 kPa and 4 MPa with a maximum temperature of 400°C. The power output from the steam turbine is 100 MW. A gas turbine provides the energy to the boiler; it accepts air at 100 kPa and 25°C, has a pressure ratio of 5, and a maximum temperature of 850°C. The exhaust gases exit the boiler at 350 K. Determine the thermal efficiency of the combined Brayton-Rankine cycle.

Solution: If we neglect the work of the pump, the enthalpy remains unchanged across the pump. Hence, $h_2 = h_1 = 192$ kJ/kg. At 400°C and 4 MPa we have $h_3 = 3214$ kJ/kg and $s_3 = 6.7698$ kJ/kg·K. State 4 is located by noting that $s_4 = s_3$ so that the quality is

$$x_4 = \frac{s_4 - s_f}{s_{fg}} = \frac{6.798 - 0.6491}{7.5019} = 0.8159$$

Thus, $h_4 = h_f + x_4 h_{fg} = 192 + (0.8159)(2393) = 2144$ kJ/kg. The steam mass flux is found using the turbine output as follows:

$$\dot{W}_{\text{ST}} = \dot{m}_s(h_3 - h_4) \qquad 100\,000 = \dot{m}_s(3214 - 2144) \qquad \dot{m}_s = 93.46 \text{ kg/s}$$

Considering the gas-turbine cycle,

$$T_6 = T_5\left(\frac{P_6}{P_5}\right)^{(k-1)/k} = (298)(5)^{0.2857} = 472.0 \text{ K}$$

Also,

$$T_8 = T_7\left(\frac{P_8}{P_7}\right)^{(k-1)/k} = (1123)\left(\frac{1}{5}\right)^{0.2857} = 709.1 \text{ K}$$

Thus we have, for the boiler,

$$\dot{m}_s(h_3 - h_2) = \dot{m}_a C_p(T_8 - T_9) \qquad (93.46)(3214 - 192) = (\dot{m}_a)(1.00)(709.1 - 350)$$

$$\dot{m}_a = 786.5 \text{ kg/s}$$

The output of the gas turbine is (note that this is not $\dot{W}_{GT}$)

$$\dot{W}_{\text{turb}} = \dot{m}_a C_p(T_7 - T_8) = (786.5)(1.00)(1123 - 709.1) = 325.5 \text{ MW}$$

The energy needed by the compressor is

$$\dot{W}_{\text{comp}} = \dot{m}_a C_p(T_6 - T_5) = (786.5)(1.00)(472 - 298) = 136.9 \text{ MW}$$

Hence, the net gas turbine output is $\dot{W}_{\text{GT}} = \dot{W}_{\text{turb}} - \dot{W}_{\text{comp}} = 325.5 - 136.9 = 188.6$ MW. The energy input by the combustor is

$$\dot{Q}_{\text{in}} = \dot{m}_a C_p(T_7 - T_6) = (786.5)(1.00)(1123 - 472) = 512 \text{ MW}$$

The above calculations allow us to determine the combined cycle efficiency as

$$\eta = \frac{\dot{W}_{\text{ST}} + \dot{W}_{\text{GT}}}{\dot{Q}_{\text{in}}} = \frac{100 + 188.6}{512} = 0.564 \quad \text{or } 56.4\%$$

Note that this efficiency is 59.3 percent higher than the Rankine cycle (see Example 9.2) and 52.8 percent higher than the Brayton cycle (see Example 8.9). Cycle efficiency could be increased even more by using steam reheaters, steam regenerators, gas intercoolers, and gas reheaters.

Solved Problems

9.1 A steam power plant is designed to operate on a Rankine cycle with a condenser outlet temperature of 80°C and boiler outlet temperature of 500°C. If the pump outlet pressure is 2 MPa, calculate the maximum possible thermal efficiency of the cycle. Compare with the efficiency of a Carnot engine operating between the same temperature limits.

To calculate the thermal efficiency we must determine the turbine work and the boiler heat transfer. The turbine work is found as follows (refer to Fig. 9-1):

At state 3: $h_3 = 3468$ kJ/kg $\quad s_3 = 7.432$ kJ/kg·K

At state 4: $s_4 = s_3 = 7.432 = 1.075 + 6.538x_4$

Thus $x_4 = 0.9723, h_4 = 335 + (0.9723)(2309) = 2580$ kJ/kg, and $w_T = h_3 - h_4 = 3468 - 2580 = 888$ kJ/kg. The boiler heat, assuming that $h_2 = h_1$ (the pump work is negligible), is $q_B = h_3 - h_2 = 3468 - 335 = 3133$ kJ/kg. The cycle efficiency is then

$$\eta = \frac{w_T}{q_B} = \frac{888}{3133} = 0.283 \quad \text{or } 28.3\%$$

The efficiency of a Carnot cycle operating between the high and low temperatures of this cycle is

$$\eta = 1 - \frac{T_L}{T_H} = 1 - \frac{353}{773} = 0.543 \quad \text{or } 54.3\%$$

9.2 For the ideal Rankine cycle shown in Fig. 9-18 determine the mass flow rate of steam and the cycle efficiency.

The turbine output is shown to be 20 MW. Referring to Fig. 9-1, we find

$$h_3 = 3422 \text{ kJ/kg}, \qquad s_3 = 6.881 \text{ kJ/kg·K}$$

$$s_4 = s_3 = 6.881 = 0.649 + 7.502x_4$$

$$\therefore x_4 = 0.8307 \qquad \therefore h_4 = 192 + (0.8307)(2393) = 2180 \text{ kJ/kg}$$

The mass flux is now calculated to be

$$\dot{m} = \frac{\dot{W}_T}{w_T} = \frac{\dot{W}_T}{h_3 - h_4} = \frac{20\,000}{3422 - 2180} = 16.1 \text{ kg/s}$$

The boiler heat transfer, neglecting the pump work so that $h_2 \cong h_1$, is

$$q_B = h_3 - h_2 = 3422 - 192 = 3230 \text{ kJ/kg}$$

The cycle efficiency is found to be

$$\eta = \frac{\dot{W}_T}{\dot{Q}_B} = \frac{\dot{W}_T}{\dot{m}q_B} = \frac{20\,000}{(16.1)(3230)} = 0.385 \quad \text{or } 38.5\%$$

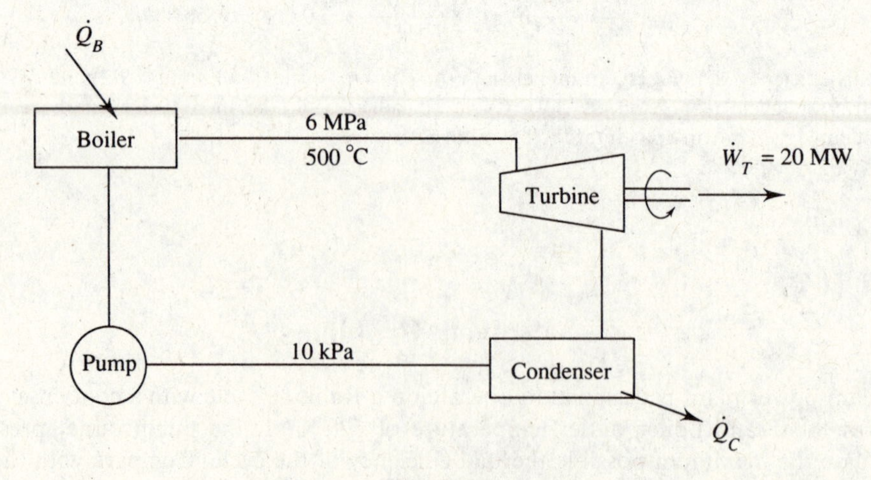

Fig. 9-18

9.3 A solar bank of collectors with an area of 8000 ft^2 supplies energy to the boiler of a Rankine cycle power plant. At peak load the collectors provide 200 Btu/ft^2-hr to the working fluid. The R134a working fluid leaves the boiler at 300 psia and 240 °F and enters the pump at 100 °F. Determine (*a*) the pump work, (*b*) the cycle efficiency, (*c*) the mass flux of the R134a, and (*d*) the maximum power output.

(*a*) The pump work requirement for this ideal cycle is (refer to Fig. 9-1)

$$w_P = (P_2 - P_1)v = [(300 - 138.8)(144)](0.01385)$$
$$= 321.5 \text{ ft-lbf/lbm} \quad \text{or } 0.413 \text{ Btu/lbm}$$

(*b*) To calculate the thermal efficiency we must know the boiler heat input. It is $q_B = h_3 - h_2 = 146.2 - (44.23 + 0.413) = 101.6$ Btu/lbm, where the enthalpy at the pump outlet, state 2, is the inlet enthalpy h_1 plus w_P.

We must also calculate the turbine work output. To locate state 4 we use the entropy as follows: $s_3 = s_4 = 0.2537$ Btu/lbm-°R. This is in the superheated region. Interpolating for the state at $P_4 = 138.8$ psia and $s_4 = 0.2537$, we find that $h_4 = 137.5$ Btu/lbm. This result requires a double interpolation, so care must be taken. The turbine work is thus

$$w_T = h_3 - h_4 = 146.2 - 137.5 = 8.7 \text{ Btu/lbm}$$

The cycle efficiency is

$$\eta = \frac{w_T - w_P}{q_B} = \frac{8.7 - 0.413}{101.6} = 0.082 \quad \text{or } 8.2\%$$

(*c*) To find the mass flux, we use the total heat flux input from the collectors: $\dot{Q}_B = (200)(8000) = \dot{m}q_B = \dot{m}(101.6)$. This results in $\dot{m} = 15{,}700$ lbm/hr or 4.37 lbm/sec.

(*d*) The maximum power output is $\dot{W}_T = \dot{m}w_T = (15{,}700)(8.7) = 137{,}000$ Btu/hr or 53.7 hp. We have used the conversion 2545 Btu/hr = 1 hp.

9.4 The steam of a Rankine cycle, operating between 4 MPa and 10 kPa, is reheated at 400 kPa to 400 °C. Determine the cycle efficiency if the maximum temperature is 600 °C.

Referring to Fig. 9-6, we find from the steam tables the following:

$$h_2 \cong h_1 = 191.8 \text{ kJ/kg}, \qquad h_3 = 3674.4 \text{ kJ/kg}, \qquad h_5 = 3273.4 \text{ kJ/kg},$$
$$s_4 = s_3 = 7.369 \text{ kJ/kg·K} \qquad s_6 = s_5 = 7.899 \text{ kJ/kg·K}$$

For the two isentropic processes we calculate the following:

$$\left.\begin{aligned} s_4 &= 7.369 \\ P_4 &= 400 \text{ kPa} \end{aligned}\right\} \quad \text{Interpolate: } h_4 = 2960 \text{ kJ/kg}$$

$$s_6 = 7.898 = 0.649 + 7.501x_6 \qquad \therefore x_6 = 0.9664 \qquad \therefore h_6 = 191.8 + 0.9664 \times 2392.8 = 2504 \text{ kJ/kg}$$

The heat transfer to the boiler is

$$q_B = h_3 - h_2 + h_5 - h_4 = 3674 - 192 + 3273 - 2960 = 3795 \text{ kJ/kg}$$

The work output from the turbine is

$$w_T = h_3 - h_4 + h_5 - h_6 = 3674 - 2960 + 3273 - 2504 = 1483 \text{ kJ/kg}$$

The cycle efficiency is finally calculated to be

$$\eta = \frac{w_T}{q_B} = \frac{1483}{3795} = 0.391 \quad \text{or } 39.1\%$$

9.5 An ideal reheat Rankine cycle operates between 8 MPa and 4 kPa with a maximum temperature of 600 °C (Fig. 9-19). Two reheat stages, each with a maximum temperature of 600 °C, are to be added at 1 MPa and 100 kPa. Calculate the resulting cycle efficiency.

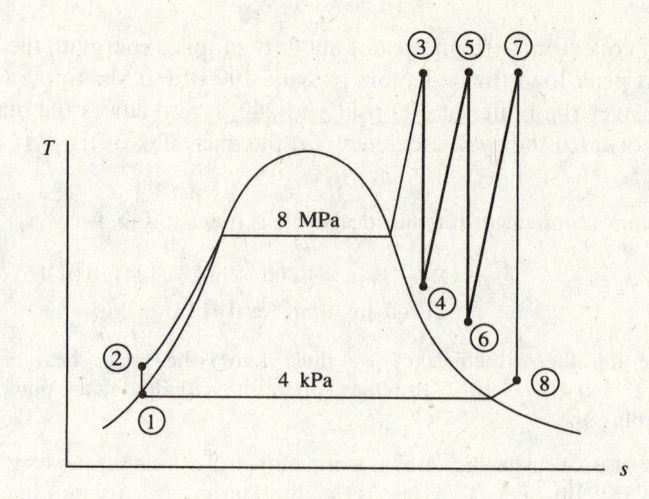

Fig. 9-19

From the steam tables we find

$$h_1 \cong h_2 = 121.5 \text{ kJ/kg} \qquad h_3 = 3642 \text{ kJ/kg} \qquad h_5 = 3698 \text{ kJ/kg} \qquad h_7 = 3705 \text{ kJ/kg}$$
$$s_3 = s_4 = 7.021 \text{ kJ/kg·K} \qquad s_5 = s_6 = 8.030 \text{ kJ/kg·K} \qquad s_7 = s_8 = 9.098 \text{ kJ/kg·K}$$

We interpolate at each of the superheated states 4, 6, and 8:

$$\left.\begin{array}{l} s_4 = 7.021 \text{ kJ/kg·K} \\ P_4 = 1 \text{ MPa} \end{array}\right\} \therefore h_4 = 2995 \text{ kJ/kg} \qquad \left.\begin{array}{l} s_6 = 8.030 \text{ kJ/kg·K} \\ P_6 = 100 \text{ kPa} \end{array}\right\} \therefore h_6 = 2972 \text{ kJ/kg}$$

$$\left.\begin{array}{l} s_8 = 9.098 \text{ kJ/kg·K} \\ P_8 = 4 \text{ kPa} \end{array}\right\} \therefore h_8 = 2762 \text{ kJ/kg}$$

The boiler heat transfer is

$$q_B = h_3 - h_2 + h_5 - h_4 + h_7 - h_6 = 3642 - 122 + 3698 - 2995 + 3705 - 2972 = 4956 \text{ kJ/kg}$$

The turbine work is

$$w_T = h_3 - h_4 + h_5 - h_6 + h_7 - h_8 = 3642 - 2995 + 3698 - 2972 + 3705 - 2762 = 2316 \text{ kJ/kg}$$

The cycle efficiency is then calculated to be

$$\eta = \frac{w_T}{q_B} = \frac{2316}{4956} = 0.467 \quad \text{or } 46.7\%$$

9.6 The condenser pressure of a regenerative cycle is 3 kPa and the feedwater pump provides a pressure of 6 MPa to the boiler. Calculate the cycle efficiency if one open feedwater heater is to be used. The maximum temperature is 600 °C.

The pressure at which the steam passing through the turbine is intercepted is estimated by selecting a saturation temperature half way between the boiler saturation temperature and the condenser saturation temperature; i.e., referring to Fig. 9-7, $T_6 = (\frac{1}{2})(275.6 + 24.1) = 149.8\,^\circ\text{C}$. The closest pressure entry to this saturation temperature is at 400 kPa. Hence, this is the selected pressure for the feedwater heater. Using the steam tables, we find

$$h_2 \cong h_1 = 101 \text{ kJ/kg} \qquad h_7 \cong h_6 = 604.3 \text{ kJ/kg}$$
$$h_3 = 3658.4 \text{ kJ/kg} \qquad s_3 = s_4 = s_5 = 7.168 \text{ kJ/kg·K}$$

For the isentropic processes we find

$$\left.\begin{array}{l} s_5 = 7.168 \text{ kJ/kg·K} \\ P_5 = 0.4 \text{ MPa} \end{array}\right\} \quad \therefore h_5 = 2859 \text{ kJ/kg}$$

$$s_4 = 7.168 = 0.3545 + 8.2231x_4 \qquad \therefore x_4 = 0.8286 \qquad \therefore h_4 = 101 + (0.8286)(2444.5) = 2126 \text{ kJ/kg}$$

If we assume $\dot{m}_6 = 1$ kg/s, we find from (*9.8*) that

$$\dot{m}_5 = \frac{h_6 - h_2}{h_5 - h_2}\dot{m}_6 = \left(\frac{640 - 101}{2859 - 101}\right)(1) = 0.195 \text{ kg/s}$$

Then we have:

$$\dot{m}_2 = \dot{m}_6 - \dot{m}_5 = 1 - 0.195 = 0.805 \text{ kg/s}$$
$$\dot{Q}_B = \dot{m}_6(h_3 - h_7) = (1)(3658 - 604) = 3054 \text{ kW}$$
$$\dot{W}_T = \dot{m}_6(h_3 - h_5) + \dot{m}_2(h_5 - h_4) = (1)(3658 - 2859) + (0.805)(2859 - 2126) = 1389 \text{ kW}$$

The cycle efficiency is finally calculated to be

$$\eta = \frac{\dot{W}_T}{\dot{Q}_B} = \frac{1389}{3054} = 0.455 \quad \text{or } 45.5\%$$

9.7 For the regenerative cycle shown in Fig. 9-20 determine the thermal efficiency, the mass flux of steam, and the ratio of rejected heat to added heat. Neglect pump work.

Referring to Fig. 9-7*b* to identify the states and using the steam tables, we find

$$h_2 \cong h_1 = 191.8 \text{ kJ/kg} \qquad h_6 \cong h_7 = 762.8 \text{ kJ/kg} \qquad h_3 = 3625.3 \text{ kJ/kg}$$

The enthalpies of states 4 and 5 are determined by assuming an isentropic process as follows:

$$\left.\begin{array}{l} s_5 = s_3 = 6.904 \text{ kJ/kg·K} \\ P_5 = 1 \text{ MPa} \end{array}\right\} \quad \therefore h_5 = 2932 \text{ kJ/kg}$$

$$s_4 = s_3 = 6.904 = 0.6491 + 7.5019x_4 \qquad \therefore x_4 = 0.8338$$
$$\therefore h_4 = 191.8 + (0.8338)(2392.8) = 2187 \text{ kJ/kg}$$

An energy balance on the heater, which is assumed insulated, is $\dot{m}_5(h_5 - h_6) = \dot{m}_2(h_7 - h_2)$. A mass balance provides (see Fig. 9-10) $\dot{m}_7 = \dot{m}_5 + \dot{m}_2$. Assuming $\dot{m}_7 = 1$ kg/s, the above two equations are combined to give

$$\dot{m}_2 = \frac{h_5 - h_6}{h_7 - h_2 + h_5 - h_6} = \frac{2932 - 763}{763 - 192 + 2932 - 763} = 0.792 \text{ kg/s}$$

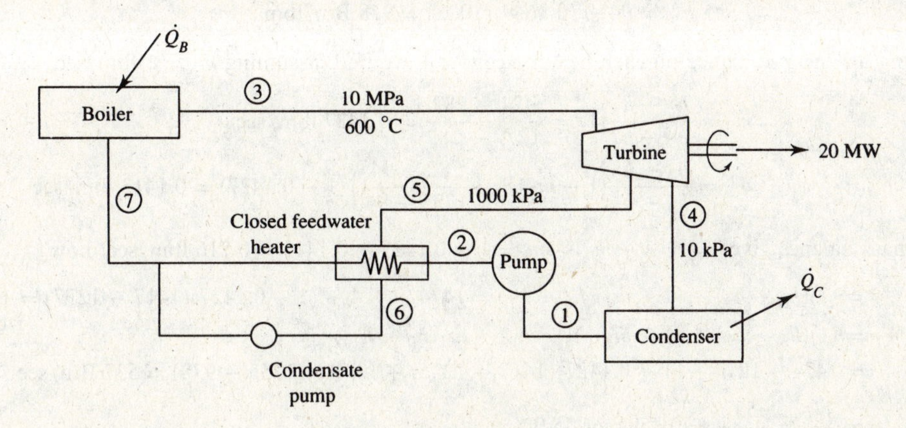

Fig. 9-20

We then have $\dot{m}_5 = 1 - \dot{m}_2 = 1 - 0.792 = 0.208$ kg/s. The turbine power (with $\dot{m}_7 = 1$ kg/s) can now be calculated to be

$$\dot{W}_T = \dot{m}_7(h_3 - h_5) + \dot{m}_2(h_5 - h_4) = (1.0)(3625 - 2932) + (0.792)(2932 - 2187) = 1283 \text{ kW}$$

The boiler heat rate is

$$\dot{Q}_B = \dot{m}_7(h_3 - h_7) = (1.0)(3625 - 763) = 2862 \text{ kW}$$

The cycle efficiency is calculated to be

$$\eta = \frac{\dot{W}_T}{\dot{Q}_B} = \frac{1283}{2862} = 0.448 \quad \text{or } 44.8\%$$

The mass flux of steam is found as

$$\dot{m}_7 = \frac{\dot{W}_T}{(\dot{W}_T)_{\text{with } \dot{m}_7 = 1}} = \frac{20}{1.283} = 15.59 \text{ kg/s}$$

The ratio of rejected heat to added heat is

$$\frac{\dot{Q}_C}{\dot{Q}_B} = \frac{\dot{Q}_B - \dot{W}_T}{\dot{Q}_B} = 1 - \frac{\dot{W}_T}{\dot{Q}_B} = 1 - \frac{1283}{2862} = 0.552$$

9.8 A power plant operates on a reheat/regenerative cycle in which steam at 1000 °F and 2000 psia enters the turbine. It is reheated at a pressure of 400 psia to 800 °F and has two open feedwater heaters, one using extracted steam at 400 psia and the other using extracted steam at 80 psia. Determine the thermal efficiency if the condenser operates at 2 psia.

Refer to the *T-s* diagram of Fig. 9-11 to identify the various states. The pump power requirements are negligible. From the steam tables the enthalpies are

$$h_2 \cong h_1 = 94 \text{ Btu/lbm} \qquad h_3 = 282 \text{ Btu/lbm} \qquad h_5 = 424 \text{ Btu/lbm}$$
$$h_6 = 1474 \text{ Btu/lbm} \qquad h_8 = 1417 \text{ Btu/lbm}$$

The enthalpies of states 7, 9, and 10 are found assuming isentropic processes as follows:

$$\left.\begin{array}{l} s_7 = s_6 = 1.560 \text{ Btu/lbm-°R} \\ P_7 = 400 \text{ psia} \end{array}\right\} \quad \therefore h_7 = 1277 \text{ Btu/lbm}$$

$$\left.\begin{array}{l} s_9 = s_8 = 1.684 \text{ Btu/lbm-°R} \\ P_9 = 80 \text{ psi} \end{array}\right\} \quad \therefore h_9 = 1235 \text{ Btu/lbm}$$

$$s_{10} = s_8 = 1.684 = 0.17499 + 1.7448x_{10} \qquad \therefore x_{10} = 0.8649$$
$$\therefore h_{10} = 94 + (0.8649)(1022) = 978 \text{ Btu/lbm}$$

Using an energy balance on each heater [see (*9.8*)], we find, assuming $\dot{m}_5 = 1$ lbm/sec,

$$\dot{m}_7 = \frac{h_5 - h_3}{h_7 - h_3}(1) = \frac{424 - 282}{1277 - 282} = 0.1427 \text{ lbm/sec}$$
$$\dot{m}_9 = \frac{h_3 - h_2}{h_9 - h_2}(1 - \dot{m}_7) = \left(\frac{282 - 94}{1235 - 94}\right)(1 - 0.1427) = 0.1413 \text{ lbm/sec}$$

A mass balance gives $\dot{m}_2 = 1 - \dot{m}_7 - \dot{m}_9 = 1 - 0.1427 - 0.1413 = 0.716$ lbm/sec; now

$$\dot{Q}_B = (1)(h_6 - h_5) + (1 - \dot{m}_7)(h_8 - h_7) = 1474 - 424 + (1 - 0.1427)(1417 - 1277) = 1170 \text{ Btu/sec}$$
$$\dot{W}_T = (1)(h_6 - h_7) + (1 - \dot{m}_7)(h_8 - h_9) + \dot{m}_2(h_9 - h_{10})$$
$$= 1474 - 1277 + (1 - 0.1427)(1417 - 1235) + (0.716)(1235 - 978) \doteq 537 \text{ Btu/sec}$$
$$\eta = \frac{\dot{W}_T}{\dot{Q}_B} = \frac{537}{1170} = 0.459 \quad \text{or } 45.9\%$$

9.9 The turbine of Prob. 9.2 is 87 percent efficient. Determine the mass flow rate and the cycle efficiency with $\dot{W}_T = 20$ MW.

Referring to Fig. 9-16 and using the steam tables, we find the following enthalpies:

$$h_3 = 3422 \text{ kJ/kg} \qquad h_2 \cong h_1 = 192 \text{ kJ/kg} \qquad s_{4'} = s_3 = 6.881 = 0.649 + 7.502x_{4'}$$

$$\therefore x_{4'} = 0.8307 \qquad \therefore h_{4'} = 192 + (0.8307)(2393) = 2180 \text{ kJ/kg}$$

The calculation is completed as follows:

$$w_s = h_3 - h_{4'} = 3422 - 2180 = 1242 \text{ kJ/kg}$$

$$w_a = \eta_T w_s = (0.87)(1242) = 1081 \text{ kJ/kg}$$

$$\dot{m} = \frac{\dot{W}_T}{w_a} = \frac{20\,000}{1081} = 18.5 \text{ kg/s}$$

$$\eta = \frac{\dot{W}_T}{\dot{Q}_B} = \frac{\dot{W}_T}{\dot{m}(h_3 - h_2)} = \frac{20\,000}{(18.5)(3422 - 192)} = 0.317 \quad \text{or } 31.7\%$$

9.10 The turbine of a Rankine cycle operating between 4 MPa and 10 kPa is 84 percent efficient. If the steam is reheated at 400 kPa to 400 °C, determine the cycle efficiency. The maximum temperature is 600 °C. Also, calculate the mass flux of condenser cooling water if it increases 10 °C as it passes through the condenser when the cycle mass flux of steam is 10 kg/s.

Referring to Figs. 9-6 and 9-16 and using the steam tables, we find the following enthalpies:

$$h_2 \cong h_1 = 192 \text{ kJ/kg} \qquad h_3 = 3674 \text{ kJ/kg} \qquad h_5 = 3273 \text{ kJ/kg}$$

$$\left.\begin{array}{l} s_{4'} = s_3 = 7.369 \text{ kJ/kg·K} \\ P_4 = 400 \text{ kPa} \end{array}\right\} \quad \therefore h_{4'} = 2960 \text{ kJ/kg}$$

$$s_{6'} = s_5 = 7.899 = 0.649 + 7.501x_{6'} \qquad \therefore x_{6'} = 0.9665$$

$$\therefore h_{6'} = 192 + (0.9665)(2393) = 2505 \text{ kJ/kg}$$

We find the actual work from the turbine to be

$$w_T = \eta_T(h_3 - h_{4'}) + \eta_T(h_5 - h_{6'}) = (0.84)(3674 - 2960) + (0.84)(3273 - 2505) = 1247 \text{ kJ/kg}$$

To find the boiler heat requirement, we must calculate the actual h_4:

$$\eta_T = \frac{w_a}{w_s} = \frac{h_3 - h_4}{h_3 - h_{4'}} \qquad 0.84 = \frac{3674 - h_4}{3674 - 2960} \qquad h_4 = 3074 \text{ kJ/kg}$$

Then

$$q_B = h_3 - h_2 + h_5 - h_4 = 3674 - 192 + 3273 - 3074 = 3681 \text{ kJ/kg}$$

$$\eta = \frac{w_T}{q_B} = \frac{1247}{3681} = 0.339 \quad \text{or } 33.9\%$$

To find the heat rejected by the condenser we must determine the actual h_6:

$$\eta_T = \frac{w_a}{w_s} = \frac{h_5 - h_6}{h_5 - h_{6'}} \qquad 0.84 = \frac{3273 - h_6}{3273 - 2505} \qquad h_6 = 2628 \text{ kJ/kg}$$

Thus $\dot{Q}_C = \dot{m}(h_6 - h_1) = (10)(2628 - 192) = 24.36$ MW. Because this heat is carried away by the cooling water,

$$\dot{Q}_w = \dot{m}_w c_p \Delta T_w \qquad 24\,360 = \dot{m}_w(4.18)(10) \qquad \dot{m}_w = 583 \text{ kg/s}$$

9.11 A gas-turbine cycle inlets 20 kg/s of atmospheric air at 15 °C, compresses it to 1200 kPa, and heats it to 1200 °C in a combustor. The gases leaving the turbine heat the steam of a Rankine cycle to 350 °C and exit the heat exchanger (boiler) at 100 °C. The pump of the Rankine cycle operates between 10 kPa and 6 MPa. Calculate the maximum power ouput of the combined cycle and the combined cycle efficiency.

The temperature of gases leaving the gas turbine is (see Fig. 9-17)

$$T_8 = T_7\left(\frac{P_8}{P_7}\right)^{(k-1)/k} = (1473)\left(\frac{100}{1200}\right)^{0.2857} = 724.2 \text{ K}$$

The temperature of the air exiting the compressor is

$$T_6 = T_5\left(\frac{P_6}{P_5}\right)^{(k-1)/k} = (288)\left(\frac{1200}{100}\right)^{0.2857} = 585.8 \text{ K}$$

The net power output of the gas turbine is then

$$\begin{aligned}\dot{W}_{GT} = \dot{W}_{turb} - \dot{W}_{comp} &= \dot{m}C_p(T_7 - T_8) - \dot{m}C_p(T_6 - T_5)\\ &= (20)(1.00)(1473 - 724.2 - 585.8 + 288) = 9018 \text{ kW}\end{aligned}$$

The temperature exiting the condenser of the Rankine cycle is 45.8 °C. An energy balance on the boiler heat exchanger allows us to find the mass flux $\dot{m}_s$ of the steam:

$$\dot{m}_a C_p(T_8 - T_9) = \dot{m}_s(h_3 - h_2) \qquad (20)(1.00)(724.2 - 100) = \dot{m}_s(3043 - 191.8)$$
$$\dot{m}_s = 3.379 \text{ kg/s}$$

The isentropic process $3 \to 4$ allows h_4 to be found:

$$s_4 = s_3 = 6.3342 = 0.6491 + 7.5019x_4 \qquad \therefore x_4 = 0.7578$$
$$\therefore h_4 = 191.8 + (0.7578)(2392.8) = 2005 \text{ kJ/kg}$$

The steam turbine output is $\dot{W}_{ST} = \dot{m}(h_3 - h_4) = (3.379)(3043 - 2005) = 3507$ kW. The maximum power output (we have assumed ideal processes in the cycles) is, finally,

$$\dot{W}_{out} + \dot{W}_{GT} + \dot{W}_{ST} = 9018 + 3507 = 12\,525 \text{ kW} \quad \text{or } 12.5 \text{ MW}$$

The energy input to this combined cycle is $\dot{Q}_{in} = \dot{m}_a C_p(T_7 - T_6) = (20)(1.00)(1473 - 585.8) = 17.74$ MW. The cycle efficiency is then

$$\eta = \frac{\dot{W}_{out}}{\dot{Q}_{in}} = \frac{12.5}{17.74} = 0.70$$

Supplementary Problems

9.12 A power plant operating on an ideal Rankine cycle has steam entering the turbine at 500 °C and 2 MPa. If the steam enters the pump at 10 kPa, calculate (*a*) the thermal efficiency with pump work included, (*b*) the thermal efficiency neglecting pump work, and (*c*) the percentage error in efficiency neglecting pump work.

9.13 An ideal Rankine cycle operates between temperatures of 500 °C and 60 °C. Determine the cycle efficiency and the quality of the turbine outlet steam if the pump outlet pressure is (*a*) 2 MPa, (*b*) 6 MPa, and (*c*) 10 MPa.

9.14 The influence of maximum temperature on the efficiency of a Rankine cycle is desired. Holding the maximum and minimum pressures constant at 1000 psia and 2 psia, respectively, what is the thermal efficiency if the boiler outlet steam temperature is (*a*) 800 °F, (*b*) 1000 °F, and (*c*) 1200 °F?

9.15 A power plant is to be operated on an ideal Rankine cycle with the superheated steam exiting the boiler at 4 MPa and 500 °C. Calculate the thermal efficiency and the quality at the turbine outlet if the condenser pressure is (*a*) 20 kPa, (*b*) 10 kPa, and (*c*) 8 kPa.

9.16 A power plant operates on a Rankine cycle between temperatures of 600 °C and 40 °C. The maximum pressure is 8 MPa and the turbine output is 20 MW. Determine the minimum mass flow rate of cooling water through the condenser if a maximum temperature differential of 10 °C is allowed.

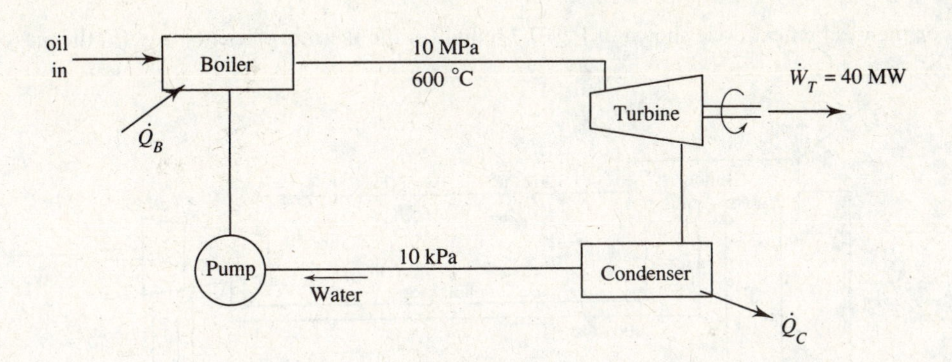

Fig. 9-21

9.17 Oil, with a heating value of 30 MJ/kg, is used in the boiler shown schematically in Fig. 9-21. If 85 percent of the energy is transferred to the working fluid, how much oil is needed per hour?

9.18 Hot geyser water at 95 °C is available to supply energy to the boiler of a Rankine cycle power plant. R134a is the working fluid. The maximum possible mass flux of hot water is 2.0 kg/s. The R134a exits the boiler as saturated vapor at 80 °C, and the condenser temperature is 40 °C. Calculate (*a*) pump work rate, (*b*) the thermal efficiency, and (*c*) the maximum possible power output. Assume that the hot water can equal the R134a temperature as it leaves the boiler.

9.19 Coal, with a heating value of 2500 Btu/lbm, is used to provide energy to the working fluid in a boiler which is 85 percent efficient. Determine the minimum mass flux of coal, in lbm/hr, that would be necessary for the turbine output to be 100 MW. The pump receives water at 2 psia, in the simple Rankine cycle, and delivers it to the boiler at 2000 psia. Superheated steam is to leave the boiler at 1000 °F.

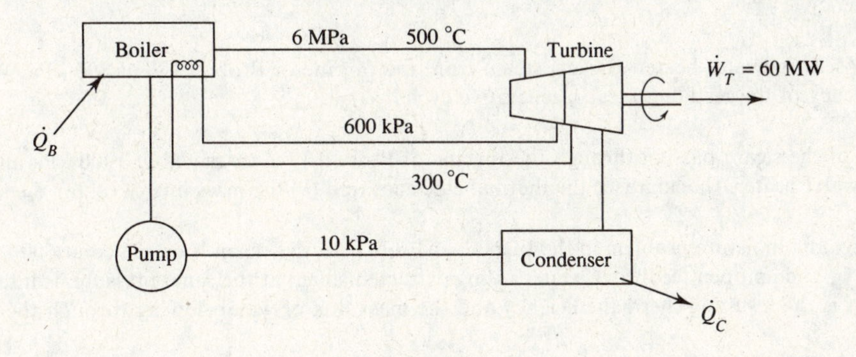

Fig. 9-22

9.20 For the ideal reheat cycle shown in Fig. 9-22, calculate the thermal efficiency and the pump mass flux.

9.21 The steam passing through the turbine of the power cycle of Prob. 9.12 is reheated at 100 kPa to 400 °C. Find the thermal efficiency.

9.22 The steam passing through the turbine of Prob. 9.13*b* is reheated to 300 °C at an extraction pressure of (*a*) 100 kPa, (*b*) 400 kPa, and (*c*) 600 kPa. Calculate the thermal efficiency.

9.23 The power cycle of Prob. 9.14*b* is proposed for reheat. Calculate the thermal efficiency if the steam is reheated to 1000 °F after being extracted at a pressure of (*a*) 400 psia, (*b*) 200 psia, and (*c*) 100 psia.

9.24 The steam passing through the turbine of Prob. 9.17 is reheated at 600 kPa to 400 °C and at 50 kPa to 400 °C. (*a*) What is the resulting thermal efficiency? (*b*) Calculate the oil needed per hour for the same power output of the turbine of Prob. 9.17.

9.25 For the ideal reheat cycle shown in Fig. 9-23, find (*a*) the thermal efficiency and (*b*) the mass flux of steam.

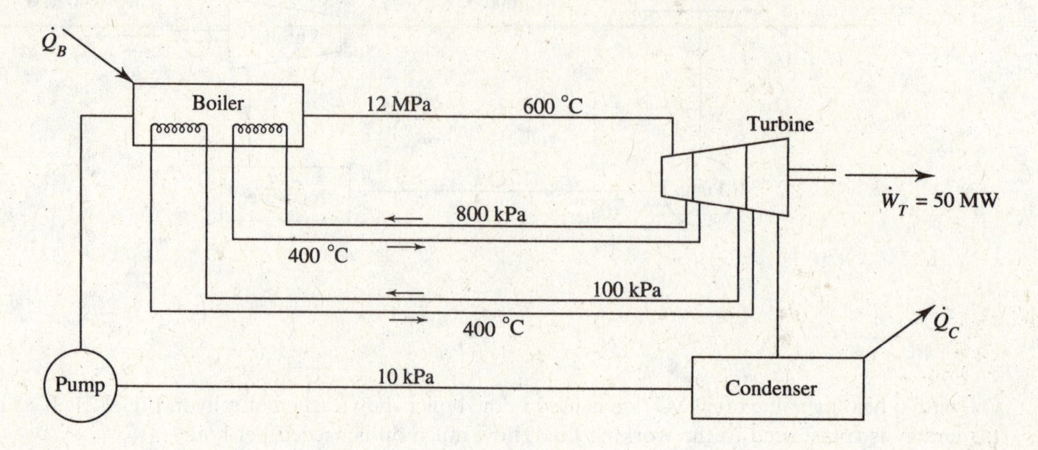

Fig. 9-23

9.26 An open feedwater heater is to be designed for the power cycle of Prob. 9.12 by extracting steam from the turbine at 400 kPa. Determine the thermal efficiency of the ideal regenerative cycle.

9.27 A portion of the steam passing through the turbine of Prob. 9.13*b* is extracted and fed into an open feedwater heater. Calculate the thermal efficiency if it is extracted at a pressure of (*a*) 600 kPa, (*b*) 800 kPa, and (*c*) 1000 kPa.

9.28 An open feedwater heater extracts steam from the turbine of Prob. 9.14*b* at 100 psia. Determine the thermal efficiency.

9.29 A closed feedwater heater extracts steam from the turbine of Prob. 9.13*b* at 800 kPa. What is the thermal efficiency of the resulting ideal regenerative cycle?

9.30 Part of the steam passing through the turbine of Prob. 9.17 is extracted at 1000 kPa and fed into a closed feedwater heater. Calculate (*a*) the thermal efficiency and (*b*) the mass flux of oil for the same power output.

9.31 To avoid a moisture problem in the turbine of Prob. 9.16 the steam is extracted at 600 kPa and reheated to 400 °C, and an open feedwater heater, using extracted steam at the same pressure, is inserted into the cycle. What is the resulting thermal efficiency and the mass flux of water flowing through the feedwater pump?

9.32 For the ideal reheat/regenerative cycle shown in Fig. 9-24 calculate (*a*) the thermal efficiency, (*b*) the mass flux of water fed to the boiler, and (*c*) the mass flux of condenser cooling water.

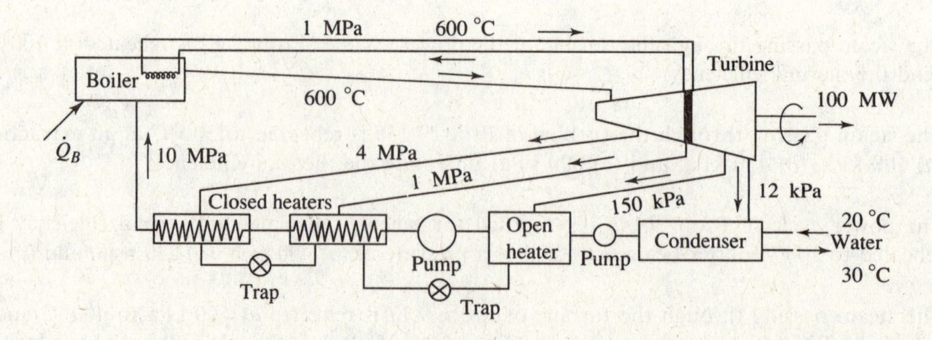

Fig. 9-24

9.33 A power plant is to operate on a supercritical steam cycle with reheat and regeneration. The steam leaves the boiler at 4000 psia and 1000 °F. It is extracted from the turbine at 400 psia; part enters an open feedwater heater and the remainder is reheated to 800 °F. The condenser pressure is 2 psia. Assuming an ideal cycle, calculate the thermal efficiency.

9.34 For the steam power cycle, operating as shown in the *T-s* diagram of Fig. 9-25, two open feedwater heaters are employed. Calculate the thermal efficiency.

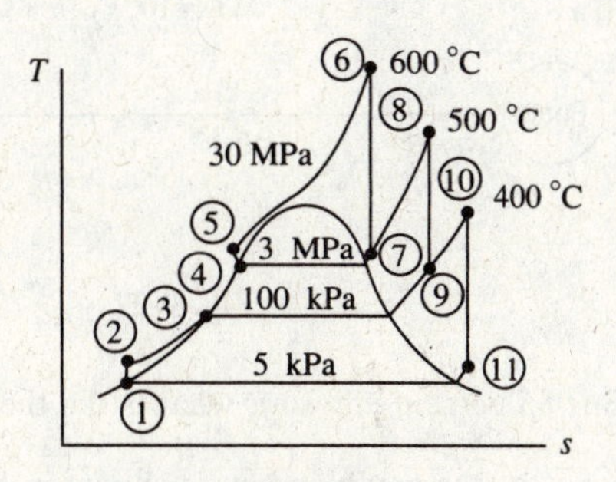

Fig. 9-25

9.35 Determine the cycle thermal efficiency if the turbine is 85 percent efficient in (*a*) Prob. 9.12, (*b*) Prob. 9.13*a*, (*c*) Prob. 9.14*b*, and (*d*) Prob. 9.16.

9.36 If the turbine of Prob. 9.17 is 80 percent efficient, determine the mass flux of oil needed to maintain the same power output.

9.37 Assume a turbine efficiency of 85 percent for Prob. 9.18 and calculate the thermal efficiency and the expected power output.

9.38 For the simple Rankine cycle shown in Fig. 9-26 the turbine efficiency is 85 percent. Determine (*a*) the thermal efficiency, (*b*) the mass flux of steam, (*c*) the diameter of the inlet pipe to the turbine if a maximum velocity of 100 m/s is allowed, and (*d*) the mass flux of condenser cooling water.

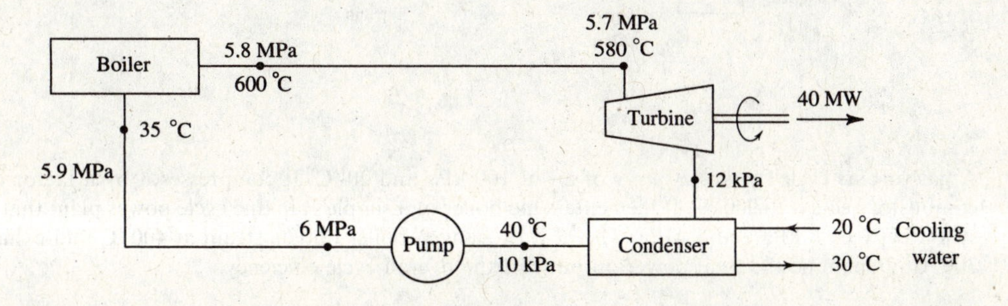

Fig. 9-26

9.39 The actual turbine of Prob. 9.20 has an efficiency of 85 percent in the high-pressure side of the turbine and 80 percent in the low-pressure side. Calculate the cycle thermal efficiency and the pump mass flux for the same power output.

9.40 Calculate the cycle thermal efficiency if the turbine is 85 percent efficient for the cycle of Prob. 9.25.

9.41 Calculate the cycle thermal efficiency if the turbine is 87 percent for the cycle of (*a*) Prob. 9.26, (*b*) Prob. 9.28*b*, and (*c*) Prob. 9.29.

9.42 Determine the thermal efficiency for the cycle shown in Fig. 9-27 if the turbine is 85 percent efficient.

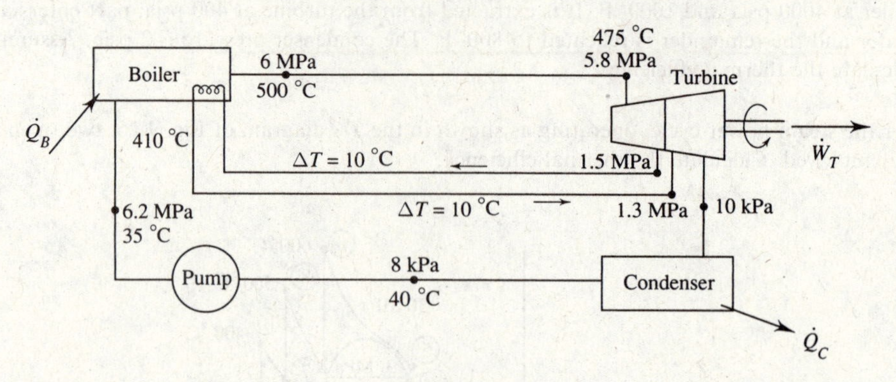

Fig. 9-27

9.43 If the turbine of Prob. 9.33 is 85 percent efficient, what is the thermal efficiency of the cycle?

9.44 Calculate the thermal efficiency of the combined cycle shown in Fig. 9-28.

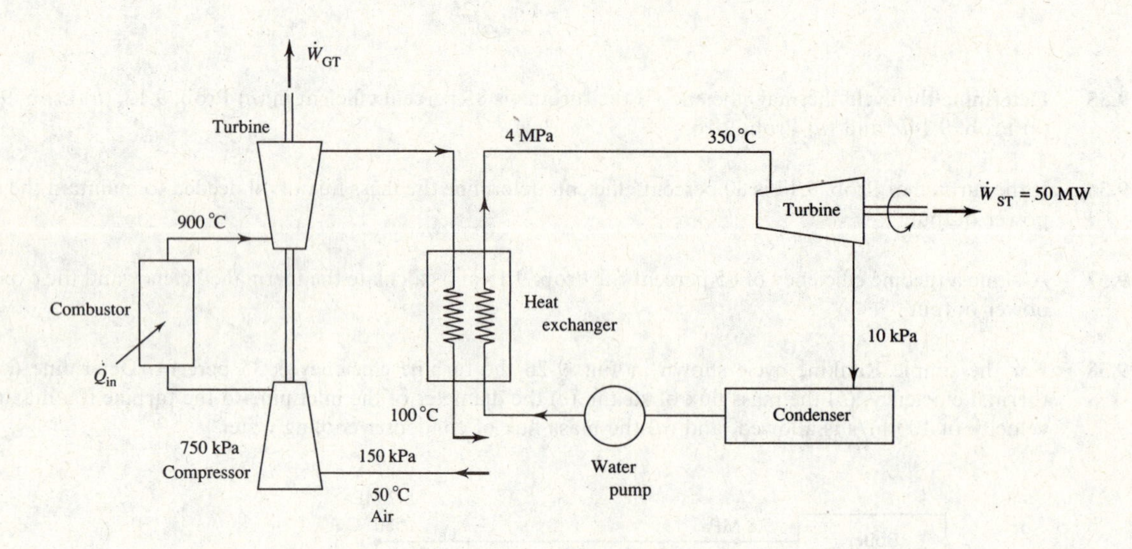

Fig. 9-28

9.45 A gas-turbine cycle intakes 50 kg/s of air at 100 kPa and 20 °C. It compresses it by a factor of 6 and the combustor heats it to 900 °C. It then enters the boiler of a simple Rankine cycle power plant that operates on steam between 8 kPa and 4 MPa. The heat exchanger–boiler outlets steam at 400 °C and exhaust gases at 300 °C. Determine the total power output and the overall cycle efficiency.

9.46 The compressor and turbine of the gas cycle of Prob. 9.45 are 85 percent efficient and the steam turbine is 87 percent efficient. Calculate the combined cycle power output and efficiency.

Review Questions for the FE Examination

9.1FE The Rankine power cycle is idealized. Which of the following is not one of the idealizations?

(A) Friction is absent.
(B) Heat transfer does not occur across a finite temperature difference.
(C) Pressure drops in pipes are neglected.
(D) Pipes connecting components are insulated.

9.2FE The component of the Rankine cycle that leads to relatively low cycle efficiency is:
(A) The pump
(B) The boiler
(C) The turbine
(D) The condenser

9.3FE Water passes through four primary components as it traverses a cycle in a power plant: a turbine, a condenser, a pump, and a boiler. Estimate the heat rejected in the condenser if $\dot{Q}_{\text{boiler}} = 60\,\text{MJ/s}$, $\dot{W}_{\text{turbine}} = 20 \times 10^6$ N·m/s, and $\dot{W}_{\text{pump}} = 100$ hp.
(A) 174 MJ/s
(B) 115 MJ/s
(C) 52 MJ/s
(D) 40 MJ/s

9.4FE Find q_B in the Rankine cycle of Fig. 9-29.
(A) 3410 kJ/kg
(B) 3070 kJ/kg
(C) 1050 kJ/kg
(D) 860 kJ/kg

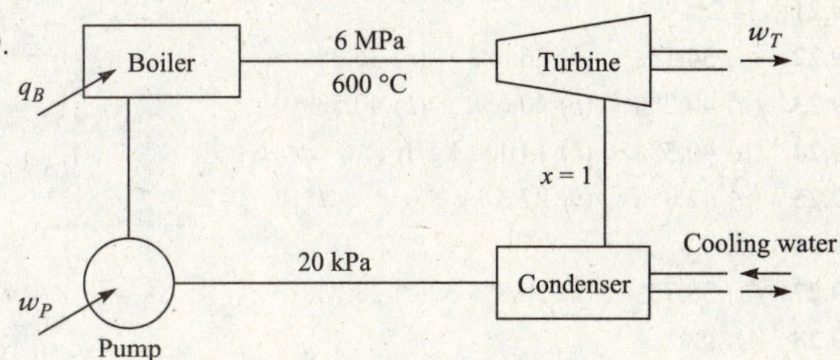

Fig. 9-29

9.5FE Find w_T of the turbine of Fig. 9-29.
(A) 3410 kJ/kg
(B) 3070 kJ/kg
(C) 1050 kJ/kg
(D) 860 kJ/kg

9.6FE Find w_P of the pump of Fig. 9-29.
(A) 4 kJ/kg
(B) 6 kJ/kg
(C) 8 kJ/kg
(D) 10 kJ/kg

9.7FE The thermal efficiency of the turbine in Fig. 9-29 is nearest:
(A) 64%
(B) 72%
(C) 76%
(D) 81%

9.8FE The cooling water in Fig. 9-29 is allowed a 10 °C increase. If $\dot{m}_{\text{steam}} = 2$ kg/s, determine $\dot{m}_{\text{cooling water}}$ if the quality of the steam leaving the condenser is just 0.0.
(A) 80 kg/s
(B) 91 kg/s
(C) 102 kg/s
(D) 113 kg/s

9.9FE An open feedwater heater accepts superheated steam at 400 kPa and 200 °C and mixes the steam with condensate at 20 kPa from an ideal condenser. If 20 kg/s flows out of the condenser, estimate $\dot{m}$ of the superheated steam.
(A) 2.26 kg/s
(B) 2.71 kg/s
(C) 3.25 kg/s
(D) 3.86 kg/s

9.10FE How much heat is transferred from the superheated steam of Question 9.9FE?
(A) 3860 kJ/s
(B) 4390 kJ/s
(C) 5090 kJ/s
(D) 7070 kJ/s

Answers to Supplementary Problems

9.12 (*a*) 33.9% (*b*) 34.0% (*c*) 0.29%

9.13 (*a*) 31.6%, 0.932 (*b*) 36.4%, 0.855 (*c*) 38.5%, 0.815

9.14 (*a*) 37.0% (*b*) 38.7% (*c*) 40.4%

9.15 (*a*) 34.9%, 0.884 (*b*) 36.8%, 0.859 (*c*) 38.7%, 0.828

9.16 664 kg/s

9.17 13 480 kg/h

9.18 (*a*) 1.17 kW (*b*) 9.2% (*c*) 12.6 kW

9.19 217,000 lbm/hr

9.20 38.4%, 44.9 kg/s

9.21 34.2%

9.22 (*a*) 34.1% (*b*) 36.0% (*c*) 36.3%

9.23 (*a*) 40.2% (*b*) 40.6% (*c*) 40.5%

9.24 (*a*) 40.3% (*b*) 14 000 kg/h

9.25 (*a*) 42.0% (*b*) 27.3 kg/s

9.26 35.6%

9.27 (*a*) 38.7% (*b*) 38.8% (*c*) 38.7%

9.28 41.2%

9.29 38.8%

9.30 (*a*) 44.8% (*b*) 12 600 kg/h

9.31 44.7%, 13.59 kg/s

9.32 (*a*) 47.2% (*b*) 67.8 kg/s (*c*) 2680 kg/s

9.33 46.6%

9.34 50.5%

9.35 (*a*) 28.8% (*b*) 26.9% (*c*) 32.9% (*d*) 35.6%

9.36 16 850 kg/h

9.37 7.7%, 10.7 W

9.38 (*a*) 32.2% (*b*) 35.6 kg/s (*c*) 17.6 cm (*d*) 1972 kg/s

9.39 34.0%, 54.6 kg/s

9.40 35.7%

9.41 (*a*) 31.0% (*b*) 34.5% (*c*) 33.8%

9.42 31.5%

9.43 41.4%

9.44 56%

9.45 16 MW, 47%

9.46 11.6 MW, 35.8%

Answers to Review Questions for the FE Examination

9.1FE (B) **9.2FE** (B) **9.3FE** (D) **9.4FE** (A) **9.5FE** (C) **9.6FE** (B) **9.7FE** (D) **9.8FE** (D) **9.9FE** (B) **9.10FE** (D)

Refrigeration Cycles

10.1 INTRODUCTION

Refrigeration involves the transfer of heat from a lower-temperature region to a higher-temperature region. If the objective of the device that accomplishes the heat transfer is to cool the lower-temperature region, the device is a *refrigerator*. If the objective is to heat the higher-temperature region, the device is referred to as a *heat pump*. Refrigeration devices most often operate on a cycle that uses a refrigerant that is vaporized in the cycle, but refrigeration can also be accomplished by a gas, as is done on aircraft. In some special cases, absorption refrigeration cycles are used, especially on large university campuses. Each of these cycles will be presented in this chapter.

10.2 THE VAPOR REFRIGERATION CYCLE

It is possible to extract heat from a space by operating a vapor cycle, similar to the Rankine cycle, in reverse. Work input is, of course, required in the operation of such a cycle, as shown in Fig. 10-1*a*.

The work is input by a compressor that increases the pressure, and thereby the temperature, through an isentropic compression process in the ideal cycle. The working fluid (often R134a) then enters a condenser in which heat is extracted, resulting in saturated liquid. The pressure is then reduced in an expansion process so that the fluid can be evaporated with the addition of heat from the refrigerated space.

The most efficient cycle, a Carnot cycle, is shown in Fig. 10-1*b*. There are, however, two major drawbacks when an attempt is made to put such a cycle into actual operation. First, it is not advisable to compress the mixture of liquid and vapor as represented by state 1 in Fig. 10-1*b* since the liquid droplets would cause excessive wear; in addition, equilibrium between the liquid phase and the vapor phase is difficult to maintain in such a process. Hence, in the ideal refrigeration cycle a saturated vapor state is assumed at the end of the evaporation process; this allows superheated vapor to exist in the compressor, as shown by process 1-2 in Fig. 10-1*c*. Second, it would be quite expensive to construct a device to be used in the expansion process that would be nearly isentropic (no losses allowed). It is much simpler to reduce the pressure irreversibly by using an expansion valve which employs a throttling process in which enthalpy remains constant, as shown by the dotted line in Fig. 10-1*c*. Even though this expansion process is characterized by losses, it is considered to be part of the "ideal" vapor refrigeration cycle. Because the

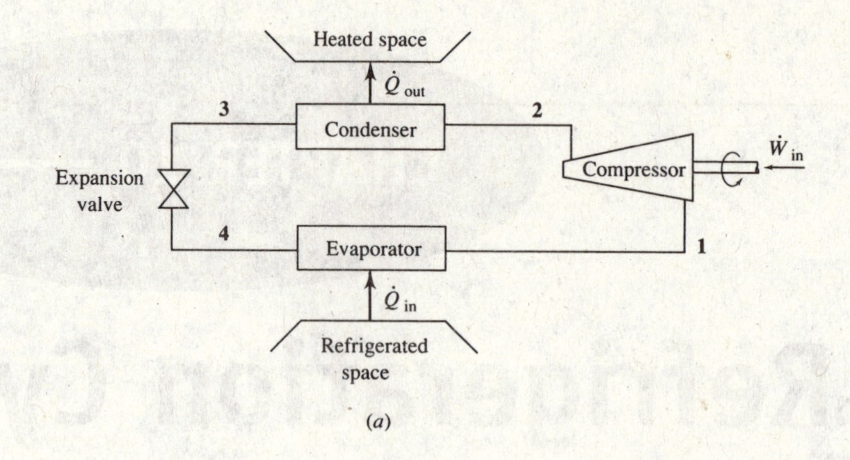

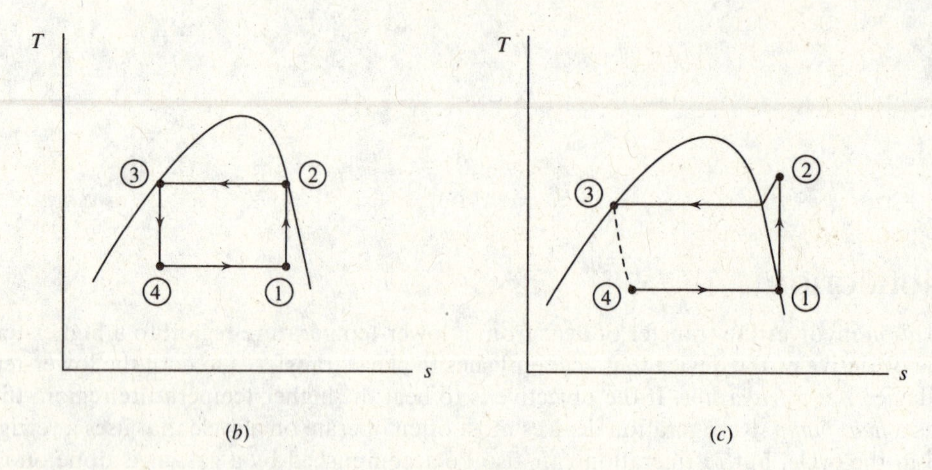

Fig. 10-1 The vapor refrigeration cycle.

expansion process is a nonequilibrium process, the area under the *T-s* diagram does not represent the net work input.

The performance of the refrigeration cycle, when used as a refrigerator, is measured by

$$\text{COP}_R = \frac{\dot{Q}_{in}}{\dot{W}_{in}} \tag{10.1}$$

When the cycle is used as a heat pump, the performance is measured by

$$\text{COP}_{HP} = \frac{\dot{Q}_{out}}{\dot{W}_{in}} \tag{10.2}$$

We do not calculate the efficiency of a refrigeration cycle since the efficiency is not of particular interest. What is of interest is the ratio of the output energy to the input energy. The coefficient of performance can attain values of perhaps 5 for properly designed heat pumps and 4 for refrigerators.

The condensation and evaporation temperatures, and hence the pressures, are established by the particular situation that motivates the design of the refrigeration unit. For example, in a home refrigerator that is designed to cool the freezer space to −18 °C (0 °F) it is necessary to design the evaporator to operate at approximately −25 °C to allow for effective heat transfer between the space and the cooling coils. The refrigerant condenses by transferring heat to air maintained at about 20 °C; consequently, to allow for effective heat transfer from the coils that transport the refrigerant, the refrigerant must be maintained at a temperature of at least 28 °C. This is shown in Fig. 10-2.

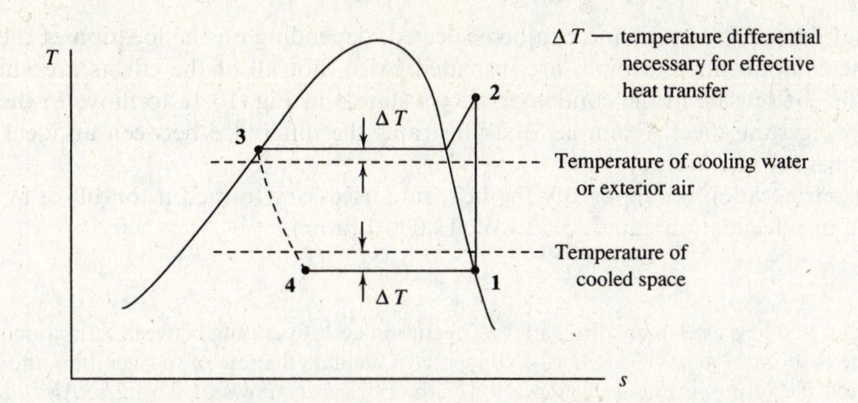

Fig. 10-2 The refrigeration cycle showing design temperatures.

To accomplish refrigeration for most spaces, it is necessary that the evaporation temperature be quite low, in the neighborhood of $-25\,^\circ\mathrm{C}$, perhaps. This, of course, rules out water as a possible refrigerant. Two common refrigerants in use today are ammonia (NH_3) and R134a (CF_3CH_2F). The thermodynamic properties of R134a are presented in Appendix D. The selection of a refrigerant depends on the two design temperatures shown in Fig. 10-2. For example, temperatures well below $-100\,^\circ\mathrm{C}$ are required to liquefy many gases. Neither ammonia nor R134a may be used at such low temperatures since they do not exist in a liquid form below $-100\,^\circ\mathrm{C}$. Also, it is desirable to operate a refrigeration cycle such that the low pressure is above atmospheric pressure, thereby avoiding air contamination should a leak occur. In addition, for most applications the refrigerant must be nontoxic, stable, and relatively inexpensive.

Deviations from the ideal vapor refrigeration cycle are shown on the *T-s* diagram of Fig. 10-3*b*. These include:

- Pressure drops due to friction in connecting pipes.
- Heat transfer occurs from or to the refrigerant through the pipes connecting the components.
- Pressure drops occur through the condenser and evaporator tubes.
- Heat transfer occurs from the compressor.
- Frictional effects and flow separation occur on the compressor blades.
- The vapor entering the compressor may be slightly superheated.
- The temperature of the liquid exiting the condenser may be below the saturation temperature.

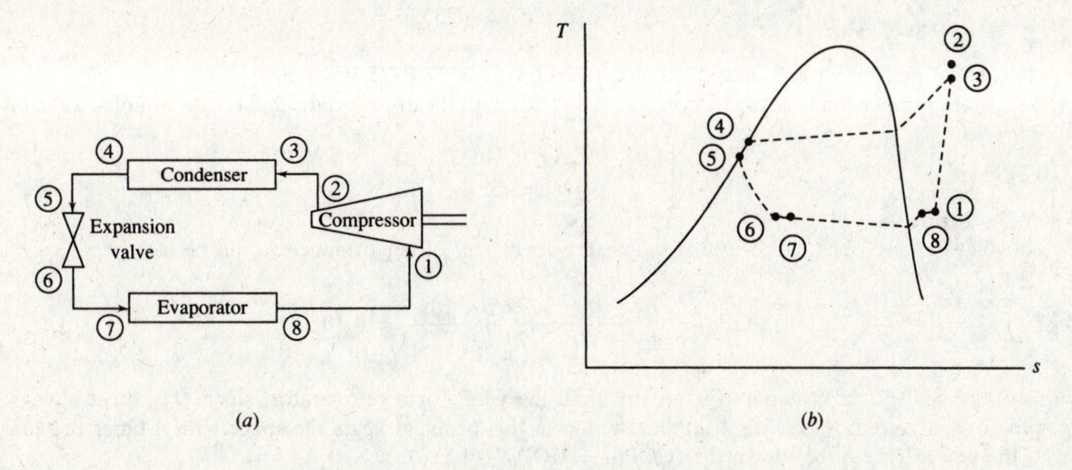

Fig. 10-3 The actual refrigeration cycle showing losses.

Some of these effects are small and can be neglected, depending on the location of the components and whether the components and pipes are insulated. Also, not all of the effects are undesirable; the subcooling of the condensate in the condenser allows state 4 in Fig. 10-1*c* to move to the left, thereby increasing the refrigerant effect. Example 10.2 illustrates the difference between an ideal refrigeration cycle and an actual refrigeration cycle.

A "ton" of refrigeration is supposedly the heat rate necessary to melt a ton of ice in 24 hours. By definition, 1 ton of refrigeration equals 3.52 kW (12,000 Btu/hr).

EXAMPLE 10.1 R134a is used in an ideal vapor refrigeration cycle operating between saturation temperatures of −20 °C in the evaporator and 39.39 °C in the condenser. Calculate the rate of refrigeration, the coefficient of performance, and the rating in horsepower per ton if the refrigerant flows at 0.6 kg/s. Also, determine the coefficient of performance if the cycle is operated as a heat pump.

Solution: The *T-s* diagram in Fig. 10-4 is drawn as an aid in the solution. The enthalpy of each state is needed. From Appendix D we find that $h_1 = 235.3$ kJ/kg, $h_3 = h_4 = 105.3$ kJ/kg, and $s_1 = 0.9332$ kJ/kg·K. Using $s_1 = s_2$, we interpolate at a pressure of 1.0 MPa, which is the pressure associated with the saturation temperature of 39.39 °C, and find that

$$h_2 = \left(\frac{0.9332 - 0.9066}{0.9428 - 0.9066}\right)(280.2 - 268.7) + 268.7 = 277.2 \text{ kJ/kg}$$

The rate of refrigeration is measured by the heat transfer rate needed in the evaporation process, namely,

$$\dot{Q}_E = \dot{m}(h_1 - h_4) = (0.6)(235.3 - 105.3) = 78.0 \text{ kW}$$

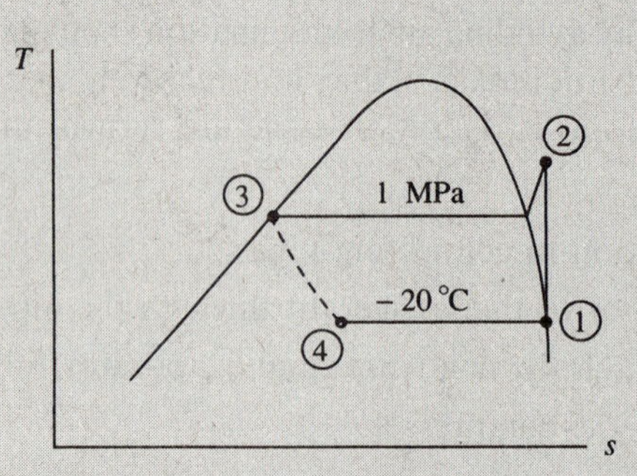

Fig. 10-4

The power needed to operate the compressor is

$$\dot{W}_C = \dot{m}(h_2 - h_1) = (0.6)(277.2 - 235.3) = 25.1 \text{ kW}$$

The coefficient of performance is then calculated to be $\text{COP}_R = 78.0/25.1 = 3.10$.

The horsepower per ton of refrigeration is determined, with the appropriate conversion of units, as follows:

$$\text{Hp/ton} = \frac{25.1/0.746}{78.0/3.52} = 1.52$$

If the above cycle were operated as a heat pump, the coefficient of performance would be

$$\text{COP}_{HP} = \frac{h_2 - h_3}{h_2 - h_1} = \frac{277.2 - 105.3}{277.2 - 235.3} = 4.10$$

Obviously, the COP for a heat pump is greater than the COP for a refrigerator, since $\dot{Q}_{out}$ must always be greater than $\dot{Q}_{in}$. Note, however, that the heat pump in this problem heats the space with 4 times the energy input to the device. (It can be shown that $\text{COP}_{HP} = \text{COP}_R + 1$.)

EXAMPLE 10.2 The ideal refrigeration cycle of Example 10.1 is used in the operation of an actual refrigerator. It experiences the following real effects:

The refrigerant leaving the evaporator is superheated to −13 °C.
The refrigerant leaving the condenser is subcooled to 38 °C.
The compressor is 80 percent efficient.

Calculate the actual rate of refrigeration and the coefficient of performance.

Solution: From Appendix D we find, using $T_3 = 38\,°\mathrm{C}$, that $h_3 = h_4 = 103.2$ kJ/kg. Also, from Table D-1 we observe that $P_1 \cong 0.18$ MPa. From Table D-3, at $P_1 = 0.18$ MPa and $T_1 = -13\,°\mathrm{C}$,

$$h_1 = 240 \text{ kJ/kg} \qquad s_1 = 0.930 \text{ kJ/kg·K}$$

If the compressor were isentropic, then, with $s_{2'} = s_1$ and $P_2 = 1.0$ MPa,

$$h_{2'} = \left(\frac{0.93 - 0.9066}{0.9428 - 0.9066}\right)(280.2 - 268.7) + 268.7 = 276 \text{ kJ/kg}$$

From the definition of efficiency, $\eta = w_s/w_a$, we have

$$0.8 = \frac{h_{2'} - h_1}{h_2 - h_1} = \frac{276 - 240}{h_2 - 240} \qquad \therefore h_2 = 285 \text{ kJ/kg}$$

The rate of refrigeration is $\dot{Q}_E = (0.6)(240 - 103.2) = 82$ kW. Note that the real effects have actually increased the capability to refrigerate a space. The coefficient of performance becomes

$$\mathrm{COP_R} = \frac{82}{(0.6)(285 - 240)} = 3.04$$

The decrease in the $\mathrm{COP_R}$ occurs because the power input to the compressor has increased.

10.3 THE MULTISTAGE VAPOR REFRIGERATION CYCLE

In Example 10.2 the subcooling of the condensate leaving the condenser resulted in increased refrigeration. Subcooling is an important consideration in designing a refrigeration system. It can be accomplished either by designing a larger condenser or by designing a heat exchanger that uses the refrigerant from the evaporator as the coolant.

Another technique that can result in increased refrigeration is to place two refrigeration cycles in series (a two-stage cycle), operating as shown in Fig. 10-5*a*; the increased refrigeration is shown in Fig. 10-5*b*. This two-stage cycle has the added advantage that the power required to compress the

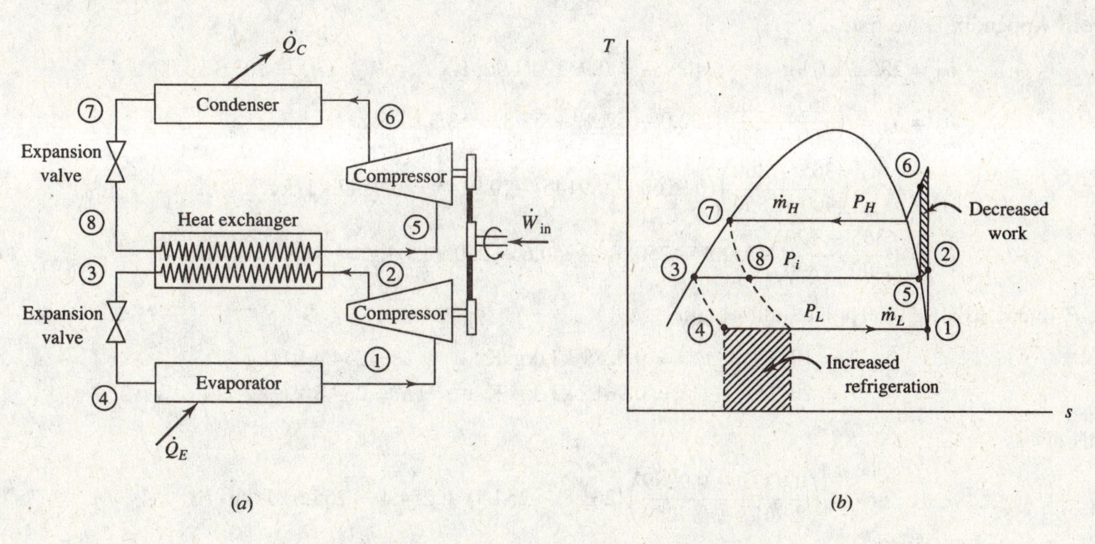

Fig. 10-5 The two-stage refrigeration cycle.

refrigerant is substantially reduced. Note that the high-temperature refrigerant leaving the low-pressure stage compressor is used to evaporate the refrigerant in the high-pressure stage. This requires a heat exchanger, and, of course, two expansion valves and two compressors. The additional costs of this added equipment must be justified by improved performance. For extremely low refrigeration temperatures several stages may be justified.

The optimal value for the intermediate pressure P_i is given by

$$P_i = (P_H P_L)^{1/2} \tag{10.3}$$

where P_H and P_L are the respective high and low absolute pressures, shown in Fig. 10-5*b*. In this discussion the same refrigerant is assumed in both systems; if different refrigerants are used, then the appropriate *T-s* diagram must be used for each fluid.

To determine the relationship between the mass fluxes of the two systems we simply apply the first law (an energy balance) to the heat exchanger. This gives

$$\dot{m}_H(h_5 - h_8) = \dot{m}_L(h_2 - h_3) \tag{10.4}$$

where $\dot{m}_H$ is the mass flux of the refrigerant in the high-pressure system and $\dot{m}_L$ is the refrigerant mass flux in the low-pressure system. This gives

$$\frac{\dot{m}_H}{\dot{m}_L} = \frac{h_2 - h_3}{h_5 - h_8} \tag{10.5}$$

The low-pressure system actually performs the desired refrigeration. Thus, in the design process, it is this system that allows us to determine $\dot{m}_L$. If X tons ($= 3.52\,X$ kilowatts) of refrigeration is required, then

$$\dot{m}_L(h_1 - h_4) = 3.52\,X \tag{10.6}$$

The mass flux is

$$\dot{m}_L = \frac{3.52\,X}{h_1 - h_4} \tag{10.7}$$

EXAMPLE 10.3 A two-stage cycle replaces the refrigeration cycle of Example 10.1. Determine the rate of refrigeration and the coefficient of performance and compare with those of Example 10.1. Use $\dot{m}_L = 0.6$ kg/s.

Solution: Refer to Fig. 10-5 for the various state designations. Using $T_1 = -20\,°\text{C}$, we find $P_L = 133$ kPa. Also, $P_H = 1000$ kPa. Then, (*10.3*) results in

$$P_i = (P_L P_H)^{1/2} = [(133)(1000)]^{1/2} = 365 \text{ kPa}$$

From Appendix D we find

$$h_1 = 235.3 \text{ kJ/kg} \qquad s_1 = s_2 = 0.9332 \text{ kJ/kg·K} \qquad h_7 = h_8 = 105.3 \text{ kJ/kg}$$

$$h_3 = h_4 = \left(\frac{365 - 360}{400 - 360}\right)(62.0 - 57.8) + 57.8 = 58.3 \text{ kJ/kg}$$

$$s_5 = s_6 = \left(\frac{365 - 360}{400 - 360}\right)(0.9160 - 0.9145) + 0.9145 = 0.9147 \text{ kJ/kg·K}$$

$$h_5 = \left(\frac{365 - 320}{400 - 320}\right)(252.3 - 250.6) + 250.6 = 250.8 \text{ kJ/kg}$$

At $P_i = 365$ kPa we interpolate and obtain

$$T = 10\,°\text{C} \qquad s = 0.9289 \text{ kJ/kg·K} \qquad h = 254.4 \text{ kJ/kg}$$
$$T = 20\,°\text{C} \qquad s = 0.9617 \text{ kJ/kg·K} \qquad h = 263.8 \text{ kJ/kg}$$

This gives

$$h_2 = \left(\frac{0.9332 - 0.9289}{0.9617 - 0.9289}\right)(263.8 - 254.4) + 254.4 = 255.6 \text{ kJ/kg}$$

Also, interpolating at $P_6 = 1.0$ MPa, we find

$$h_6 = \left(\frac{0.9147 - 0.9066}{0.9428 - 0.9066}\right)(280.2 - 268.7) + 268.7 = 271.3 \text{ kJ/kg}$$

From the above, $\dot{Q}_E = \dot{m}_L(h_1 - h_4) = (0.6)(235.3 - 58.3) = 106.2$ kW. This compares with a value of 78.0 kW from the simple refrigeration cycle of Example 10.1. That represents a 36 percent increase in the rate of refrigeration. The mass flux in the high-pressure stage is found from (*10.5*) to be

$$\dot{m}_H = \dot{m}_L \frac{h_2 - h_3}{h_5 - h_8} = (0.6)\left(\frac{255.6 - 58.3}{250.8 - 105.3}\right) = 0.814 \text{ kg/s}$$

The power input to the compressors is

$$\dot{W}_{\text{in}} = \dot{m}_L(h_2 - h_1) + \dot{m}_H(h_6 - h_5) = (0.6)(255.6 - 235.3) + (0.814)(271.3 - 250.8) = 28.9 \text{ kW}$$

The coefficient of performance is now calculated to be

$$\text{COP}_\text{R} = \frac{\dot{Q}_E}{\dot{W}_{in}} = \frac{106.2}{28.9} = 3.68$$

This compares with a value of 3.10 from the refrigeration cycle of Example 10.1, a 19 percent increase. The advantages of using two stages is obvious when considering the increased refrigeration and performance; the equipment is much more expensive, however, and must be justified economically.

10.4 THE HEAT PUMP

The heat pump utilizes the vapor refrigeration cycle discussed in Sec. 10.2. It can be used to heat a house in cool weather or cool a house in warm weather, as shown schematically in Fig. 10-6. Note that in the heating mode the house gains heat from the condenser, whereas in the cooling mode the house loses heat to the evaporator. This is possible since the evaporator and the condenser are similar heat exchangers. In an actual situation, valving is used to perform the desired switching of the heat exchangers.

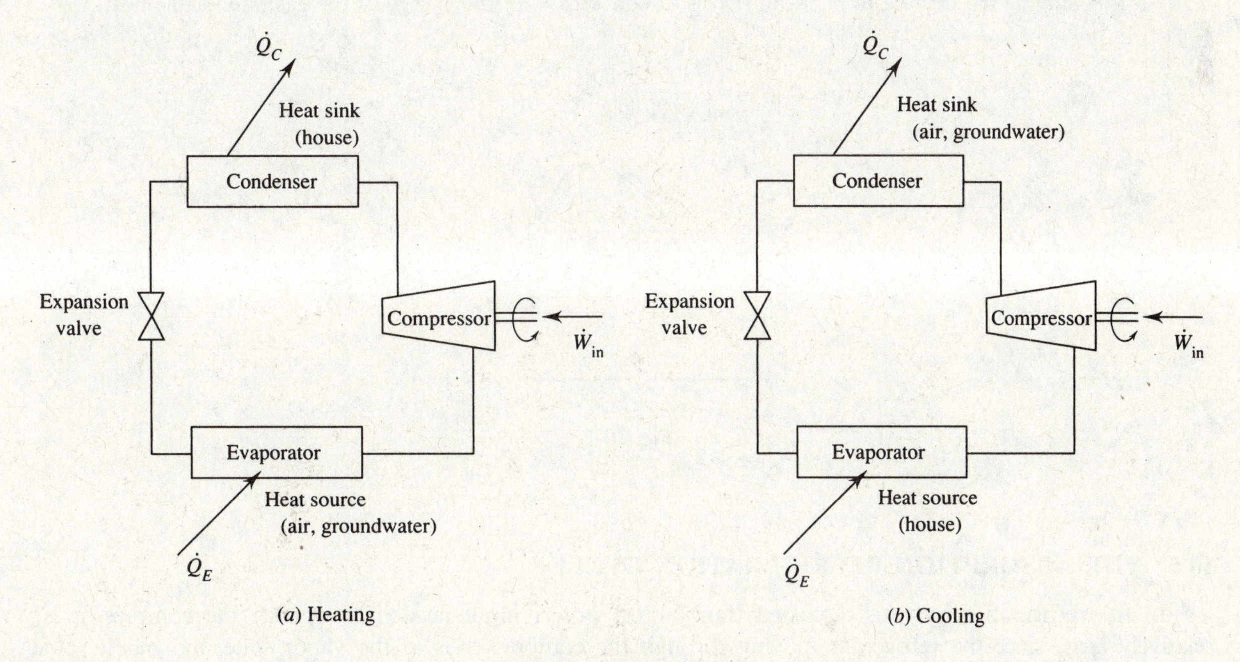

Fig. 10-6 The heat pump.

The heat pump system is sized to meet the heating load or the cooling load, whichever is greater. In southern areas where the cooling loads are extremely large, the system may be oversized for the small heating demand of a chilly night; an air conditioner with an auxiliary heating system may be advisable in those cases. In a northern area where the large heating load demands a relatively large heat pump, the cooling load on a warm day may be too low for effective use of the heat pump; the large cooling capacity would quickly reduce the temperature of the house without a simultaneous reduction in the humidity, a necessary feature of any cooling system. In that case, a furnace which provides the heating with an auxiliary cooling system is usually advisable. Or, the heat pump could be designed based on the cooling load, with an auxiliary heater for times of heavy heating demands.

EXAMPLE 10.4 A heat pump using R134a is proposed for heating a home that requires a maximum heating load of 300 kW. The evaporator operates at $-12\,^{\circ}\mathrm{C}$ and the condenser at 900 kPa. Assume an ideal cycle.

(*a*) Determine the COP.
(*b*) Determine the cost of electricity at \$0.08/kWh.
(*c*) Compare the R134a system with the cost of operating a furnace using natural gas at \$0.75/therm if there are 100 000 kJ/therm of natural gas.

Solution:

(*a*) The *T-s* diagram (Fig. 10-7) is sketched for reference. From Appendix D we find $h_1 = 240.2$ kJ/kg, $s_1 = s_2 = 0.9267$ kJ/kg·K, and $h_3 = h_4 = 99.56$ kJ/kg. Interpolating, there results

$$h_2 = \left(\frac{0.9267 - 0.9217}{0.9566 - 0.9217}\right)(282.3 - 271.2) + 271.2 = 272.8 \text{ kJ/kg}$$

The heat rejected by the condenser is

$$\dot{Q}_c = \dot{m}(h_2 - h_3) \qquad 300 = \dot{m}(272.8 - 99.56)$$

This gives the refrigerant mass flux as $\dot{m} = 1.732$ kg/s. The required power by the compressor is then $\dot{W}_{\text{in}} = \dot{m}(h_2 - h_1) = (1.732)(272.8 - 240.2) = 56.5$ kW. This results in a coefficient of performance of

$$\text{COP}_\text{R} = \frac{\dot{Q}_C}{\dot{W}_{in}} = \frac{300}{56.5} = 5.31$$

(*b*) Cost of electricity (56.5 kW)(\$0.08/kWh) = \$4.52/h.

(*c*) Assuming the furnace to be ideal, that is, it converts all of the energy of the gas into usable heat, we have

$$\text{Cost of gas} = \left[\frac{(300)(3600)}{100\,000}\right](0.75) = \$8.10/\text{h}$$

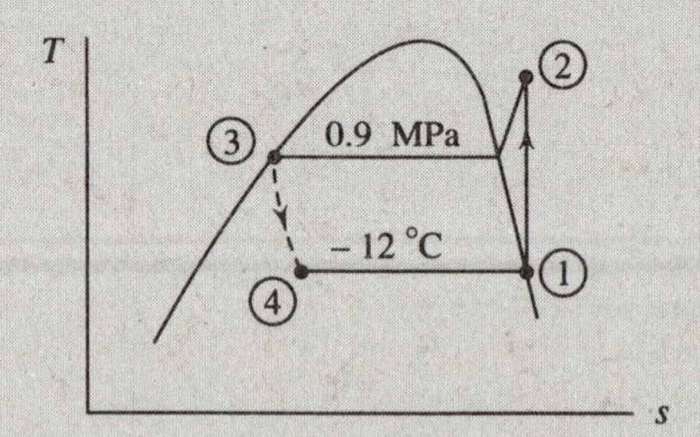

Fig. 10-7

10.5 THE ABSORPTION REFRIGERATION CYCLE

In the refrigeration systems discussed thus far the power input needed to operate the compressor is relatively large since the refrigerant moving through the compressor is in the vapor state and has a very large specific volume when compared with that of a liquid. We can markedly reduce this power if we increase

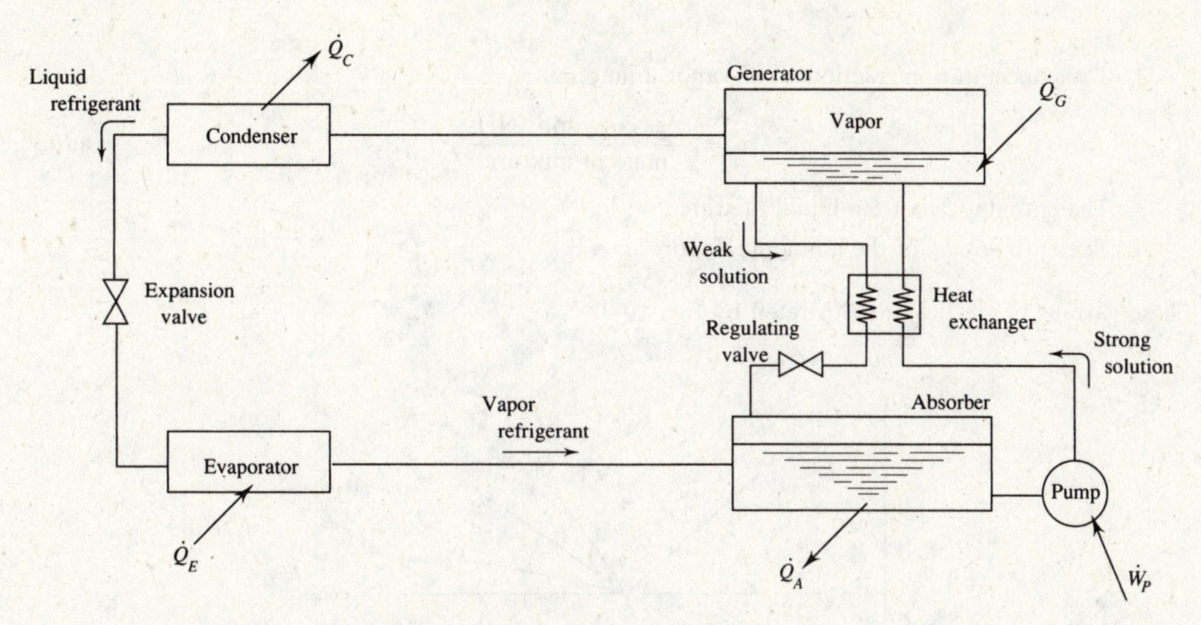

Fig. 10-8 The absorption refrigeration cycle.

the pressure with a pump operating with a liquid. Such a refrigeration cycle exists; it is the *absorption refrigeration cycle*, shown schematically in Fig. 10-8. Note that the compressor of the conventional refrigeration cycle has been replaced with the several pieces of equipment shown on the right of the cycle. The absorber, the pump, the heat exchanger, and the generator are the major additional components that replace the compressor.

Saturated, low-pressure refrigerant vapor leaves the evaporator and enters the absorber where it is absorbed into the weak carrier solution. Heat is released in this absorption process, and to aid the process the temperature is maintained at a relatively low value by removing heat $\dot{Q}_A$. The much stronger liquid solution leaves the absorber and is pumped to the higher condenser pressure, requiring very little pump power. It passes through a heat exchanger, which increases its temperature, and enters the generator where the added heat boils off the refrigerant which then passes on to the condenser. The remaining weak carrier solution is then returned from the generator to the absorber to be recharged with refrigerant; on its way to the absorber the temperature of the carrier solution is reduced in the heat exchanger and its pressure is reduced with a regulating valve.

The primary disadvantage of the absorption cycle is that a relatively high-temperature energy source must be available to supply the heat transfer $\dot{Q}_G$; this is typically supplied by a source that would otherwise be wasted, such as rejected steam from a power plant. The additional heat $\dot{Q}_G$ must be inexpensive, or the additional cost of the extra equipment cannot be justified.

For applications in which the refrigerated space is maintained at temperatures below 0 °C, the refrigerant is normally ammonia and the carrier is water. For air-conditioning applications the refrigerant can be water and the carrier either lithium bromide or lithium chloride. With water as the refrigerant a vacuum of 0.001 MPa must be maintained in the evaporator and absorber to allow for an evaporator temperature of 7 °C. Since the evaporator temperature must be about 10 °C below the temperature of the air that is cooling the space, such a low pressure is not unreasonable.

To analyze the absorption cycle we must know the amount of refrigerant contained in a mixture, both in liquid form and in vapor form. This can be found with the aid of an equilibrium diagram, such as that for an ammonia-water mixture. At a given temperature and pressure the equilibrium diagram displays the following properties:

1. The concentration fraction x' of liquid ammonia:

$$x' = \frac{\text{mass of liquid NH}_3}{\text{mass of mixture}} \tag{10.8}$$

2. The concentration fraction x'' of vapor ammonia:

$$x'' = \frac{\text{mass of vapor NH}_3}{\text{mass of mixture}} \tag{10.9}$$

3. The enthalpy h_L of the liquid mixture.
4. The enthalpy h_v of the ammonia vapor.

These various properties are illustrated by Fig. 10-9.

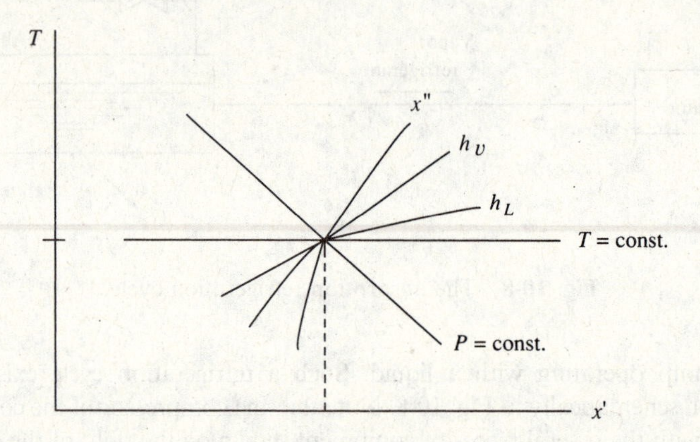

Fig. 10-9 An equilibrium diagram.

Finally, in the absorber and the generator two streams enter and one stream leaves. To determine the properties of the leaving stream, it is necessary to use a mass balance and an energy balance on each device; mass balances on both the refrigerant and the mixture are necessary.

10.6 THE GAS REFRIGERATION CYCLE

If the flow of the gas is reversed in the Brayton cycle of Sec. 8.9, the gas undergoes an isentropic expansion process as it flows through the turbine, resulting in a substantial reduction in temperature, as shown in Fig. 10-10. The gas with low turbine exit temperature can be used to refrigerate a space to temperature T_2 by extracting heat at rate $\dot{Q}_{in}$ from the refrigerated space.

Figure 10-10 illustrates a *closed* refrigeration cycle. (An *open* cycle system is used in aircraft; air is extracted from the atmosphere at state 2 and inserted into the passenger compartment at state 1. This provides both fresh air and cooling.) An additional heat exchanger may be used, like the regenerator of the Brayton power cycle, to provide an even lower turbine exit temperature, as illustrated in Fig. 10-11 of Example 10.6. The gas does not enter the expansion process (the turbine) at state 5; rather, it passes through an internal heat exchanger (it does not exchange heat with the surroundings). This allows the temperature of the gas entering the turbine to be much lower than that of Fig. 10-10. The temperature T_1 after the expansion is so low that gas liquefication is possible. It should be noted, however, that the coefficient of performance is actually reduced by the inclusion of an internal heat exchanger.

A reminder: when the purpose of a thermodynamic cycle is to cool a space, we do not define a cycle's efficiency; rather, we define its *coefficient of performance*:

$$\text{COP}_\text{R} = \frac{\text{desired effect}}{\text{energy that costs}} = \frac{\dot{Q}_{in}}{\dot{W}_{in}} \tag{10.10}$$

where $\dot{W}_{in} = \dot{m}(w_{comp} - w_{turb})$.

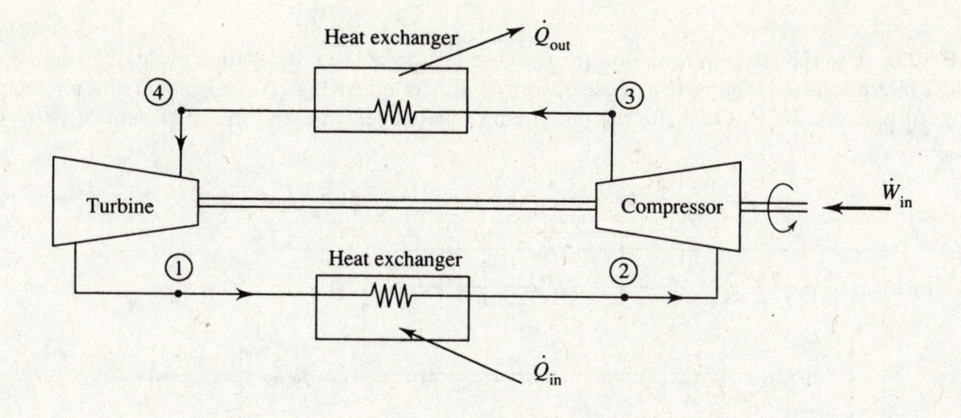

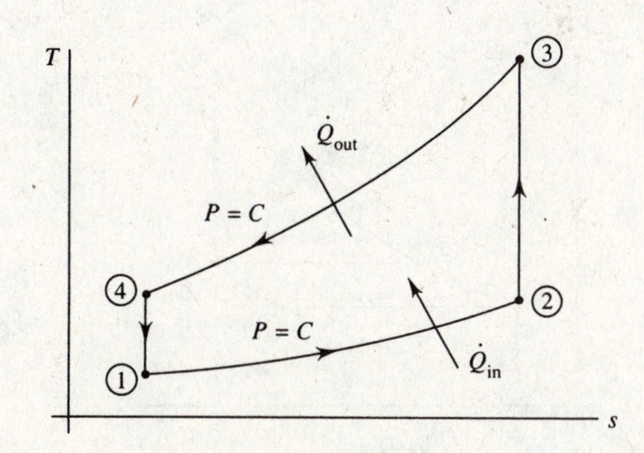

Fig. 10-10 The gas refrigeration cycle.

EXAMPLE 10.5 Air enters the compressor of a simple gas refrigeration cycle at $-10\,^\circ$C and 100 kPa. For a compression ratio of 10 and a turbine inlet temperature of $30\,^\circ$C, calculate the minimum cycle temperature and the coefficient of performance.

Solution: Assuming isentropic compression and expansion processes we find

$$T_3 = T_2\left(\frac{P_3}{P_2}\right)^{(k-1)/k} = (263)(10)^{0.2857} = 508\text{ K}$$

$$T_1 = T_4\left(\frac{P_1}{P_4}\right)^{(k-1)/k} = (303)\left(\frac{1}{10}\right)^{0.2857} = 157\text{ K} = -116\,^\circ\text{C}$$

The COP is now calculated as follows:

$$q_{\text{in}} = C_p(T_2 - T_1) = (1.00)(263 - 157) = 106\text{ kJ/kg}$$

$$w_{\text{comp}} = C_p(T_3 - T_2) = (1.00)(508 - 263) = 245\text{ kJ/kg}$$

$$w_{\text{turb}} = C_p(T_4 - T_1) = (1.00)(303 - 157) = 146\text{ kJ/kg}$$

$$\therefore \text{COP}_\text{R} = \frac{q_{\text{in}}}{w_{\text{comp}} - w_{\text{turb}}} = \frac{106}{245 - 146} = 1.07$$

This coefficient of performance is quite low when compared with that of a vapor refrigeration cycle. Thus gas refrigeration cycles are used only for special applications.

EXAMPLE 10.6 Use the given information for the compressor of the refrigeration cycle of Example 10.5 but add an ideal internal heat exchanger, a regenerator, as illustrated in Fig. 10-11, so that the air temperature entering the turbine is $-40\,^\circ\mathrm{C}$. Calculate the minimum cycle temperature and the coefficient of performance.

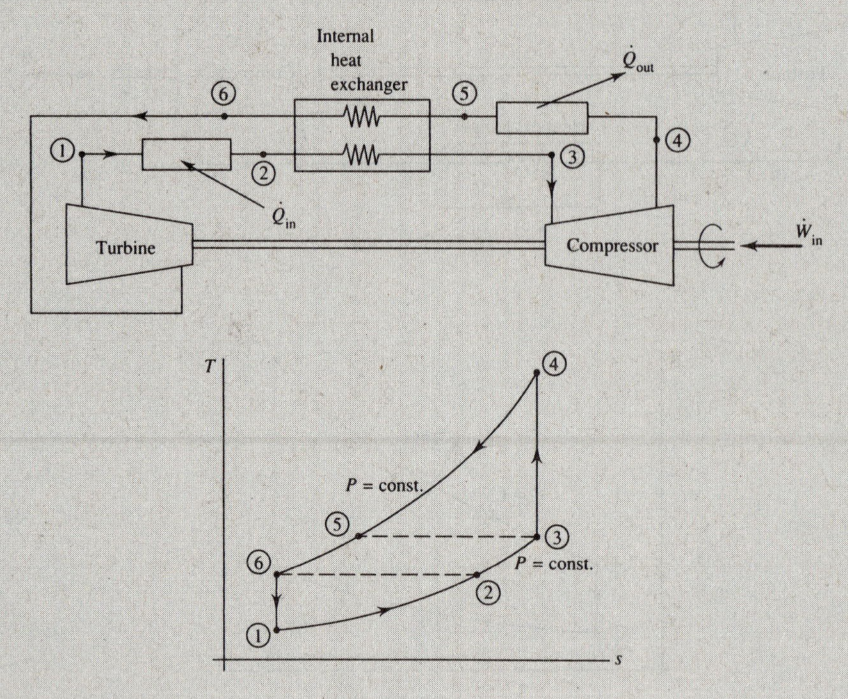

Fig. 10-11

Solution: Assuming isentropic compression we again have $T_4 = T_3(P_4/P_3)^{(k-1)/k} = (263)(10)^{0.2857} = 508$ K. For an ideal internal heat exchanger we would have $T_5 = T_3 = 263$ K and $T_6 = T_2 = 233$ K. The minimum cycle temperature is

$$T_1 = T_6\left(\frac{P_1}{P_6}\right)^{(k-1)/k} = (233)\left(\frac{1}{10}\right)^{0.2857} = 121\ \mathrm{K} = -152\,^\circ\mathrm{C}$$

For the COP:

$$q_{in} = C_p(T_2 - T_1) = (1.00)(233 - 121) = 112\ \mathrm{kJ/kg}$$
$$w_{comp} = C_p(T_4 - T_3) = (1.00)(508 - 263) = 245\ \mathrm{kJ/kg}$$
$$w_{turb} = C_p(T_6 - T_1) = (1.00)(233 - 121) = 112\ \mathrm{kJ/kg}$$
$$\therefore \mathrm{COP_R} = \frac{q_{in}}{w_{comp} - w_{turb}} = \frac{112}{245 - 112} = 0.842$$

Obviously, the COP is lower than that of the cycle with no internal heat exchanger. The objective is not to increase the COP but to provide extremely low refrigeration temperatures.

Solved Problems

10.1 An ideal refrigeration cycle uses R134a as the working fluid between saturation temperatures of $-20\,^\circ\mathrm{F}$ and $50\,^\circ\mathrm{F}$. If the refrigerant mass flux is 2.0 lbm/sec, determine the rate of refrigeration and the coefficient of performance.

Referring to Fig. 10-1*c*, we find from Appendix D that

$$h_1 = 98.81 \text{ Btu/lbm} \qquad h_3 = h_4 = 27.28 \text{ Btu/lbm} \qquad s_1 = 0.2250 \text{ Btu/lbm-°R}$$

Recognizing that the R134a is compressed isentropically in the ideal cycle, state 2 is located as follows:

$$\left.\begin{array}{r} s_2 = s_1 = 0.2250 \text{ Btu/lbm-°R} \\ P_2 \cong 60 \text{ psia} \end{array}\right\} \therefore h_2 = 112.2 \text{ Btu/lbm}$$

where P_2 is the saturation pressure at 50 °F. We can now calculate the desired information:

$$\dot{Q}_E = \dot{m}(h_1 - h_4) = (2)(98.81 - 27.28) = 143.1 \text{ Btu/sec (42.9 tons)}$$

$$\dot{W}_{\text{in}} = \dot{m}(h_2 - h_1) = (2)(112.2 - 98.81) = 26.8 \text{ Btu/sec}$$

$$\text{COP} = \frac{\dot{Q}_E}{\dot{W}_{in}} = \frac{143.1}{26.8} = 5.34$$

10.2 R134a is compressed from 200 kPa to 1.0 MPa in an 80 percent efficient compressor (Fig. 10-12). The condenser exiting temperature is 40 °C. Calculate the COP and the refrigerant mass flux for 100 tons (352 kW) of refrigeration.

From the R134a table we find that

$$h_1 = 241.3 \text{ kJ/kg} \qquad h_3 = h_4 = 105.29 \text{ kJ/kg} \qquad s_1 = 0.9253 \text{ kJ/kg·K}$$

State 2′ is located, assuming an isentropic process, as follows:

$$\left.\begin{array}{r} s_{2'} = s_1 = 0.9253 \text{ kJ/kg·K} \\ P_2 = 1.0 \text{ MPa} \end{array}\right\} \therefore h_{2'} = 274.6 \text{ kJ/kg}$$

The efficiency of the compressor allows us to determine the actual compressor work. It is

$$w_a = \frac{w_s}{\eta_c} = \frac{h_{2'} - h_1}{\eta_c} = \frac{274.6 - 241.3}{0.8} = 41.7 \text{ kJ/kg}$$

The cycle COP_R is calculated to be

$$\text{COP}_\text{R} = \frac{h_1 - h_4}{w_a} = \frac{241.3 - 105.29}{41.7} = 3.26$$

The mass flux of refrigerant is found from $\dot{Q}_E$:

$$\dot{Q}_E = \dot{m}(h_1 - h_4) \qquad 352 = \dot{m}(241.3 - 105.29) \qquad \dot{m} = 2.59 \text{ kg/s}$$

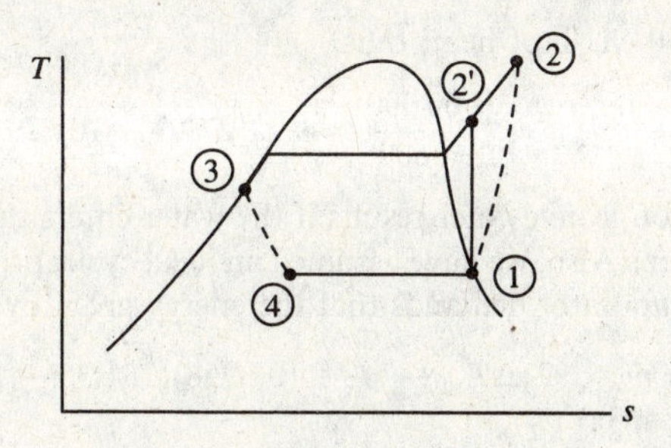

Fig. 10-12

10.3 A two-stage refrigeration system operates between high and low pressures of 1.6 MPa and 100 kPa, respectively. If the mass flux of R134a in the low-pressure stage is 0.6 kg/s, find (*a*) the tons of refrigeration, (*b*) the coefficient of performance, and (*c*) the mass flux of cooling water used to cool the R134a in the condenser if $\Delta T_w = 15$ °C.

The intermediate pressure is $P_i = (P_H P_L)^{1/2} = [(1.6)(0.1)]^{1/2} = 0.4$ MPa. Referring to Fig. 10-5, the R134a tables provide us with

$$h_1 = 231.35 \text{ kJ/kg} \qquad h_5 = 252.32 \text{ kJ/kg} \qquad h_7 = h_8 = 134.02 \text{ kJ/kg}$$
$$h_3 = h_4 = 62.0 \text{ kJ/kg} \qquad s_1 = 0.9395 \text{ kJ/kg·K} \qquad s_5 = 0.9145 \text{ kJ/kg·K}$$

Assuming the compressors to be isentropic, the enthalpies of states 2 and 6 are found by extrapolation as follows:

$$\left.\begin{array}{r} s_2 = s_1 = 0.9395 \text{ kJ/kg·K} \\ P_2 = 0.4 \text{ MPa} \end{array}\right\} \quad h_2 = 259.5 \text{ kJ/kg}$$

$$\left.\begin{array}{r} s_6 = s_5 = 0.9145 \text{ kJ/kg·K} \\ P_2 = 1.6 \text{ MPa} \end{array}\right\} \quad h_6 = 280.8 \text{ kJ/kg}$$

The mass flux of the R134a in the high-pressure stage is

$$\dot{m}_H = \dot{m}_L \frac{h_2 - h_3}{h_5 - h_8} = (0.6)\left(\frac{259.5 - 62.0}{252.32 - 134.02}\right) = 1.00 \text{ kg/s}$$

(*a*) $\dot{Q}_E = \dot{m}_L(h_1 - h_4) = (0.6)(231.35 - 62.0) = 101.6 \text{ kW} = 28.9 \text{ tons}$

(*b*) $\dot{W}_m = \dot{m}_L(h_2 - h_1) + \dot{m}_H(h_6 - h_5) = (0.6)(259.5 - 231.35) + (1.00)(280.8 - 252.3) = 45.4 \text{ kW}$

$$\text{COP}_R = \frac{\dot{Q}_E}{\dot{W}_m} = \frac{101.6}{45.4} = 2.24$$

(*c*) Cooling water is used to cool the R134a in the condenser. As energy balance on the condenser provides

$$\dot{m}_w C_p \Delta T_w = \dot{m}_H(h_6 - h_7) \qquad \dot{m}_w = \frac{(1.00)(280.8 - 134.02)}{(4.18)(15)} = 2.34 \text{ kg/s}$$

10.4 A heat pump uses groundwater at 12 °C as an energy source. If the energy delivered by the heat pump is to be 60 MJ/h, estimate the minimum mass flux of groundwater if the compressor operates with R134a between pressures of 100 kPa and 1.0 MPa. Also, calculate the minimum compressor horsepower.

Referring to Fig. 10-1*c*, the R134a tables provide

$$h_1 = 231.35 \text{ kJ/kg} \qquad h_3 = h_4 = 105.29 \text{ kJ/kg} \qquad s_1 = 0.9395 \text{ kJ/kg·K}$$

State 2 is located assuming an isentropic process as follows:

$$\left.\begin{array}{r} s_2 = s_1 = 0.9395 \text{ kJ/kg·K} \\ P_2 = 1.0 \text{ MPa} \end{array}\right\} \quad \therefore h_2 = 279.3 \text{ kJ/kg}$$

The condenser delivers 60 MJ/h of heat; thus,

$$\dot{Q}_C = \dot{m}_{\text{ref}}(h_2 - h_3) \qquad \frac{60\,000}{3600} = \dot{m}_{\text{ref}}(279.3 - 105.29) \qquad \therefore \dot{m}_{\text{ref}} = 0.0958 \text{ kg/s}$$

The minimum mass flux of groundwater results if the water enters the evaporator at 12 °C and leaves at 0 °C (the freezing point of water). Also, we have assumed an ideal cycle, providing us with a minimum mass flux. An energy balance on the evaporator demands that the energy given by the R134a be lost by the groundwater:

$$\dot{m}_{\text{ref}}(h_1 - h_4) = \dot{m}_{\text{water}} C_p \Delta T_{\text{water}} \qquad (0.0958)(231.35 - 105.29) = \dot{m}_{\text{water}}(4.18)(12 - 0)$$
$$\dot{m}_{\text{water}} = 0.241 \text{ kg/s}$$

Finally, the minimum compressor power is

$$\dot{W}_{in} = \dot{m}_{\text{ref}}(h_2 - h_1) = (0.0958)(279.3 - 231.35) = 4.59 \text{ kW} = 6.16 \text{ hp}$$

10.5 A simple gas cycle produces 10 tons of refrigeration by compressing air from 200 kPa to 2 MPa. If the maximum and minimum temperatures are 300 °C and −90 °C, respectively, find the compressor power and the cycle COP. The compressor is 82 percent efficient and the turbine is 87 percent efficient.

The ideal compressor inlet temperature (see Fig. 10-10) is $T_{2'} = T_3(P_2/P_3)^{(k-1)/k} = (573)(200/2000)^{0.2857} = 296.8$ K. Because the compressor is 82 percent efficient, the actual inlet temperature T_2 is found as follows:

$$\eta_{\text{comp}} = \frac{w_s}{w_a} = \frac{C_p(T_3 - T_{2'})}{C_p(T_3 - T_2)} \qquad \therefore T_2 = \left(\frac{1}{0.82}\right)[(0.82)(573) - 573 + 296.8] = 236.2 \text{ K}$$

The low-temperature heat exchanger produces 10 tons = 35.2 kW of refrigeration:

$$\dot{Q}_{\text{in}} = \dot{m}C_p(T_2 - T_1) \qquad 35.2 = \dot{m}(1.00)(236.2 - 183) \qquad \therefore \dot{m} = 0.662 \text{ kg/s}$$

The compressor power is then $\dot{W}_{\text{comp}} = \dot{m}C_p(T_3 - T_2) = (0.662)(1.00)(573 - 236.2) = 223$ kW. The turbine produces power to help drive the compressor. The ideal turbine inlet temperature is

$$T_{4'} = T_1\left(\frac{P_4}{P_1}\right)^{(k-1)/k} = (183)\left(\frac{2000}{200}\right)^{0.2857} = 353.3 \text{ K}$$

The turbine power output is $\dot{W}_{\text{turb}} = \dot{m}\eta_{\text{turb}}C_p(T_{4'} - T_1) = (0.662)(0.87)(1.00)(353.3 - 183) = 98.1$ kW. The cycle COP is now calculated to be

$$\text{COP} = \frac{\dot{Q}_{\text{in}}}{\dot{W}_{\text{net}}} = \frac{(10)(3.52)}{223 - 98.1} = 0.282$$

10.6 Air enters the compressor of a gas refrigeration cycle at −10 °C and is compressed from 200 kPa to 800 kPa. The high-pressure air is then cooled to 0 °C by transferring energy to the surroundings and then to −30 °C with an internal heat exchanger before it enters the turbine. Calculate the minimum possible temperature of the air leaving the turbine, the coefficient of performance, and the mass flux for 8 tons of refrigeration. Assume ideal components.

Refer to Fig. 10-11 for designation of states. The temperature at the compressor outlet is

$$T_4 = T_3\left(\frac{P_4}{P_3}\right)^{(k-1)/k} = (283)\left(\frac{800}{200}\right)^{0.2857} = 420.5 \text{ K}$$

The minimum temperature at the turbine outlet follows from an isentropic process:

$$T_1 = T_6\left(\frac{P_1}{P_6}\right)^{(k-1)/k} = (243)\left(\frac{200}{800}\right)^{0.2857} = 163.5 \text{ K}$$

The coefficient of performance is calculated as follows:

$$q_{\text{in}} = C_p(T_2 - T_1) = (1.00)(243 - 163.5) = 79.5 \text{ kJ/kg}$$
$$w_{\text{comp}} = C_p(T_4 - T_3) = (1.00)(420.5 - 283) = 137.5 \text{ kJ/kg}$$
$$w_{\text{turb}} = C_p(T_6 - T_1) = (1.00)(243 - 163.5) = 79.5 \text{ kJ/kg}$$
$$\therefore \text{COP} = \frac{q_{\text{in}}}{w_{\text{comp}} - w_{\text{turb}}} = \frac{79.5}{137.5 - 79.5} = 1.37$$

We find the mass flux as follows:

$$\dot{Q}_{\text{in}} = \dot{m}q_{\text{in}} \qquad (8)(3.52) = (\dot{m})(79.5) \qquad \dot{m} = 0.354 \text{ kg/s}$$

Supplementary Problems

10.7 An ideal vapor refrigeration cycle utilizes R134a as the working fluid between saturation temperatures of −20 °C and 40 °C. For a flow of 0.6 kg/s, determine (*a*) the rate of refrigeration, (*b*) the coefficient of performance, and (*c*) the coefficient of performance if used as a heat pump.

10.8 R134a is used in an ideal refrigeration cycle between pressures of 120 kPa and 1000 kPa. If the compressor requires 10 hp, calculate (*a*) the rate of refrigeration, (*b*) the coefficient of performance, and (*c*) the coefficient of performance if used as a heat pump.

10.9 An ideal refrigeration cycle using R134a produces 10 tons of refrigeration. If it operates between saturation temperatures of $-10\,^\circ$F and $110\,^\circ$F, determine (*a*) the COP, (*b*) the power input needed for the compressor, and (*c*) the volume rate of flow into the compressor.

10.10 If the low pressure is 200 kPa in Prob. 10.8, rework the problem.

10.11 For 20 tons of refrigeration calculate the minimum work input to the compressor for the cycle shown in Fig. 10-13 if the working fluid is (*a*) R134a, and (*b*) water.

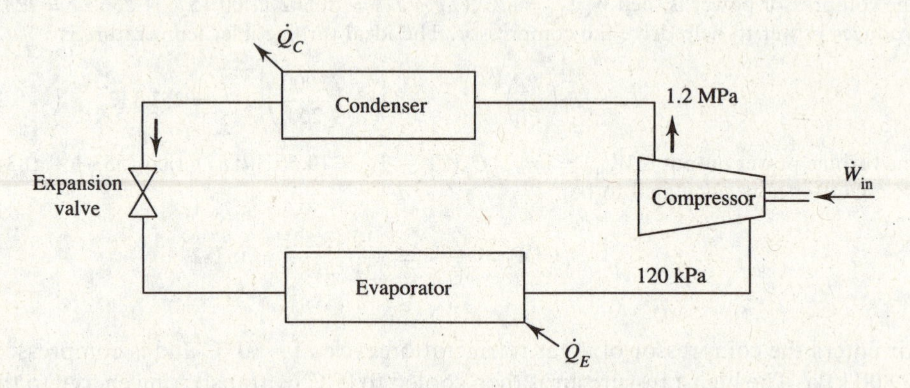

Fig. 10-13

10.12 The turbine shown in Fig. 10-14 produces just enough power to operate the compressor. The R134a is mixed in the condenser and is then separated into mass fluxes $\dot{m}_p$ and $\dot{m}_r$. Determine $\dot{m}_p/\dot{m}_r$ and $\dot{Q}_B/\dot{Q}_E$.

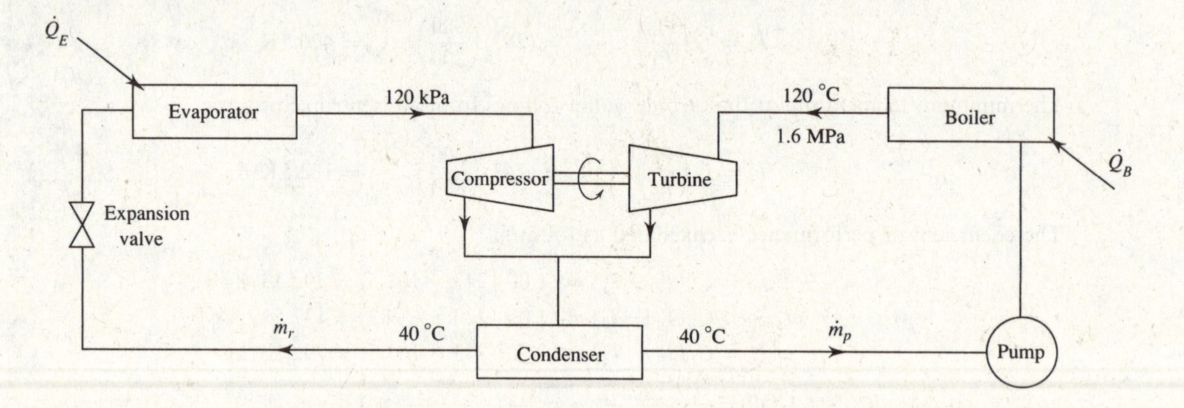

Fig. 10-14

10.13 Assume that the refrigerant leaving the condenser of Prob. 10.7 is subcooled to 36 °C. Calculate the coefficient of performance.

10.14 The compressor of a refrigeration cycle accepts R134a as saturated vapor at 200 kPa and compresses it to 1200 kPa; it is 80 percent efficient. The R134a leaves the condenser at 40 °C. Determine (*a*) the COP and (*b*) the mass flux of R134a for 10 tons of refrigeration.

10.15 R134a enters a compressor at 15 psia and 0 °F and leaves at 180 psia and 200 °F. If it exits the condenser as saturated liquid and the system produces 12 tons of refrigeration, calculate (*a*) the COP, (*b*) the mass flux of refrigerant, (*c*) the power input to the compressor, (*d*) the compressor efficiency, and (*e*) the volume rate of flow entering the compressor.

10.16 A refrigeration cycle circulates 0.2 kg/s of R134a. Saturated vapor enters the compressor at 140 kPa and leaves at 1200 kPa and 80 °C. The temperature at the condenser exit is 44 °C. Determine (*a*) the COP, (*b*) the tons of refrigeration, (*c*) the required power input, (*d*) the efficiency of the compressor, and (*e*) the mass flux of condenser cooling water if a temperature rise of 10 °C is allowed.

10.17 A refrigeration cycle utilizes a compressor which is 80 percent efficient; it accepts R134a as saturated vapor at −24 °C. The liquid leaving the condenser is at 800 kPa and 30 °C. For a mass flux of 0.1 kg/s calculate (*a*) the COP, (*b*) the tons of refrigeration, and (*c*) the mass flux of condenser cooling water for a temperature rise of 10 °C.

10.18 An ideal two-stage refrigeration cycle with $P_H = 1$ MPa and $P_L = 120$ kPa has an intermediate pressure of the R134a of 320 kPa. If 10 tons of refrigeration is produced, calculate the mass fluxes in both loops and the COP.

10.19 An ideal two-stage cycle with an intermediate pressure of 400 kPa replaces the refrigeration cycle of Prob. 10-9*a*. Determine the COP and the necessary power input.

10.20 (*a*) For a 20-ton refrigeration cycle like that shown in Fig. 10-15, operating with R134a between pressures of 1000 and 160 kPa, determine the maximum coefficient of performance and the minimum power input. (*b*) Determine the maximum COP and the minimum power input for a single-stage system operating between the same pressures.

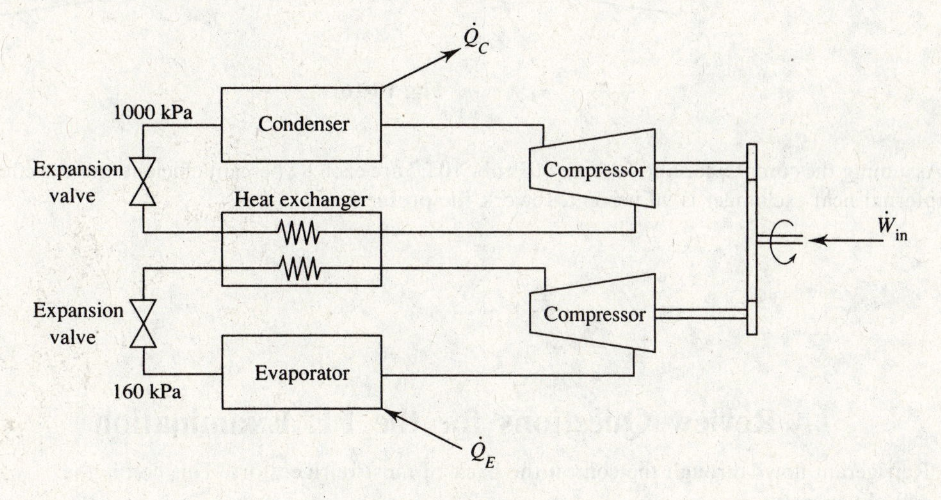

Fig. 10-15

10.21 A two-stage refrigeration system using R134a operates between pressures of 1.0 MPa and 90 kPa with a mass flux of 0.5 kg/s in the high-pressure stage. Assuming ideal cycles, calculate (*a*) the tons of refrigeration, (*b*) the power input, (*c*) the rating in compressor horsepower per ton of refrigeration, and (*d*) the mass flux of condenser cooling water if a 20 °C temperature rise is allowed.

10.22 A heat pump using R134a as the refrigerant provides 80 MJ/h to a building. The cycle operates between pressures of 1000 and 200 kPa. Assuming an ideal cycle, determine (*a*) the COP, (*b*) the compressor horsepower, and (*c*) the volume flow rate into the compressor.

10.23 A home heating system uses a heat pump with R134a as the refrigerant. The maximum heating load results when the temperature of 1000 ft^3/min of circulation air is raised 45 °F. If the compressor increases the pressure from 30 to 160 psia, calculate (*a*) the COP, (*b*) the compressor power needs, and (*c*) the mass flux of R134a. Assume an ideal cycle.

10.24 Air flows at the rate of 2.0 kg/s through the compressor of an ideal gas refrigeration cycle where the pressure increases to 500 kPa from 100 kPa. The maximum and minimum cycle temperatures are 300 °C and – 20 °C, respectively. Calculate the COP and the power needed to drive the compressor.

10.25 Rework Prob. 10.24 assuming the efficiencies of the compressor and turbine are 84 percent and 88 percent, respectively.

10.26 An ideal internal heat exchanger is added to the cycle of Prob. 10.24 (see Fig. 10-11 of Example 10.6) so that the low temperature is reduced to – 60 °C while the maximum temperature remains at 300 °C. Determine the COP and the compressor power requirement.

10.27 What is the COP for the ideal air cycle shown in Fig. 10-16 if it is (*a*) used to refrigerate a space, and (*b*) used to heat a space?

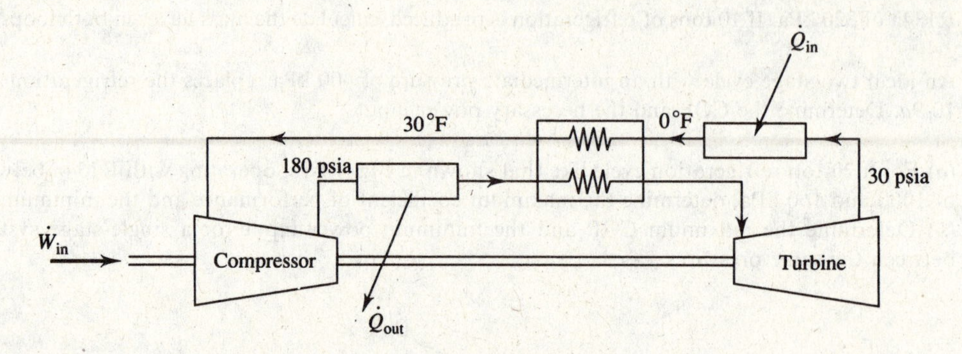

Fig. 10-16

10.28 Assuming the compressor and turbine of Prob. 10.27 are each 87 percent efficient and the effectiveness of the internal heat exchanger is 90 percent, rework the problem.

Review Questions for the FE Examination

10.1FE Refrigerant flows through the coil on the back of most refrigerators. This coil is the:
(A) Evaporator
(B) Intercooler
(C) Reheater
(D) Condenser

10.2FE An ideal refrigeration cycle operates between a low temperature of – 12 °C and a high pressure of 800 kPa. If the mass flow rate is 2 kg/s of R134a, the compressor power requirement is nearest:
(A) 38 kW
(B) 34 kW
(C) 50 kW
(D) 60 kW

10.3FE The quality of the refrigerant of Question 10.2FE entering the evaporator is nearest:
(A) 14%
(B) 26%
(C) 29%
(D) 34%

10.4FE How much heat could the cycle of Question 10.2FE supply?
(A) 1270 MJ/hr
(B) 1060 MJ/hr
(C) 932 MJ/hr
(D) 865 MJ/hr

10.5FE The COP of the cycle of Question 10.2FE, when used as a refrigerator, is nearest:
(A) 2.4
(B) 3.4
(C) 4.0
(D) 4.8

10.6FE Which of the following refrigeration cycle deviations from the ideal is considered desirable?
(A) Pressure drops through the connecting tubes.
(B) Friction on the compressor blades.
(C) Subcooling of the refrigerant in the condenser.
(D) Superheating of the vapor entering the compressor.

10.7FE Air enters the compressor of an ideal gas-refrigeration cycle at 10 °C and 80 kPa. If the maximum and minimum temperatures are 250 °C and −50 °C, the compressor work is nearest:
(A) 170 kJ/kg
(B) 190 kJ/kg
(C) 220 kJ/kg
(D) 2400 kJ/kg

10.8FE The pressure ratio across the compressor in the cycle of Question 10.7FE is nearest:
(A) 8.6
(B) 8.2
(C) 7.8
(D) 7.4

10.9FE The COP for the refrigeration cycle of Question 10.7FE is nearest:
(A) 1.04
(B) 1.18
(C) 1.22
(D) 1.49

Answers to Supplementary Problems

10.7 (*a*) 77.5 kW (*b*) 3.08 (*c*) 4.08

10.8 (*a*) 21.8 kW (*b*) 2.94 (*c*) 3.94

10.9 (*a*) 2.60 (*b*) 18.1 hp (*c*) 1.71 ft^3/sec

10.10 (*a*) 30.5 kW (*b*) 4.08 (*c*) 5.08

10.11 (*a*) 28.6 kW (*b*) 19.4 kW

10.12 3.35, 1.91

10.13 3.22

10.14 (*a*) 2.91 (*b*) 0.261 kg/s

10.15 (*a*) 1.40 (*b*) 0.772 lbm/sec (*c*) 40.6 hp (*d*) 62.5% (*e*) 2.38 ft^3/sec

10.16 (*a*) 1.65 (*b*) 6.95 tons (*c*) 14.8 kW (*d*) 60.2% (*e*) 0.940 kg/s

10.17 (*a*) 2.78 (*b*) 4.01 tons (*c*) 0.460 kg/s

10.18 0.195 kg/s, 0.272 kg/s, 3.42

10.19 3.02, 23.4 kW

10.20 (*a*) 4.04, 17.4 kW (*b*) 3.49, 20.2 kW

10.21 (*a*) 17.8 tons (*b*) 20.9 kW (*c*) 1.58 hp/ton (*d*) 2.93 kg/s

10.22 (*a*) 5.06 (*b*) 5.88 hp (*c*) 0.013 m^3/s

10.23 (*a*) 4.79 (*b*) 4.06 hp (*c*) 0.193 lbm/sec

10.24 1.73, 169 hp

10.25 0.57, 324 hp

10.26 1.43, 233 hp

10.27 (*a*) 1.28 (*b*) 2.28

10.28 (*a*) 0.737 (*b*) 1.74

Answers to Review Questions for the FE Examination

10.1FE (D) **10.2FE** (D) **10.3FE** (C) **10.4FE** (A) **10.5FE** (D) **10.6FE** (C) **10.7FE** (D) **10.8FE** (A) **10.9FE** (B)

Thermodynamic Relations

11.1 THREE DIFFERENTIAL RELATIONSHIPS

Let us consider a variable z which is a function of x and y. Then we may write

$$z = f(x, y) \qquad dz = \left(\frac{\partial z}{\partial x}\right)_y dx + \left(\frac{\partial z}{\partial y}\right)_x dy \tag{11.1}$$

This relationship is an exact mathematical formulation for the differential z. Let us write dz in the form

$$dz = M\,dx + N\,dy \tag{11.2}$$

where

$$M = \left(\frac{\partial z}{\partial x}\right)_y \qquad N = \left(\frac{\partial z}{\partial y}\right)_x \tag{11.3}$$

If we have exact differentials (and we will when dealing with thermodynamic properties), then we have the first important relationship:

$$\left(\frac{\partial M}{\partial y}\right)_x = \left(\frac{\partial N}{\partial x}\right)_y \tag{11.4}$$

This is proved by substituting in for M and N from our previous equations:

$$\frac{\partial^2 z}{\partial y\,\partial x} = \frac{\partial^2 z}{\partial x\,\partial y} \tag{11.5}$$

which is true providing the order of differentiation makes no difference in the result, which it does not for the functions of interest in our study of thermodynamics.

To find our second important relationship, first consider that x is a function of y and z, that is, $x = f(y, z)$. Then we may write

$$dx = \left(\frac{\partial x}{\partial y}\right)_z dy + \left(\frac{\partial x}{\partial z}\right)_y dz \tag{11.6}$$

Substituting for dz from (*11.1*), we have

$$dx = \left(\frac{\partial x}{\partial y}\right)_z dy + \left(\frac{\partial x}{\partial z}\right)_y \left[\left(\frac{\partial z}{\partial x}\right)_y dx + \left(\frac{\partial z}{\partial y}\right)_x dy\right] \tag{11.7}$$

or, rearranging,

$$0 = \left[1 - \left(\frac{\partial x}{\partial z}\right)_y \left(\frac{\partial z}{\partial x}\right)_y\right] dx - \left[\left(\frac{\partial x}{\partial z}\right)_y \left(\frac{\partial z}{\partial y}\right)_x + \left(\frac{\partial x}{\partial z}\right)_z\right] dy \tag{11.8}$$

The two independent variables x and y can be varied independently, that is, we can fix x and vary y, or fix y and vary x. If we fix x, then $dx = 0$; hence the bracketed coefficient of dy must be zero. If we fix y, then $dy = 0$ and the bracketed coefficient of dx is zero. Consequently,

$$1 - \left(\frac{\partial x}{\partial z}\right)_y \left(\frac{\partial z}{\partial x}\right)_y = 0 \tag{11.9}$$

and

$$\left(\frac{\partial x}{\partial z}\right)_y \left(\frac{\partial z}{\partial y}\right)_x + \left(\frac{\partial x}{\partial y}\right)_z = 0 \tag{11.10}$$

The first equation gives

$$\left(\frac{\partial x}{\partial z}\right)_y \left(\frac{\partial z}{\partial x}\right)_y = 1 \tag{11.11}$$

which leads to our second important relationship:

$$\left(\frac{\partial x}{\partial z}\right)_y = \frac{1}{(\partial z/\partial x)_y} \tag{11.12}$$

Now rewrite (*11.10*) as

$$\left(\frac{\partial x}{\partial z}\right)_y \left(\frac{\partial z}{\partial y}\right)_x = -\left(\frac{\partial x}{\partial y}\right)_z \tag{11.13}$$

Dividing through by $(\partial x/\partial y)_z$ and using (*11.12*),

$$\left[\left(\frac{\partial x}{\partial y}\right)_z\right]^{-1} = \left(\frac{\partial y}{\partial x}\right)_z \tag{11.14}$$

we obtain the *cyclic formula*

$$\left(\frac{\partial x}{\partial z}\right)_y \left(\frac{\partial z}{\partial y}\right)_x \left(\frac{\partial y}{\partial x}\right)_z = -1 \tag{11.15}$$

EXAMPLE 11.1 Estimate the change in the specific volume of air, assuming an ideal gas, using the differential form for dv, if the temperature and pressure change from 25 °C and 122 kPa to 29 °C and 120 kPa. Compare with the change calculated directly from the ideal-gas law.

Solution: Using $v = RT/P$, we find

$$dv = \left(\frac{\partial v}{\partial T}\right)_P dT + \left(\frac{\partial v}{\partial P}\right)_T dp = \frac{R}{P}\,dT - \frac{RT}{P^2}\,dP = \left(\frac{0.287}{121}\right)(4) - \frac{(0.287)(300)}{(121)^2}(-2) = 0.02125\ \text{m}^3/\text{kg}$$

where we have used average values for P and T.

The ideal-gas law provides

$$\Delta v = \frac{RT_2}{P_2} - \frac{RT_1}{P_1} = \frac{(0.287)(302)}{120} - \frac{(0.287)(298)}{122} = 0.02125\,\text{m}^3/\text{kg}$$

Obviously the change in state of 4 °C and −2 kPa is sufficiently small that the differential change dv approximates the actual change Δv.

11.2 THE MAXWELL RELATIONS

For small (differential) changes in the internal energy and the enthalpy of a simple compressible system, we may write the differential forms of the first law as [see (*6.9*) and (*6.12*)]

$$du = T\,ds - P\,dv \tag{11.16}$$

$$dh = T\,ds + v\,dP \tag{11.17}$$

We introduce two other properties: the Helmholtz function a and the Gibbs function g:

$$a = u - Ts \tag{11.18}$$

$$g = h - Ts \tag{11.19}$$

In differential form, using (*11.16*) and (*11.17*), we can write

$$da = -P\,dv - s\,dT \tag{11.20}$$

$$dg = v\,dP - s\,dT \tag{11.21}$$

Applying our first important relationship from calculus [see (*11.4*)] to the four exact differentials above, we obtain the *Maxwell relations*:

$$\left(\frac{\partial T}{\partial v}\right)_s = -\left(\frac{\partial P}{\partial s}\right)_v \tag{11.22}$$

$$\left(\frac{\partial T}{\partial P}\right)_s = \left(\frac{\partial v}{\partial s}\right)_P \tag{11.23}$$

$$\left(\frac{\partial P}{\partial T}\right)_v = \left(\frac{\partial s}{\partial v}\right)_T \tag{11.24}$$

$$\left(\frac{\partial v}{\partial T}\right)_P = -\left(\frac{\partial s}{\partial P}\right)_T \tag{11.25}$$

Through the Maxwell relations, changes in entropy (an immeasurable quantity) can be expressed in terms of changes in v, T, and P (measurable quantities). By extension, the same can be done for internal energy and enthalpy (see Sec. 11.4).

EXAMPLE 11.2 Assuming that $h = h(s, P)$, what two differential relationships does this imply? Verify one of the relationships using the steam tables at 400 °C and 4 MPa.

Solution: If $h = h(s, P)$ we can write

$$dh = \left(\frac{\partial h}{\partial s}\right)_P ds + \left(\frac{\partial h}{\partial P}\right)_s dP$$

But the first law can be written as [see (*11.17*)] $dh = T\,ds + v\,dP$. Equating coefficients of ds and dP, there results

$$T = \left(\frac{\partial h}{\partial s}\right)_P \qquad v = \left(\frac{\partial h}{\partial P}\right)_s$$

Let's verify the constant-pressure relationships. At $P = 4$ MPa and using central differences (use entries on either side of the desired state) at $T = 400\,°\text{C}$, we have

$$\left(\frac{\partial h}{\partial s}\right)_P = \frac{3330 - 3092}{6.937 - 6.583} = 672 \text{ K or } 399\,°\text{C}$$

This compares favorably with the specified temperature of 400 °C.

11.3 THE CLAPEYRON EQUATION

We may use the Maxwell relations in a variety of ways. For example, (*11.24*) allows us to express the quantity h_{fg} (the enthalpy of vaporization) using P, v, and T data alone. Suppose we desire h_{fg} at the point (v_0, T_0) of Fig. 11-1. Since the temperature remains constant during the phase change, we can write

$$\left(\frac{\partial s}{\partial v}\right)_{T=T_0} = \frac{s_g - s_f}{v_g - v_f} \tag{11.26}$$

Consequently, (*11.24*) gives

$$\left(\frac{\partial P}{\partial T}\right)_{v=v_0} = \frac{s_{fg}}{v_{fg}} \tag{11.27}$$

But, we can integrate (*11.17*), knowing that P and T are constant during a phase change:

$$\int dh = \int T_0\,ds - \int v\,dP^{\nearrow 0} \qquad \text{or} \qquad h_{fg} = T_0 s_{fg} \tag{11.28}$$

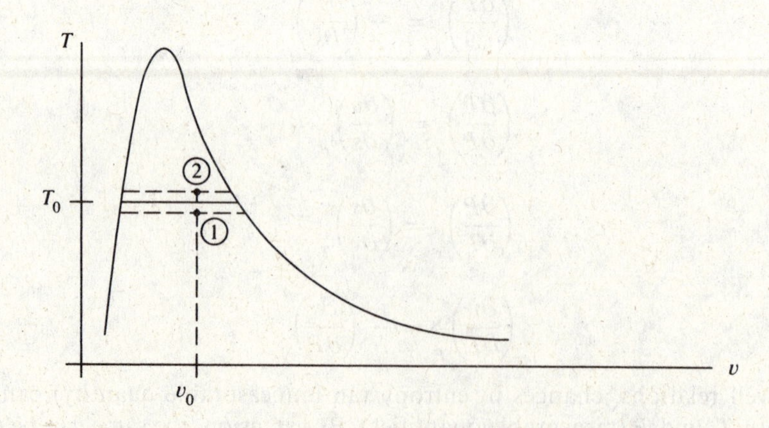

Fig. 11-1 A T-v diagram for a phase-change substance.

This is substituted into (*11.27*) to give the *Clapeyron equation*:

$$\left(\frac{\partial P}{\partial T}\right)_{v=v_0} = \frac{h_{fg}}{T_0 v_{fg}} \qquad \text{or} \qquad h_{fg} = T_0 v_{fg}\left(\frac{\partial P}{\partial T}\right)_{v=v_0} \tag{11.29}$$

The partial derivative $(\partial P/\partial T)_{v=v_0}$ can be evaluated from the saturated-state tables using the central-difference approximation

$$\left(\frac{\partial P}{\partial T}\right)_{v=v_0} \cong \frac{P_2 - P_1}{T_2 - T_1} \tag{11.30}$$

where T_2 and T_1 are selected at equal intervals above and below T_0. (See Fig. 11-1 and Example 11.3.)

For relatively low pressures, the Clapeyron equation can be modified when $v_g \gg v_f$. We may treat the saturated vapor as an ideal gas, so that

$$v_{fg} = v_g - v_f \cong v_g = \frac{RT}{P} \tag{11.31}$$

Then (*11.29*) becomes (dropping the subscript 0)

$$\left(\frac{\partial P}{\partial T}\right)_v = \frac{P h_{fg}}{RT^2} \tag{11.32}$$

This is often referred to as the *Clausius–Clapeyron equation.* It may also be used for the sublimation process involving a solid to vapor phase change.

During a phase change, the pressure depends only on the temperature; hence, we may use an ordinary derivative so that

$$\left(\frac{\partial P}{\partial T}\right)_v = \left(\frac{dP}{dT}\right)_{\text{sat}} \tag{11.33}$$

Then (*11.32*) can be rearranged as

$$\left(\frac{dP}{P}\right)_{\text{sat}} = \frac{h_{fg}}{R}\left(\frac{dT}{T^2}\right)_{\text{sat}} \tag{11.34}$$

This is integrated between two saturation states to yield

$$\ln\left(\frac{P_2}{P_1}\right)_{\text{sat}} \cong \frac{h_{fg}}{R}\left(\frac{1}{T_1} - \frac{1}{T_2}\right)_{\text{sat}} \tag{11.35}$$

where we have assumed h_{fg} to be constant between state 1 and state 2 (hence the "approximately equal to" symbol). Relationship (*11.35*) may be used to approximate the pressure or temperature below the limits of tabulated values (see Example 11.4).

EXAMPLE 11.3 Predict the value for the enthalpy of vaporization for water at 200 °C assuming steam to be an ideal gas. Calculate the percent error.

Solution: At 200 °C and 155.4 kPa the specific volume of the saturated steam is, in the ideal-gas approximation, $v_g = RT/P = (0.462)(473)/155 = 0.1406\ \text{m}^3/\text{kg}$. For liquid water the density is approximately 1000 kg/m^3 so that $v_f \cong 0.001\ \text{m}^3/\text{kg}$ (or we can use v_f from the steam tables). Hence we find

$$h_{fg} = Tv_{fg}\left(\frac{\partial P}{\partial T}\right)_v = (473)(0.1406 - 0.001)\left(\frac{1906 - 1254}{210 - 190}\right) = 2153\ \text{kJ/kg}$$

This compares with $h_{fg} = 1941$ kJ/kg from the steam tables, the error being

$$\%\ \text{error} = \left(\frac{2153 - 1941}{1941}\right)(100) = 10.9\%$$

This error is due to the inaccuracy of the value for v_g.

EXAMPLE 11.4 Suppose the steam tables started at $P_{\text{sat}} = 2$ kPa ($T_{\text{sat}} = 17.5\,^\circ\text{C}$) and we desired T_{sat} at $P_{\text{sat}} = 1$ kPa. Predict T_{sat} and compare with the value from the steam tables.

Solution: Since the pressure is quite low, we will assume that $v_g \gg v_f$ and that v_g is given by the ideal-gas law. Using values for h_{fg} at $P_{\text{sat}} = 4$ kPa, 3 kPa, and 2 kPa we assume that at $P_{\text{sat}} = 1$ kPa, $h_{fg} = 2480$ kJ/kg. Then (*11.35*) provides us with

$$\ln\left(\frac{P_2}{P_1}\right)_{\text{sat}} = \frac{h_{fg}}{R}\left(\frac{1}{T_1} - \frac{1}{T_2}\right)_{\text{sat}} \qquad \ln\left(\frac{1}{2}\right) = \left(\frac{2480}{0.462}\right)\left(\frac{1}{290.5} - \frac{1}{T_2}\right) \qquad \therefore T_2 = 280\,\text{K or } 7.0\,^\circ\text{C}$$

This is very close to the value of 6.98 °C from the steam tables.

11.4 FURTHER CONSEQUENCES OF THE MAXWELL RELATIONS

Internal Energy

Considering the internal energy to be a function of T and v, we can write

$$du = \left(\frac{\partial u}{\partial T}\right)_v dT + \left(\frac{\partial u}{\partial v}\right)_T dv = C_v\,dT + \left(\frac{\partial u}{\partial v}\right)_T dv \tag{11.36}$$

where we have used the definition $C_v = (\partial u/\partial T)_v$. The differential form of the first law is

$$du = T\,ds - P\,dv \tag{11.37}$$

Assuming $s = f(T, v)$, the above relationship can be written as

$$du = T\left[\left(\frac{\partial s}{\partial T}\right)_v dT + \left(\frac{\partial s}{\partial v}\right)_T dv\right] - P\,dv = T\left(\frac{\partial s}{\partial T}\right)_v dT + \left[T\left(\frac{\partial s}{\partial v}\right)_T - P\right]dv \tag{11.38}$$

When this expression for du is equated to that of (*11.36*), one obtains

$$C_v = T\left(\frac{\partial s}{\partial T}\right)_v \tag{11.39}$$

$$\left(\frac{\partial u}{\partial v}\right)_T = T\left(\frac{\partial s}{\partial v}\right)_T - P = T\left(\frac{\partial P}{\partial T}\right)_v - P \tag{11.40}$$

where we have used the Maxwell relation (*11.24*). We can now relate du to the properties P, v, T, and C_v by substituting (*11.40*) into (*11.36*):

$$du = C_v\,dT + \left[T\left(\frac{\partial P}{\partial T}\right)_v - P\right]dv \tag{11.41}$$

This can be integrated to provide $(u_2 - u_1)$ if we have an equation of state that provides the relationship between P, v, and T so that $(\partial P/\partial T)_v$ is known.

Enthalpy

Considering enthalpy to be a function of T and P, steps similar to those above result in

$$C_p = T\left(\frac{\partial s}{\partial T}\right)_P \tag{11.42}$$

$$dh = C_p\,dT + \left[v - T\left(\frac{\partial v}{\partial T}\right)_P\right]dP \tag{11.43}$$

which can be integrated to give $(h_2 - h_1)$ if an equation of state is known.

Since we know that $h = u + Pv$, we have

$$h_2 - h_1 = u_2 - u_1 + P_2v_2 - P_1v_1 \tag{11.44}$$

Hence, if we know $P = f(T, v)$, we can find $(u_2 - u_1)$ from (*11.41*) and $(h_2 - h_1)$ from (*11.44*). If we know $v = f(P, T)$, we can find $(h_2 - h_1)$ from (*11.43*) and $(u_2 - u_1)$ from (*11.44*). In the first case we know P explicitly as a function of T and v; in the second case we know v explicitly as a function of P and T. For an ideal gas, $Pv = RT$ so that the bracketed quantities in (*11.41*) and (*11.43*) are zero, as we have assumed earlier in our study of an ideal gas in which $u = u(T)$ and $h = h(T)$. For a nonideal gas an equation of state will be provided so that one of the bracketed quantities can be evaluated.

Entropy

Finally, let us find an expression for ds. Consider $s = s(T, v)$. Then, using (*11.39*) and (*11.24*), we have

$$ds = \left(\frac{\partial s}{\partial T}\right)_v dT + \left(\frac{\partial s}{\partial v}\right)_T dv = \frac{C_v}{T} dT + \left(\frac{\partial P}{\partial T}\right)_v dv \tag{11.45}$$

Alternatively, we can let $s = s(T, P)$. Then, using (*11.42*) and (*11.25*), we find

$$ds = \frac{C_p}{T} dT - \left(\frac{\partial v}{\partial T}\right)_P dP \tag{11.46}$$

These two equations can be integrated to yield

$$s_2 - s_1 = \int_{T_1}^{T_2} \frac{C_v}{T} dT + \int_{v_1}^{v_2} \left(\frac{\partial P}{\partial T}\right)_v dv = \int_{T_1}^{T_2} \frac{C_p}{T} dT - \int_{P_1}^{P_2} \left(\frac{\partial v}{\partial T}\right)_P dP \tag{11.47}$$

For an ideal gas these equations simplify to the equations of Chap. 6. See Sec. 11.7 for actual calculations involving real gases.

EXAMPLE 11.5 Derive an expression for the enthalpy change in an isothermal process of a gas for which the equation of state is $P = RT/(v - b) - (a/v^2)$.

Solution: Since P is given explicitly, we find an expression for Δu and then use (*11.44*). For a process in which $dT = 0$, (*11.41*) provides

$$\Delta u = \int_{v_1}^{v_2} \left[T\left(\frac{\partial P}{\partial T}\right)_v - P\right] dv = \int_{v_1}^{v_2} \left(\frac{TR}{v - b} - \frac{RT}{v - b} + \frac{a}{v^2}\right) dv = -a\left(\frac{1}{v_2} - \frac{1}{v_1}\right)$$

The expression for Δh is then

$$h_2 - h_1 = \Delta u + P_2v_2 - P_1v_1 = a\left(\frac{1}{v_1} - \frac{1}{v_2}\right) + P_2v_2 - P_1v_1$$

EXAMPLE 11.6 We know that $C_p = A + BT$ along a low-pressure isobar $P = P^*$ (see Fig. 11-2). If the equation of state is $P = RT/(v - b) - (a/v^2)$ find an expression for Δs.

Solution: Since we know P explicitly, we use (*11.47*) to find Δs:

$$\Delta s = \int_{T_1}^{T_2} \frac{C_v}{T} dT + \int_{v_1}^{v_2} \left(\frac{\partial P}{\partial T}\right)_v dv$$

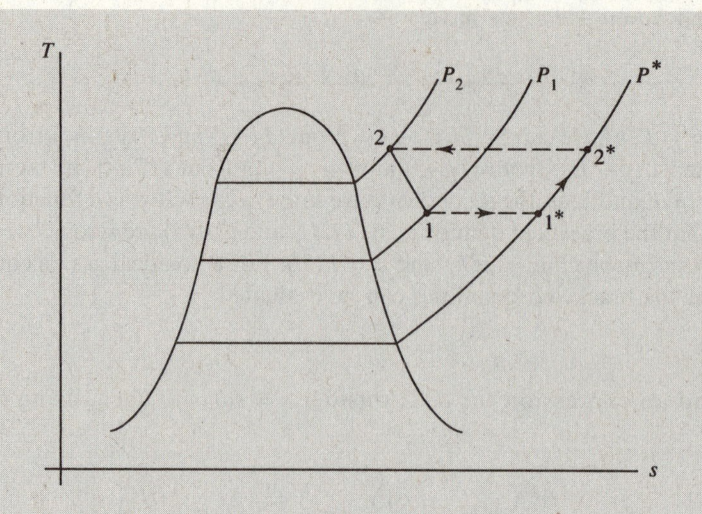

Fig. 11-2

Our expression for C_p holds only along $P = P^*$. Rather than integrating directly from 1 to 2, as shown in Fig. 11-2, we proceed isothermally from 1 to 1^*, then along $P = P^*$ from 1^* to 2^*, and finally isothermally from 2^* to 2. This results in

$$\begin{aligned}\Delta s &= -\int_{v_1}^{v_1^*}\left(\frac{\partial P}{\partial T}\right)_v dv + \int_{T_1^*}^{T_2^*}\frac{C_p}{T}\,dT + \int_{v_2^*}^{v_2}\left(\frac{\partial P}{\partial T}\right)_v dv \\ &= -\int_{v_1}^{v_1^*}\frac{R}{v-b}\,dv + \int_{T_1^*}^{T_2^*}\left(\frac{A}{T}+B\right)dT + \int_{v_2^*}^{v_2}\frac{R}{v-b}\,dv \\ &= R\ln\frac{v_1-b}{v_1^*-b} + A\ln\frac{T_2}{T_1} + B(T_2-T_1) + R\ln\frac{v_2-b}{v_2^*-b}\end{aligned}$$

We could calculate a numerical value for Δs if the initial and final states, A, B, P^*, a, and b, were provided for a particular gas.

11.5 RELATIONSHIPS INVOLVING SPECIFIC HEATS

If we can relate the specific heats to P, v, and T, we will have completed our objective of relating the "hidden" thermodynamic quantities to the three measurable properties.

The exact differential $ds = M\,dT + N\,dP$ was written in (*11.46*) as

$$ds = \frac{C_p}{T}\,dT - \left(\frac{\partial v}{\partial T}\right)_P dP \tag{11.48}$$

Using (*11.4*), we can write

$$\left[\frac{\partial}{\partial P}(C_p/T)\right]_T = -\left[\frac{\partial}{\partial T}\left(\frac{\partial v}{\partial T}\right)_P\right]_P \tag{11.49}$$

or, rearranging,

$$\left(\frac{\partial C_p}{\partial P}\right)_T = -T\left(\frac{\partial^2 v}{\partial T^2}\right)_P \tag{11.50}$$

If we start with (*11.45*), we obtain

$$\left[\frac{\partial}{\partial v}(C_v/T)\right]_T = \left[\frac{\partial}{\partial T}\left(\frac{\partial P}{\partial T}\right)_v\right]_v \tag{11.51}$$

resulting in

$$\left(\frac{\partial C_v}{\partial v}\right)_T = T\left(\frac{\partial^2 P}{\partial T^2}\right)_v \tag{11.52}$$

Consequently, knowing an equation of state, the quantities $(\partial C_p/\partial P)_T$ and $(\partial C_v/\partial v)_T$ can be found for an isothermal process.

A third useful relation can be found by equating (*11.48*) and (*11.45*):

$$\frac{C_p}{T}\,dT - \left(\frac{\partial v}{\partial T}\right)_P dP = \frac{C_v}{T}\,dT + \left(\frac{\partial P}{\partial T}\right)_v dv \tag{11.53}$$

so that

$$dT = \frac{T(\partial v/\partial T)_P}{C_p - C_v}\,dP + \frac{T(\partial P/\partial T)_v}{C_p - C_v}\,dv \tag{11.54}$$

But, since $T = T(P, v)$, we can write

$$dT = \left(\frac{\partial T}{\partial P}\right)_v dP + \left(\frac{\partial T}{\partial v}\right)_P dv \tag{11.55}$$

Equating the coefficients of dP in the above two expressions for dT gives

$$C_p - C_v = T\left(\frac{\partial v}{\partial T}\right)_P \left(\frac{\partial P}{\partial T}\right)_v = -T\left(\frac{\partial v}{\partial T}\right)_P^2 \left(\frac{\partial P}{\partial v}\right)_T \tag{11.56}$$

where we have used both (*11.12*) and (*11.15*). The same relationship would have resulted had we equated the coefficients of dv in (*11.54*) and (*11.55*). We can draw three important conclusions from (*11.56*):

1. $C_p = C_v$ for a truly incompressible substance ($v = \text{const.}$). Since $(\partial v/\partial T)_P$ is quite small for a liquid or soild, we usually assume that $C_p \cong C_v$.
2. $C_p \to C_v$ as $T \to 0$ (absolute zero).
3. $C_p \geq C_v$ since $(\partial P/\partial v)_T < 0$ for all known substances.

Equation (*11.56*) can be written in terms of the *volume expansivity*

$$\beta = \frac{1}{v}\left(\frac{\partial v}{\partial T}\right)_P \tag{11.57}$$

and the *bulk modulus*

$$B = -v\left(\frac{\partial P}{\partial v}\right)_T \tag{11.58}$$

as

$$C_p - C_v = vT\beta^2 B \tag{11.59}$$

Values for β and B can be found in handbooks of material properties.

EXAMPLE 11.7 Find an expression for $C_p - C_v$ if the equation of state is $P = RT/(v - b) - (a/v^2)$.

Solution: Equation (*11.56*) provides us with

$$C_p - C_v = T\left(\frac{\partial v}{\partial T}\right)_P \left(\frac{\partial P}{\partial T}\right)_v$$

Our given equation of state can be written as

$$T = \frac{1}{R}\left[P(v-b) + \frac{a}{v^2}(v-b)\right]$$

so that

$$(\partial T/\partial v)_P = (P - a/v^2 + 2ab/v^3)/R = 1/(\partial v/\partial T)_P$$

Hence

$$C_p - C_v = TR^2 \Big/ [(P + a/v^2 + 2ab/v^3)(v-b)]$$

This reduces to $C_p - C_v = R$ if $a = b = 0$, the ideal-gas relationship.

EXAMPLE 11.8 Calculate the entropy change of a 10-kg block of copper if the pressure changes from 100 kPa to 50 MPa while the temperature remains constant. Use $\beta = 5 \times 10^{-5}\ \text{K}^{-1}$ and $\rho = 8770\ \text{kg/m}^3$.

Solution: Using one of Maxwell's equations and *(11.57)*, the entropy differential is

$$ds = \left(\frac{\partial s}{\partial P}\right)_T dP + \left(\frac{\partial s}{\partial T}\right)_P \cancelto{0}{dT} = -\left(\frac{\partial v}{\partial T}\right)_P dP = -v\beta\, dP$$

Assuming v and β to be relatively constant over this pressure range, the entropy change is

$$s_2 - s_1 = -\frac{1}{\rho}\beta(P_2 - P_1) = -\frac{1}{8770}(5 \times 10^{-5})[(50 - 0.1) \times 10^6] = -0.285\ \text{J/kg·K}$$

If we had considered the copper to be incompressible ($dv = 0$) the entropy change would be zero, as observed from *(11.47)*. The entropy change in this example results from the small change in volume of the copper.

11.6 THE JOULE–THOMSON COEFFICIENT

When a fluid passes through a throttling device (a valve, a porous plug, a capillary tube, or an orifice) the enthalpy remains constant, the result of the first law. In the refrigeration cycle such a device was used to provide a sudden drop in the temperature. A drop does not always occur: the temperature may remain constant or the temperature may increase. Which situation occurs depends on the value of the *Joule–Thomson coefficient*,

$$\mu_j \equiv \left(\frac{\partial T}{\partial P}\right)_h \tag{11.60}$$

If μ_j is positive, a temperature decrease follows the pressure decrease across the device; if μ_j is negative, a temperature increase results; for $\mu_j = 0$, a zero temperature change results. Let us express μ_j in terms of P, v, T, and C_p as we did with the other properties in Sec. 11.4. The differential expression for dh is given in *(11.43)* as

$$dh = C_p\, dT + \left[v - T\left(\frac{\partial v}{\partial T}\right)_P\right] dP \tag{11.61}$$

If we hold h constant, as demanded by *(11.60)*, we find

$$0 = C_p\, dT + \left[v - T\left(\frac{\partial v}{\partial T}\right)_P\right] dP \tag{11.62}$$

or, in terms of partial derivatives,

$$\mu_j = \left(\frac{\partial T}{\partial P}\right)_h = \frac{1}{C_p}\left[T\left(\frac{\partial v}{\partial T}\right)_P - v\right] \tag{11.63}$$

Since μ_j is quite easy to measure, this relationship provides us with a relatively easy method to evaluate C_p. For an ideal gas, $h = h(T)$ or $T = T(h)$. Therefore, when h is held constant, T is held constant, so $\partial T/\partial P = \mu_j = 0$.

EXAMPLE 11.9 Find the Joule–Thomson coefficient for steam at 400 °C and 1 MPa using both expressions given in (*11.63*).

Solution: We can use (*11.42*) and find C_p:

$$C_p = T\left(\frac{\partial s}{\partial T}\right)_P \cong T\left(\frac{\Delta s}{\Delta T}\right)_P = 673\frac{7.619 - 7.302}{450 - 350} = 2.13 \text{ kJ/kg·K}$$

Then (*11.63*) gives, using $C_p = 2130$ J/kg·K,

$$\mu_j = \frac{1}{C_p}\left[T\left(\frac{\partial v}{\partial T}\right)_P - v\right] = \left(\frac{1}{2130}\right)\left[(673)\left(\frac{0.3304 - 0.2825}{450 - 350}\right) - 0.3066\right] = 7.40 \times 10^{-6} \text{ K/Pa}$$

Using the other expression in (*11.64*) we find (we hold enthalpy constant at 3264 kJ/kg)

$$\mu_j = \left(\frac{\partial T}{\partial P}\right)_h = \frac{403.7 - 396.2}{(1.5 - 0.5) \times 10^6} = 7.50 \times 10^{-6} \text{ K/Pa}$$

Since μ_j is positive, the temperature decreases due to the sudden decrease in pressure across a throttling device.

11.7 ENTHALPY, INTERNAL-ENERGY, AND ENTROPY CHANGES OF REAL GASES

Gases at relatively low pressure can usually be treated as an ideal gas so that $Pv = RT$. For ideal gases, the relations of the previous sections reduce to the simplified relations of the earlier chapters in this book. In this section we will evaluate the changes in enthalpy, internal energy, and entropy of real (nonideal) gases using the generalized relations of Sec. 11.4.

The general relation for the enthalpy change is found by integrating (*11.43*):

$$h_2 - h_1 = \int_{T_1}^{T_2} C_p\,dT + \int_{P_1}^{P_2}\left[v - T\left(\frac{\partial v}{\partial T}\right)_P\right]dP \tag{11.64}$$

The change in a property is independent of the path selected. Rather than going directly from 1 to 2, let us select the path shown in Fig. 11-3 that takes us to such a low pressure P^* that the process from 1^* to 2^* involves an ideal gas. Certainly $P^* = 0$ will work, so let's set $P^* = 0$ (actually, a small pressure ε may be used). The processes from 1 to 1^* and from 2^* to 2 are isothermal, so that

$$h_1^* - h_1 = \int_{P_1}^{0}\left[v - T\left(\frac{\partial v}{\partial T}\right)_P\right]_{T=T_1} dP \tag{11.65}$$

$$h_2 - h_2^* = \int_{0}^{P_2}\left[v - T\left(\frac{\partial v}{\partial T}\right)_P\right]_{T=T_2} dP \tag{11.66}$$

For the ideal process from 1^* to 2^* we have

$$h_2^* - h_1^* = \int_{T_1}^{T_2} C_p\,dT \tag{11.67}$$

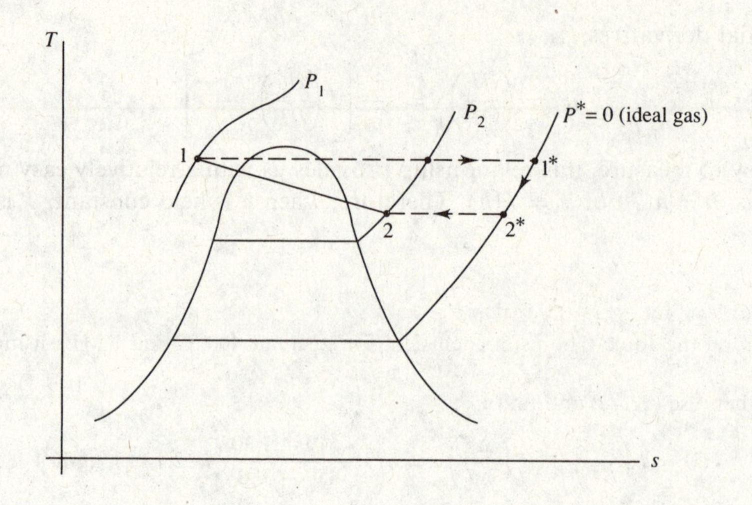

Fig. 11-3 Schematic used to find Δh in a real gas.

The enthalpy change is then

$$h_2 - h_1 = (h_1^* - h_1) + (h_2^* - h_1^*) + (h_2 - h_2^*) \tag{11.68}$$

The ideal-gas change $(h_2^* - h_1^*)$ is found using the $C_p(T)$ relationship or the gas tables. For the isothermal changes of the real gas we introduce the equation of state $Pv = ZRT$, where Z is the compressibility factor. Using $v = ZRT/P$, the integrals of *(11.65)* and *(11.66)* can be put in the form

$$\frac{h^* - h}{T_c} = -RT_R^2 \int_0^{P_R} \left(\frac{\partial Z}{\partial T_R}\right)_{P_R} \frac{dP_R}{P_R} \tag{11.69}$$

where the reduced temperature $T_R = T/T_c$ and the reduced pressure $P_R = P/P_c$ have been used. The quantity $(h^* - h)/T_c$ is called the *enthalpy departure* and has been determined numerically using a graphical integration of the compressibility chart. The result is presented in Fig. H-1 using molar units. Obviously, $h^* - h = 0$ for an ideal gas, since $h = h(T)$ and the process is isothermal.

The internal-energy change is found from the definition of enthalpy [see *(11.44)*] and is

$$u_2 - u_1 = h_2 - h_1 - R(Z_2T_2 - Z_1T_1) \tag{11.70}$$

where we have used $Pv = ZRT$.

The change in entropy of a real gas can be found using a technique similar to that used for the enthalpy change. For an isothermal process *(11.47)* provides the entropy change as

$$s_2 - s_1 = -\int_{P_1}^{P_2} \left(\frac{\partial v}{\partial T}\right)_P dP \tag{11.71}$$

We again integrate from the given state along an isothermal path to a low pressure where an ideal gas can be assumed, resulting in

$$s - s_0^* = -\int_{P_{\text{low}}}^{P} \left(\frac{\partial v}{\partial T}\right)_P dP \tag{11.72}$$

where the asterisk denotes an ideal-gas state. The above equation, integrated along an isotherm from the ideal-gas state to any state that is approximated as an ideal gas, takes the form

$$s^* - s_0^* = -\int_{P_{\text{low}}}^{P} \frac{R}{P}\, dP \tag{11.73}$$

Subtracting the above two equations provides, for an isothermal process,

$$s^* - s = -\int_{P_{\text{low}}}^{P}\left[\frac{R}{P} - \left(\frac{\partial v}{\partial T}\right)_P\right]dP \qquad (11.74)$$

Introducing the nonideal-gas equation of state $Pv = ZRT$, we have

$$s^* - s = R\int_{P_{\text{low}}}^{P}\left[(Z-1) - T_R\left(\frac{\partial Z}{\partial T_R}\right)_{P_R}\right]\frac{dP_R}{P_R} \qquad (11.75)$$

which is called the *entropy departure*. This has also been determined numerically from the compressibility charts and is presented in Appendix I using molar units. We can now find the entropy change between any two states using

$$s_2 - s_1 = -(s_2^* - s_2) + (s_2^* - s_1^*) + (s_1^* - s_1) \qquad (11.76)$$

In this equation the quantity $s_2^* - s_1^*$ represents the entropy change between the two given states, on the assumption that the gas behaves as an ideal gas; it does not represent a change along the $P^* = 0$ path illustrated in Fig. 11-3.

EXAMPLE 11.10 Calculate the enthalpy change, the internal-energy change, and the entropy change of nitrogen as it undergoes a process from $-50\,^\circ\text{C}$, 2 MPa, to $40\,^\circ\text{C}$, 6 MPa. Use (*a*) the equations for an ideal gas with constant specific heats, (*b*) the ideal-gas tables, and (*c*) the equations of this section.

Solution:

(*a*) $\Delta h = C_p\Delta T = (1.042)[40 - (-50)] = 93.8\text{ kJ/kg} \quad \Delta u = C_v\Delta T = (0.745)\,[40 - (-50)] = 67.0\text{ kJ/kg}$

$$\Delta s = C_p \ln\frac{T_2}{T_1} - R\ln\frac{P_2}{P_1} = 1.042\ln\frac{313}{223} - 0.297\ln\frac{6}{2} = 0.0270\text{ kJ/kg·K}$$

(*b*) Interpolating in the ideal-gas table (Table E-2) gives

$\Delta h = h_2 - h_1 = (9102 - 6479)/28 = 93.7\text{ kJ/kg} \qquad \Delta u = u_2 - u_1 = (6499 - 4625)/28 = 66.9\text{ kJ/kg}$

$\Delta s = s_2^o - s_1^o - R\ln\frac{P_2}{P_1} = (192.9 - 183.0)/28 - 0.297\ln(6/2) = 0.0273\text{ kJ/kg·K}$

(*c*) Using (*11.69*) and the enthalpy departure chart in Appendix H we find

$$T_{R1} = \frac{T_1}{T_{cr}} = \frac{223}{126.2} = 1.77 \qquad P_{R1} = \frac{P_1}{P_{cr}} = \frac{2}{3.39} = 0.590$$

$$T_{R2} = \frac{T_2}{T_{cr}} = \frac{313}{126.2} = 2.48 \qquad P_{R2} = \frac{P_2}{P_{cr}} = \frac{6}{3.39} = 1.77$$

The enthalpy departure chart (Appendix H) provides us with

$$\frac{\bar{h}_1^* - \bar{h}_1}{T_c} = 1.6\text{ kJ/kmol·K} \qquad \therefore h_1^* - h_1 = \frac{(1.6)(126.2)}{28} = 7.21\text{ kJ/kg}$$

$$\frac{\bar{h}_2^* - \bar{h}_2}{T_c} = 2.5\text{ kJ/kmol·K} \qquad \therefore h_2^* - h_2 = \frac{(2.5)(126.2)}{28} = 11.27\text{ kJ/kg}$$

Consequently,

$$\Delta h = (h_2 - h_2^*) + (h_1^* - h_1) + (h_2^* - h_1^*) = -11.27 + 7.21 + (1.042)[40 - (-50)] = 90\text{ kJ/kg}$$

To find the internal energy change we use (*11.70*). The Z values are found, using the compressibility chart with the above T_R and P_R values, to be $Z_1 = 0.99$ and $Z_2 = 0.985$. Then

$$\Delta u = \Delta h - R(Z_2T_2 - Z_1T_1) = 90 - (0.297)[(0.985)(313) - (0.99)(223)] = 64\text{ kJ/kg}$$

To find the entropy change we first find $s_1^* - s_1$ and $s_2^* - s_2$ using the entropy departure chart in Appendix I.

$$\bar{s}_1^* - \bar{s}_1 = 1.0\text{ kJ/kmol·K} \qquad \therefore s_1^* - s_1 = 1.0/28 = 0.036\text{ kJ/kg·K}$$

$$\bar{s}_2^* - \bar{s}_2 = 1.2\text{ kJ/kmol·K} \qquad \therefore s_2^* - s_2 = 1.2/28 = 0.043\text{ kJ/kg·K}$$

The entropy change is then

$$\Delta s = (s_2 - s_2^*) + (s_1^* - s_1) + (s_2^* - s_1^*) = -0.043 + 0.036 + 1.042 \ln \frac{313}{223} - 0.297 \ln \frac{6}{2} = 0.02 \text{ kJ/kg·K}$$

Note that the real-gas effects in this example were not very pronounced. The temperatures were quite high compared to T_c and the pressures were not excessively large. Also, accuracy using the small charts is quite difficult.

Solved Problems

11.1 Verify (*11.15*) using the equation of state for an ideal gas.

The equation of state for an ideal gas is $Pv = RT$. Let the three variables be P, v, T. Relationship (*11.15*) takes the form

$$\left(\frac{\partial P}{\partial T}\right)_v \left(\frac{\partial T}{\partial v}\right)_P \left(\frac{\partial v}{\partial P}\right)_T = -1$$

The partial derivatives are

$$\left(\frac{\partial P}{\partial T}\right)_v = \frac{\partial}{\partial T}\left(\frac{RT}{v}\right)_v = \frac{R}{v} \qquad \left(\frac{\partial T}{\partial v}\right)_P = \frac{\partial}{\partial v}\left(\frac{Pv}{R}\right)_P = \frac{P}{R} \qquad \left(\frac{\partial v}{\partial P}\right)_T = \frac{\partial}{\partial P}\left(\frac{RT}{P}\right)_T = -\frac{RT}{P^2}$$

Form the product and simplify:

$$\left(\frac{\partial P}{\partial T}\right)_v \left(\frac{\partial T}{\partial v}\right)_P \left(\frac{\partial v}{\partial P}\right)_T = \frac{R}{v}\frac{P}{R}\left(-\frac{RT}{P^2}\right) = -\frac{RT}{Pv} = -1$$

The relationship is verified.

11.2 Derive the Maxwell relation (*11.23*) from (*11.22*) using (*11.15*).

The right side of the Maxwell relation (*11.23*) involves v, s, and P so that

$$\left(\frac{\partial v}{\partial s}\right)_P \left(\frac{\partial P}{\partial v}\right)_s \left(\frac{\partial s}{\partial P}\right)_v = -1 \qquad \text{or} \qquad \left(\frac{\partial v}{\partial s}\right)_P = -\left(\frac{\partial v}{\partial P}\right)_s \left(\frac{\partial P}{\partial s}\right)_v \tag{1}$$

From calculus,

$$\left(\frac{\partial T}{\partial v}\right)_s \left(\frac{\partial v}{\partial P}\right)_s = \left(\frac{\partial T}{\partial P}\right)_s$$

Using (*11.22*) the above relation is written as

$$-\left(\frac{\partial P}{\partial s}\right)_v \left(\frac{\partial v}{\partial P}\right)_s = \left(\frac{\partial T}{\partial P}\right)_s$$

Substituting this into (*1*) provides

$$\left(\frac{\partial v}{\partial s}\right)_P = \left(\frac{\partial T}{\partial P}\right)_s$$

which is the Maxwell relation (*11.23*).

11.3 Verify the third Maxwell relation (*11.24*) using the steam table at 600 °F and 80 psia.

We approximate the first derivative using central differences if possible:

$$\left(\frac{\partial P}{\partial T}\right)_{v=7.794} = \frac{(100-60)(144)}{857.6-348.2} = 11.3 \text{ lbf/ft}^2\text{-°F}$$

$$\left(\frac{\partial s}{\partial v}\right)_{T=600} = \frac{1.7582-1.8165}{6.216-10.425} = 0.0139 \text{ Btu/ft}^3\text{-°R} \quad \text{or } 10.8 \text{ lbf/ft}^2\text{-°R}$$

The difference in the above is less than 5 percent, which is due primarily to the fact that the entries in the steam table are relatively far apart. A table with more entries would result in less error.

11.4 Verify the Clapeyron equation for R134a at 500 kPa.

The Clapeyron equation is $(\partial P/\partial T)_v = h_{fg}/Tv_{fg}$. From Table D-2 for R134a we find, at 500 kPa using central differences,

$$\left(\frac{\partial P}{\partial T}\right)_v = \frac{600-400}{21.58-8.93} = 15.81 \text{ kPa/°C}$$

We also observe that at $P = 500$ kPa, $T = 15.74$°C, $h_{fg} = 184.74$ kJ/kg, and $v_{fg} = 0.0409 - 0.0008056 = 0.04009$ m^3/kg.

Checking the above Clapeyron equation, we have

$$15.81 \stackrel{?}{=} \frac{184.74}{(15.74+273)(0.04009)} = 15.96$$

This is less than 1 percent difference, verifying the Clapeyron equation.

11.5 Find an expression for the change in internal energy if $P = RT/(v-b) - (a/v^2)$ and $C_v = A + BT$. Simplify the expression for an ideal gas with constant specific heats.

We integrate (*11.41*) as follows:

$$\begin{aligned}\Delta u &= \int C_v\,dT + \int\left[T\left(\frac{\partial P}{\partial T}\right)_v - P\right]dv \\ &= \int (A+BT)\,dT + \int\left[T\frac{R}{v-b} - \frac{RT}{v-b} + \frac{a}{v^2}\right]dv \\ &= \int_{T_1}^{T_2}(A+BT)\,dT + \int_{v_1}^{v_2}\frac{a}{v^2}\,dv \\ &= A(T_2-T_1) + \frac{1}{2}B(T_2^2-T_1^2) - a\left(\frac{1}{v_2}-\frac{1}{v_1}\right)\end{aligned}$$

For an ideal gas $P = RT/v$ so that $a = b = 0$, and if $C_v =$ const., we set $B = 0$. Then the above expression simplifies to $\Delta u = A(T_2 - T_1) = C_v(T_2 - T_1)$.

11.6 Find an expression for $C_p - C_v$ if the equation of state is

$$v = \frac{RT}{P} - \frac{a}{RT} + b$$

From the equation of state we find $(\partial v/\partial T)_P = (R/P) + (a/RT^2)$. To find $(\partial P/\partial T)_v$ we first write the equation of state as

$$P = RT\left(v - b + \frac{a}{RT}\right)^{-1}$$

so that

$$\left(\frac{\partial P}{\partial T}\right)_v = \frac{(v-b)R + 2a/T}{(v-b+a/RT)^2}$$

Using (*11.56*) the desired expression is

$$C_p - C_v = \left(\frac{TR}{P} + \frac{a}{RT}\right)\frac{(v-b)R + 2a/T}{(v-b+a/RT)^2}$$

This reduces to $C_p - C_v = R$ for an ideal gas; that is, for $a = b = 0$.

11.7 The specific heat C_v of copper at 200 °C is desired. If C_v is assumed to be equal to C_p estimate the error. Use $\beta = 5 \times 10^{-5}\ \text{K}^{-1}$, $B = 125$ GPa, and $\rho = 8770\ \text{kg/m}^3$.

Equation (*11.59*) provides the relation

$$C_p - C_v = vT\beta^2 B = \left(\frac{1}{8770}\right)(473)(5\times10^{-5})^2(125\times10^9) = 16.85\ \text{J/kg·K}$$

From Table B-4 the specific heat of copper is approximated at 200 °C to be about 0.40 kJ/kg·K. Hence,

$$C_v = C_p - 0.01685 = 0.4 - 0.01685 = 0.383\ \text{kJ/kg·K}$$

Assuming $C_v = 0.4$ kJ/kg·K,

$$\%\ \text{error} = \left(\frac{0.4 - 0.383}{0.383}\right)(100) = 4.4\%$$

This error may be significant in certain calculations.

11.8 The Joule–Thomson coefficient is measured to be 0.001 °R-ft²/lbf for steam at 600 °F and 100 psia. Calculate the value of C_p.

Equation (*11.63*) is used to evaluate C_p. With values from the steam table at 600 °F and 160 psia we find

$$C_p = \frac{1}{\mu_j}\left[T\left(\frac{\partial v}{\partial T}\right)_P - v\right] = \left(\frac{1}{0.001}\right)\left[(1060)\left(\frac{4.243 - 3.440}{700 - 500}\right) - 3.848\right]$$

$$= 408\ \text{ft-lbf/lbm-°R} \quad \text{or } 0.524\ \text{Btu/lbm-°R}$$

11.9 Calculate the change in enthalpy of air which is heated from 300 K and 100 kPa to 700 K and 2000 kPa using the enthalpy departure chart. Compare with Prob. 4.10(*c*).

The reduced temperatures and pressures are

$$T_{R1} = \frac{T_1}{T_c} = \frac{300}{133} = 2.26 \qquad P_{R1} = \frac{P_1}{P_c} = \frac{100}{3760} = 0.027$$

$$T_{R2} = \frac{700}{133} = 5.26 \qquad P_{R2} = \frac{2000}{3760} = 0.532$$

The enthalpy departure chart provides $h_2^* - h_2 \cong 0$ and $h_1^* - h_1 \cong 0$, so that

$$h_2 - h_1 = h_2^* - h_1^* = 713.27 - 300.19 = 413.1\ \text{kJ/kg}$$

where we have used the ideal-gas tables for the ideal-gas enthalpy change $h_2^* - h_1^*$. Obviously, the real-gas effects in this problem are negligible and the result is the same as that of Prob. 4.10(*c*).

11.10 Nitrogen is compressed in a steady-flow device from 1.4 MPa and 20 °C to 20 MPa and 200 °C. Calculate (*a*) the change in enthalpy, (*b*) the change in entropy, and (*c*) the heat transfer if the work input is 200 kJ/kg.

The reduced temperatures and pressures are

$$T_{R1} = \frac{T_1}{T_{cr}} = \frac{293}{126.2} = 2.32 \qquad P_{R1} = \frac{P_1}{P_{cr}} = \frac{1.4}{3.39} = 0.413$$

$$T_{R2} = \frac{473}{126.2} = 3.75 \qquad P_{R2} = \frac{20}{3.39} = 5.90$$

(*a*) The enthalpy departure chart allows us to find

$$h_1^* - h_1 = \frac{\bar{h}_1^* - \bar{h}_1}{T_{cr}} \frac{T_{cr}}{M} = (0.3)\left(\frac{126.2}{28}\right) = 1.4 \text{ kJ/kg}$$

$$h_2^* - h_2 = \frac{\bar{h}_2^* - \bar{h}_2}{T_{cr}} \frac{T_{cr}}{M} = (2.5)\left(\frac{126.2}{28}\right) = 6.8 \text{ kJ/kg}$$

The enthalpy change is found to be

$$h_2 - h_1 = (h_1^* - h_1) + (h_2 - h_2^*) + (h_2^* - h_1^*) = 1.4 - 6.8 + (1.04)(200 - 20)$$
$$= 182 \text{ kJ/kg}$$

(*b*) The entropy departure chart provides

$$s_1^* - s_1 = \frac{\bar{s}_1^* - \bar{s}_1}{M} = \frac{0.1}{28} = 0.004 \text{ kJ/kg·K}$$

$$s_2^* - s_2 = \frac{\bar{s}_2^* - \bar{s}_2}{M} = \frac{0.5}{28} = 0.02 \text{ kJ/kg·K}$$

The entropy change is then

$$s_2 - s_1 = (s_1^* - s_1) + (s_2 - s_2^*) + (s_2^* - s_1^*) = 0.004 - 0.02 + 1.04 \ln \frac{473}{293} - 0.297 \ln \frac{20}{1.4}$$
$$= -0.308 \text{ kJ/kg·K}$$

(*c*) From the first law, $q = \Delta h + w = 182 - 200 = -18$ kJ/kg. The negative sign means that heat is leaving the device.

11.11 **Methane is compressed isothermally in a steady-flow compressor from 100 kPa and 20 °C to 20 MPa. Calculate the minimum power required if the mass flux is 0.02 kg/s.**

The reduced temperatures and pressures are

$$T_{R2} = T_{R1} = \frac{T_1}{T_{cr}} = \frac{293}{191.1} = 1.53 \qquad P_{R1} = \frac{0.1}{4.64} = 0.02 \qquad P_{R2} = \frac{20}{4.64} = 4.31$$

Minimum power is required for an isothermal process if the process is reversible, so that the heat transfer is given by $q = T\Delta s$. The entropy change is

$$\Delta s = \cancel{(s_1^* - s_1)}^0 + (s_2 - s_2^*) + (s_2^* - s_1^*) = 0 - \frac{7}{16} + 2.25 \ln 1 - 0.518 \ln \frac{20}{0.1} = -3.18 \text{ kJ/kg·K}$$

so that $q = T\Delta s = (293)(-3.18) = 932$ kJ/kg. The first law, $q - w = \Delta h$, requires that we find Δh. We find $\bar{h}_2^* - \bar{h}_2 = 14$ kJ/kmol·K, so that

$$\Delta h = \cancel{(h_1^* - h_1)}^0 + (h_2 - h_2^*) + \cancel{(h_2^* - h_1^*)}^0 = (-14)\left(\frac{191.1}{16}\right) = -167 \text{ kJ/kg}$$

Finally, the required power is

$$\dot{W} = (q - \Delta h)\dot{m} = [932 - (-167)](0.02) = 22 \text{ kW}$$

11.12 **Estimate the minimum power needed to compress carbon dioxide in a steady-flow insulated compressor from 200 kPa and 20 °C to 10 MPa. The inlet flow rate is 0.8 m^3/min.**

Minimum power is associated with a reversible process. Insulation results in negligible heat transfer. Consequently, an isentropic process is assumed. First, the reduced pressures and temperature are

$$P_{R1} = \frac{P_1}{P_{cr}} = \frac{0.2}{7.39} = 0.027 \qquad P_{R2} = \frac{10}{7.37} = 1.37 \qquad T_{R1} = \frac{T_1}{T_{cr}} = \frac{293}{304.2} = 0.963$$

For the isentropic process $\Delta s = 0$:

$$\Delta s = 0 = (\bar{s}_1^* - \bar{s}_1)^0 + (s_2 - s_2^*) + (s_2^* - s_1^*) = 0 + \frac{\bar{s}_2 - \bar{s}_2^*}{44} + 0.842 \ln \frac{T_2}{293} - 0.189 \ln \frac{10}{0.2}$$

Since $\bar{s}_2 - \bar{s}_2^*$ depends on T_2, this equation has T_2 as the only unknown. A trial-and-error procedure provides the solution. First, let $\bar{s}_2 - \bar{s}_2^* = 0$ and find $T_2 = 705$ K. Since $\bar{s}_2^* - \bar{s}_2 > 0$, we try the following:

$$T_2 = 750 \text{ K}, \quad T_{R2} = 2.47: \qquad 0 \stackrel{?}{=} -\frac{2}{44} + 0.842 \ln \frac{750}{293} - 0.189 \ln \frac{10}{0.2} = 0.0066$$

$$T_2 = 730 \text{ K}, \quad T_{R2} = 2.03: \qquad 0 \stackrel{?}{=} -\frac{2}{44} + 0.842 \ln \frac{730}{293} - 0.189 \ln \frac{10}{0.2} = -0.016$$

Interpolating results in $T_2 = 744$ K or 471 °C. The work for this steady-flow process can now be found to be

$$w = -\Delta h = h_1 - h_1^{*0} + h_2^* - h_2 + h_1^* - h_2^* = 0 + (2.0)\left(\frac{304.2}{44}\right) + (0.842)(20 - 471)$$
$$= -366 \text{ kJ/kg}$$

To find $\dot{W}$ we must know $\dot{m} = (\rho_1)(0.8/60)$. The density is found using

$$\rho_1 = \frac{P_1}{Z_1 R T_1} = \frac{200}{(0.99)(0.189)(293)} = 3.65 \text{ kg/m}^3$$

Finally

$$\dot{W} = \dot{m}w = \left[\frac{(3.65)(0.8)}{60}\right](-366) = -17.8 \text{ kW}$$

11.13 Calculate the maximum work that can be produced by steam at 30 MPa and 600 °C if it expands through the high-pressure stage of a turbine to 6 MPa. Use the charts and compare with tabulated values from the steam tables.

Maximum work occurs for an adiabatic reversible process, i.e., for $\Delta s = 0$. The reduced temperature and pressures are

$$T_{R1} = \frac{T_1}{T_{cr}} = \frac{873}{647} = 1.35 \qquad P_{R1} = \frac{30}{22.1} = 1.36 \qquad P_{R2} = \frac{6}{22.1} = 0.27$$

The isentropic process provides us with T_2 by a trial-and-error procedure:

$$\Delta s = 0 = (s_1^* - s_1) + (s_2 - s_2^*) + (s_2^* - s_1^*) = \frac{4}{18} + s_2 - s_2^* + 1.872 \ln \frac{T_2}{873} - 0.462 \ln \frac{6}{30}$$

If $s_2 - s_2^* = 0$, we find $T_2 = 521$ K or 248 °C. Since $s_2 - s_2^* < 0$, we try $T_2 > 521$ K:

$$T_2 = 600 \text{ K}, \quad T_{R2} = 0.93: \qquad 0 \stackrel{?}{=} \frac{4}{18} - \frac{3}{18} + 1.872 \ln \frac{600}{873} - 0.462 \ln \frac{6}{30} = 0.097$$

$$T_2 = 560 \text{ K}, \quad T_{R2} = 0.87: \qquad 0 \stackrel{?}{=} \frac{4}{18} - \frac{3.5}{18} + 1.872 \ln \frac{560}{873} - 0.462 \ln \frac{6}{30} = -0.06$$

Interpolation gives $T_2 = 575$ K or 302 °C. The work produced is then

$$w = -\Delta h = (h_1 - h_1^*) + (h_2^* - h_2) + (h_1^* - h_2^*)$$
$$= (-8)\left(\frac{647.4}{18}\right) + (4)\left(\frac{647.4}{18}\right) + \frac{30\,750 - 19\,500}{18} = 481 \text{ kJ/kg}$$

where we have used the ideal-gas Table E-6 to find $h_1^* - h_2^*$. A less accurate value would be found using $C_p \, \Delta T$.

To compare with valves obtained directly from the steam tables we use

$$\left.\begin{array}{r} s_2 = s_1 = 6.2339 \text{ kJ/kg·K} \\ P_2 = 6 \text{ MPa} \end{array}\right\} \qquad \therefore h_2 = 2982 \text{ kJ/kg}$$

The work is $w = -\Delta h = h_1 - h_2 = 3444 - 2982 = 462$ kJ/kg.

Supplementary Problems

11.14 Using (*11.1*), estimate the increase in pressure needed to decrease the volume of 2 kg of air 0.04 m^3 if the temperature changes from 30 °C to 33 °C. The initial volume is 0.8 m^3.

11.15 Using (*11.1*), estimate the temperature change if the pressure changes from 14.7 to 15 psia while the volume changes from 2.2 to 2.24 ft^3. There is 4 lbm of air.

11.16 Show that the slope of a constant-pressure line on a T-v diagram of an ideal gas increases with temperature.

11.17 Find an expression for the slope of a constant-pressure line on a T-v diagram, if $(P + a/v^2)(v - b) = RT$.

11.18 Write two relationships that result from the differential forms of the first law and the relationship $u = u(s, v)$. Verify the two relationships for steam at 300 °C and 2 MPa.

11.19 Derive Maxwell relation (*11.24*) from (*11.22*) using (*11.15*).

11.20 Verify (*11.25*) using the R134a tables at 100 kPa and 0 °C.

11.21 Verify (*11.23*) using the steam tables at 200 kPa and 400 °C.

11.22 Verify the Clapeyron equation using steam at 40 psia.

11.23 Use the Clapeyron equation to predict the enthalpy of vaporization h_{fg} of steam at 50 °C, (*a*) assuming that steam is an ideal gas; (*b*) taking v_g from the steam table. (*c*) What is h_{fg} in the steam table?

11.24 Using the Clausius–Clapeyron equation, predict T_{sat} for $P_{sat} = 2.0$ psia using the values in Table C-2E. Compare this value with that found from interpolation in Table C-1E.

11.25 (*a*) Derive the relationship $C_p = T(\partial s/\partial T)_P$ and verify the expression for dh given by (*11.43*). (*b*) For an ideal gas what is the value of the quantity in brackets in (*11.43*)?

11.26 Assume an ideal gas with constant C_p and C_v and derive simplified relationships for $s_2 - s_1$. Refer to (*11.47*).

11.27 Show that (*a*) $C_p = T(\partial P/\partial T)_s(\partial v/\partial T)_p$ and (*b*) $C_v = -T(\partial P/\partial T)_v(\partial v/\partial T)_s$.

11.28 (*a*) Use Problem 11.27(*a*) to estimate the value of C_p for steam at 3 MPa and 400 °C and compare with an estimate using $C_p = (\partial h/\partial T)_p$ at the same state. (*b*) Do the same for steam at 4000 psia and 1000 °F.

11.29 (*a*) Use Problem 11.27(*b*) to estimate the value of C_v for steam at 2 MPa and 400 °C and compare with an estimate using $C_v = (\partial u/\partial T)_v$ at the same state. (*b*) Do the same for steam at 4000 psia and 1000 °F.

11.30 Using $P = RT/v - a/v^2$ and assuming an isothermal process, find expressions for (*a*) Δh, (*b*) Δu, and (*c*) Δs.

11.31 Using $P = RT/(v - b)$ and assuming an isothermal process, find expressions for (*a*) Δh, (*b*) Δu, and (*c*) Δs.

11.32 Air undergoes a change from 20 °C and 0.8 m^3/kg to 200 °C and 0.03 m^3/kg. Calculate the enthalpy change assuming (*a*) the van der Waals equation of state and constant specific heats, (*b*) the ideal-gas tables, and (*c*) an ideal gas with constant specific heats.

11.33 Nitrogen undergoes a change from 100 °F and 5 ft^3/lbm to 600 °F and 0.8 ft^3/lbm. Calculate the enthalpy change assuming (*a*) the van der Waals equation of state and constant specific heats, (*b*) the ideal-gas tables, and (*c*) an ideal gas with constant specific heats.

11.34 Find an expression for $C_p - C_v$ if $P = RT/v - a/v^2$.

11.35 Calculate β and B for water at 5 MPa and 60 °C. Then estimate the difference $C_p - C_v$.

11.36 Calculate β and B for water at 500 psia and 100 °F. Then estimate the difference $C_p - C_v$.

11.37 Find an expression for the Joule–Thomson coefficient for a gas if $P = RT/v - a/v^2$. What is the *inversion temperature* (the temperature where $\mu_j = 0$)?

11.38 Estimate the Joule–Thomson coefficient for steam at 6 MPa and 600 °C using both expressions in (*11.63*). Approximate the value of C_p using $(\partial h/\partial T)_p$.

11.39 Estimate the temperature change of steam that is throttled from 8 MPa and 600 °C to 4 MPa.

11.40 Estimate the temperature change of R134a that is throttled from 170 psia and 200 °F to 80 psia.

11.41 Calculate the change in the enthalpy of air if its state is changed from 200 K and 900 kPa to 700 K and 6 MPa using (*a*) the enthalpy departure chart and (*b*) the ideal-gas tables.

11.42 Calculate the change in entropy of nitrogen if its state is changed from 300 °R and 300 psia to 1000 °R and 600 psia using (*a*) the entropy departure chart and (*b*) the ideal-gas tables.

11.43 Estimate the power needed to compress 2 kg/s of methane in a reversible adiabatic process from 400 kPa and 20 °C to 4 MPa in a steady-flow device (*a*) assuming ideal-gas behavior and (*b*) accounting for real-gas behavior.

11.44 An adiabatic reversible turbine changes the state of 10 kg/min of carbon dioxide from 10 MPa and 700 K to 400 kPa. Estimate the power produced (*a*) assuming ideal-gas behavior and (*b*) accounting for real-gas behavior.

11.45 Air is contained in a rigid tank and the temperature is changed from 20 °C to 800 °C. If the initial pressure is 1600 kPa, calculate the final pressure and the heat transfer (*a*) using the enthalpy departure chart and (*b*) assuming ideal-gas behavior.

11.46 Air undergoes an isothermal compression in a piston-cylinder arrangement from 100 °F and 14.7 psia to 1000 psia. Estimate the work required and the heat transfer (*a*) assuming ideal-gas behavior and (*b*) accounting for real-gas effects.

11.47 Nitrogen expands in a turbine from 200 °C and 20 MPa to 20 °C and 2 MPa. Estimate the power produced if the mass flux is 3 kg/s.

Answers to Supplementary Problems

11.14 13.0 kPa

11.15 0.843 °F

11.16 Slope $= T/v$

11.17 $(P - a/v^2 + 2ab/v^3)/R$

11.18 $T = (\partial u/\partial s)_v$, $P = -(\partial u/\partial v)_s$

11.23 (*a*) 2319 kJ/kg (*b*) 2397 kJ/kg (*c*) 2383 kJ/kg

11.24 147 °F, 126 °F

11.25 (*b*) zero

11.26 $C_v \ln T_2/T_1 + R \ln v_2/v_1$, $C_p \ln T_2/T_1 - R \ln P_2/P_1$

11.28 (*a*) 2.30 kJ/kg·K vs. 2.29 kJ/kg·K (*b*) 0.84 Btu/lbm-°R vs. 0.860 Btu/lbm-°R

11.29 (*a*) 1.75 kJ/kg·K vs. 1.71 kJ/kg·K (*b*) 0.543 Btu/lbm-°R vs. 0.500 Btu/lbm-°R

11.30 (*a*) $P_2v_2 - P_1v_1 + a(1/v_1 - 1/v_2)$ (*b*) $a\ (1/v_1 - 1/v_2)$ (*c*) $R \ln v_2/v_1$

11.31 (*a*) $P_2v_2 - P_1v_1$ (*b*) 0 (*c*) $R \ln [(v_2 - b)/(v_1 - b)]$

11.32 (*a*) 176 kJ/kg (*b*) 170 kJ/kg (*c*) 181 kJ/kg

11.33 (*a*) 123 Btu/lbm (*b*) 126 Btu/lbm (*c*) 124 Btu/lbm

11.34 $TR^2v/(Pv^2 - a)$

11.35 $5.22 \times 10^{-4}\ \text{K}^{-1}$, 2.31×10^6 kPa, 0.212 kJ/kg·K

11.36 $1.987 \times 10^{-4}\ {}^\circ\text{R}^{-1}$, 48.3×10^6 psf, 0.0221 Btu/lbm-°R

11.37 $2av/[C_p(RvT - 2a)], (Pv^2 - a)/Rv$

11.38 3.45°C/MPa, 3.55°C/MPa

11.39 −14°C

11.40 −12.6°F

11.41 (*a*) 524 kJ/kg (*b*) 513 kJ/kg

11.42 (*a*) 0.265 Btu/lbm-°R (*b*) 0.251 Btu/lbm-°R

11.43 (*a*) 923 kW (*b*) 923 kW

11.44 (*a*) 61.2 kW (*b*) 55 kW

11.45 (*a*) 617 kJ/kg (*b*) 612 kJ/kg

11.46 (*a*) −162 Btu/lbm, −162Btu/lbm

11.47 544 kW

Mixtures and Solutions

12.1 BASIC DEFINITIONS

Thus far in our thermodynamic analyses we have considered only single-component systems. In this chapter we develop methods for determining thermodynamic properties of a mixture for applying the first law to systems involving mixtures.

We begin by defining two terms which describe and define a mixture. The *mole fraction* y is defined as

$$y_i = \frac{N_i}{N} \tag{12.1}$$

where N_i is the number of moles of the ith component and N is the total number of moles. The *mass fraction mf* is defined as

$$mf_i = \frac{m_i}{m} \tag{12.2}$$

where m_i is the mass of the ith component and m is the total mass of the mixture. Clearly, the total number of moles and the total mass of a mixture are given, respectively, by

$$N = N_1 + N_2 + N_3 + \cdots \qquad m = m_1 + m_2 + m_3 + \cdots \tag{12.3}$$

Dividing the above equations by N and m, respectively, we see that

$$\Sigma y_i = 1 \qquad \Sigma mf_i = 1 \tag{12.4}$$

The (mean) molecular weight of a mixture is given by

$$M = \frac{m}{N} = \frac{\Sigma N_i M_i}{N} = \Sigma y_i M_i \tag{12.5}$$

The mixture's gas constant is then

$$R = \frac{R_u}{M} \tag{12.6}$$

where R_u denotes, as in Chapter 2, the universal molar gas constant.

Analyzing a mixture on the basis of mass (or weight) is *gravimetric analysis*. Analyzing a mixture on the basis of moles (or volume) is *volumetric analysis*. The type of analysis must be stated.

EXAMPLE 12.1 Molar analysis of air indicates that it is composed primarily of nitrogen (78%) and oxygen (22%). Determine (*a*) the mole fractions, (*b*) the gravimetric analysis, (*c*) its molecular weight, and (*d*) its gas constant. Compare with values from Appendix B.

Solution:

(*a*) The mole fractions are given as $y_1 = 0.78$ and $y_2 = 0.22$, where the subscript 1 refers to nitrogen and 2 to oxygen.

(*b*) If there are 100 mol of the mixture, the mass of each component is

$$\left.\begin{aligned} m_1 &= N_1 M_1 = (78)(28) = 2184\,\text{kg} \\ m_2 &= N_2 M_2 = (22)(32) = 704\,\text{kg} \end{aligned}\right\} \qquad \therefore\ m = 2888\,\text{kg}$$

Gravimetric analysis yields

$$mf_1 = \frac{m_1}{m} = \frac{2184}{2888} = 0.756 \qquad mf_2 = \frac{m_2}{m} = \frac{704}{2888} = 0.244$$

or, by mass, the mixture is 75.6% N_2 and 24.4% O_2.

(*c*) The molecular weight of the mixture is $M = m/N = 2888/100 = 28.9$ kg/kmol. This compares with 28.97 kg/kmol from the appendix, an error of -0.24%.

(*d*) The gas constant for air is calculated to be $R = R_u/M = 8.314/28.9 = 0.288$ kJ/kg·K. This compares with 0.287 kJ/kg·K from the appendix, an error of 0.35%.

By including argon as a component of air, the above calculations could be improved. However, it's obvious that the above analysis is quite acceptable.

12.2 IDEAL-GAS LAW FOR MIXTURES

Two models are used to obtain the *P-v-T* relation for a mixture of ideal gases. The *Amagat model* treats each component as though it exists separately at the same pressure and temperature of the mixture; the total volume is the sum of the volumes of the components. In this chapter we use the *Dalton model*, in which each component occupies the same volume and has the same temperature as the mixture; the total pressure is the sum of the component pressures, termed the *partial pressures*. For the Dalton model

$$P = P_1 + P_2 + P_3 + \cdots \tag{12.7}$$

For any component of a mixture of ideal gases the ideal-gas law is

$$P_i = \frac{N_i R_u T}{V} \tag{12.8}$$

For the mixture as a whole we have

$$P = \frac{N R_u T}{V} \tag{12.9}$$

so that

$$\frac{P_i}{P} = \frac{N_i R_u T/V}{N R_u T/V} = \frac{N_i}{N} = y_i \tag{12.10}$$

EXAMPLE 12.2 A rigid tank contains 2 kg of N_2 and 4 kg of CO_2 at a temperature of 25 °C and 2 MPa. Find the partial pressures of the two gases and the gas constant of the mixture.

Solution: To find the partial pressures we need the mole fractions. The moles of N_2 and CO_2 are, respectively,

$$\left.\begin{aligned} N_1 &= \frac{m_1}{M_1} = \frac{2}{28} = 0.0714 \text{ mol} \\ N_2 &= \frac{m_2}{M_2} = \frac{4}{44} = 0.0909 \text{ mol} \end{aligned}\right\} \quad \therefore N = 0.1623 \text{ mol}$$

The mole fractions are

$$y_1 = \frac{N_1}{N} = \frac{0.0714}{0.1623} = 0.440 \qquad y_2 = \frac{N_2}{N} = \frac{0.0909}{0.1623} = 0.560$$

The partial pressures are

$$P_1 = y_1 P = (0.44)(2) = 0.88 \text{ MPa} \qquad P_2 = y_2 P = (0.56)(2) = 1.12 \text{ MPa}$$

The molecular weight is $M = M_1 y_1 + M_2 y_2 = (28)(0.44) + (44)(0.56) = 36.96$ kg/kmol. The gas constant of the mixture is then

$$R = \frac{R_u}{M} = \frac{8.314}{36.96} = 0.225 \text{ kJ/kg·K}$$

12.3 PROPERTIES OF A MIXTURE OF IDEAL GASES

The extensive properties of a mixture, such as H, U, and S, can be found by simply adding the contributions of each component. For example, the total enthalpy of a mixture is

$$H = \Sigma H_i = H_1 + H_2 + H_3 + \cdots \tag{12.11}$$

In terms of the specific enthalpy h,

$$H = mh = \Sigma m_i h_i \qquad \text{and} \qquad H = N\bar{h} = \Sigma N_i \bar{h}_i \tag{12.12}$$

where the overbar denotes a mole basis. Dividing the above two equations by m and N, respectively, we see that

$$h = \Sigma mf_i h_i \qquad \text{and} \qquad \bar{h} = \Sigma y_i \bar{h}_i \tag{12.13}$$

Since the specific heat C_p is related to the change in the enthalpy, we may write

$$\Delta h = C_p \, \Delta T \qquad \text{and} \qquad \Delta h_i = C_{p,i} \, \Delta T \tag{12.14}$$

so that

$$\Delta h = C_p \, \Delta T = \Sigma mf_i (C_{p,i} \, \Delta T) \tag{12.15}$$

Dividing both sides by ΔT, there results

$$C_p = \Sigma mf_i C_{p,i} \tag{12.16}$$

The molar specific heat is

$$\bar{C}_p = \Sigma y_i \bar{C}_{p,i} \tag{12.17}$$

Likewise, using internal energy we would find

$$C_v = \Sigma mf_i C_{v,i} \qquad \bar{C}_v = \Sigma y_i \bar{C}_{v,i} \tag{12.18}$$

EXAMPLE 12.3 Gravimetric analysis of a mixture of three gases indicates 20% N_2, 40% CO_2, and 40% O_2. Find the heat transfer needed to increase the temperature of 20 lbm of the mixture from 80 °F to 300 °F in a rigid tank.

Solution: The heat transfer needed is given by the first law as (the work is zero for a rigid tank) $Q = \Delta U = m\,\Delta u = mC_v\,\Delta T$. We must find C_v. It is given by (12.18*a*) as

$$C_v = mf_1C_{v,1} + mf_2C_{v,2} + mf_3C_{v,3} = (0.2)(0.177) + (0.4)(0.158) + (0.4)(0.157)$$
$$= 0.161 \text{ Btu/lbm-°R}$$

The heat transfer is then $Q = mC_v\,\Delta T = (20)(0.161)(300 - 80) = 708$ Btu.

EXAMPLE 12.4 A mixture is composed of 2 mol CO_2 and 4 mol N_2. It is compressed adiabatically in a cylinder from 100 kPa and 20 °C to 2 MPa. Assuming constant specific heats, calculate (*a*) the final temperature, (*b*) the work required, and (*c*) the change in entropy.

Solution:

(*a*) The temperature is found using the isentropic relationship $T_2 = T_1(P_2/P_1)^{(k-1)/k}$. Let's find k for the mixture. The mass is $m = N_1M_1 + N_2M_2 = (2)(44) + (4)(28) = 200$ kg. The specific heats are

$$C_v = mf_1C_{v,1} + mf_2C_{v,2} = \left(\frac{88}{200}\right)(0.653) + \left(\frac{112}{200}\right)(0.745) = 0.705\,\text{kJ/kg·K}$$

$$C_p = mf_1C_{p,1} + mf_2C_{p,2} = \left(\frac{88}{200}\right)(0.842) + \left(\frac{112}{200}\right)(1.042) = 0.954\,\text{kJ/kg·K}$$

The ratio of specific heats is $k = C_p/C_v = 0.954/0.705 = 1.353$. Consequently, the final temperature is

$$T_2 = T_1\left(\frac{P_2}{P_1}\right)^{(k-1)/k} = (293)\left(\frac{2000}{1000}\right)^{0.353/1.353} = 640 \text{ K} \quad \text{or } 367\,°\text{C}$$

(*b*) The work is found using the first law with $Q = 0$:

$$W = -\Delta U = -m\,\Delta u = -mc_v\,\Delta T = (-200)(0.705)(367 - 20) = -48.9\,\text{MJ}$$

(*c*) The entropy change is

$$\Delta s = C_p \ln\frac{T_2}{T_1} - R\ln\frac{P_2}{P_1}$$
$$= 0.954\ln\frac{640}{293} - \frac{8.314}{(\frac{2}{6})(44) + (\frac{4}{6})(28)}\ln\frac{2000}{100} = -0.00184\text{ kJ/kg·K}$$

Obviously, the entropy change should be zero for this isentropic process. The above small value is a measure of the error in our calculations.

12.4 GAS-VAPOR MIXTURES

Air is a mixture of nitrogen, oxygen, and argon plus traces of some other gases. When water vapor is not included, we refer to it as *dry air*. If water vapor is included, as in *atmospheric air*, we must be careful to properly account for it. At the relatively low atmospheric temperature we can treat dry air as an ideal gas with constant specific heats. It is also possible to treat the water vapor in the air as an ideal gas, even though the water vapor may be at the saturation state; for example, at 1 kPa we find (using $R = 0.462$ kJ/kg·K from Table B-2) $v = RT/P = 129$ m^3/kg, the same value as v_g in Table C-2. Consequently, we can consider atmospheric air to be a mixture of two ideal gases. By (*12.7*), the total pressure is the sum of the partial pressure P_a of the dry air and the partial pressure P_v of the water vapor (called the *vapor pressure*):

$$P = P_a + P_v \tag{12.19}$$

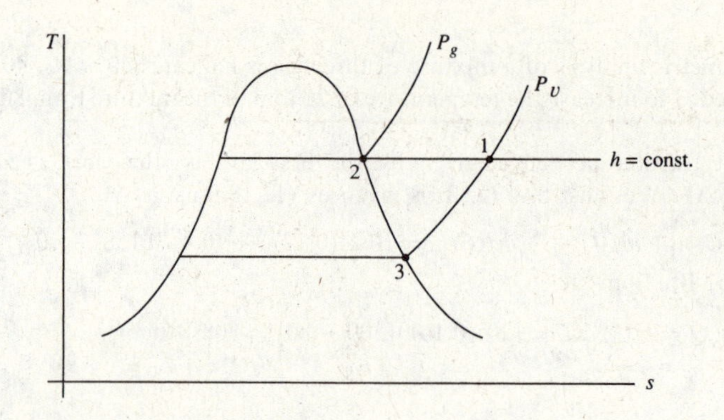

Fig. 12-1 *T-s* diagram for the water-vapor component.

Since we assume that the water vapor is an ideal gas, its enthalpy is dependent on temperature only. Hence we use the enthalpy of the water vapor to be the enthalpy of saturated water vapor at the temperature of the air, expressed as

$$h_v(T) = h_g(T) \tag{12.20}$$

In Fig. 12-1 this means that $h_1 = h_2$ where $h_2 = h_g$ from the steam tables at $T = T_1$. This is acceptable for situations in which the pressure is relatively low (near atmospheric pressure) and the temperature is below about 60 °C (140 °F).

The amount of water vapor in the air is related to the relative humidity and the humidity ratio. The *relative humidity* ϕ is defined as the ratio of the mass of the water vapor m_v to the maximum amount of water vapor m_g the air can hold at the same temperature:

$$\phi = \frac{m_v}{m_g} \tag{12.21}$$

Using the ideal-gas law we find

$$\phi = \frac{P_v V/R_v T}{P_g V/R_v T} = \frac{P_v}{P_g} \tag{12.22}$$

where the constant-pressure lines for P_v and P_g are shown in Fig. 12-1.

The *humidity ratio* ω (also referred to as *specific humidity*) is the ratio of the mass of water vapor to the mass of dry air:

$$\omega = \frac{m_v}{m_a} \tag{12.23}$$

Using the ideal-gas law for air and water vapor, this becomes

$$\begin{aligned} \omega &= \frac{P_v V/R_v T}{P_a V/R_a T} = \frac{P_v/R_v}{P_a/R_a} \\ &= \frac{P_v/0.4615}{P_a/0.287} = 0.622\frac{P_v}{P_a} \end{aligned} \tag{12.24}$$

Combining (*12.24*) and (*12.22*), we relate the above two quantities as

$$\omega = 0.622\frac{\phi P_g}{P_a} \qquad \phi = 1.608\frac{\omega P_a}{P_g} \tag{12.25}$$

Note that at state 3 in Fig. 12-1 the relative humidity is 1.0 (100%). Also note that for a given mass of water vapor in the air, ω remains constant but ϕ varies depending on the temperature.

The temperature of the air as measured by a conventional thermometer is referred to as the *dry-bulb temperature T* (T_1 in Fig. 12-1). The temperature at which condensation begins if air is cooled at constant pressure is the *dew-point temperature* $T_{d.p.}$ (T_3 in Fig. 12-1). If the temperature falls below the dew-point temperature, condensation occurs and the amount of water vapor in the air decreases. This may occur on a cool evening; it also may occur on the cool coils of an air conditioner.

EXAMPLE 12.5 The air at 25 °C and 100 kPa in a 150-m^3 room has a relative humidity of 60%. Calculate (*a*) the humidity ratio, (*b*) the dew point, (*c*) the mass of water vapor in the air, and (*d*) the mole fraction of the water vapor.

Solution:

(*a*) By (*12.22*), $P_v = P_g\phi = (3.169)(0.6) = 1.90$ kPa, where P_g is the saturation pressure at 25 °C found in Table C-1. The partial pressure of the air is then $P_a = P - P_v = 100 - 1.9 = 98.1$ kPa, where we have used the total pressure of the air in the room to be at 100 kPa. The humidity ratio is then

$$\omega = 0.622\frac{P_v}{P_a} = (0.622)\left(\frac{1.9}{98.1}\right) = 0.01205 \text{ kg H}_2\text{O/kg dry air}$$

(*b*) The dew point is the temperature T_3 of Fig. 12-1 associated with the partial pressure P_v. It is found by interpolation in Table C-l or Table C-2, whichever appears to be easier: $T_{d.p.} = 16.6$ °C.

(*c*) From the definition of the humidity ratio the mass of water vapor is found to be

$$m_v = \omega m_a = \omega\frac{P_a V}{R_a T} = (0.01205)\left[\frac{(98.1)(150)}{(0.287)(298)}\right] = 2.07 \text{ kg}$$

(*d*) To find the mole fraction of the water vapor, we first find the total moles:

$$N_v = \frac{m_v}{M_v} = \frac{2.07}{18} = 0.1152 \text{ mol} \qquad N_a = \frac{m_a}{M_a} = \frac{(98.1)(150)/(0.287)(298)}{28.97} = 5.94 \text{ mol}$$

The mole fraction of the water vapor is

$$y_v = \frac{0.1152}{5.94 + 0.1152} = 0.0194$$

This demonstrates that air with 60% humidity is about 2% water vapor by volume. We usually ignore this when analyzing air, as in Example 12.1, and consider air to be dry air. Ignoring the water vapor does not lead to significant error in most engineering applications. It must be included, however, when considering problems involving, for example, combustion and air-conditioning.

EXAMPLE 12.6 The air in Example 12.5 is cooled below the dew point to 10 °C. (*a*) Estimate the amount of water vapor that will condense. (*b*) Reheat the air back to 25 °C and calculate the relative humidity.

Solution:

(*a*) At 10 °C the air is saturated, with $\phi = 100\%$. In Fig. 12-1 we are at a state on the saturation line that lies below state 3. At 10 °C we find from Table C-1 that $P_v = 1.228$ kPa, so that

$$P_a = P - P_v = 100 - 1.228 = 98.77 \text{ kPa}$$

The humidity ratio is then $\omega = (0.622)(P_v/P_a) = (0.622)(1.228/98.77) = 0.00773$ kg H_2O/kg dry air. The difference in the humidity ratio just calculated and the humidity ratio of Example 12.5 is $\Delta\omega = 0.01205 - 0.00773 = 0.00432$ kg H_2O/kg dry air. The mass of water vapor removed (condensed) is found to be

$$\Delta m_v = \Delta\omega\, m_a = (0.00432)\left[\frac{(98.1)(150)}{(0.287)(298)}\right] = 0.743 \text{ kg H}_2\text{O}$$

where we have used the initial mass of dry air.

(*b*) As we reheat the air back to 25 °C, the ω remains constant at 0.00773. Using (*12.25*), the relative humidity is then reduced to

$$\phi = 1.608\frac{\omega P_a}{P_g} = 1.608\frac{(0.00773)(98.77)}{3.169} = 0.387 \quad \text{or } 38.7\%$$

where P_g is used as the saturation pressure at 25 °C from Table C-1.

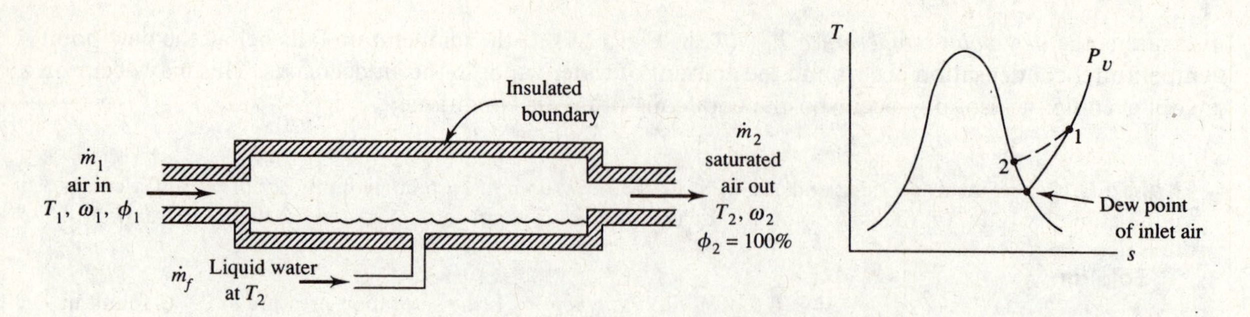

Fig. 12-2 Setup used to find ω of air.

12.5 ADIABATIC SATURATION AND WET-BULB TEMPERATURES

It is quite difficult to measure the relative humidity and the humidity ratio directly, at least with any degree of accuracy. This section presents two indirect methods for determining these quantities accurately.

Consider a relatively long insulated channel, shown in Fig. 12-2; air with an unknown relative humidity enters, moisture is added to the air by the pool of water, and saturated air exits. This process involves no heat transfer because the channel is insulated and hence it is called the *adiabatic saturation process*. The exit temperature is the *adiabatic saturation temperature*. Let us find an expression for the humidity ratio. Consider that the liquid water added is at temperature T_2. An energy balance on this control volume, neglecting kinetic and potential energy changes, is done considering the dry air and the water vapor components. With $\dot{Q} = \dot{W} = 0$ we have

$$\dot{m}_{v1}h_{v1} + \dot{m}_{a1}h_{a1} + \dot{m}_f h_{f2} = \dot{m}_{a2}h_{a2} + \dot{m}_{v2}h_{v2} \tag{12.26}$$

But, from conservation of mass for both the dry air and the water vapor,

$$\dot{m}_{a1} = \dot{m}_{a2} = \dot{m}_a \qquad \dot{m}_{v1} + \dot{m}_f = \dot{m}_{v2} \tag{12.27}$$

Using the definition of ω in (*12.23*), the above equations allow us to write

$$\dot{m}_a\omega_1 + \dot{m}_f = \omega_2\dot{m}_a \tag{12.28}$$

Substituting this into (*12.26*) for $\dot{m}_f$, there results, using $h_v \cong h_g$,

$$\dot{m}_a\omega_1 h_{g1} + \dot{m}_a h_{a1} + (\omega_2 - \omega_1)\,\dot{m}_a h_{f2} = \dot{m}_a h_{a2} + \omega_2\dot{m}_a h_{g2} \tag{12.29}$$

At state 2 we know that $\phi_2 = 1.0$ and, using (*12.25*),

$$\omega_2 = 0.622\frac{P_{g2}}{P - P_{g2}} \tag{12.30}$$

Thus, (*12.29*) becomes

$$\omega_1 = \frac{\omega_2 h_{fg2} + C_p(T_2 - T_1)}{h_{g1} - h_{f2}} \tag{12.31}$$

where $h_{a2} - h_{a1} = C_p(T_2 - T_1)$ for the dry air and $h_{fg2} = h_{g2} - h_{f2}$. Consequently, if we measure the temperatures T_2 and T_1 and the total pressure P we can find ω_2 from (*12.30*) with the remaining quantities in (*12.31*) given in Appendix C.

Because T_2 is significantly less than T_1, the apparatus sketched in Fig. 12-2 can be used to cool an airstream. This is done in relatively dry climates so that T_2 is reduced but usually not to the saturation temperature. Such a device is often referred to as a "swamp cooler." A fan blowing air through a series of wicks that stand in water is quite effective at cooling low-humidity air.

Using the device of Fig. 12-2 to obtain the adiabatic saturation temperature is a rather involved process. A much simpler approach is to wrap the bulb of a thermometer with a cotton wick saturated with water, and then either to blow air over the wick or to swing the thermometer through the air until the temperature reaches a steady-state value. This *wet-bulb temperature* $T_{\text{w.b.}}$ and the adiabatic saturation temperature are essentially the same for water if the pressure is approximately atmospheric.

EXAMPLE 12.7 The dry-bulb and wet-bulb temperatures of a 14.7-psia airstream are 100 °F and 80 °F, respectively. Determine (*a*) the humidity ratio, (*b*) the relative humidity, and (*c*) the specific enthalpy of the air.

Solution:

(*a*) We use (*12.31*) to find ω_1. But first ω_2 is found using (*12.30*):

$$\omega_2 = 0.622\frac{P_{g2}}{P - P_{g2}} = (0.622)\left(\frac{0.5073}{14.7 - 0.5073}\right) = 0.0222$$

where P_{g2} is the saturation pressure at 80 °F. Now (*12.30*) gives

$$\omega_1 = \frac{\omega_2 h_{fg2} + C_p(T_2 - T_1)}{h_{g1} - h_{f2}} = \frac{(0.0222)(1048) + (0.24)(80 - 100)}{1105 - 48.09}$$
$$= 0.01747 \text{ H}_2\text{O/lbm dry air}$$

(*b*) The partial pressure of the water vapor is found using (*12.24*):

$$\omega_1 = 0.622\frac{P_{v1}}{P_{a1}} \qquad 0.01747 = 0.622\frac{P_{v1}}{14.7 - P_{v1}} \qquad \therefore P_{v1} = 0.402 \text{ psia}$$

The relative humidity is obtained from (*12.22*): $\phi = P_{v1}/P_{g1} = 0.402/0.9503 = 0.423$ or 42.3%.

(*c*) The specific enthalpy is found by assuming a zero value for air at $T = 0$ °F. The enthalpy for the mixture is $H = H_a + H_v = m_a h_a + m_v h_v$. Dividing by m_a, we find that

$$h = h_a + \omega h_v = C_p T + \omega h_g$$
$$= (0.24)(100) + (0.01747)(1105)$$
$$= 43.3 \text{ Btu/lbm dry air}$$

where we have used $h_v = h_g$ (see Fig. 12-2). The enthalpy is always expressed per mass unit of dry air.

12.6 THE PSYCHROMETRIC CHART

A convenient way of relating the various properties associated with a water vapor–air mixture is to plot these quantities on a *psychrometric chart* such as Fig. 12-3 or (for standard atmospheric pressure) Fig. F-1 in the Appendix. Any two of the properties establishes a state from which the other properties are determined. As an example, consider a state *A* that is located by specifying the dry-bulb temperature and the relative humidity. The wet-bulb temperature would be read at 1, the dew-point temperature at 2, the enthalpy at 3, and the humidity ratio at 4. Referring to Fig. F-1, a dry-bulb temperature of 30 °C and a relative humidity of 80% would provide the following: $T_{\text{d.p.}} = 26$ °C, $T_{\text{w.b.}} = 27$ °C, $h = 85$ kJ/kg dry air, and $\omega = 0.0215$ kg H_2O/kg dry air. The chart provides us with a quick, relatively accurate method for finding the quantities of interest. If the pressure is significantly different from standard atmospheric pressure, the equations presented in the preceding sections must be used.

EXAMPLE 12.8 Using Fig. F-1E, rework Example 12.7 ($T_{\text{d.b.}} = 100$ °F, $T_{\text{w.b.}} = 80$ °F) to find ω, ϕ, and h.

Solution: Using the chart, the intersection of $T_{\text{d.b.}} = 100$ °F and $T_{\text{w.b.}} = 80$ °F gives

$$\omega = 0.0175 \text{ lbm H}_2\text{O/lbm dry air} \qquad \phi = 42\% \qquad h = 44 \text{ Btu/lbm dry air}$$

These values are less accurate than those calculated in Example 12.7, but certainly are acceptable.

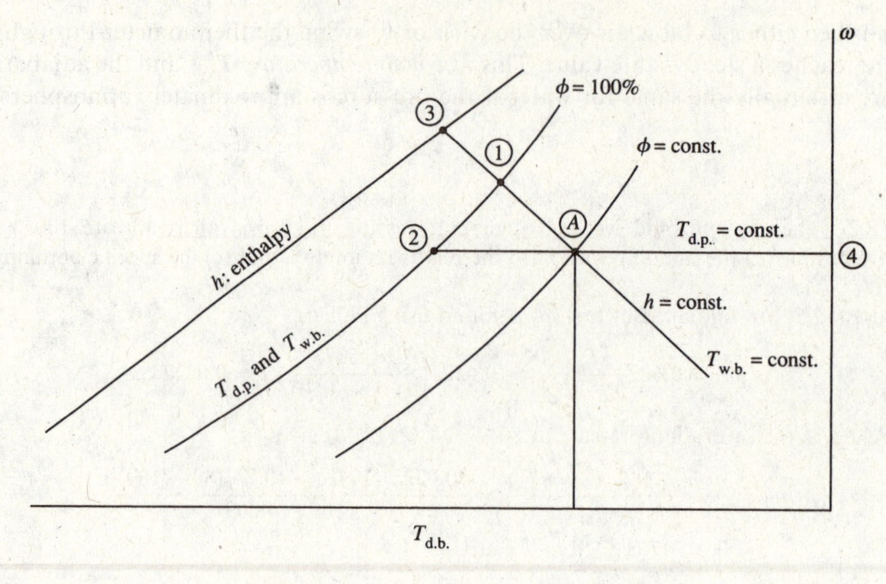

Fig. 12-3 The psychrometric chart.

12.7 AIR-CONDITIONING PROCESSES

Generally, people feel most comfortable when the air is in the "comfort zone": the temperature is between 22 °C (72 °F) and 27 °C (80 °F) and the relative humidity is between 40% and 60%. In Fig. 12-4, the area enclosed by the heavy dotted lines represents the *comfort zone*. There are several situations in which air must be conditioned to put it in the comfort zone:

- The air is too cold or too hot. Heat is simply added or extracted. This is represented by *A–C* and *B–C* in Fig. 12-4.
- The air is too cold and the humidity is too low. The air can first be heated, and then moisture added, as in *D–E–C*.
- The temperature is acceptable but the humidity is too high. The air is first cooled from *F* to *G*. Moisture is removed from *G* to *H*. Heat is added from *H* to *I*.
- The air is too hot and the humidity is low. Moisture is added, and the process represented by *J–K* results.
- An airstream from the outside is mixed with an airstream from the inside to provide natural cooling or fresh air. Process *I–M* represents the warmer inside air mixed with the outside air represented by *L–M*. State *M* represents the mixed air.

Each of these situations will be considered in the following examples. The first law will be used to predict the heating or cooling needed or to establish the final state.

EXAMPLE 12.9 Outside air at 5 °C and 70% relative humidity is heated to 25 °C. Calculate the rate of heat transfer needed if the incoming volume flow rate is 50 m^3/min. Also, find the final relative humidity. Assume $P = 100$ kPa.

Solution: The density of dry air is found using the partial pressure P_{a1} in the ideal-gas law:

$$P_{a1} = P - P_{v1} = P - \phi P_{g1} = 100 - (0.7)(0.872) = 99.4\,\text{kPa}$$

$$\therefore \rho_{a1} = \frac{P_{a1}}{R_a T_1} = \frac{99.4}{(0.287)(278)} = 1.246\,\text{kg/m}^3$$

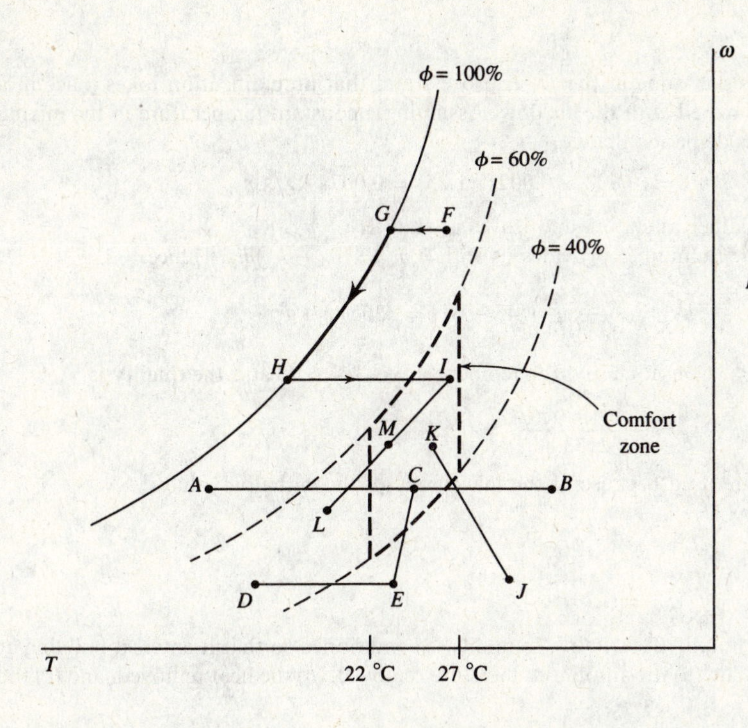

Fig. 12-4 The conditioning of air.

The mass flux of dry air is then $\dot{m}_a = (50/60)(1.246) = 1.038$ kg/s. Using the psychrometric chart at state 1 ($T_1 = 5\,°\text{C}, \phi_1 = 70\%$), we find $h_1 = 14$ kJ/kg air. Since ω remains constant (no moisture is added or removed), we follow curve A–C in Fig. 12-4; at state 2 we find that $h_2 = 35$ kJ/kg air. Hence,

$$\dot{Q} = \dot{m}_a(h_2 - h_1) = 1.038(35 - 14) = 11.4\,\text{kJ/s}$$

At state 2 we also note from the chart that $\phi_2 = 19\%$.

EXAMPLE 12.10 Outside air at 5 °C and 40% relative humidity is heated to 25 °C and the final relative humidity is raised to 40% while the temperature remains constant by introducing steam at 400 kPa into the airstream. (*a*) Find the needed rate of heat transfer if the incoming volume flow rate of air is 60 m^3/min. (*b*) Calculate the rate of steam supplied. (*c*) Calculate the state of the steam introduced.

Solution:

(*a*) The process we must follow is first simple heating and then humidification; the latter is sketched as D–E–C in Fig. 12-4, except the E–C leg is vertical at constant temperature. The partial pressure of dry air is

$$P_{a1} = P - P_{v1} = P - \phi P_{g1} = 100 - (0.4)(0.872) = 99.7\,\text{kPa}$$

where we have assumed standard atmospheric pressure. The dry air density is

$$\rho_{a1} = \frac{P_{a1}}{R_a T_1} = \frac{99.7}{(0.287)(278)} = 1.25\,\text{kg/m}^3$$

so that the mass flux of dry air is $\dot{m}_a = (60/60)(125) = 1.25$ kg/s. The rate of heat addition is found using h_1 and h_2 from the psychrometric chart:

$$\dot{Q} = \dot{m}_a(h_2 - h_1) = (1.25)(31 - 10) = 26.2\,\text{kJ/s}$$

(*b*) We assume that all the heating is done in the D–E process and that humidification takes place in a process in which the steam is mixed with the air flow. Assuming a constant temperature in the mixing process, conservation of mass demands that

$$\dot{m}_s = (\omega_3 - \omega_2)\dot{m}_a = (0.008 - 0.0021)(1.25) = 0.0074 \text{ kg/s}$$

where the air enters the humidifier at state 2 and leaves at state 3.

(*c*) An energy balance around the humidifier provides us with $h_s\dot{m}_s = (h_3 - h_2)\dot{m}_a$. Hence,

$$h_s = \frac{\dot{m}_a}{\dot{m}_s}(h_3 - h_2) = \left(\frac{1.25}{0.0074}\right)(45 - 31) = 2365 \text{ kJ/kg}$$

This is less than h_g at 400 kPa. Consequently, the temperature is 143.6 °C and the quality is

$$x_s = \frac{2365 - 604.7}{2133.8} = 0.82$$

Only two significant figures are used because of the inaccuracy of the enthalpy values.

EXAMPLE 12.11 Outside air at 80 °F and 90% relative humidity is conditioned so that it enters a building at 75 °F and 40% relative humidity. Estimate (*a*) the amount of moisture removed, (*b*) the heat removed, and (*c*) the necessary added heat.

Solution:

(*a*) The overall process is sketched as F–G–H–I in Fig. 12-4. Heat is removed during the F–H process, moisture is removed during the G–H process, and heat is added during the H–I process. Using the psychrometric chart, we find the moisture removed to be

$$\Delta\omega = \omega_3 - \omega_2 = 0.0075 - 0.0177 = -0.010 \text{ lbm } H_2O\text{/lbm dry air}$$

where states 2 and 3 are at G and H, respectively.

(*b*) The heat that must be removed to cause the air to follow the F–G–H process is $q = h_3 - h_1 = 20 - 39.5 = -18.5$ Btu/lbm dry air.

(*c*) The heat that must be added to change the state of the air from the saturated state at H to the desired state at I is

$$q = h_4 - h_3 = 26.5 - 20 = 6.5 \text{ Btu/lbm dry air}$$

EXAMPLE 12.12 Hot, dry air at 40 °C and 10% relative humidity passes through an evaporative cooler. Water is added as the air passes through a series of wicks and the mixture exits at 27 °C. Find (*a*) the outlet relative humidity, (*b*) the amount of water added, and (*c*) the lowest temperature that could be realized.

Solution:

(*a*) The heat transfer is negligible in an evaporative cooler, so that $h_2 \cong h_1$. A constant enthalpy line is shown in Fig. 12-4 and is represented by J–K. From the psychrometric chart we find, at 27 °C, $\phi_2 = 45\%$.

(*b*) The added water is found to be $\omega_2 - \omega_1 = 0.010 - 0.0046 = 0.0054$ kg H_2O/kg dry air.

(*c*) The lowest possible temperature occurs when $\phi = 100\%$: $T_{min} = 18.5$ °C.

EXAMPLE 12.13 Outside cool air at 15 °C and 40% relative humidity (airstream 1) is mixed with inside air taken near the ceiling at 32 °C and 70% relative humidity (airstream 2). Determine the relative humidity and temperature of the resultant airstream 3 if the outside flow rate is 40 m^3/min and the inside flow rate is 20 m^3/min.

Solution: An energy and mass balance of the mixing of airstream 1 with airstream 2 to produce airstream 3 would reveal the following facts relative to the psychrometric chart:

State 3 lies on a straight line connecting state 1 and state 2.

The ratio of the distance 2–3 to the distance 3–1 is equal to $\dot{m}_{a1}/\dot{m}_{a2}$.

State 1 and state 2 can be located on the psychrometric chart. We must determine $\dot{m}_{a1}$ and $\dot{m}_{a2}$:

$$P_{a1} = P - P_{v1} = 100 - 1.7 = 98.3 \text{ kPa} \qquad P_{a2} = P - P_{v2} = 100 - 4.8 = 95.2 \text{ kPa}$$

$$\therefore \rho_{a1} = \frac{98.3}{(0.287)(288)} = 1.19 \text{ kg/m}^3 \qquad \rho_{a2} = \frac{95.2}{(0.287)(305)} = 1.09 \text{ kg/m}^3$$

$$\therefore \dot{m}_{a1} = (40)(1.19) = 47.6 \text{ kg/min} \qquad \dot{m}_{a2} = (20)(1.09) = 21.8 \text{ kg/min}$$

State 3 is located by the ratio $d_{2-3}/d_{3-1} = \dot{m}_{a1}/\dot{m}_{a2} = 47.6/21.8 = 2.18$, where d_{2-3} is the distance from state 2 to state 3. State 3 is positioned on the psychrometric chart, and we find $\phi_3 = 63\%$ and $T_3 = 20.2\,^\circ\text{C}$.

EXAMPLE 12.14 Water is used to remove the heat from the condenser of a power plant. 10 000 kg per minute of 40 °C water enters a cooling tower, as shown in Fig. 12-5. Water leaves at 25 °C. Air enters at 20 °C and leaves at 32 °C. Estimate (*a*) the volume flow rate of air into the cooling tower, and (*b*) the mass flux of water that leaves the cooling tower from the bottom.

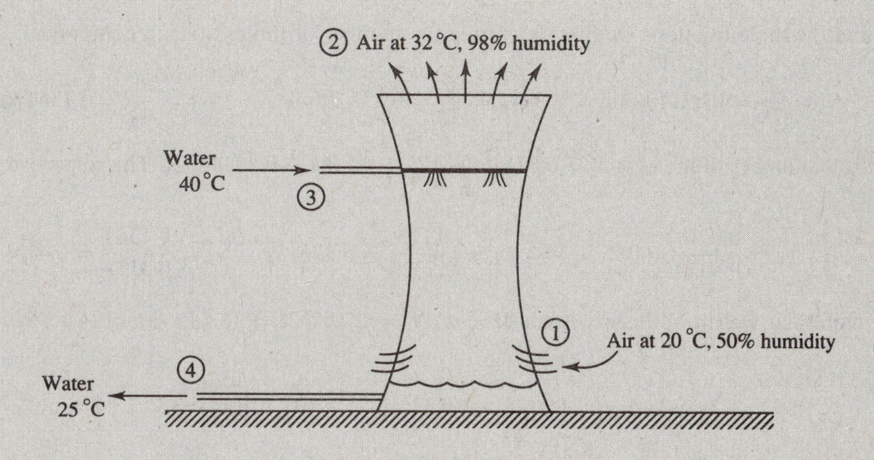

Fig. 12-5

Solution:

(*a*) An energy balance for the cooling tower provides $\dot{m}_{a1}h_1 + \dot{m}_{w3}h_3 = \dot{m}_{a2}h_2 + \dot{m}_{w4}h_4$, where $\dot{m}_{a1} = \dot{m}_{a2} = \dot{m}_a$ is the mass flux of dry air. From the psychrometric chart we find

$$h_1 = 37 \text{ kJ/kg dry air} \qquad h_2 = 110 \text{ kJ/kg dry air} \qquad \omega_1 = 0.0073 \text{ kg H}_2\text{O/kg dry air}$$
$$\omega_2 = 0.0302 \text{ kg H}_2\text{O/kg dry air}$$

From the steam tables we use h_f at the given temperature and find $h_3 = 167.5$ kJ/kg and $h_4 = 104.9$ kJ/kg. A mass balance on the water results in $\dot{m}_{w4} = \dot{m}_{w3} - (\omega_2 - \omega_1)\dot{m}_a$. Substituting this into the energy balance, with $\dot{m}_{a1} = \dot{m}_{a2} = \dot{m}_a$,

$$\dot{m}_a = \frac{\dot{m}_{w3}(h_4 - h_3)}{h_1 - h_2 + (\omega_2 - \omega_1)h_4} = \frac{(10\,000)(104.9 - 167.5)}{37 - 110 + (0.0302 - 0.0073)(104.9)} = 8870 \text{ kg/min}$$

From the psychrometric chart we find that $v_1 = 0.84\ m^3/kg$ dry air. This allows us to find the volume flow rate:

$$\text{Volume flow rate} = \dot{m}_a v_1 = (8870)(0.84) = 7450\ m^3/min$$

This air flow rate requires fans, although there is some "chimney effect" since the hotter air wants to rise.

(*b*) $\dot{m}_4 = \dot{m}_{w3} - (\omega_2 - \omega_1)\dot{m}_a = 10\,000 - (0.0302 - 0.0073)(8870) = 9800$ kg/min

If the exiting water is returned to the condenser, it must be augmented by 200 kg/min so that 10 000 kg/min is furnished. The added water is called *makeup water.*

Solved Problems

12.1 Gravimetric analysis of a mixture indicates 2 kg N_2, 4 kg O_2, and 6 kg CO_2. Determine (*a*) the mass fraction of each component, (*b*) the mole fraction of each component, (*c*) the molecular weight of the mixture, and (*d*) its gas constant.

(*a*) The total mass of the mixture is $m = 2 + 4 + 6 = 12$ kg. The respective mass fractions are

$$mf_1 = \frac{2}{12} = 0.1667 \qquad mf_2 = \frac{4}{12} = 0.3333 \qquad mf_3 = \frac{6}{12} = 0.5$$

(*b*) To find the mole fractions we first determine the number of moles of each component:

$$N_1 = \frac{2}{28} = 0.0714\ \text{kmol} \qquad N_2 = \frac{4}{32} = 0.125\ \text{kmol} \qquad N_3 = \frac{6}{44} = 0.1364\ \text{kmol}$$

The total number of moles is $N = 0.0714 + 0.125 + 0.1364 = 0.3328$ mol. The respective mole fractions are

$$y_1 = \frac{0.0714}{0.3328} = 0.215 \qquad y_2 = \frac{0.125}{0.3328} = 0.376 \qquad y_3 = \frac{0.1364}{0.3328} = 0.410$$

(*c*) The molecular weight of the mixture is $M = m/N = 12/0.3328 = 36.1$ kg/kmol. Alternatively, we could write

$$M = \sum y_i M_i = (0.215)(28) + (0.376)(32) + (0.410)(44) = 36.1\ \text{kg/kmol}$$

(*d*) The gas constant is $R = \dfrac{R_u}{M} = \dfrac{8.314}{36.1} = 0.230$ kJ/kg·K.

12.2 The partial pressure of each component of a mixture of N_2 and O_2 is 10 psia. If the temperature is 80 °F find the specific volume of the mixture.

The mole fractions are equal since the partial pressures are equal [see (*12.10*)]: $y_1 = 0.5$ and $y_2 = 0.5$. The molecular weight is then

$$M = \sum y_i M_i = (0.5)(28) + (0.5)(32) = 40\ \text{lbm/lbmol}$$

and the gas constant is $R = R_u/M = 1545/40 = 38.6$ ft-lbf/lbm-°R. Hence,

$$v = \frac{RT}{P} = \frac{(38.6)(540)}{(20)(144)} = 7.24\ \text{ft}^3/\text{lbm}$$

12.3 A mixture of ideal gases consists of 2 kmol CH_4, 1 kmol N_2, and 1 kmol CO_2, all at 20 °C and 20 kPa. Heat is added until the temperature increases to 400 °C while the pressure remains constant. Calculate (*a*) the heat transfer, (*b*) the work done, and (*c*) the change in entropy.

To find the quantities of interest we first calculate the specific heats of the mixture, using Table B-2:

$$\left.\begin{aligned} m_1 &= (2)(16) = 32\text{ kg} \\ m_2 &= (1)(14) = 14\text{ kg} \\ m_3 &= (1)(44) = 44\text{ kg} \end{aligned}\right\} \qquad \therefore\ m = 90\text{ kg}$$

$$mf_1 = \frac{32}{90} = 0.356 \qquad mf_2 = \frac{14}{90} = 0.1556 \qquad mf_3 = \frac{44}{90} = 0.489$$

$$\therefore C_p = \sum mf_i C_{p,i} = (0.356)(2.254) + (0.1556)(1.042) + (0.489)(0.842) = 1.376\text{ kJ/kg·K}$$

$$C_v = \sum mf_i C_{v,i} = (0.356)(1.735) + (0.1556)(0.745) + (0.489)(0.653) = 1.053\text{ kJ/kg·K}$$

(*a*) For a constant-pressure process, $Q = \Delta H = mC_p\,\Delta T = (90)(1.376)(400 - 20) = 47\,060$ kJ.

(*b*) $$W = Q - \Delta U = Q - mC_v\,\Delta T = 47\,060 - (90)(1.053)(400 - 20) = 11\,050\text{ kJ}$$

(*c*) $$\Delta S = m(C_p \ln T_2/T_1 - R \ln 1) = (90)(1.376 \ln 673/293) = 103.0\text{ kJ/K}$$

12.4 An insulated, rigid tank contains 2 mol N_2 at 20 °C and 200 kPa separated by a membrane from 4 mol CO_2 at 100 °C and 100 kPa. The membrane ruptures and the mixture reaches equilibrium. Calculate the final temperature and pressure.

The first law, with $Q = W = 0$, requires $0 = \Delta U = N_1\overline{C}_{v,1}(T - 20) + N_2\overline{C}_{v,2}(T - 100)$. The specific heat $\overline{C}_{v,i} = M_i C_{v,i}$. Using values from Table B-2, we have

$$0 = (2)(28)(0.745)(T - 20) + (4)(44)(0.653)(T - 100)$$

This equation can be solved to yield the equilibrium temperature as $T = 78.7$ °C. The initial volumes occupied by the gases are

$$V_1 = \frac{N_1 R_u T_1}{P_1} = \frac{(2)(8.314)(293)}{200} = 24.36\text{ m}^3 \qquad V_2 = \frac{N_2 R_u T_2}{P_2} = \frac{(4)(8.314)(373)}{100} = 124\text{ m}^3$$

The total volume remains fixed at $124.0 + 24.4 = 148.4\text{ m}^3$. The pressure is then

$$P = \frac{NR_uT}{V} = \frac{(6)(8.314)(273 + 78.7)}{148.4} = 118.2\text{ kPa}$$

12.5 A mixture of 40% N_2 and 60% O_2 by weight is compressed from 70 °F and 14.7 psia to 60 psia. Estimate the horsepower required by an 80 percent-efficient compressor and the entropy change, if the mass flux is 10 lbm/min.

The efficiency of a compressor is based on an isentropic process. Let us find k and C_p:

$$C_p = mf_1 C_{p,1} + mf_2 C_{p,2} = (0.4)(0.248) + (0.6)(0.219) = 0.231\text{ Btu/lbm-°R}$$

$$C_v = mf_1 C_{v,1} + mf_2 C_{v,2} = (0.4)(0.177) + (0.6)(0.157) = 0.165\text{ Btu/lbm-°R}$$

$$k = \frac{C_p}{C_v} = \frac{0.231}{0.165} = 1.4$$

The isentropic relation provides

$$T_2 = T_1\left(\frac{P_2}{P_1}\right)^{(k-1)/k} = (530)\left(\frac{60}{14.7}\right)^{(1.4-1)/1.4} = 792\text{ °R}$$

For an ideal compressor, $\dot{W}_{\text{comp}} = \dot{m}\Delta h = \dot{m}C_p\,\Delta T = (\frac{10}{60})(0.231)(778)(792 - 530) = 7850$ ft-lbf/sec, where the factor 778 ft-lbf/Btu provides us with the desired units. If the compressor is 80 percent efficient, the actual power is $\dot{W}_{\text{comp}} = 7850/0.8 = 9810$ ft-lbf/sec or 17.8 hp.

To find the entropy change we need the actual outlet temperature. Using the definition of compressor efficiency,

$$\eta_{\text{comp}} = \frac{w_s}{w_a} = \frac{C_p(\Delta T)_s}{C_p(\Delta T)_a}$$

we find $0.8 = (792 - 530)/(T_2 - 530)$ and $T_2 = 857.5\,^\circ\text{R}$. The change in entropy is then

$$\Delta s = C_p \ln \frac{T_2}{T_2} - R \ln \frac{P_2}{P_1} = 0.231 \ln \frac{857.5}{530} - 0.066 \ln \frac{60}{14.7} = 0.0183 \text{ Btu/lbm-}^\circ\text{R}$$

where we have used $R = C_p - C_v = 0.231 - 0.165 = 0.066$ Btu/lbm-°R.

12.6 Outside air at 30 °C and 100 kPa is observed to have a dew point of 20 °C. Find the relative humidity, the partial pressure of the dry air, and the humidity ratio using equations only.

At 30 °C we find the saturation pressure from Table C-1 (see Fig. 12-1) to be $P_g = 4.246$ kPa. At 20 °C the partial pressure of the water vapor is $P_v = 2.338$ kPa. Consequently, the relative humidity is

$$\phi = \frac{P_v}{P_g} = \frac{2.338}{4.246} = 0.551 \quad \text{or } 55.1\%$$

The partial pressure of the dry air is $P_a = P - P_v = 100 - 2.338 = 97.66$ kPa. The humidity ratio is found to be

$$\omega = 0.622 \frac{P_v}{P_a} = (0.622)\left(\frac{2.338}{97.66}\right) = 0.01489 \text{ kg H}_2\text{O/kg dry air}$$

12.7 Outside air at 25 °C has a relative humidity of 60%. What would be the expected wet-bulb temperature?

We assume an atmospheric pressure of 100 kPa. The saturation pressure at 25 °C is $P_g = 3.169$ kPa, so that

$$P_v = \phi P_g = (0.6)(3.169) = 1.901 \text{ kPa}$$

and

$$P_a = P - P_v = 100 - 1.901 = 98.1 \text{ kPa}$$

The humidity ratio of the outside air is

$$\omega_1 = 0.622 \frac{P_v}{P_a} = (0.622)\left(\frac{1.901}{98.1}\right) = 0.01206 \text{ kg H}_2\text{O/kg dry air}$$

Using ω_2 from *(12.30)* we can write *(12.31)* as

$$(h_{g1} - h_{f2})\omega_1 = 0.622 \frac{P_{g2}}{P - P_{g2}} h_{fg2} + c_p(T_2 - T_1)$$

Substituting in the known values, we must solve

$$(2547.2 - h_{f2})(0.01206) = 0.622 \frac{P_{g2}}{100 - P_{g2}} h_{fg2} + (1.00)(T_2 - 25)$$

This is solved by trial and error:

$$T_2 = 20\,^\circ\text{C}: \quad 29.7 \stackrel{?}{=} 32.2 \qquad T_2 = 15\,^\circ\text{C}: \quad 30.0 \stackrel{?}{=} 16.6$$

Interpolation yields $T_2 = 19.3\,^\circ\text{C}$.

12.8 Rework Prob. 12.7 using the psychrometric chart.

The wet-bulb or adiabatic saturation temperature is found by first locating the intersection of a vertical line for which $T = 25\,^\circ\text{C}$ and the curved line for which $\phi = 60\%$ humidity. Follow the line for which $T_{\text{w.b.}} =$ const. that slopes upward to the left and read $T_{\text{w.b.}} = 19.4\,^\circ\text{C}$.

12.9 Air at 90 °F and 20% relative humidity is cooled to 75 °F. Assuming standard atmospheric pressure, calculate the required rate of energy transfer, if the inlet flow rate is 1500 ft^3/min, and find the final humidity, using *(a)* the psychrometric chart and *(b)* the equations.

(*a*) The partial pressure is $P_{a1} = P - P_{v1} = P - \phi P_{g1} = 14.7 - (0.2)(0.6988) = 14.56$ psia; hence,

$$\rho_{a1} = \frac{P_{a1}}{R_a T_1} = \frac{(14.56)(144)}{(53.3)(550)} = 0.0715\ \text{lbm/ft}^3$$

and $\dot{m}_a = (1500/60)(0.0715) = 1.788$ lbm/sec. The psychrometric chart at state 1 provides $h_1 = 28.5$ Btu/lbm dry air. With $\omega = \text{const.}$, state 2 is located by following an *A–C* curve in Fig. 12-4; we find $h_2 = 24.5$ Btu/lbm dry air. This gives

$$\dot{Q}_a = \dot{m}(h_2 - h_1) = (1.788)(24.5 - 28.5) = -7.2\ \text{Btu/sec}$$

The relative humidity is found on the chart at state 2 to be $\phi_2 = 32.5\%$.

(*b*) The equations provide more accurate results than can be obtained by using the psychrometric chart. The value for $\dot{m}_a$ from part (*a*) has been calculated so we'll simply use that number. The rate of heat transfer is

$$\dot{Q} = \dot{m}_a(h_{a2} - h_{a1}) + \dot{m}_v(h_{v2} - h_{v1}) = \dot{m}_a C_p(T_2 - T_1) + \dot{m}_v(h_{v2} - h_{v1})$$

We find $\dot{m}_v$ as follows:

$$\dot{m}_v = \omega \dot{m}_a = 0.622\frac{\phi P_g}{P_a}\dot{m}_a = (0.622)(0.2)\left(\frac{0.6988}{14.56}\right)(1.788) = 0.01067\ \text{lbm/sec}$$

Thus $\dot{Q} = (1.788)(0.24)(75 - 90) + (0.01067)(1094.2 - 1100.7) = -6.51$ Btu/sec.

To find the relative humidity using (*12.22*) we must find P_{v2} and P_{g2}. The final temperature is 75 °F; Table C-1E gives, by interpolation, $P_{g2} = 0.435$ psia. Since the mass of vapor and the mass of dry air remain constant, the partial pressure of vapor and dry air remain constant. Hence,

$$P_{v2} = P_{v1} = \phi P_{g1} = (0.2)(0.6988) = 0.1398\ \text{psia}$$

The final relative humidity is $\phi_2 = P_{v2}/P_{g2} = 0.1398/0.435 = 0.321$ or 32.1%. The values found in part (*b*) are more accurate than those of part (*a*), especially for $\dot{Q}$, since it is difficult to read h_1 and h_2 accurately.

12.10 Air at 90 °F and 90% relative humidity is cooled to 75 °F. Calculate the required rate of energy transfer if this inlet flow rate is 1500 ft³/min. Also, find the final humidity. Compare with the results of Prob. 12.9. Use the psychrometric chart.

The first step is to find the mass flux of dry air. It is found as follows:

$$P_{a1} = P - P_{v1} = P - \phi P_{g1} = 14.7 - (0.9)(0.6988) = 14.07\ \text{psia}$$

$$\therefore \rho_{a1} = \frac{P_{a1}}{R_a T_1} = \frac{(14.07)(144)}{(53.3)(550)} = 0.0691\ \text{lbm/ft}^3 \quad \text{and}$$

$$\dot{m}_a = \left(\frac{1500}{60}\right)(0.0691) = 1.728\ \text{lbm/sec}$$

State 1 is located on the psychrometric chart by $T_{\text{d.b.}} = 90\,°\text{F}$, $\phi = 90\%$. Hence, by extrapolation, $h_1 = 52$ Btu/lbm dry air. To reduce the temperature to 75 °F, it is necessary to remove moisture, following curve *F–G–H* in Fig. 12-4. State 2 ends on the saturation line, and $h_2 = 38.6$ Btu/lbm dry air. This gives

$$\dot{Q} = \dot{m}(h_2 - h_1) = (1.728)(38.6 - 52) = -23.2\ \text{Btu/sec}$$

The relative humidity is $\phi_2 = 100\%$.

12.11 A rigid 2-m³ tank contains air at 160 °C and 400 kPa and a relative humidity of 20%. Heat is removed until the final temperature is 20 °C. Determine (*a*) the temperature at which condensation begins, (*b*) the mass of water condensed during the process, and (*c*) the heat transfer.

(*a*) The pressure in this problem is not atmospheric, so the psychrometric chart is not applicable. The initial partial pressure of the vapor is $P_{v1} = \phi P_{g1} = (0.2)(617.8) = 123.6$ kPa. The specific volume of the water vapor is

$$v_{v1} = \frac{R_v T_1}{P_{v1}} = \frac{(0.462)(433)}{123.6} = 1.62\ \text{m}^3/\text{kg}$$

At this specific volume (the volume remains constant), the temperature at which condensation begins is $T_{cond} = 92.5\,°C$.

(*b*) The partial pressure of the dry air is $P_{a1} = P - P_{v1} = 400 - 123.6 = 276.4$ kPa. The mass of dry air is

$$m_a = \frac{P_{a1}V_1}{R_aT_1} = \frac{(276.4)(2)}{(0.287)(433)} = 4.45\text{ kg}$$

The initial humidity ratio is

$$\omega_1 = 0.622\frac{P_{v1}}{P_{a1}} = (0.622)\left(\frac{123.6}{276.4}\right) = 0.278\text{ kg H}_2\text{O/kg dry air}$$

The final relative humidity is $\phi_2 = 1.0$, so that $P_{v2} = 2.338$ kPa. The final partial pressure of the dry air results from $P_{a1}/T_1 = P_{a2}/T_2$, so that $P_{a2} = (P_{a1})(T_2/T_1) = (276.4)(293/433) = 187$ kPa. The final humidity ratio becomes

$$\omega_2 = 0.622\frac{P_{v2}}{P_{a2}} = (0.622)\left(\frac{2.338}{187}\right) = 0.00778\text{ kg H}_2\text{O/kg dry air}$$

The moisture removed is $m_{cond} = m_a(\omega_1 - \omega_2) = (4.45)(0.278 - 0.00778) = 1.20$ kg.

(*c*) The heat transfer results from the first law:

$$\begin{aligned} Q &= m_a(u_{a2} - u_{a1}) + m_{v2}u_{v2} - m_{v1}u_{v1} + \Delta m_w(h_{fg})_{avg} \\ &= m_a\left[C_p(T_2 - T_1) + \omega_2 u_{v2} - \omega_1 u_{v1} + (\omega_2 - \omega_1)(h_{fg})_{avg}\right] \end{aligned}$$

Treating the vapor as an ideal gas, that is, $u_v = u_g$ at the given temperatures, we have

$$\begin{aligned} Q = (4.45)[&(0.717)(20 - 160) + (0.00778)(2402.9) \\ &- (0.278)(2568.4) + (0.00778 - 0.278)(2365)] = -6290\text{ kJ} \end{aligned}$$

12.12 Hot, dry air at 40 °C, 1 atm, and 20% humidity passes through an evaporative cooler until the humidity is 40%; it is then cooled to 25 °C. For an inlet airflow of 50 m^3/min, (*a*) how much water is added per hour and (*b*) what is the rate of cooling?

(*a*) The psychrometric chart is used with $h_1 = h_2$, providing us with

$$\omega_1 = 0.0092\text{ kg H}_2\text{O/kg dry air} \qquad \omega_2 = 0.0122\text{ kg H}_2\text{O/kg dry air}$$

We find the mass flux $\dot{m}_a$ of dry air as follows:

$$P_{a1} = P - P_{v1} = P - \phi P_{g1} = 100 - (0.2)(7.383) = 98.52\text{ kPa}$$

$$\therefore \rho_{a1} = \frac{P_{a1}}{R_aT_1} = \frac{98.52}{(0.287)(313)} = 1.097\text{ kg/m}^3$$

and

$$\therefore \dot{m}_a = (\rho_{a1})(50) = (1.097)(50) = 54.8\text{ kg/min}$$

The water addition rate is

$$(\dot{m}_w)_{added} = \dot{m}_a(\omega_2 - \omega_1) = (54.8)(0.0122 - 0.0092) = 0.1644\text{ kg/min} = 9.86\text{ kg/h}$$

(*b*) No heat is transferred during the process from 1 to 2. From 2 to 3 the humidity ratio remains constant and the psychrometric chart yields

$$h_2 = 64\text{ kJ/kg dry air} \qquad h_3 = 64\text{ kJ/kg dry air}$$

The rate of heat transfer is $\dot{Q} = \dot{m}_a(h_3 - h_2) = (54.8)(56 - 64) = -440$ kJ/min.

12.13 Outside air at 10 °C and 30% relative humidity is available to mix with inside air at 30 °C and 60% humidity. The inside flow rate is 50 m^3/min. Use the equations to determine what the outside flow rate should be to provide a mixed stream at 22 °C.

Mass balances and an energy balance provide

$$\text{Dry air:}\ \dot m_{a1} + \dot m_{a2} = \dot m_{a3}$$
$$\text{Vapor:}\ \dot m_{a1}\omega_1 + \dot m_{a2}\omega_2 = \dot m_{a3}\omega_3$$
$$\text{Energy:}\ \dot m_{a1}h_1 + \dot m_{a2}h_2 = \dot m_{a3}h_3$$

Using the given quantities we find, assuming a pressure of 100 kPa:

$$P_{a1} = P - P_{v1} = P - \phi_1 P_{g1} = 100 - (0.3)(1.228) = 99.6\ \text{kPa}$$
$$P_{a2} = P - \phi_2 P_{g2} = 100 - (0.6)(4.246) = 97.5\ \text{kPa}$$
$$\rho_{a1} = \frac{P_{a1}}{R_a T_1} = \frac{99.6}{(0.287)(283)} = 1.226\ \text{kg/m}^3 \qquad \rho_{a2} = \frac{P_{a2}}{R_a T_2} = \frac{97.5}{(0.287)(303)} = 1.121\ \text{kg/m}^3$$
$$\omega_1 = \frac{0.622 P_{v1}}{P_{a1}} = \frac{(0.622)(0.3)(1.228)}{99.6} = 0.00230\ \text{kg H}_2\text{O/kg dry air}$$
$$\omega_2 = \frac{(0.622)(0.6)(4.246)}{97.5} = 0.01625\ \text{kg H}_2\text{O/kg dry air}$$
$$h_1 = C_p T_1 + \omega_1 h_{g1} = (1.00)(10) + (0.0023)(2519.7) = 15.8\ \text{kJ/kg dry air}$$
$$h_2 = C_p T_2 + \omega_2 h_{g2} = (1.00)(30) + (0.01625)(2556.2) = 71.5\ \text{kJ/kg dry air}$$
$$h_3 = C_p T_3 + \omega_3 h_{g3} = (1.00)(22) + (\omega_3)(2542) = 22 + 2542\,\omega_3$$

Substituting the appropriate values in the energy equation and choosing the outside flow rate as $\dot V_1$ gives

$$(1.226\dot V_1)(15.8) + (1.121)(50)(71.5) = [1.226\dot V_1 + (1.121)(50)](22 + 2542\omega_3)$$

The vapor mass balance is $(1.226\dot V_1)(0.0023) + (1.121)(50)(0.01625) = [1.226\dot V_1 + (1.121)(50)]\omega_3$. Solving for ω_3 in terms of $\dot V_1$ from the above equation and substituting into the energy equation, we find $\dot V_1 = 31.1\ \text{m}^3/\text{min}$.

Supplementary Problems

12.14 For the following mixtures calculate the mass fraction of each component and the gas constant of the mixture. (*a*) 2 kmol CO_2, 3 kmol N_2, 4 kmol O_2; (*b*) 2 lbmol N_2, 3 lbmol CO, 4 lbmol O_2; (*c*) 3 kmol N_2, 2 kmol O_2, 5 kmol H_2; (*d*) 3 kmol CH_4; 2 kmol air, 1 kmol CO_2; and (*e*) 21 lbmol O_2, 78 lbmol N_2, 1 lbmol Ar.

12.15 For the following mixtures calculate the mole fraction of each component and the gas constant of the mixture. (*a*) 2 kg CO_2, 3 kg N_2, 4 kg O_2; (*b*) 2 lbm N_2, 3 lbm CO, 4 lbm O_2; (*c*) 3 kg N_2, 2 kg O_2, 5 kg H_2; (*d*) 3 kg CH_4, 2 kg air, 1 kg CO_2; and (*e*) 21 lbm O_2, 78 lbm N_2, 1 lbm Ar.

12.16 A mixture of gases consists of 21% N_2, 32% O_2, 16% CO_2, and 31% H_2, by volume. Determine: (*a*) the mass fraction of each component, (*b*) the mixture's molecular weight, and (*c*) its gas constant.

12.17 Gravimetric analysis of a mixture of gases indicates 21% O_2, 30% CO_2, and 49% N_2. Calculate (*a*) its volumetric analysis and (*b*) its gas constant.

12.18 Volumetric analysis of a mixture of gases shows 60% N_2, 20% O_2, and 20% CO_2. (*a*) How many kilograms would be contained in 10 m^3 at 200 kPa and 40 °C? (*b*) How many pounds would be contained in 300 ft^3 at 39 psia and 100 °F?

12.19 A mixture of gases contains 2 kmol O_2, 3 kmol CO_2, and 4 kmol N_2. If the mixture is contained in a 10-m^3 tank at 50 °C, estimate (*a*) the pressure in the tank and (*b*) the partial pressure of the N_2.

12.20 Gravimetric analysis of a mixture of gases indicates 60% N_2, 20% O_2, and 20% CO_2. (*a*) What volume is needed to contain 100 kg of the mixture at 25 °C and 200 kPa? (*b*) What volume is needed to contain 200 lbm of the mixture at 80 °F and 30 psia?

12.21 Volumetric analysis of a mixture of gases contained in a 10-m^3 tank at 400 kPa indicates 60% H_2, 25% N_2, and 15% CO_2. Determine the temperature of the mixture if its total mass is 20 kg.

12.22 The partial pressures of a mixture of gases are 20 kPa (N_2), 60 kPa (O_2), and 80 kPa (CO_2). If 20 kg of the mixture is contained in a tank at 60 °C, what is the volume of the tank?

12.23 A mixture of oxygen and hydrogen has the same molecular weight (molar mass) as does air. (*a*) What is its volumetric analysis? (*b*) What is its gravimetric analysis?

12.24 A rigid tank contains 10 kg of a mixture of 20% CO_2 and 80% N_2 by volume. The initial pressure and temperature are 200 kPa and 60 °C. Calculate the heat transfer needed to increase the pressure to 600 kPa using (*a*) constant specific heats and (*b*) the ideal-gas tables.

12.25 Twenty pounds of a mixture of gases is contained in a 30-ft^3 rigid tank at 30 psia and 70 °F. Volumetric analysis indicates 20% CO_2, 30% O_2, and 50% N_2. Calculate the final temperature if 400 Btu of heat is added. Assume constant specific heats.

12.26 An insulated cylinder contains a mixture of gases initially at 100 kPa and 25 °C with a volumetric analysis of 40% N_2 and 60% CO_2. Calculate the work needed to compress the mixture to 400 kPa assuming a reversible process. Use constant specific heats.

12.27 A mixture of gases is contained in a cylinder at an initial state of 0.2 m^3, 200 kPa, and 40 °C. Gravimetric analysis is 20% CO_2 and 80% air. Calculate (*a*) the heat transfer needed to maintain the temperature at 40 °C while the pressure is reduced to 100 kPa and (*b*) the entropy change. Assume constant specific heats.

12.28 A mixture of gases with a volumetric analysis of 30% H_2, 50% N_2, and 20% O_2 undergoes a constant-pressure process in a cylinder at an initial state of 30 psia, 100 °F, and 0.4 ft^3. If the volume increases to 1.2 ft^3 determine (*a*) the heat transfer and (*b*) the entropy change. Assume constant specific heats.

12.29 A tank containing 3 kg of CO_2 at 200 kPa and 140 °C is connected to a second tank containing 2 kg of N_2 at 400 kPa and 60 °C. A valve is opened and the two tanks are allowed to equalize in pressure. If the final temperature is 50 °C, find (*a*) the heat transfer, (*b*) the final pressure, and (*c*) the entropy change.

12.30 A stream of nitrogen at 150 kPa and 50 °C mixes with a stream of oxygen at 150 kPa and 20 °C. The mass flux of nitrogen is 2 kg/min and that of oxygen is 4 kg/min. The mixing occurs in a steady-flow insulated chamber. Calculate the temperature of the exiting stream.

12.31 A mixture of gases with a volumetric analysis of 20% CO_2, 30% N_2, and 50% O_2 is cooled from 1000 °R to 500 °R in a steady-flow heat exchanger. Estimate the heat transfer using (*a*) constant specific heats and (*b*) the ideal-gas tables.

12.32 A mixture of gases with a gravimetric analysis of 20% CO_2, 30% N_2, and 50% O_2 is cooled from 400 °C to 50 °C by transferring 1MW of heat from the steady-flow heat exchanger. Find the mass flux, assuming constant specific heats.

12.33 A mixture of 40% O_2 and 60% CO_2 by volume enters a nozzle at 40 m/s, 200 °C, and 200 kPa. It passes through an adiabatic nozzle and exits at 20 °C. Find the exit velocity and pressure. Assume constant specific heats.

12.34 If the inlet diameter of the nozzle of Prob. 12.33 is 20 cm, find the exit diameter.

12.35 A mixture of 40% N_2 and 60% CO_2 by volume enters a nozzle at negligible velocity and 80 psia and 1000 °F. If the mixture exits at 20 psia, what is the maximum possible exit velocity? Assume constant specific heats.

12.36 A mixture of 40% N_2 and 60% CO_2 by volume enters a supersonic diffuser at 1000 m/s and 20 °C and exits at 400 m/s. Find the exit temperature. Assume constant specific heats.

12.37 A mixture of 60% air and 40% CO_2 by volume at 600 kPa and 400 °C expands through a turbine to 100 kPa. Estimate the maximum power output if the mass flux is 4 kg/min. Assume constant specific heats.

12.38 If the turbine of Prob. 12.37 is 85 percent efficient, estimate the exit temperature.

12.39 A compressor increases the pressure of a mixture of gases from 100 to 400 kPa. If the mixture enters at 25 °C, find the minimum power requirement if the mass flux is 0.2 kg/s. Assume constant specific heats for the following gravimetric analyses of the mixture: (*a*) 10% H_2 and 90% O_2; (*b*) 90% H_2 and 10% O_2; and (*c*) 20% N_2, 30% CO_2, and 50% O_2.

12.40 Atmospheric air at 30 °C and 100 kPa has a relative humidity of 40%. Determine (*a*) the humidity ratio, (*b*) the dew-point temperature, and (*c*) the specific volume of the dry air.

12.41 Atmospheric air at 90 °F and 14.2 psia has a humidity ratio of 0.02. Calculate (*a*) the relative humidity, (*b*) the dew-point temperature, (*c*) the specific volume of the dry air, and (*d*) the enthalpy ($h = 0$ at 0 °F) per unit mass of dry air.

12.42 The air in a 12 × 15 × 3 m room is at 25 °C and 100 kPa, with a 50% relative humidity. Estimate (*a*) the humidity ratio, (*b*) the mass of dry air, (*c*) the mass of water vapor in the room, and (*d*) the enthalpy in the room ($h = 0$ at 0 °C).

12.43 A tank contains 0.4 kg of dry air and 0.1 kg of saturated water vapor at 30 °C. Calculate (*a*) the volume of the tank and (*b*) the pressure in the tank.

12.44 The partial pressure of water vapor is 1 psia in atmospheric air at 14.5 psia and 110 °F. Find (*a*) the relative humidity, (*b*) the humidity ratio, (*c*) the dew-point temperature, (*d*) the specific volume of the dry air, and (*e*) the enthalpy per unit mass of dry air.

12.45 A person wearing glasses comes from outside, where the temperature is 10 °C, into a room with 40% relative humidity. At what room temperature will the glasses start to fog up?

12.46 The outer surface temperature of a glass of cola, in a room at 28 °C, is 5 °C. At what relative humidity will water begin to collect on the outside of the glass?

12.47 A cold-water pipe at 50 °F runs through a basement where the temperature is 70 °F. At what relative humidity will water begin to condense on the pipe?

12.48 On a cold winter day the temperature on the inside of a thermopane window is 10 °C. If the inside temperature is 27 °C, what relative humidity is needed to just cause condensation on the window?

12.49 Atmospheric air has a dry-bulb temperature of 30 °C and a wet-bulb temperature of 20 °C. Calculate (*a*) the humidity ratio, (*b*) the relative humidity, and (*c*) the enthalpy per kg of dry air ($h = 0$ at 0 °C).

12.50 Use the psychrometric chart (Appendix F) to provide the missing values in Table 12-1.

Table 12-1

	Dry-Bulb Temperature	Wet-Bulb Temperature	Relative Humidity	Humidity Ratio	Dew-Point Temperature	Specific Enthalpy
(*a*)	20 °C			0.012		
(*b*)	20 °C	10 °C				
(*c*)	70 °F		60%			
(*d*)		60 °F	70%			
(*e*)		25 °C			15 °C	
(*f*)		70 °C		0.015		

12.51 Atmospheric air at 10 °C and 60% relative humidity is heated to 27 °C. Use the psychrometric chart to estimate the final humidity and the rate of heat transfer needed if the mass flux of dry air is 50 kg/min.

12.52 Heat is removed from a room without condensing out any of the water vapor. Use the psychrometric chart to calculate the final relative humidity if the air is initially at 35 °C and 50% relative humidity and the temperature is reduced to 25 °C.

12.53 Outside air at 40 °F and 40% relative humidity enters through the cracks in a house and is heated to 75 °F. Estimate the final relative humidity of the air if no other sources of water vapor are available.

12.54 Atmospheric air at 10 °C and 40% relative humidity is heated to 25 °C in the heating section of an air-conditioning device and then steam is introduced to increase the relative humidity to 50% while the temperature increases to 26 °C. Calculate the mass flux of water vapor added and the rate of heat transfer needed in the heating section if the volume flow rate of inlet air is 50 m^3/min.

12.55 Atmospheric air at 40 °F and 50% relative humidity enters the heating section of an air-conditioning device at a volume flow rate of 100 ft^3/min. Water vapor is added to the heated air to increase the relative humidity to 55%. Estimate the rate of heat transfer needed in the heating section and the mass flux of water vapor added if the temperature after the heating section is 72 °F and the temperature at the exit is 74 °F.

12.56 Outside air in a dry climate enters an air conditioner at 40 °C and 10% relative humidity and is cooled to 22 °C. (*a*) Calculate the heat removed. (*b*) Calculate the total energy required to condition outside (humid) air at 30 °C and 90% relative humidity to 22 °C and 10% relative humidity. (*Hint:* Sum the energy removed and the energy added.)

12.57 One hundred m^3/min of outside air at 36 °C and 80% relative humidity is conditioned for an office building by cooling and heating. Estimate both the rate of cooling and the rate of heating required if the final state of the air is 25 °C and 40% relative humidity.

12.58 Room air at 29 °C and 70% relative humidity is cooled by passing it over coils through which chilled water at 5 °C flows. The mass flux of the chilled water is 0.5 kg/s and it experiences a 10 °C temperature rise. If the room air exits the conditioner at 18 °C and 100% relative humidity, estimate (*a*) the mass flux of the room air and (*b*) the heat transfer rate.

12.59 Atmospheric air at 100 °F and 15% relative humidity enters an evaporative cooler at 900 ft^3/min and leaves with a relative humidity of 60%. Estimate (*a*) the exit temperature and (*b*) the mass flux at which water must be supplied to the cooler.

12.60 Outside air at 40 °C and 20% relative humidity is to be cooled by using an evaporative cooler. If the flow rate of the air is 40 m^3/min, estimate (*a*) the minimum possible temperature of the exit stream and (*b*) the maximum mass flux needed for the water supply.

12.61 Thirty m^3/min of outside air at 0 °C and 40% relative humidity is first heated and then passed through an evaporative cooler so that the final state is 25 °C and 50% relative humidity. Determine the temperature of the air when it enters the cooler, the heat transfer rate needed during the heating process, and the mass flux of water required by the cooler.

12.62 Outside air at 10 °C and 60% relative humidity mixes with 50 m^3/min of inside air at 28 °C and 40% relative humidity. If the outside flow rate is 30 m^3/min, estimate the relative humidity, the temperature, and the mass flux of the exiting stream.

12.63 Inside air at 80 °F and 80% relative humidity is mixed with 900 ft^3/min of outside air at 40 °F and 20% relative humidity. If the relative humidity of the exiting stream is 60%, estimate (*a*) the flow rate of the inside air, (*b*) the temperature of the exiting stream, and (*c*) the heat transfer rate from the outside air to the inside air.

12.64 Cooling water leaves the condenser of a power plant at 38 °C with a mass flux of 40 kg/s. It is cooled to 24 °C in a cooling tower that receives atmospheric air at 25 °C and 60% relative humidity. Saturated air exits the tower at 32 °C. Estimate (*a*) the required volume flow rate of entering air and (*b*) the mass flux of the makeup water.

12.65 A cooling tower cools 40 lbm/sec of water from 80 °F to 60 °F by moving 800 ft^3/sec of atmospheric air with dry-bulb and wet-bulb temperatures of 75 °F and 55 °F, respectively, through the tower. Saturated air exits the tower. Find (*a*) the temperature of the exiting air stream and (*b*) the mass flux of the makeup water.

12.66 A cooling tower cools water from 35 °C to 27 °C. The tower receives 200 m^3/s of atmospheric air at 30 °C and 40% relative humidity. The air exits the tower at 33 °C and 95% relative humidity. Estimate (*a*) the mass flux of water that is cooled and (*b*) the mass flux of makeup water.

Review Questions for the FE Examination

12.1FE The air in a conference room is to be conditioned from its present state of 18 °C and 40% humidity. Select the appropriate conditioning strategy.
(A) Heat and dehumidify.
(B) Cool, dehumidify, and then heat.
(C) Heat, dehumidify, and then heat.
(D) Heat and humidify.

12.2FE For which of the following situations would you expect condensation not to occur?
(A) On a glass of ice water sitting on the kitchen table.
(B) On the grass on a cool evening in Tucson, Arizona.
(C) On your glasses when you come into a warm room on a cold winter day.
(D) On the inside of a window pane in an apartment that has a high infiltration rate on a cold winter day.

12.3FE Which temperature is most different from the outside temperature as measured with a conventional thermometer?
(A) Wet-bulb temperature
(B) Dry-bulb temperature
(C) Dew-point temperature
(D) Ambient temperature

12.4FE An evaporative cooler inlets air at 40 °C and 20% humidity. If the exiting air is at 80% humidity, the exiting temperature is nearest:
(A) 29 °C
(B) 27 °C
(C) 25 °C
(D) 23 °C

12.5FE The dry-bulb temperature is 35 °C and the wet-bulb temperature is 27 °C. The relative humidity is nearest:
(A) 65%
(B) 60%
(C) 55%
(D) 50%

12.6FE Estimate the liters of water in the air in a 3 m × 10 m × 20 m room if the temperature is 20 °C and the humidity is 70%.
(A) 7.3 L
(B) 6.2 L
(C) 5.1 L
(D) 4.0 L

12.7FE Air at 5 °C and 80% humidity is heated to 25 °C in an enclosed room. The final humidity is nearest:
(A) 36%
(B) 32%
(C) 26%
(D) 22%

12.8FE Air at 30 °C and 80% humidity is to be conditioned to 20 °C and 40% humidity. How much heating is required?
(A) 30 kJ/kg
(B) 25 kJ/kg
(C) 20 kJ/kg
(D) 15 kJ/kg

12.9FE Air at 35 °C and 70% humidity in a 3 m × 10 m × 20 m classroom is cooled to 25 °C and 40% humidity. Estimate the amount of water removed.
(A) 4 kg
(B) 6 kg
(C) 11 kg
(D) 16 kg

12.10FE Outside air at 15 °C and 40% humidity is mixed with inside air at 32 °C and 70% humidity taken near the ceiling. Estimate the humidity of the mixed stream if the outside flow rate is 40 m^3/min and the inside flow rate is 20 m^3/min.
(A) 63%
(B) 58%
(C) 53%
(D) 49%

12.11FE The temperature of the mixed stream of air of Question 12.10FE is nearest:
(A) 28 °C
(B) 25 °C
(C) 23 °C
(D) 20 °C

Answers to Supplementary Problems

12.14 (*a*) 0.293, 0.28, 0.427, 0.249 kJ/kg·K (*b*) 0.209, 0.313, 0.478, 51.9 ft-lbf/lbm-°R (*c*) 0.532, 0.405, 0.063, 0.526 kJ/kg·K (*d*) 0.32, 0.386, 0.293, 0.333 kJ/kg·K (*e*) 0.232, 0.754, 0.014, 53.4 ft-lbf/lbm-°R

12.15 (*a*) 0.164, 0.386, 0.450, 0.256 kJ/kg·K (*b*) 0.235, 0.353, 0.412, 52.1 ft-lbf/lbm-°R (*c*) 0.0401, 0.0234, 0.9365, 2.22 kJ/kg·K (*d*) 0.671, 0.247, 0.0813, 0.387 kJ/kg·K (*e*) 0.189, 0.804, 0.0072, 53.6 ft-lbf/lbm-°R.

12.16 (*a*) 0.247, 0.431, 0.296, 0.026 (*b*) 23.78 (*c*) 0.350 kJ/kg·K

12.17 (*a*) 0.212, 0.221, 0.567 (*b*) 0.257 kJ/kg·K

12.18 (*a*) 24.59 kg (*b*) 62.3 lbm

12.19 (*a*) 2420 kPa (*b*) 1074 kPa

12.20 (*a*) 39.9 m^3 (*b*) 1202 ft^3

12.21 83.0 °C

12.22 9.23 m^3

12.23 (*a*) 89.9%, 10.1% (*b*) 0.993, 0.00697

12.24 (*a*) 4790 kJ (*b*) 5520 kJ

12.25 190 °F

12.26 82.3 kJ/kg

12.27 (*a*) 27.7 kJ (*b*) 88.6 J/K

12.28 (*a*) 15.5 Btu (*b*) 0.0152 Btu/°R

12.29 (*a*) −191 kJ (*b*) 225 kPa (*c*) − 0.410 kJ/K

12.30 30.8 °C

12.31 (*a*) − 111 Btu/lbm (*b*) − 116 Btu/lbm

12.32 3.03 kg/s

12.33 560 m/s, 28.1 kPa

12.34 11.2 cm

12.35 2130 ft/sec

12.36 484 °C

12.37 15.2 kW

12.38 189 °C

12.39 (*a*) 65.6 kW (*b*) 380 kW (*c*) 24.6 kW

12.40 (*a*) 0.01074 kg H_2O/kg dry air (*b*) 14.9 °C (*c*) 0.885 m^3/kg

12.41 (*a*) 63.3% (*b*) 75.5 °F (*c*) 14.8 ft^3/lbm (*d*) 43.6 Btu/lbm dry air

12.42 (*a*) 0.0102 kg H_2O/kg dry air (*b*) 621 kg (*c*) 6.22 kg (*d*) 31.4 MJ

12.43 (*a*) 3.29 m^3 (*b*) 14.82 kPa

12.44 (*a*) 78.4% (*b*) 0.0461 (*c*) 101.7 °F (*d*) 15.4 ft^3/lbm (*e*) 77.5 Btu/lbm dry air

12.45 24.2 °C

12.46 22.9%

12.47 49%

12.48 34.1%

12.49 (*a*) 0.01074 (*b*) 40.2% (*c*) 57.5 kJ/kg dry air

12.50 (*a*) 17.9 °C, 82%, 16.9 °C, 50.5 kJ/kg (*b*) 25%, 0.0035, − 1 °C, 29 kJ/kg (*c*) 61 °F, 0.0095, 55.7 °F, 27 Btu/lbm (*d*) 66 °F, 0.0097, 56 °F, 26.5 Btu/lbm (*e*) 47.5 °C, 17%, 0.0107, 76 kJ/kg (*f*) 73.5 °F, 85%, 68.5 °F, 34 Btu/lbm

12.51 20%, 14 kW

12.52 88%

12.53 12%

12.54 0.458 kg/min, 19.33 kW

12.55 609 Btu/min, 0.514 lbm/min

12.56 (*a*) 19 kJ/kg dry air (*b*) 98 kJ/kg dry air

12.57 152 kW, 26.8 kW

12.58 (*a*) 0.91 kg/s (*b*) 20.9 kW

12.59 (*a*) 76 °F (*b*) 0.354 lbm/min

12.60 (*a*) 21.7 °C (*b*) 0.329 kg/min

12.61 45 °C, 30 kW, 0.314 kg/min

12.62 49%, 20.7 °C, 94.2 kg/min

12.63 (*a*) 228 ft^3/min (*b*) 47.5 °F (*c*) 367 Btu/min

12.64 (*a*) 37 m^3/s (*b*) 0.8 kg/s

12.65 (*a*) 73 °F (*b*) 0.78 lbm/sec

12.66 (*a*) 530 kg/s (*b*) 5.9 kg/s

Answers to Review Questions for the FE Examination

12.1FE (D) **12.2FE** (B) **12.3FE** (C) **12.4FE** (C) **12.5FE** (B) **12.6FE** (A) **12.7FE** (D) **12.8FE** (D) **12.9FE** (C) **12.10FE** (D) **12.11FE** (D)

Combustion

13.1 COMBUSTION EQUATIONS

Let us begin our review of this particular variety of chemical-reaction equations by considering the combustion of propane in a pure oxygen environment. The chemical reaction is represented by

$$C_3H_8 + 5O_2 \rightarrow 3CO_2 + 4H_2O \tag{13.1}$$

Note that the number of moles of the elements on the left-hand side may not equal the number of moles on the right-hand side. However, the number of atoms of an element must remain the same before, after, and during a chemical reaction; this demands that the mass of each element be conserved during combustion.

In writing the equation we have demonstrated some knowledge of the products of the reaction. Unless otherwise stated we will assume *complete combustion*: the products of the combustion of a hydrocarbon fuel will be H_2O and CO_2. *Incomplete combustion* results in products that contain H_2, CO, C, and/or OH.

For a simple chemical reaction, such as (*13.1*), we can immediately write down a balanced chemical equation. For more complex reactions the following systematic method proves useful:

1. Set the number of moles of fuel equal to 1.
2. Balance CO_2 with number of C from the fuel.
3. Balance H_2O with H from the fuel.
4. Balance O_2 from CO_2 and H_2O.

For the combustion of propane we assumed that the process occurred in a pure oxygen environment. Actually, such a combustion process would normally occur in air. For our purposes we assume that air consists of 21% O_2 and 79% N_2 by volume so that for each mole of O_2 in a reaction we will have

$$\frac{79}{21} = 3.76 \ \frac{\text{mol } N_2}{\text{mol } O_2} \tag{13.2}$$

Thus, on the (simplistic) assumption that N_2 will not undergo any chemical reaction, (*13.1*) is replaced by

$$C_3H_8 + 5(O_2 + 3.76N_2) \rightarrow 3CO_2 + 4H_2O + 18.8N_2 \tag{13.3}$$

The minimum amount of air that supplies sufficient O_2 for the complete combustion of the fuel is called *theoretical air* or *stoichiometric air*. When complete combustion is achieved with theoretical air, the products contain no O_2, as in the reaction of (*13.3*). In practice, it is found that if complete combustion is to occur, air must be supplied in an amount greater than theoretical air. This is due to the chemical kinetics and molecular activity of the reactants and products. Thus we often speak in terms of *percent theoretical air* or *percent excess air*, where

$$\%\ \text{theoretical air} = 100\% + \%\ \text{excess air} \tag{13.4}$$

Slightly insufficient air results in CO being formed; some hydrocarbons may result from larger deficiencies.

The parameter that relates the amount of air used in a combustion process is the *air-fuel ratio* (AF), which is the ratio of the mass of air to the mass of fuel. The reciprocal is the *fuel-air ratio* (FA). Thus

$$AF = \frac{m_{\text{air}}}{m_{\text{fuel}}} \qquad FA = \frac{m_{\text{fuel}}}{m_{\text{air}}} \tag{13.5}$$

Again, considering propane combustion with theoretical air as in (*13.3*), the air-fuel ratio is

$$AF = \frac{m_{\text{air}}}{m_{\text{fuel}}} = \frac{(5)(4.76)(29)}{(1)(44)} = 15.69\,\frac{\text{kg air}}{\text{kg fuel}} \tag{13.6}$$

where we have used the molecular weight of air as 29 kg/kmol and that of propane as 44 kg/kmol. If, for the combustion of propane, $AF > 15.69$, a *lean mixture* occurs; if $AF < 15.69$, a *rich mixture* results.

The combustion of hydrocarbon fuels involves H_2O in the products of combustion. The calculation of the dew point of the products is often of interest; it is the saturation temperature at the partial pressure of the water vapor. If the temperature drops below the dew point, the water vapor begins to condense. The condensate usually contains corrosive elements, and thus it is often important to ensure that the temperature of the products does not fall below the dew point.

EXAMPLE 13.1 Butane is burned with dry air at an air-fuel ratio of 20. Calculate (*a*) the percent excess air, (*b*) the volume percentage of CO_2 in the products, and (*c*) the dew-point temperature of the products.

Solution: The reaction equation for theoretical air is

$$C_4H_{10} + 6.5(O_2 + 3.76N_2) \rightarrow 4CO_2 + 5H_2O + 24.44N_2$$

(*a*) The air-fuel ratio for theoretical air is

$$AF_{\text{th}} = \frac{m_{\text{air}}}{m_{\text{fuel}}} = \frac{(6.5)(4.76)(29)}{(1)(58)} = 15.47\,\frac{\text{kg air}}{\text{kg fuel}}$$

This represents 100% theoretical air. The actual air-fuel ratio is 20. The excess air is then

$$\%\ \text{excess air} = \left(\frac{AF_{\text{act}} - AF_{\text{th}}}{AF_{\text{th}}}\right)(100\%) = \frac{20 - 15.47}{15.47}(100\%) = 29.28\%$$

(*b*) The reaction equation with 129.28% theoretical air is

$$C_4H_{10} + (6.5)(1.2928)(O_2 + 3.76N_2) \rightarrow 4CO_2 + 5H_2O + 1.903O_2 + 31.6N_2$$

The volume percentage is obtained using the total moles in the products of combustion. For CO_2 we have

$$\%\ CO_2 = \left(\frac{4}{42.5}\right)(100\%) = 9.41\%$$

(*c*) To find the dew-point temperature of the products we need the partial pressure of the water vapor. It is found using the mole fraction to be

$$P_v = y_{H_2O}P_{\text{atm}} = \left(\frac{5}{42.5}\right)(100) = 11.76\ \text{kPa}$$

where we have assumed an atmospheric pressure of 100 kPa. Using Table C-2 we find the dew-point temperature to be $T_{\text{d.p.}} = 49\,^\circ\text{C}$.

EXAMPLE 13.2 Butane is burned with 90% theoretical air. Calculate the volume percentage of CO in the products and the air-fuel ratio. Assume no hydrocarbons in the products.

Solution: For incomplete combustion we add CO to the products of combustion. Using the reaction equation from Example 13.1,

$$C_4H_{10} + (0.9)(6.5)(O_2 + 3.76N_2) \rightarrow aCO_2 + 5H_2O + 22N_2 + bCO$$

With atomic balances on the carbon and oxygen we find:

$$\left.\begin{array}{ll} \text{C:} & 4 = a + b \\ \text{O:} & 11.7 = 2a + 5 + b \end{array}\right\} \quad \therefore a = 2.7,\ b = 1.3$$

The volume percentage of CO is then

$$\%\,CO = \left(\frac{1.3}{31}\right)(100\%) = 4.19\%$$

The air-fuel ratio is

$$AF = \frac{m_{air}}{m_{fuel}} = \frac{(0.9)(6.5)(4.76)(29)}{(1)(58)} = 13.92\ \frac{\text{lbm air}}{\text{lbm fuel}}$$

EXAMPLE 13.3 Butane is burned with dry air, and volumetric analysis of the products on a dry basis (the water vapor is not measured) gives 11.0% CO_2, 1.0% CO, 3.5% O_2, and 84.5% N_2. Determine the percent theoretical air.

Solution: The problem is solved assuming that there is 100 moles of dry products. The chemical equation is

$$aC_4H_{10} + b(O_2 + 3.76N_2) \rightarrow CO_2 + 1CO + 3.5O_2 + 84.5N_2 + cH_2O$$

We perform the following balances:

$$\begin{array}{lll} \text{C:} & 4a = 11 + 1 & \therefore a = 3 \\ \text{H:} & 10a = 2c & \therefore c = 15 \\ \text{O:} & 2b = 22 + 1 + 7 + c & \therefore b = 22.5 \end{array}$$

A balance on the nitrogen allows a check: $3.76b = 84.5$, or $b = 22.47$. This is quite close, so the above values are acceptable. Dividing through the chemical equation by the value of a so that we have 1 mol fuel,

$$C_4H_{10} + 7.5(O_2 + 3.76N_2) \rightarrow 3.67CO_2 + 0.33CO + 1.17O_2 + 28.17N_2 + 5H_2O$$

Comparing this with the combustion equation of Example 13.1 using theoretical air, we find

$$\%\text{ theoretical air} = \left(\frac{7.5}{6.5}\right)(100\%) = 107.7\%$$

EXAMPLE 13.4 Volumetric analysis of the products of combustion of an unknown hydrocarbon, measured on a dry basis, gives 10.4% CO_2, 1.2% CO, 2.8% O_2, and 85.6% N_2. Determine the composition of the hydrocarbon and the percent theoretical air.

Solution: The chemical equation for 100 mol dry products is

$$C_aH_b + c(O_2 + 3.76N_2) \rightarrow 10.4CO_2 + 1.2CO + 2.8O_2 + 85.6N_2 + dH_2O$$

Balancing each element,

$$\begin{aligned}
&\text{C:} \quad a = 10.4 + 1.2 \qquad \therefore a = 11.6 \\
&\text{N:} \quad 3.76c = 85.6 \qquad \therefore c = 22.8 \\
&\text{O:} \quad 2c = 20.8 + 1.2 + 5.6 + d \qquad \therefore d = 18.9 \\
&\text{H:} \quad b = 2d \qquad \therefore b = 37.9
\end{aligned}$$

The chemical formula for the fuel is $C_{11.6}H_{37.9}$. This could represent a mixture of hydrocarbons, but it is not any species listed in Appendix B, since the ratio of hydrogen atoms to carbon atoms is $3.27 \simeq 13/4$.

To find the percent theoretical air we must have the chemical equation using 100% theoretical air:

$$C_{11.6}H_{37.9} + 21.08(O_2 + 3.76N_2) \rightarrow 11.6CO_2 + 18.95H_2O + 79.26N_2$$

Using the number of moles of air from the actual chemical equation, we find

$$\%\ \text{theoretical air} = \left(\frac{22.8}{21.08}\right)(100\%) = 108\%$$

13.2 ENTHALPY OF FORMATION, ENTHALPY OF COMBUSTION, AND THE FIRST LAW

When a chemical reaction occurs, there may be considerable change in the chemical composition of a system. The problem this creates is that for a control volume the mixture that exits is different from the mixture that enters. Since various tables use different zeros for the enthalpy, it is necessary to establish a standard reference state, which we shall choose as 25 °C (77 °F) and 1 atm and which shall be denoted by the superscript "°", for example, h°.

Consider the combustion of H_2 with O_2, resulting in H_2O:

$$H_2 + \tfrac{1}{2}\, O_2 \rightarrow H_2O(l) \tag{13.7}$$

If H_2 and O_2 enter a combustion chamber at 25 °C (77 °F) and 1 atm and $H_2O(l)$ leaves the chamber at 25 °C (77 °F) and 1 atm, the measured heat transfer will be $-285\,830$ kJ for each kmol of $H_2O(l)$ formed. [The symbol (l) after a chemical compound implies the liquid phase and (g) implies the gaseous phase. If no symbol is given, a gas is implied.] The negative sign on the heat transfer means energy has left the control volume, as shown schematically in Fig. 13-1.

The first law applied to a combustion process in a control volume is

$$Q = H_P - H_R \tag{13.8}$$

where H_P is the enthalpy of the *products of combustion* that leave the combustion chamber and H_R is the enthalpy of the *reactants* that enter. If the reactants are stable elements, as in our example in Fig. 13-1, and the process is at constant temperature and constant pressure, then the enthalpy change is called the

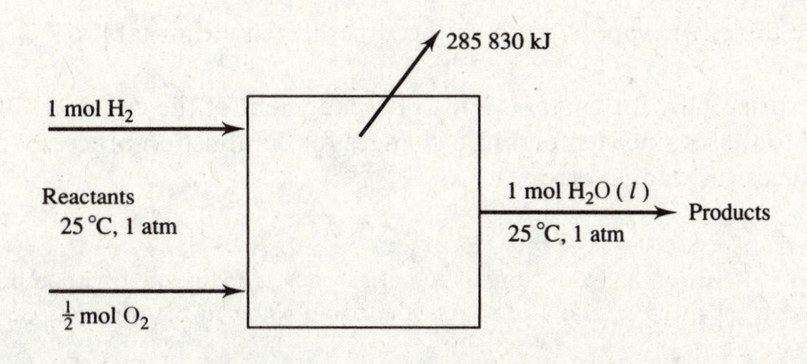

Fig. 13-1 The control volume used during combustion.

enthalpy of formation, denoted by $\bar{h}_f^\circ$. The enthalpies of formation of numerous compounds are listed in Table B-6. Note that some compounds have a positive $\bar{h}_f^\circ$, indicating that they require energy to form (an *endothermic reaction*); others have a negative $\bar{h}_f^\circ$, indicating that they give off energy when they are formed (an *exothermic reaction*).

The enthalpy of formation is the enthalpy change when a compound is formed. The enthalpy change when a compound undergoes complete combustion at constant temperature and pressure is called the *enthalpy of combustion.* For example, the enthalpy of formation of H_2 is zero, yet when 1 mol H_2 undergoes complete combustion to $H_2O(l)$, it gives off 285 830 kJ heat; the enthalpy of combustion of H_2 is 285 830 kJ/kmol. Values are listed for several compounds in Table B-7. If the products contain liquid water, the enthalpy of combustion is the *higher heating value* (HHV); if the products contain water vapor, the enthalpy of combustion is the *lower heating value*. The difference between the higher heating value and the lower heating value is the heat of vaporization $\bar{h}_{fg}$ at standard conditions.

For any reaction the first law, represented by *(13.8)*, can be applied to a control volume. If the reactants and products consist of several components, the first law is, neglecting kinetic and potential energy changes,

$$Q - W_S = \sum_{\text{prod}} N_i(\bar{h}_f^\circ + \bar{h} - \bar{h}^\circ)_i - \sum_{\text{react}} N_i(\bar{h}_f^\circ + \bar{h} - \bar{h}^\circ)_i \tag{13.9}$$

where N_i represents the number of moles of substance i. The work is often zero, but not in, for example, a combustion turbine.

If combustion occurs in a rigid chamber, for example, a bomb calorimeter, the first law is

$$Q = U_P - U_R = \sum_{\text{prod}} N_i(\bar{h}_f^\circ + \bar{h} - \bar{h}^\circ - Pv)_i - \sum_{\text{react}} N_i(\bar{h}_f^\circ + \bar{h} - \bar{h}^\circ - Pv)_i \tag{13.10}$$

where we have used enthalpy since the $\bar{h}_f^\circ$ values are tabulated. Since the volume of any liquid or solid is negligible compared to the volume of the gases, we write *(13.10)* as

$$Q = \sum_{\text{prod}} N_i(\bar{h}_f^\circ + \bar{h} - \bar{h}^\circ - R_u T)_i - \sum_{\text{react}} N_i(\bar{h}_f^\circ + \bar{h} - \bar{h}^\circ - R_u T)_i \tag{13.11}$$

If $N_{\text{prod}} = N_{\text{react}}$, the Q for the rigid volume is equal to Q for the control volume for the isothermal process.

In the above relations we employ one of the following methods to find $(\bar{h} - \bar{h}^\circ)$:

For a solid or liquid

Use $\bar{C}_p \Delta T$.

For gases

Method 1: Assume an ideal gas with constant specific heat so that $\bar{h} - \bar{h}^\circ = \bar{C}_p\, \Delta T$.

Method 2: Assume an ideal gas and use tabulated values for $\bar{h}$.

Method 3: Assume nonideal-gas behavior and use the generalized charts.

Method 4: Use tables for vapors, such as the superheated steam tables.

Which method to use (especially for gases) is left to the judgment of the engineer. In our examples we'll usually use method 2 for gases since temperature changes for combustion processes are often quite large and method 1 introduces substantial error.

EXAMPLE 13.5 Calculate the enthalpy of combustion of gaseous propane and of liquid propane assuming the reactants and products to be at 25 °C and 1 atm. Assume liquid water in the products exiting the steady-flow combustion chamber.

Solution: Assuming theoretical air (the use of excess air would not influence the result since the process is isothermal), the chemical equation is

$$C_3H_8 + 5(O_2 + 3.76N_2) \rightarrow 3CO_2 + 4H_2O(l) + 18.8N_2$$

where, for the HHV, a liquid is assumed for H_2O. The first law becomes, for the isothermal process $h = h^\circ$,

$$Q = H_P - H_R = \sum_{\text{prod}} N_i(\bar{h}_f^\circ)_i - \sum_{\text{react}} N_i(\bar{h}_f^\circ)_i$$

$$= (3)(-393\,520) + (4)(-285\,830) - (-103\,850) = -2\,220\,000 \text{ kJ/kmol fuel}$$

This is the enthalpy of combustion; it is stated with the negative sign. The sign is dropped for the HHV; for gaseous propane it is 2220 MJ for each kmol of fuel.

For liquid propane we find

$$Q = (3)(-393\,520) + (4)(-285\,830) - (-103\,850 - 15\,060) = -2\,205\,000 \text{ kJ/kmol fuel}$$

This is slightly less than the HHV for gaseous propane, because some energy is needed to vaporize the liquid fuel.

EXAMPLE 13.6 Calculate the heat transfer required if propane and air enter a steady-flow combustion chamber at 25 °C and 1 atm and the products leave at 600 K and 1 atm. Use theoretical air.

Solution: The combustion equation is written using H_2O in the vapor form due to the high exit temperature:

$$C_3H_8 + 5(O_2 + 3.76\,N_2) \rightarrow 3CO_2 + 4H_2O(g) + 18.8N_2$$

The first law takes the form [see (*13.9*)]

$$Q = \sum_{\text{prod}} N_i(\bar{h}_f^\circ + \bar{h} - \bar{h}^\circ)_i - \sum_{\text{react}} N_i(\bar{h}_f^\circ + \bar{h} - \bar{h}^\circ)_i$$

$$= (3)(-393\,520 + 22\,280 - 9360) + (4)(-241\,810 + 20\,400 - 9900)$$
$$+ (18.8)(17\,560 - 8670) - (-103\,850) = -1\,796\,000 \text{ kJ/kmol fuel}$$

where we have used method 2 listed for gases. This heat transfer is less than the enthalpy of combustion of propane, as it should be, since some energy is needed to heat the products to 600 K.

EXAMPLE 13.7 Liquid octane at 25 °C fuels a jet engine. Air at 600 K enters the insulated combustion chamber and the products leave at 1000 K. The pressure is assumed constant at 1 atm. Estimate the exit velocity using theoretical air.

Solution: The equation is $C_8H_{18}(l) + 12.5(O_2 + 3.76N_2) \rightarrow 8CO_2 + 9H_2O + 47N_2$. The first law, with $Q = W_s = 0$ and including the kinetic energy change (neglect V_{inlet}), is

$$0 = H_P - H_R + \frac{V^2}{2}M_P \quad \text{or} \quad V^2 = \frac{2}{M_P}(H_R - H_P)$$

where M_P is the mass of the products per kmol fuel. For the products,

$$H_P = (8)(-393\,520 + 42\,770 - 9360) + (9)(-241\,810 + 35\,880 - 9900)$$
$$+ (47)(30\,130 - 8670) = -3\,814\,700 \text{ kJ/kmol fuel}$$

For the reactants, $H_R = (-249\,910) + (12.5)(17\,930 - 8680) + (47)(17\,560 - 8670) = 283\,540$ kJ/kmol.

The mass of the products is $M_P = (8)(44) + (9)(18) + (47)(28) = 1830$ kg/kmol-fuel and so

$$V^2 = \frac{2}{1830}\left[(0.28354 + 3.8147)10^9\right] \qquad \therefore V = 2120 \text{ m/s}$$

EXAMPLE 13.8 Liquid octane is burned with 300% excess air. The octane and air enter the steady-flow combustion chamber at 25°C and 1 atm and the products exit at 1000 K and 1 atm. Estimate the heat transfer.

Solution: The reaction with theoretical air is $C_8H_{18} + 12.5(O_2 + 3.76N_2) \rightarrow 8CO_2 + 9H_2O + 47N_2$. For 300% excess air (400% theoretical air) the reaction is

$$C_8H_{18}(l) + 50(O_2 + 3.76N_2) \rightarrow 8CO_2 + 9H_2O + 37.5O_2 + 188N_2$$

The first law applied to the combustion chamber is

$$\begin{aligned} Q = H_P - H_R &= (8)(-393\,520 + 42\,770 - 9360) + (9)(-241\,810 + 35\,880 - 9900) \\ &\quad + (37.5)(31\,390 - 8680) + (188)(30\,130 - 8670) - (-249\,910) \\ &= 312\,500\ \text{kJ/kmol fuel} \end{aligned}$$

In this situation heat must be added to obtain the desired exit temperature.

EXAMPLE 13.9 A constant-volume bomb calorimeter is surrounded by water at 77°F. Liquid propane is burned with pure oxygen in the calorimeter, and the heat transfer is determined to be −874,000 Btu/lbmol. Calculate the enthalpy of formation and compare with that given in Table B-6.

Solution: The complete combustion of propane follows $C_3H_8 + 5O_2 \rightarrow 3CO_2 + 4H_2O(g)$. The surrounding water sustains a constant-temperature process, so that *(13.11)* becomes

$$Q = \sum_{\text{prod}} N_i(\bar{h}_f^\circ)_i - \sum_{\text{react}} N_i(\bar{h}_f^\circ)_i + (N_R - N_P)R_uT = -874{,}000$$

$$-874{,}000 = (3)(-169{,}300) + (4)(-104{,}040) - (\bar{h}_f^\circ)_{C_3H_8} + (6 - 7)(1.986)(537)$$

$$\therefore\ (\bar{h}_f^\circ)_{C_3H_8} = -51{,}130\ \text{Btu/lbmol}$$

This compares with $\bar{h}_f^\circ$ from the Table B-6 of $(-44{,}680 - 6480) = -51{,}160$ Btu/lbmol.

13.3 ADIABATIC FLAME TEMPERATURE

If we consider a combustion process that takes place adiabatically, with no work or changes in kinetic and potential energy, then the temperature of the products is referred to as the *adiabatic flame temperature*. We find that the maximum adiabatic flame temperature that can be achieved occurs at theoretical air. This fact allows us to control the adiabatic flame temperature by the amount of excess air involved in the process: The greater the amount of excess air the lower the adiabatic flame temperature. If the blades in a turbine can withstand a certain maximum temperature, we can determine the excess air needed so that the maximum allowable blade temperature is not exceeded. We will find that an iterative (trial-and-error) procedure is needed to find the adiabatic flame temperature. A quick approximation to the adiabatic flame temperature is found by assuming the products to be completely N_2. An example will illustrate.

The adiabatic flame temperature is calculated assuming complete combustion, no heat transfer from the combustion chamber, and no dissociation of the products into other chemical species. Each of these effects tends to reduce the adiabatic flame temperature. Consequently, the adiabatic flame temperature that we will calculate represents the maximum possible flame temperature for the specified percentage of theoretical air.

If a significant amount of heat transfer does occur, we can account for it by including the following term in the energy equation:

$$\dot{Q} = UA(T_P - T_E) \qquad (13.12)$$

where U = overall heat-transfer coefficient (specified),
T_E = temperature of environment,
T_P = temperature of products,
A = surface area of combustion chamber.
[Note that the units on U are kW/m^2·K or Btu/sec-ft^2-°R.]

EXAMPLE 13.10 Propane is burned with 250% theoretical air; both are at 25°C and 1 atm. Predict the adiabatic flame temperature in the steady-flow combustion chamber.

Solution: The combustion with theoretical air is $C_3H_8 + 5(O_2 + 3.76N_2) \rightarrow 3CO_2 + 4H_2O + 18.8N_2$. For 250% theoretical air we have

$$C_3H_8 + 12.5(O_2 + 3.76N_2) \rightarrow 3CO_2 + 4H_2O + 7.5O_2 + 47N_2$$

Since $Q = 0$ for an adiabatic process we demand that $H_R = H_P$. The enthalpy of the reactants, at 25°C, is $H_R = -103\,850$ kJ/kmol fuel.

The temperature of the products is the unknown; and we cannot obtain the enthalpies of the components of the products from the tables without knowing the temperatures. This requires a trial-and-error solution. To obtain an initial guess, we assume the products to be composed entirely of nitrogen:

$$H_R = H_P = -103\,850 = (3)(-393\,520) + (4)(-241\,820) + (61.5)(\bar{h}_P - 8670)$$

where we have noted that the products contain 61.5 mol of gas. This gives $\bar{h}_P = 43\,400$ kJ/kmol, which suggests a temperature of about 1380 K (take T_P a little less than that predicted by the all-nitrogen assumption). Using this temperature we check using the actual products:

$$\begin{aligned} -103\,850 \stackrel{?}{=} &(3)(-393\,520 + 64\,120 - 9360) + (4)(-241\,820 + 52\,430 - 9900) \\ &+ (7.5)(44\,920 - 8680) + (47)(42\,920 - 8670) = 68\,110 \end{aligned}$$

The temperature is obviously too high. We select a lower value, $T_P = 1300$ K. There results:

$$\begin{aligned} -103\,850 \stackrel{?}{=} &(3)(-393\,520 + 59\,520 - 9360) + (4)(-241\,820 + 48\,810 - 9900) \\ &+ (7.5)(44\,030 - 8680) + (47)(40\,170 - 8670) = -96\,100 \end{aligned}$$

We use the above two results for 1380 K and 1300 K and, assuming a linear relationship, predict that T_P is

$$T_P = 1300 - \left[\frac{103\,850 - 96\,100}{68\,110 - (-96\,100)}\right](1380 - 1300) = 1296\text{ K}$$

EXAMPLE 13.11 Propane is burned with theoretical air; both are at 25°C and 1 atm in a steady-flow combustion chamber. Predict the adiabatic flame temperature.

Solution: The combustion with theoretical air is $C_3H_8 + 5(O_2 + 3.76N_2) \rightarrow 3CO_2 + 4H_2O + 18.8N_2$. For the adiabatic process the first law takes the form $H_R = H_P$. Hence, assuming the products to be composed entirely of nitrogen,

$$-103\,850 = (3)(-393\,520) + (4)(-241\,820) + (25.8)(\bar{h}_P - 8670)$$

where the products contain 25.8 mol gas. This gives $\bar{h}_P = 87\,900$ kJ/kmol, which suggests a temperature of about 2600 K. With this temperature we find, using the actual products:

$$\begin{aligned} -103\,850 \stackrel{?}{=} &(3)(-393\,520 + 137\,400 - 9360) + (4)(-241\,820 + 114\,300 - 9900) \\ &+ (18.8)(86\,600 - 8670) = 119\,000 \end{aligned}$$

At 2400 K there results:

$$-103\,850 \stackrel{?}{=} (3)(-393\,520 + 125\,200 - 9360) + (4)(-241\,820 + 103\,500 - 9900) + (18.8)(79\,320 - 8670) = -97\,700$$

A straight line extrapolation gives $T_P = 2394$ K.

EXAMPLE 13.12 The overall heat-transfer coefficient of a steady-flow combustion chamber with a 2-m^2 surface area is determined to be 0.5 kW/m^2·K. Propane is burned with theoretical air, both at 25 °C and 1 atm. Predict the temperature of the products of combustion if the propane mass flow rate is 0.2 kg/s.

Solution: The molar influx is $\dot{m}_{\text{fuel}} = 0.2/44 = 0.004545$ kmol/s, where the molecular weight of propane, 44 kg/kmol, is used. Referring to the chemical reaction given in Example 13.11, the mole fluxes of the products are given by:

$$\dot{M}_{CO_2} = (3)(0.004545) = 0.01364 \text{ kmol/s} \qquad \dot{M}_{H_2O} = (4)(0.004545) = 0.02273 \text{ kmol/s}$$
$$\dot{M}_{N_2} = (18.8)(0.004545) = 0.1068 \text{ kmol/s}$$

We can write the energy equation (the first law) as

$$\dot{Q} + \dot{H}_R = \dot{H}_P$$

Using (*13.12*), the energy equation becomes

$$-(0.5)(2)(T_P - 298) + (0.004545)(-103\,850) = (0.01364)(-393\,520 + \bar{h}_{CO_2} - 9360) + (0.02273)(-241\,820 + \bar{h}_{H_2O} - 9900) + (0.1068)(\bar{h}_{N_2} - 8670)$$

For a first guess at T_P let us assume a somewhat lower temperature than that of Example 13.11, since energy is leaving the combustion chamber. The guesses follow:

$$T_P = 1600 \text{ K:} \quad -1774 \stackrel{?}{=} -4446 - 4295 + 4475 = -4266$$
$$T_P = 2000 \text{ K:} \quad -2174 \stackrel{?}{=} -4120 - 3844 + 5996 = -1968$$
$$T_P = 1900 \text{ K:} \quad -2074 \stackrel{?}{=} -4202 - 3960 + 5612 = -2550$$

Interpolation between the last two entries gives $T_P = 1970$ K. Checking,

$$T_P = 1970 \text{ K:} \quad -2144 \stackrel{?}{=} -4145 - 3879 + 5881 = -2143$$

Hence, $T_P = 1970$ K. If we desire the temperature of the products to be less than this, we can increase the overall heat-transfer coefficient or add excess air.

Solved Problems

13.1 Ethane (C_2H_6) is burned with dry air which contains 5 mol O_2 for each mole of fuel. Calculate (*a*) the percent of excess air, (*b*) the air-fuel ratio, and (*c*) the dew-point temperature.

The stoichiometric equation is $C_2H_6 + 3.5(O_2 + 3.76N_2) \rightarrow 2CO_2 + 3H_2O + 6.58N_2$. The required combustion equation is

$$C_2H_6 + 5(O_2 + 3.76N_2) \rightarrow 2CO_2 + 3H_2O + 1.5O_2 + 18.8N_2$$

(*a*) There is excess air since the actual reaction uses 5 mol O_2 rather than 3.5 mol. The percent of excess air is

$$\% \text{ excess air} = \left(\frac{5 - 3.5}{3.5}\right)(100\%) = 42.9\%$$

(*b*) The air-fuel ratio is a mass ratio. Mass is found by multiplying the number of moles by the molecular weight:

$$AF = \frac{(5)(4.76)(29)}{(1)(30)} = 23.0 \text{ kg air/kg fuel}$$

(*c*) The dew-point temperature is found using the partial pressure of the water vapor in the combustion products. Assuming atmospheric pressure of 100 kPa, we find

$$P_v = y_{H_2O} P_{atm} = \left(\frac{3}{25.3}\right)(100) = 1.86 \text{ kPa}$$

Using Table C-2, we interpolate and find $T_{d.p.} = 49\,°C$.

13.2 A fuel mixture of 60% methane, 30% ethane, and 10% propane by volume is burned with stoichiometric air. Calculate the volume flow rate of air required if the fuel mass is 12 lbm/hr assuming the air to be at 70 °F and 14.7 psia.

The reaction equation, assuming 1 mol fuel, is

$$0.6CH_4 + 0.3C_2H_6 + 0.1C_3H_8 + a(O_2 + 3.76N_2) \rightarrow bCO_2 + cH_2O + dN_2$$

We find a, b, c, and d by balancing the various elements as follows:

$$\begin{aligned} &\text{C:} \quad 0.6 + 0.6 + 0.3 = b && \therefore b = 1.5 \\ &\text{H:} \quad 2.4 + 1.8 + 0.8 = 2c && \therefore c = 2.5 \\ &\text{O:} \quad 2a = 2b + c && \therefore a = 2.75 \\ &\text{N:} \quad (2)(3.76a) = 2d && \therefore d = 10.34 \end{aligned}$$

The air-fuel ratio is

$$AF = \frac{(2.75)(4.76)(29)}{(0.6)(16) + (0.3)(30) + (0.1)(44)} = \frac{379.6}{23} = 16.5 \frac{\text{lbm air}}{\text{lbm fuel}}$$

and $\dot{m}_{air} = (AF)\dot{m}_{fuel} = (16.5)(12) = 198$ lbm/h. To find the volume flow rate we need the air density. It is

$$\rho_{air} = \frac{P}{RT} = \frac{(14.7)(144)}{(53.3)(530)} = 0.0749 \text{ lbm/ft}^3$$

whence

$$AV = \frac{\dot{m}}{\rho_{air}} = \frac{198/60}{0.0749} = 44.1 \text{ ft}^3\text{/min}$$

(The volume flow rate is usually given in ft^3/min (cfm).)

13.3 Butane (C_4H_{10}) is burned with 20 °C air at 70% relative humidity. The air-fuel ratio is 20. Calculate the dew-point temperature of the products. Compare with Example 13.1.

The reaction equation using dry air (the water vapor in the air does not react, but simply tags along; it will be included later) is

$$C_4H_{10} + a(O_2 + 3.76N_2) \rightarrow 4CO_2 + 5H_2O + bO_2 + cN_2$$

The air-fuel ratio of 20 allows us to calculate the constant a, using $M_{fuel} = 58$ kg/kmol, as follows:

$$AF = \frac{m_{dry\ air}}{m_{fuel}} = \frac{(a)(4.76)(29)}{(1)(58)} = 20 \qquad \therefore a = 8.403$$

We also find that $b = 1.903$ and $c = 31.6$. The partial pressure of the moisture in the 20 °C air is

$$P_v = \phi P_g = (0.7)(2.338) = 1.637 \text{ kPa}$$

The ratio of the partial pressure to the total pressure (100 kPa) equals the mole ratio, so that

$$N_v = N\frac{P_v}{P} = (8.403 \times 4.76 + N_v)\left(\frac{1.637}{100}\right) \qquad \text{or} \qquad N_v = 0.666 \text{ kmol/kmol fuel}$$

We simply add N_v to each side of the reaction equation:

$$C_4H_{10} + 8.403(O_2 + 3.76N_2) + 0.666H_2O \rightarrow 4CO_2 + 5.666H_2O + 1.903O_2 + 31.6N_2$$

The partial pressure of water vapor in the products is $P_v = Py_{H_2O} = (100)(5.666/43.17) = 13.1$ kPa. From Table C-2 we find the dew-point temperature to be $T_{d.p.} = 51\,°C$, which compares with 49 °C using dry air as in Example 13.1. Obviously the moisture in the combustion air does not significantly influence the products. Consequently, we usually neglect the moisture.

13.4 Methane is burned with dry air, and volumetric analysis of the products on a dry basis gives 10% CO_2, 1% CO, 1.8% O_2, and 87.2% N_2. Calculate (*a*) the air-fuel ratio, (*b*) the percent excess air, and (*c*) the percentage of water vapor that condenses if the products are cooled to 30 °C.

Assume 100 mol dry products. The reaction equation is

$$aCH_4 + b(O_2 + 3.76N_2) \rightarrow 10CO_2 + CO + 1.8O_2 + 87.2N_2 + cH_2O$$

A balance on the atomic masses provides the following:

$$\text{C:} \quad a = 10 + 1 \qquad \therefore a = 11$$
$$\text{H:} \quad 4a = 2c \qquad \therefore c = 22$$
$$\text{O:} \quad 2b = 20 + 1 + 3.6 + c \qquad \therefore b = 23.3$$

Dividing the reaction equation by a so that we have 1 mol fuel:

$$CH_4 + 2.12(O_2 + 3.76N_2) \rightarrow 0.909CO_2 + 0.091CO + 0.164O_2 + 7.93N_2 + 2H_2O$$

(*a*) The air-fuel ratio is calculated from the reaction equation to be

$$AF = \frac{m_{air}}{m_{fuel}} = \frac{(2.12)(4.76)(29)}{(1)(16)} = 18.29 \text{ kg air/kg fuel}$$

(*b*) The stoichiometric reaction is $CH_4 + 2(O_2 + 3.76N_2) \rightarrow CO_2 + 2H_2O + 7.52N_2$. This gives the excess air as

$$\% \text{ excess air} = \left(\frac{2.12 - 2}{2}\right)(100\%) = 6\%$$

(*c*) There are 2 mol water vapor in the combustion products before condensation. If N_w represents moles of water vapor that condense when the products reach 30 °C, then $2 - N_w$ is the number of water vapor moles and $11.09 - N_w$ is the total number of moles in the combustion products at 30 °C. We find N_w as follows:

$$\frac{N_v}{N} = \frac{P_v}{P} \qquad \frac{2 - N_w}{11.09 - N_w} = \frac{4.246}{100} \qquad \therefore N_w = 1.597 \text{ mol } H_2O$$

The percentage of water vapor that condenses out is

$$\% \text{ condensate} = \left(\frac{1.597}{2}\right)(100) = 79.8\%$$

13.5 An unknown hydrocarbon fuel combusts with dry air; the resulting products have the following dry volumetric analysis: 12% CO_2, 15% CO, 3% O_2, and 83.5% N_2. Calculate the percent excess air.

The reaction equation for 100 mol dry products is

$$C_aH_b + c(O_2 + 3.76N_2) \rightarrow 12CO_2 + 1.5CO + 3O_2 + 83.5N_2 + dH_2O$$

A balance on each element provides the following:

$$\begin{array}{lll} \text{C:} & a = 12 + 1.5 & \therefore a = 13.5 \\ \text{N:} & 3.76c = 83.5 & \therefore c = 22.2 \\ \text{O:} & 2c = 24 + 1.5 + 6 + d & \therefore d = 12.9 \\ \text{H:} & b = 2d & \therefore b = 25.8 \end{array}$$

The fuel mixture is represented by $C_{13.5}H_{25.8}$. For theoretical air with this fuel, we have

$$C_{13.5}H_{25.8} + 19.95(O_2 + 3.76N_2) \rightarrow 13.5CO_2 + 12.9H_2O + 75.0N_2$$

Comparing this with the actual equation above, we find

$$\%\text{ excess air} = \left(\frac{22.2 - 19.95}{19.95}\right)(100\%) = 11.3\%$$

13.6 Carbon reacts with oxygen to form carbon dioxide in a steady-flow chamber. Calculate the energy involved and state the type of reaction. Assume the reactants and products are at 25 °C and 1 atm.

The reaction equation is $C + O_2 \rightarrow CO_2$. The first law and Table B-6 give

$$Q = H_P - H_R = \sum_{\text{prod}} N_i(\bar{h}_f^\circ)_i - \sum_{\text{react}} N_i(\bar{h}_f^\circ)_i$$
$$= (1)(-393\,520) - 0 - 0 = -393\,520 \text{ kJ/kmol}$$

The reaction is exothermic (negative Q).

13.7 Methane enters a steady-flow combustion chamber at 77 °F and 1 atm with 80% excess air which is at 800 °R and 1 atm. Calculate the heat transfer if the products leave at 1600 °R and 1 atm.

The reaction equation with 180% theoretical air and with the water in vapor form is

$$CH_4 + 3.6(O_2 + 3.76N_2) \rightarrow CO_2 + 2H_2O(g) + 1.6O_2 + 13.54N_2$$

The first law, with zero work, provides the heat transfer:

$$Q = \sum_{\text{prod}} N_i(\bar{h}_f^\circ + \bar{h} - \bar{h}^\circ)_i - \sum_{\text{react}} N_i(\bar{h}_f^\circ + \bar{h} - \bar{h}^\circ)_i$$
$$= (1)(-169{,}300 + 15{,}829 - 4030) + (2)(-104{,}040 + 13{,}494 - 4258) + (1.6)(11{,}832 - 3725)$$
$$+ (13.54)(11{,}410 - 3730) - (-32{,}210) - (3.6)(5602 - 3725) - (13.54)(5564 - 3730)$$
$$= -229{,}500 \text{ Btu/lbmol fuel}$$

13.8 Ethane at 25 °C is burned in a steady-flow combustion chamber with 20% excess air at 127 °C, but only 95% of the carbon is converted to CO_2. If the products leave at 1200 K, calculate the heat transfer. The pressure remains constant at 1 atm.

The stoichiometric reaction equation is

$$C_2H_6 + 3.5(O_2 + 3.76N_2) \rightarrow 2CO_2 + 3H_2O + 11.28N_2$$

With 120% theoretical air and the product CO, the reaction equation becomes

$$C_2H_6 + 4.2(O_2 + 3.76N_2) \rightarrow 1.9CO_2 + 0.1CO + 3H_2O + 0.75O_2 + 11.28N_2$$

The first law with zero work is $Q = H_P - H_R$. The enthalpy of the products is [see (*13.9*)]

$$H_P = (1.9)(-393\,520 + 53\,850 - 9360) + (0.1)(-110\,530 + 37\,100 - 8670)$$
$$+ (3)(-241\,820 + 44\,380 - 9900) + (0.75)(38\,450 - 8680) + (11.28)(36\,780 - 8670)$$
$$= -1\,049\,000 \text{ kJ/kmol fuel}$$

The enthalpy of the reactants is

$$H_R = -84\,680 + (4.2)(11\,710 - 8680) + (15.79)(11\,640 - 8670) = -25\,060 \text{ kJ/kmol fuel}$$

Then $Q = -1\,049\,000 - (-25\,060) = -1\,024\,000$ kJ/kmol fuel.

13.9 A rigid volume contains 0.2 lbm of propane gas and 0.8 lbm of oxygen at 77 °F and 30 psia. The propane burns completely, and the final temperature, after a period of time, is observed to be 1600 °R. Calculate (*a*) the final pressure and (*b*) the heat transfer.

The moles of propane and oxygen are $N_{\text{propane}} = 0.2/44 = 0.004545$ lbmol and $N_{\text{oxygen}} = 0.8/32 = 0.025$ lbmol. For each mole of propane there is $0.025/0.004545 = 5.5$ mol O_2. The reaction equation for complete combustion is then

$$C_3H_8 + 5.5O_2 \rightarrow 3CO_2 + 4H_2O(g) + 0.5O_2$$

(*a*) We use the ideal-gas law to predict the final pressure. Since the volume remains constant, we have

$$V = \frac{N_1 R_u T_1}{P_1} = \frac{N_2 R_u T_2}{P_2} \qquad \frac{(6.5)(537)}{30} = \frac{(7.5)(1600)}{P_2} \qquad \therefore P_2 = 103.1 \text{ psia}$$

(*b*) By (*13.11*), with $R_u = 1.986$ Btu/lbmol-°R, we have for each mole of propane:

$$\begin{aligned} Q &= \sum_{\text{prod}} N_i(\bar{h}_f^\circ + \bar{h} - \bar{h}^\circ - R_u T)_i - \sum_{\text{react}} N_i(\bar{h}_f^\circ + \bar{h} - \bar{h}^\circ - R_u T)_i \\ &= (3)[-169{,}300 + 15{,}830 - 4030 - (1.986)(1600)] \\ &\quad + (4)[-104{,}040 + 13{,}490 - 4260 - (1.986)(1600)] \\ &\quad + (0.5)[11{,}830 - 3720 - (1.986)(1600)] \\ &\quad - (1)[-44{,}680 - (1.986)(537)] - (5.5)[(-1.986)(537)] \\ &= -819{,}900 \text{ Btu/lbmol fuel} \end{aligned}$$

Thus $Q = (-819{,}900)(0.004545) = 3730$ Btu.

13.10 Propane is burned in a steady-flow combustion chamber with 80% theoretical air, both at 25 °C and 1 atm. Estimate the adiabatic flame temperature and compare with that of Examples 13.10 and 13.11.

Using the stoichiometric reaction equation of Example 13.11 and assuming production of CO, the combustion with 80% theoretical air follows

$$C_3H_8 + 4(O_2 + 3.76N_2) \rightarrow CO_2 + 4H_2O + 2CO + 15.04N_2$$

A mass balance of the elements is required to obtain this equation. For an adiabatic process, the first law takes the form $H_R = H_P$, where H_R for propane is $-103\,850$ kJ/kmol. Assuming the temperature close to but less than that of Example 13.11, we try $T_P = 2200$ K:

$$\begin{aligned} -103\,850 &\stackrel{?}{=} (-393\,520 + 112\,940 - 9360) + (4)(-241\,820 + 92\,940 - 9900) \\ &\quad + (2)(-110\,530 + 72\,690 - 8670) + (15.04)(72\,040 - 8670) = -65\,000 \end{aligned}$$

At 2100 K:

$$\begin{aligned} -103\,850 &\stackrel{?}{=} (-393\,520 + 106\,860 - 9360) + (4)(-241\,820 + 87\,740 - 9900) \\ &\quad + (2)(-110\,530 + 69\,040 - 8670) + (15.04)(68\,420 - 8670) = -153\,200 \end{aligned}$$

A straight-line interpolation provides the adiabatic flame temperature $T_P = 2156$ K. Note that this temperature is less than that of the stoichiometric reaction of Example 13.11, as was the temperature for Example 13.10 where excess air was used. The stoichiometric reaction provides the maximum adiabatic flame temperature.

13.11 An insulated, rigid 0.7-m³ tank contains 0.05 kg of ethane and 100% theoretical air at 25 °C. The fuel is ignited and complete combustion occurs. Estimate (*a*) the final temperature and (*b*) the final pressure.

With 100% theoretical air, $C_2H_6 + 3.5(O_2 + 3.76N_2) \rightarrow 2CO_2 + 3H_2O + 13.16N_2$.

(*a*) The first law, with $Q = W = 0$, is written for this constant-volume process using (*13.11*):

$$\sum_{\text{react}} N_i(\bar{h}_f^\circ + \bar{h} - \bar{h}^\circ - R_uT)_i = \sum_{\text{prod}} N_i(\bar{h}_f^\circ + \bar{h} - \bar{h}^\circ - R_uT)_i$$

The reactants are at 25 °C (the initial pressure is unimportant if not extremely large) and the products are at T_P; therefore,

$$\text{L.H.S.} = (1)[-84\,680 - (8.314)(298)] + (3.5)[(-8.314)(298)] + (13.16)[(-8.314)(298)]$$
$$\text{R.H.S.} = (2)[-393\,520 + \bar{h}_{CO_2} - 9360 - 8.314T_P]$$
$$+ (3)[(-241\,820 + \bar{h}_{H_2O} - 9900 - 8.314T_P) + (13.16)(\bar{h}_{N_2} - 8670 - 8.314T_P)]$$

or

$$1\,579\,000 = 2\bar{h}_{CO_2} + 3\bar{h}_{H_2O} + 13.16\bar{h}_{N_2} - 151T_P$$

We solve for T_P by trial and error:

$$T_P = 2600\text{ K}: 1\,579\,000 \stackrel{?}{=} (2)(137\,400) + (3)(114\,300) + (13.16)(86\,850) - (151)(2600) = 1\,365\,000$$

$$T_P = 2800\text{ K}: 1\,579\,000 \stackrel{?}{=} (2)(149\,810) + (3)(125\,200) + (13.16)(94\,010) - (151)(2800) = 1\,490\,000$$

$$T_P = 3000\text{ K}: 1\,579\,000 \stackrel{?}{=} (2)(162\,230) + (3)(136\,260) + (13.16)(101\,410) - (151)(3000) = 1\,615\,000$$

Interpolation provides a temperature between 2800 K and 3000 K: $T_P = 2942\text{K}$.

(*b*) We have $N_{\text{fuel}} = 0.05/30 = 0.001667$ kmol; therefore, $N_{\text{prod}} = (18.16)(0.001667) = 0.03027$ kmol. The pressure in the products is then

$$P_{\text{prod}} = \frac{N_{\text{prod}}R_uT_{\text{prod}}}{V} = \frac{(0.03027)(8.314)(2942)}{0.7} = 1058 \text{ kPa}$$

Supplementary Problems

13.12 The following fuels combine with stoichiometric air: (*a*) C_2H_4, (*b*) C_3H_6, (*c*) C_4H_{10}, (*d*) C_5H_{12}, (*e*) C_8H_{18}, and (*f*) CH_3OH. Provide the correct values for x, y, z in the reaction equation

$$C_aH_b + w(O_2 + 3.76N_2) \rightarrow xCO_2 + yH_2O + zN_2$$

13.13 Methane (CH_4) is burned with stoichiometric air and the products are cooled to 20 °C assuming complete combustion at 100 kPa. Calculate (*a*) the air-fuel ratio, (*b*) the percentage of CO_2 by weight of the products, (*c*) the dew-point temperature of the products, and (*d*) the percentage of water vapor condensed.

13.14 Repeat Prob. 13.13 for ethane (C_2H_6).

13.15 Repeat Prob. 13.13 for propane (C_3H_8).

13.16 Repeat Prob. 13.13 for butane (C_4H_{10}).

13.17 Repeat Prob. 13.13 for octane (C_4H_{18}).

13.18 Ethane (C_2H_6) undergoes complete combustion at 95 kPa with 180% theoretical air. Find (*a*) the air-fuel ratio, (*b*) the percentage of CO_2 by volume in the products, and (*c*) the dew-point temperature.

13.19 Repeat Prob. 13.18 for propane (C_3H_8).

13.20 Repeat Prob. 13.18 for butane (C_4H_{10}).

13.21 Repeat Prob. 13.18 for octane (C_5H_{18}).

13.22 Calculate the mass flux of fuel required if the inlet air flow rate is 20 m^3/min at 20 °C and 100 kPa using stoichiometric air with (*a*) methane (CH_4), (*b*) ethane (C_2H_6), (*c*) propane (C_3H_8), (*d*) butane (C_4H_{10}), and (*e*) octane (C_5H_{18}).

13.23 Propane (C_3H_8) undergoes complete combustion at 90 kPa and 20 °C with 130% theoretical air. Calculate the air-fuel ratio and the dew-point temperature if the relative humidity of the combustion air is (*a*) 90%, (*b*) 80%, (*c*) 60%, and (*d*) 40%.

13.24 An air-fuel ratio of 25 is used in an engine that burns octane (C_8H_{18}). Find the percentage of excess air required and the percentage of CO_2 by volume in the products.

13.25 Butane (C_4H_{10}) is burned with 50% excess air. If 5% of the carbon in the fuel is converted to CO, calculate the air-fuel ratio and the dew-point of the products. Combustion takes place at 100 kPa.

13.26 A fuel which is 60% ethane and 40% octane by volume undergoes complete combustion with 200% theoretical air. Find (*a*) the air-fuel ratio, (*b*) the percent by volume of N_2 in the products, and (*c*) the dew-point temperature of the products if the pressure is 98 kPa.

13.27 One lbm of butane, 2 lbm of methane, and 2 lbm of octane undergo complete combustion with 20 lbm of air. Calculate (*a*) the air-fuel ratio, (*b*) the percent excess air, and (*c*) the dew-point temperature of the products if the combustion process occurs at 14.7 psia.

13.28 Each minute 1 kg of methane, 2 kg of butane, and 2 kg of octane undergo complete combustion with stoichiometric 20 °C air. Calculate the flow rate of air required if the process takes place at 100 kPa.

13.29 A volumetric analysis of the products of butane (C_4H_{10}) on a dry basis yields 7.6% CO_2, 8.2% O_2, 82.8% N_2, and 1.4% CO. What percent excess air was used?

13.30 A volumetric analysis of the products of combustion of octane (C_8H_{18}) on a dry basis yields 9.1% CO_2, 7.0% O_2, 83.0% N_2, and 0.9% CO. Calculate the air-fuel ratio.

13.31 Three moles of a mixture of hydrocarbon fuels, denoted by C_xH_y, is burned and a volumetric analysis on a dry basis of the products yields 10% CO_2, 8% O_2, 1.2% CO, and 80.8% N_2. Estimate the values for x and y and the percent theoretical air utilized.

13.32 Producer gas, created from coal, has a volumetric analysis of 3% CH_4, 14% H_2, 50.9% N_2, 0.6% O_2, 27% CO, and 4.5% CO_2. Complete combustion occurs with 150% theoretical air at 100 kPa. What percentage of the water vapor will condense out if the temperature of the products is 20 °C?

13.33 Using the enthalpy of formation data from Table B-6 calculate the enthalpy of combustion for a steady-flow process, assuming liquid water in the products. Inlet and outlet temperatures are 25 °C and the pressure is 100 kPa. (Compare with the value listed in Table B-7.) The fuel is (*a*) methane, (*b*) acetylene, (*c*) propane gas, and (*d*) liquid pentane.

13.34 Propane gas (C_3H_8) undergoes complete combustion with stoichiometric air; both are at 77 °F and 1 atm. Calculate the heat transfer if the products from a steady-flow combustor are at (*a*) 77 °F, (*b*) 1540 °F, and (*c*) 2540 °F.

13.35 Liquid propane (C_3H_8) undergoes complete combustion with air; both are at 25 °C and 1 atm. Calculate the heat transfer if the products from a steady-flow combustor are at 1000 K and the percent theoretical air is (*a*) 100%, (*b*) 150%, and (*c*) 200%.

13.36 Ethane gas (C_2H_6) at 25 °C is burned with 150% theoretical air at 500 K and 1 atm. Find the heat transfer from a steady-flow combustor if the products are at 1000 K and (*a*) complete combustion occurs; (*b*) 95% of the carbon is converted to CO_2 and 5% to CO.

13.37 Complete combustion occurs between butane gas (C_4H_{10}) and air; both are at 25 °C and 1 atm. If the steady-flow combustion chamber is insulated, what percent theoretical air is needed to maintain the products at (*a*) 1000 K and (*b*) 1500 K?

13.38 Complete combustion occurs between ethylene gas (C_2H_4) and air; both are at 77 °F and 1 atm. If 150,000 Btu of heat is removed per lbmol of fuel from the steady-flow combustor, estimate the percent theoretical air required to maintain the products at 1500 °R.

13.39 Butane gas (C_4H_{10}) at 25 °C is burned in a steady-flow combustion chamber with 150% theoretical air at 500 K and 1 atm. If 90% of the carbon is converted to CO_2 and 10% to CO, estimate the heat transfer if the products are at 1200 K.

13.40 Butane gas (C_4H_{10}) undergoes complete combustion with 40% excess air; both are at 25 °C and 100 kPa. Calculate the heat transfer from the steady-flow combustor if the products are at 1000 K and the humidity of the combustion air is (*a*) 90%, (*b*) 70%, and (*c*) 50%.

13.41 A rigid tank contains a mixture of 0.2 kg of ethane gas (C_2H_6) and 1.2 kg of O_2 at 25 °C and 100 kPa. The mixture is ignited and complete combustion occurs. If the final temperature is 1000 K, find the heat transfer and the final pressure.

13.42 A mixture of 1 lbmol methane gas (CH_4) and stoichiometric air at 77 °F and 20 psia is contained in a rigid tank. If complete combustion occurs, calculate the heat transfer and the final pressure if the final temperature is 1540 °F.

13.43 A mixture of octane gas (C_8H_{18}) and 20% excess air at 25 °C and 200 kPa is contained in a 50-liter cylinder. Ignition occurs and the pressure remains constant until the temperature reaches 800 K. Assuming complete combustion, estimate the heat transfer during the expansion process.

13.44 A mixture of butane gas (C_4H_{10}) and stoichiometric air is contained in a rigid tank at 25 °C and 100 kPa. If 95% of the carbon is burned to CO_2 and the remainder to CO, calculate the heat transfer from the tank and the volume percent of the water that condenses out if the final temperature is 25 °C.

13.45 Butane gas (C_4H_{10}) mixes with air, both at 25 °C and 1 atm, and undergoes complete combustion in a steady-flow insulated combustion chamber. Calculate the adiabatic flame temperature for (*a*) 100% theoretical air, (*b*) 150% theoretical air, and (*c*) 100% excess air.

13.46 Ethane (C_2H_6) at 25 °C undergoes complete combustion with air at 400 K and 1 atm in a steady-flow insulated combustor. Determine the exit temperature for 50% excess air.

13.47 Hydrogen gas and air, both of 400 K and 1 atm, undergo complete combustion in a steady-flow insulated combustor. Estimate the exit temperature for 200% theoretical air.

13.48 Liquid methyl alcohol (CH_3OH) at 25 °C reacts with 150% theoretical air. Find the exit temperature, assuming complete combustion, from a steady-flow insulated combustor if the air enters at (*a*) 25 °C, (*b*) 400 K, and (*c*) 600 K. Assume atmospheric pressure.

13.49 Ethene (C_2H_4) at 77 °F undergoes complete combustion with stoichiometric air at 77 °F and 70% humidity in an insulated steady-flow combustion chamber. Estimate the exit temperature assuming a pressure of 14.5 psia.

13.50 Ethane (C_2H_6) at 25 °C combusts with 90% theoretical air at 400 K and 1 atm in an insulated steady-flow combustor. Determine the exit temperature.

13.51 A mixture of liquid propane (C_3H_8) and stoichiometric air at 25 °C and 100 kPa undergoes complete combustion in a rigid container. Determine the maximum temperature and pressure (the *explosion pressure*) immediately after combustion.

Answers to Supplementary Problems

13.12 (*a*) 2, 2, 11.28 (*b*) 3, 3, 16.92 (*c*) 4, 5, 24.44 (*d*) 5, 6, 30.08 (*e*) 8, 9, 47 (*f*) 1, 2, 5.64

13.13 (*a*) 17.23 (*b*) 15.14% (*c*) 59 °C (*d*) 89.8%

13.14 (*a*) 16.09 (*b*) 17.24% (*c*) 55.9 °C (*d*) 87.9%

13.15 (*a*) 15.67 (*b*) 18.07% (*c*) 54.6 °C (*d*) 87.0%

13.16 (*a*) 15.45 (*b*) 18.52% (*c*) 53.9 °C (*d*) 86.4%

13.17 (*a*) 17.78 (*b*) 14.3% (*c*) 60.1 °C (*d*) 90.4%

13.18 (*a*) 28.96 (*b*) 6.35% (*c*) 43.8 °C

13.19 (*a*) 28.21 (*b*) 6.69% (*c*) 42.5 °C

13.20 (*a*) 27.82 (*b*) 6.87% (*c*) 41.8 °C

13.21 (*a*) 30.23 (*b*) 10.48% (*c*) 45.7 °C

13.22 (*a*) 1.38 kg/min (*b*) 1.478 kg/min (*c*) 1.518 kg/min (*d*) 1.539 kg/min (*e*) 1.392 kg/min

13.23 (*a*) 20.67, 50.5 °C (*b*) 20.64, 50.2 °C (*c*) 20.57, 49.5 °C (*d*) 20.50, 48.9 °C

13.24 65.4%, 7.78%

13.25 23.18, 46.2 °C

13.26 (*a*) 30.8 (*b*) 76.0% (*c*) 40.3 °C

13.27 (*a*) 19.04 (*b*) 118.7% (*c*) 127 °F

13.28 65.92 m^3/min

13.29 59%

13.30 21.46

13.31 3.73, 3.85, 152.6%

13.32 76.8%

13.33 (*a*) − 890 300 kJ/kmol (*b*) − 1 299 600 kJ/kmol (*c*) − 2 220 000 kJ/kmol (*d*) − 3 505 000 kJ/kmol

13.34 (*a*) − 955,100 Btu/lbmol (*b*) − 572,500 Btu/lbmol (*c*) − 13,090 Btu/lbmol

13.35 (*a*) − 1 436 000 kJ/kmol (*b*) − 1 178 000 kJ/kmol (*c*) − 919 400 kJ/kmol

13.36 (*a*) − 968 400 kJ/kmol (*b*) − 929 100 kJ/kmol

13.37 (*a*) 411% (*b*) 220%

13.38 1045%

13.39 − 1 298 700 kJ/kmol

13.40 (*a*) − 1 871 000 kJ/kmol (*b*) − 1 801 000 kJ/kmol (*c*) − 1 735 000 kJ/kmol

13.41 − 12 780 kJ, 437 kPa

13.42 − 220,600 Btu, 74.5 psia

13.43 − 216 kJ

13.44 − 2 600 400 kJ/kmol fuel, 81.3%

13.45 (*a*) 2520 K (*b*) 1830 K (*c*) 1510 K

13.46 1895 K

13.47 1732 K

13.48 (*a*) 1520 K (*b*) 1585 K (*c*) 1690 K

13.49 4740 °R

13.50 2410 K

13.51 2965 K, 1035 kPa

Sample Exams for a Semester Course for Engineering Students

We are including a set of sample exams that we used in one of our courses at Michigan State University. They are four-part, multiple-choice exams that contain questions similar to those found on the Fundamentals of Engineering exam and the Graduate Record Exam/Engineering. Thermodynamics provides an excellent course to expose engineering students to the nuances of a multiple-choice exam, a rarity in engineering courses.

Research has shown that grades are independent of the type of exam given but some experience is desirable so that students have a chance to try out different strategies in taking such an exam. It is very different from the four-problem, partial credit exam so popular in engineering courses. Students typically do not like multiple-choice exams in engineering courses since a simple mistake makes the question completely wrong, but that is the type of exam they will have to take to gain admission to the profession of engineering or a graduate school, law school, or medical school. Such exams are long and difficult and care must be taken to not let the time slip away attempting to answer a question.

In the following exams, a passing grade was given if a student exceeded the score of a random guesser, so that 6 correct answers out of 20 questions would be just a "pass," the lowest "D." Usually, a 16 would be the top grade in the class, so that would be equivalent to a "100," the top "A." To obtain 16 correct answers to these multiple-choice exams is indeed a feat. Try it, if you think that is not true, even if you are a professor!

These exams cover Chapters 1, 2, 3, 5, and 6 and only portions of Chapters 4, 8, 9, 10, and 12. A second course could cover the material that was not covered in the first course. Professors are encouraged to make more use of multiple-choice exams even in other engineering courses so that students will be experienced in taking such exams. Even the more advanced design-oriented exams are all multiple choice in the process of becoming a Professional Engineer. Such exams are more difficult to create but the grading of the exams is extremely efficient, allowing professors to always return the exam the following class period.

If you have comments or questions about these exams, please contact me at MerleCP@sbcglobal.net.

The answers and sketchy solutions to the exams are included after each exam.

Exam No. 1 (This exam covers Chapter 1 through Section 4.7)

Name ____________

1. Thermodynamics is a science that does not include the study of energy
   (A) storage (B) utilization (C) transfer (D) transformation

2. The units of joules can be converted to which of the following?
   (A) Pa·m^2 (B) N·kg (C) Pa/m^2 (D) Pa·m^3

3. In thermodynamics we focus our attention on which form of energy?
   (A) kinetic (B) potential (C) internal (D) total

4. The standard atmospheric pressure in meters of ammonia ($\rho = 600$ kg/m^3) is:
   (A) 17 m (B) 19 m (C) 23 m (D) 31 m

5. A gage pressure of 400 kPa acting on a 4-cm-diameter piston is resisted by an 800-N/m spring. How much is the spring compressed? Neglect the weight of the spring.
   (A) 0.628 m (B) 0.951 m (C) 1.32 m (D) 1.98 m

6. Select the correct T-v diagram if steam at 0.005 m^3/kg is heated to 0.5 m^3/kg while maintaining $P = 500$ kPa.

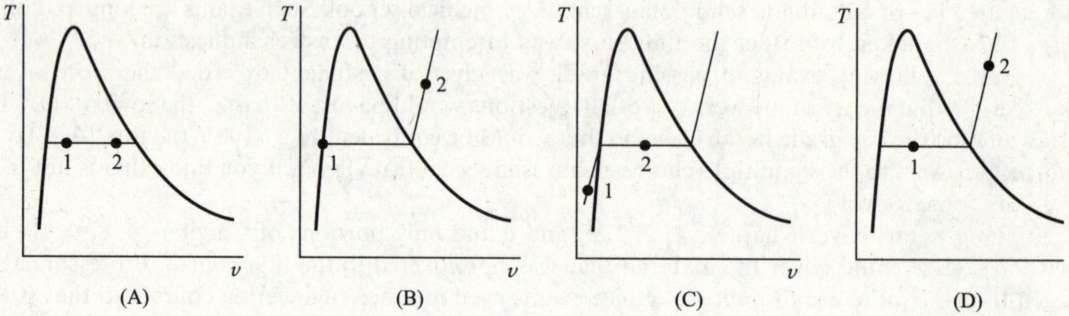

T v 1 2
(A)
T v 1 2
(B)
T v 1 2
(C)
T v 1 2
(D)

7. Estimate the enthalpy h of steam at 288 °C and 2 MPa.
   (A) 2931 kJ/kg (B) 2957 kJ/kg (C) 2972 kJ/kg (D) 2994 kJ/kg

8. Estimate v of steam at 200 °C and $u = 2000$ kJ/kg.
   (A) 0.0762 m^3/kg (B) 0.0801 m^3/kg (C) 0.0843 m^3/kg (D) 0.0921 m^3/kg

9. Saturated water vapor is heated in a rigid volume from 200 °C to 600 °C. The final pressure is nearest:
   (A) 3.30 MPa (B) 3.25 MPa (C) 3.20 MPa (D) 3.15 MPa

10. A 0.2-m^3 tire is filled with 25 °C air until $P = 280$ kPa. The mass of air is nearest:
    (A) 7.8 kg (B) 0.888 kg (C) 0.732 kg (D) 0.655 kg

11. Calculate the mass of nitrogen at 150 K and 2 MPa contained in 0.2 m^3.
(A) 10.2 kg (B) 8.98 kg (C) 6.23 kg (D) 2.13 kg

12. Find the work produced if air expands from 0.2 m^3 to 0.8 m^3 if $P = 0.2 + 0.4V$ kPa.
(A) 0.48 kJ (B) 0.42 kJ (C) 0.36 kJ (D) 0.24 kJ

13. Ten kilograms of saturated liquid water expands until $T_2 = 200\,°C$ while $P = \text{const.} = 500$ kPa. Find W_{1-2}.
(A) 1920 kJ (B) 2120 kJ (C) 2340 kJ (D) 2650 kJ

14. If $P_1 = 400$ kPa and $T_1 = 400\,°C$, what is T_2 when the piston hits the stops?
(A) 314 °C (B) 315 °C (C) 316 °C (D) 317 °C

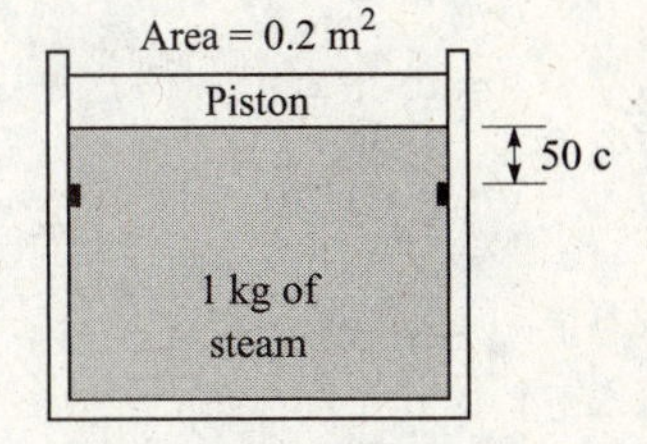

15. Calculate the heat released for the steam of Question 14.
(A) 190 kJ (B) 185 kJ (C) 180 kJ (D) 175 kJ

16. After the piston hits the stops in Question 14, how much additional heat is released until $P_3 = 100$ kPa?
(A) 1580 kJ (B) 1260 kJ
(C) 930 kJ (D) 730 kJ

17. Find the temperature rise in the volume shown after 5 minutes.
(A) 423 °C (B) 378 °C
(C) 313 °C (D) 287 °C

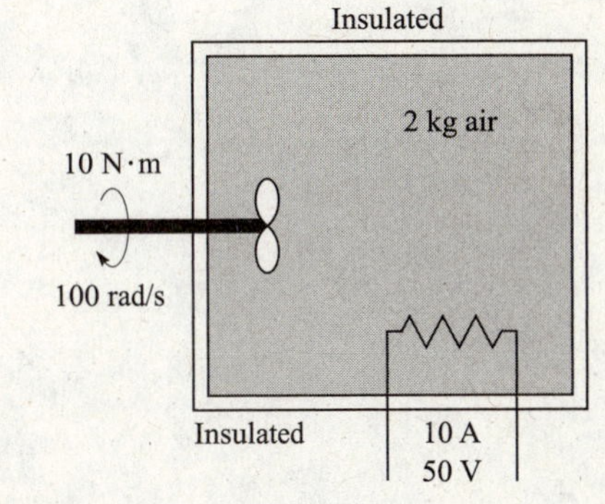

18. One kilogram of air is compressed at $T = \text{const.} = 100\,°C$ until the volume is halved. The rejected heat is:
(A) 42 kJ (B) 53 kJ
(C) 67 kJ (D) 74 kJ

19. Energy is added to 5 kg of air with a paddle wheel so that $\Delta T = 100\,°C$ while $P = \text{const.}$ in an insulated container. The paddle-wheel work is:
(A) 500 kJ (B) 423 kJ (C) 402 kJ (D) 383 kJ

20. Five kilograms of copper at 200 °C is submerged in 20 kg of water at 10 °C in an insulated container. After a long time the temperature is nearest:
(A) 14 °C (B) 16 °C (C) 18 °C (D) 20 °C

Equations of use in the closed-book exam (use this page for your work)

$W = mg$

$PV = mRT$

$F = Kx$

$T = \_\_\ ^\circ\text{C} + 273$

$P_{abs} = P_{gage} + P_{atm}$

$P = \dfrac{F}{A}$

$Z = \dfrac{Pv}{RT}$

$v = \dfrac{V}{m}$

$h = u + Pv, \qquad \Delta h = C_p \Delta T$

$v = v_f + x(v_g - v_f)$

$Q - W = \Delta U, \qquad \Delta u = C_v \Delta T$

$W = \int P\,dV$

$\dot{W} = T\omega$

$= mRT \ln \dfrac{V_2}{V_1} \quad \text{if } T = \text{const.}$

$\dot{W} = VI$

Solution to Exam No. 1

1. (B)

2. (D) $J = N \cdot m = (N/m^2)m^3 = Pa \cdot m^3$

3. (C)

4. (A) $P = \rho gh$. $100\,000 = 600 \times 9.8h$. $\therefore h = 17.0$ m

5. (A) $PA = Kx$. $400\,000\pi \times 0.02^2 = 800x$. $\therefore x = 0.628$ m

6. (D) $v_g > v_1 > v_f$. $v_2 > v_g$. $\therefore$ State 1 is in the quality region, state 2 is superheat.

7. (D) $h = \frac{38}{50}(3023 - 2902) = 2994$ kJ/kg

8. (C) $2000 = 851 + x(2595 - 851)$. $\therefore x = 0.659$. $v = 0.0012 + 0.659(0.1274 - 0.0012) = 0.0843\ m^3/kg$

9. (D) $v_1 = v_g = 0.1274 = v_2$. $T_2 = 600\,°C$. $P_2 = \frac{0.1324 - 0.1274}{0.1324 - 0.09885}(1) + 3 = 3.15$ MPa

10. (B) $m = PV/RT = 380 \times 0.2/(0.287 \times 298) = 0.888$ kg

11. (A) $P_R = 2/3.39 = 0.59$. $T_R = 150/126.2 = 1.19$. $\therefore Z = 0.875$.

$$m = PV/ZRT = 2000 \times 0.2/(0.875 \times 0.297 \times 150) = 10.2 \text{ kg}$$

12. (D) $W = \int P\,dV = \int_{0.2}^{0.8}(0.2 + 0.4V)\,dV = 0.2(0.8 - 0.2) + 0.2(0.8^2 - 0.2^2) = 0.24$ kJ

13. (B) $W = mP\Delta v = 10 \times 500(0.443 - 0.0012) = 2120$ kJ

14. (B) $v_1 = 0.7726$. $v_2 = v_1 - 0.5(0.2) = 0.6726$. $P_2 = 400$ kPa. $\therefore T_2 = 315\,°C$

15. (D) $Q = m\Delta h = 1(3098 - 3273) = 175$ kJ

16. (A) $Q = m\Delta u = 1(2829 - 1246) = 1580$ kJ
since $v_3 = v_2 = 0.6726 = 0.001 + x_3(1.694 - 0.001)$. $\therefore x_3 = 0.397$ and $u_3 = 1246$

17. (C) $W = (10 \times 100 + 50 \times 100)5 \times 60 = Q = mC_v\Delta T = 2 \times 717\Delta T$. $\therefore \Delta T = 313\,°C$

18. (D) $Q = mRT \ln \frac{v_2}{v_1} = 1 \times 0.287 \times 373 \ln \frac{1}{2} = -74$ kJ

19. (A) $-W = m\Delta h = 5 \times C_p\Delta T = 5 \times 1.0 \times 100 = 500$ kJ

20. (A) $m_{cu}C_{cu}\Delta T_{cu} = m_w C_w \Delta T_w$. $5 \times 0.39(200 - T_2) = 20 \times 4.18(T_2 - 10)$. $\therefore T_2 = 14\,°C$

Exam No. 2 (This exam covers Section 4.8 through Chapter 6)

Name ____________

1. Which of the following is not a control volume?
   (A) insulated tank (B) car radiator (C) compressor (D) turbine

2. Flow in a pipe is assumed to be uniform. The x-coordinate is along the pipe and r is the radial coordinate. Which of the following is not true?
   (A) $v = v(r)$ (B) $v = v(x)$ (C) $v = v(t)$ (D) $v =$ const.

3. Liquid water flows in the pipe at 12 m/s. The pipe's diameter is reduced by a factor of 2. What is the velocity in the reduced section?
   (A) 36 m/s (B) 48 m/s (C) 24 m/s (D) 3 m/s

4. A nozzle accelerates steam at 4 MPa and 500 °C to 1 MPa and 300 °C. If $V_1 = 20$ m/s, the exiting velocity is nearest:
   (A) 575 m/s (B) 750 m/s (C) 825 m/s (D) 890 m/s

5. Calculate the mass flow rate at the nozzle exit of Question 4 if $d_1 = 10$ cm.
   (A) 2.9 kg/s (B) 2.3 kg/s (C) 1.8 kg/s (D) 1.1 kg/s

6. 100 kg/min of air enters a tube at 25 °C and leaves at 20 °C. The heat loss is nearest:
   (A) 750 kJ/min (B) 670 kJ/min (C) 500 kJ/min (D) 480 kJ/min

7. Calculate the power needed to raise the pressure of water 4 MPa assuming the temperature remains constant. The mass flux is 5 kg/s and $\dot{Q} = 0$.
   (A) 20 kW (B) 32 kW (C) 48 kW (D) 62 kW

8. Which of the following is usually assumed to be reversible?
   (A) paddle wheel (B) compressing a gas (C) melting an ice cube (D) a burst membrane

9. A Carnot engine operates between reservoirs at 20 °C and 200 °C. If 10 kW of power is produced, the rejected heat rate is nearest:
   (A) 26.3 kJ/s (B) 20.2 kJ/s (C) 16.3 kJ/s (D) 12 kJ/s

10. An inventor invents a thermal engine that operates between ocean layers at 27 °C and 10 °C. It produces 10 kW and discharges 9900 kJ/min of heat. Such an engine is:
    (A) impossible (B) reversible (C) possible (D) probable

11. Select the Kelvin–Planck statement of the second law.
(A) An engine cannot produce more heat than the heat it receives.
(B) A refrigerator cannot transfer heat from a low-temperature reservoir to a high-temperature reservoir without work.
(C) An engine cannot produce work without discharging heat.
(D) An engine discharges heat if the work is less than the heat it receives.

12. Select the incorrect statement relating to a Carnot cycle.
(A) There are two adiabatic processes.
(B) There are two constant pressure processes.
(C) Each process is a reversible process.
(D) Work occurs on all four processes.

13. A Carnot refrigerator requires 10 kW to remove 20 kJ/s from a 20 °C reservoir. What is T_H?
(A) 440 K (B) 400 K (C) 360 K (D) 340 K

14. Find ΔS_{net} if a 10-kg copper mass at 100 °C is submerged in a 20 °C lake.
(A) 1.05 kJ/K (B) 0.73 kJ/K (C) 0.53 kJ/K (D) 0.122 kJ/K

15. Find the maximum work output from a steam turbine if $P_1 = 10$ MPa, $T_1 = 600$ °C, and $P_2 = 40$ kPa.
(A) 1625 kJ/kg (B) 1545 kJ/kg (C) 1410 kJ/kg (D) 0.1225 kJ/kg

16. Find the minimum work input to an air compressor if $P_1 = 100$ kPa, $T_1 = 27$ °C, and $P_2 = 800$ kPa.
(A) 276 kJ/kg (B) 243 kJ/kg (C) 208 kJ/kg (D) 187 kJ/kg

17. The entropy change while expanding air in a cylinder from 800 kPa is 0.2 kJ/kg·K. If the temperature remains constant at 27 °C, what is the heat transfer?
(A) 140 kJ/kg (B) 115 kJ/kg (C) 100 kJ/kg (D) 60 kJ/kg

18. Find the maximum outlet velocity for an air nozzle that operates between 400 kPa and 100 kPa if $T_1 = 27$ °C.
(A) 570 m/s (B) 440 m/s (C) 380 m/s (D) 320 m/s

19. The efficiency of an adiabatic steam turbine operating between $P_1 = 10$ MPa, $T_1 = 600$ °C, and $P_2 = 40$ kPa is 80%. What is u_2?
(A) 2560 kJ/kg (B) 2484 kJ/kg (C) 2392 kJ/kg (D) 2304 kJ/kg

20. The efficiency of an adiabatic air compressor operating between $P_1 = 100$ kPa, $T_1 = 27$ °C, and $T_2 = 327$ °C is 80%. What is P_2?
(A) 540 kPa (B) 680 kPa (C) 720 kPa (D) 780 kPa

Equations of use in the closed-book exam (use this page for your work)

$\rho A V = \dot{m} = \text{const.}$

$Q - W = \Delta U$

$$\dot{Q} - \dot{W}_s = \dot{m}\left(h_2 - h_1 + \frac{V_2^2 - V_1^2}{2}\right) \quad \text{(gas or vapor)}$$

$$= \dot{m}\left(\frac{P_2 - P_1}{\rho} + u_2 - u_1 + \frac{V_2^2 - V_1^2}{2}\right) \quad \text{(liquid)}$$

$$\eta_{\max} = 1 - \frac{T_L}{T_H} \qquad (\text{COP}_\text{R})_{\max} = \frac{T_L}{T_H - T_L}$$

$$\Delta S = \int \frac{\delta Q}{T} \qquad W = \int P\,dV$$

$$= mRT \ln \frac{V_2}{V_1} \text{ if } T = \text{const.}$$

$$\Delta s = C_p \ln \frac{T_2}{T_1} - R \ln \frac{P_2}{P_1} \quad \text{(gas)}$$

$$= \text{C} \ln \frac{T_2}{T_1} \quad \text{(solid or liquid)}$$

$$\frac{T_2}{T_1} = \left(\frac{P_2}{P_1}\right)^{\frac{k-1}{k}} \qquad k_{\text{air}} = 1.4$$

$P = \rho RT$ $\qquad R_{\text{air}} = 0.287 \text{ kJ/kg·K}$ $\qquad v = v_f + x v_{fg}$

Solution to Exam No. 2

1. (A)

2. (A)

3. (B) $V_2 = V_1/A_1 = 12 \times 2^2 = 48$ m/s

4. (D) $0 = h_1 - h_1 + (V_2^2 - V_1^2)/2 = 3051 - 3445 + (V_2^2 - 20^2)/2 \times 1000.$ $\quad \therefore V_2 = 888$ m/s

5. (C) $\dot{m} = A_1 V_1 / v_1 = \pi \times 0.05^2 \times 0/0.08643 = 1.817$ kg/s

6. (C) $\dot{Q} = \dot{m}\Delta h = \dot{m}C_p \Delta T = 100 \times 1.0(20 - 25) = -500$ kJ/min

7. (A) $-\dot{W}_S = \dot{m}\Delta h = \dot{m}(\Delta u + v\Delta P) = 5(0 + 0.001 \times 4000) = 20$ kW

8. (B) All others operate in one direction, i.e., heat cannot flow into a heater.

9. (C) $\eta = \dfrac{\dot{W}}{\dot{Q}_{in}} = 1 - \dfrac{T_L}{T_H} = 1 - \dfrac{293}{473} = \dfrac{10}{\dot{Q}_{in}}$ $\quad \therefore \dot{Q}_{in} = 26.3 = \dot{Q}_{out} + \dot{W} = \dot{Q}_{out} + 10.$ $\quad \therefore \dot{Q}_{out} = 16.3$ kJ/s

10. (A) $\eta_{max} = 1 - \dfrac{T_L}{T_H} = 1 - \dfrac{283}{300} = 0.0567.$ $\quad \eta_{engine} = \dfrac{10}{10 + 9900/60} = 0.0571 > \eta_{max}.$ $\quad \therefore$Impossible

11. (C)

12. (B)

13. (A) COP $= \dfrac{20}{10} = \dfrac{1}{T_H/T_L - 1} = \dfrac{T_L}{T_H - T_L} = \dfrac{293}{T_H - 293}.$ $\quad \therefore T_H = 439.5$ K

14. (D) $\Delta S_{net} = \Delta S_{cu} + \Delta S_{lake} = 0.39 \times 10 \ln \dfrac{293}{373} + \dfrac{0.39 \times 10 \times (100 - 20)}{293} = 0.122$ kJ/K

15. (C) $h_1 - h_2 = 3870 - 2462 = 1408$ kJ/kg where we used $s_2 - s_1 = 7.1696 = 1.026 - 6.6449x_2$.
$\therefore x_2 = 0.9246$, and $h_2 = 2462$

16. (B) $T_2 = 300(800/100)^{0.2857} = 0.543.$ $\quad w_{min} = C_p \Delta T = 1.00(543 - 300) = 243$ kJ/kg

17. (D) $q = T\Delta s = 300 \times 0.2 = 60$ kJ/kg

18. (B) $T_2 = 300(100/400)^{0.2857} = 201.9$ K. $\quad 0 = \dfrac{V_2^2}{2 \times 1000} + 1.0(201.9 - 300).$ $\quad \therefore V_2 = 443$ m/s

19. (A) $\eta = \dfrac{h_1 - h_2}{h_1 - h_{2'}}.$ $\quad 0.8 = \dfrac{3870 - h_2}{3870 - 2462}.$ $\quad \therefore h_2 = 2744$ and $u_2 = 2560$ where we used
$s_{2'} = s_1 = 7.1696 = 1.026x_{2'}.$ $\quad \therefore x_{2'} = 0.9246$ and $h_{2'} = 2462.$

20. (D) $\eta = \dfrac{h_{1'} - h_2}{h_1 - h_2}.$ $\quad 0.8 = \dfrac{1.0(T_{2'} - 300)}{1.0(600 - 300)}.$ $\quad \therefore T_{2'} = 540$ K. $P_2 = 100(540/300)^{3.5} = 782$ kPa

Exam No. 3 (This exam covers Chapters 8, 9, and 12)

Name ____________

1. Power cycles are idealized. Which of the following is not an idealization?
(A) Friction is absent.
(B) Heat transfer does not occur across a finite temperature difference.
(C) Pipes connecting components are insulated.
(D) Pressure drops in pipes are neglected.

2. In the analysis of the Otto power cycle, which do we not assume?
(A) The combustion process is a constant-pressure process.
(B) All processes are quasiequilibrium processes.
(C) The exhaust process is replaced by a heat rejection process.
(D) The specific heats are assumed constant.

3. The temperatures in a Carnot cycle using air are 300 K and 900 K. A pressure of 100 kPa exists at state 1, 4677 kPa at state 2, and 2430 kPa at state 3. Find the pressure at state 4 assuming constant specific heats.
(A) 82 kPa (B) 75 kPa (C) 52 kPa (D) 47 kPa

4. Find the net work for the Carnot cycle of Question 3.
(A) 113 kJ/kg (B) 168 kJ/kg (C) 198 kJ/kg (D) 236 kJ/kg

5. Find the heat addition for the Carnot cycle of Question 3.
(A) 169 kJ/kg (B) 198 kJ/kg (C) 232 kJ/kg (D) 277 kJ/kg

6. The four temperatures in an Otto cycle using air are $T_1 = 300$ K, $T_2 = 754$ K, $T_3 = 1600$ K, $T_4 = 637$ K. If $P_1 = 100$ kPa, find P_2. Assume constant specific heats.
(A) 2970 kPa (B) 2710 kPa (C) 2520 kPa (D) 2360 kPa

7. Find the heat added for the Otto cycle of Question 6.
(A) 930 kJ/kg (B) 850 kJ/kg (C) 760 kJ/kg (D) 610 kJ/kg

8. Find the compression ratio for the Otto cycle of Question 6.
(A) 11 (B) 10 (C) 9 (D) 8

9. The four temperatures of a Brayton cycle are $T_1 = 300$ K, $T_2 = 500$ K, $T_3 = 1000$ K, $T_4 = 600$ K. Find the heat added. Assume constant specific heats.
(A) 200 kJ/kg (B) 300 kJ/kg (C) 400 kJ/kg (D) 500 kJ/kg

10. Find the pressure ratio of the Brayton cycle of Question 9.
(A) 2 (B) 4 (C) 6 (D) 8

11. Find the back-work ratio w_C/w_T of the Brayton cycle of Question 9.
(A) 0.8 (B) 0.7 (C) 0.6 (D) 0.5

12. An ideal regenerator is added to the Brayton cycle of Question 9. Find η_R/η_B where η_R is the cycle efficiency with the regenerator and η_B is the cycle efficiency without the regenerator.
(A) 1.2 (B) 1.25 (C) 1.3 (D) 1.35

13. The high and low pressure in an ideal Rankine cycle are 10 MPa and 20 kPa. If the high temperature of the steam is 600 °C, the heat input is nearest:
(A) 3375 kJ/kg (B) 3065 kJ/kg (C) 2875 kJ/kg (D) 2650 kJ/kg

14. Find the turbine output for the Rankine cycle of Question 13.
(A) 1350 kJ/kg (B) 1400 kJ/kg (C) 1450 kJ/kg (D) 1500 kJ/kg

15. The steam in the Rankine cycle of Question 13 is reheated at 4 MPa to 600 °C. The total heat input is nearest:
(A) 3850 kW/kg (B) 3745 kW/kg (C) 3695 kW/kg (D) 3625 kW/kg

16. If $\dot{m} = 2400$ kg/min, what is the pump power requirement for the Rankine cycle of Question 13?
(A) 1580 kW (B) 1260 kW (C) 930 kW (D) 730 kW

17. Given: $T = 25\,°C$, $\phi = 100\%$. Find ω. Assume $P_{atm} = 100$ kPa. (Use equations.)
(A) 0.0208 (B) 0.0206 (C) 0.0204 (D) 0.0202

18. Given: $T = 25\,°C$, $\omega = 0.015$. Find ϕ. Assume $P_{atm} = 100$ kPa. (Use equations.)
(A) 72.3% (B) 72.9% (C) 73.6% (D) 74.3%

19. Air at 5 °C and 90% humidity is heated to 25 °C. Estimate the final humidity.
(A) 20% (B) 25% (C) 30% (D) 35%

20. Air at 35 °C and 70% humidity in a 3 m by 10 m by 20 m room is cooled to 25 °C and 40% humidity. How much water is removed?
(A) 14 kg (B) 12 kg (C) 10 kg (D) 8 kg

Information of use in the closed-book exam (use this page for your work)

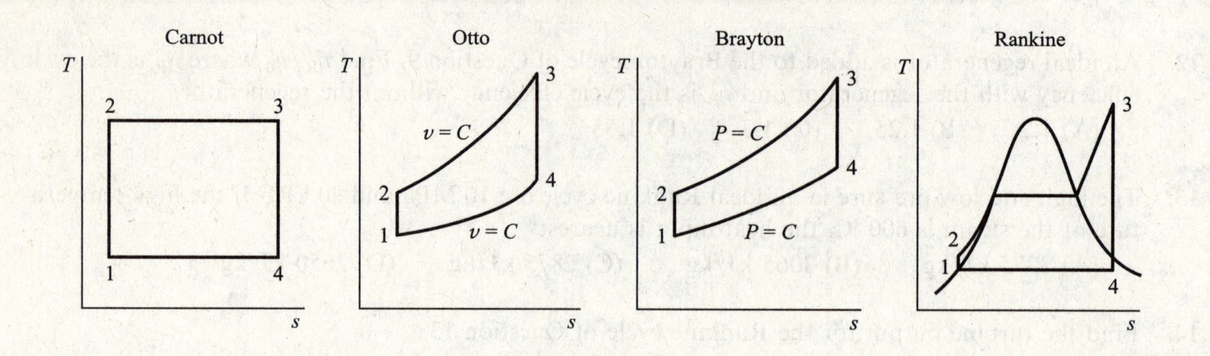

$PV = mRT$ $\quad R_{\text{air}} = 0.287 \text{ kJ/kg·K}$ $\quad C_{p,\text{air}} = 1.0 \text{ kJ/kg·K}$ $\quad C_{v,\text{air}=0.718} \text{ kJ/kg·K}$

$\frac{T_2}{T_1} = \left(\frac{P_2}{P_1}\right)^{\frac{k-1}{k}}$ $\quad \frac{T_2}{T_1} = \left(\frac{V_1}{V_2}\right)^{k-1}$ $\quad \frac{P_2}{P_1} = \left(\frac{V_1}{V_2}\right)^{k}$ $\quad k_{\text{air}} = 1.4$

$Q - W = m\Delta u$ $\quad \Delta u = C_v \Delta T$ $\quad \Delta s = C_p \ln\frac{T_2}{T_1} - R\ln\frac{P_2}{P_1}$

$\dot{Q} - \dot{W} = \dot{m}\Delta h$ $\quad \Delta h = C_p \Delta T$ $\quad \Delta h = \Delta u + \Delta(Pv)$

$\phi = \frac{P_v}{P_g}$ $\quad \omega = 0.622\frac{P_v}{P_g}$ $\quad P = P_v + P_g$

Solution to Exam No. 3

1. (B)

2. (A)

3. (C) $P_4 = P_3(T_4/T_3)^{k/(k-1)} = 2430(300/900)^{3.5} = 52$ kPa

4. (A) $w = \Delta T \Delta s = 600(-C_p \ln P_3/P_2) = 600(-1.0 \ln 2430/4677) = 113$ kJ/kg

5. (A) $q = T\Delta s = 900(-C_p \ln P_3/P_2) = 900(-1.0 \ln 2430/4677) = 169$ kJ/kg

6. (C) $P_2 = P_1(T_2/T_1)^{k/(k-1)} = 100(754/300)^{3.5} = 2520$ kPa

7. (D) $q = C_v \Delta T = 0.717(1600 - 754) = 607$ kJ/kg

8. (B) $v_1/v_2 = (T_2/T_1)^{1/(k-1)} = (754/300)^{2.5} = 10$

9. (D) $q = \Delta h = C_p \Delta T = 1.0(1000 - 500) = 500$ kJ/kg

10. (C) $P_2/P_1 = (T_2/T_1)^{k/(k-1)} = (500/300)^{3.5} = 6$

11. (D) $w_C/w_T = \Delta h_C/\Delta h_T = C_p \Delta T_C / C_p \Delta T_T = (500 - 300)/(1000 - 600) = 0.5$

12. (B) $\eta_B = \dfrac{w_T - w_C}{q_{\text{in}}} = \dfrac{1000 - 600 - (500 - 300)}{1000 - 500} = 0.4. \qquad \eta_R = \dfrac{1000 - 600 - (500 - 300)}{1000 - 600} = 0.5.$

$\therefore \dfrac{\eta_R}{\eta_B} = 1.25$

13. (A) $q = h_3 - h_2 = 3625 - 251 = 3374$ kJ/kg

14. (A) $s_3 = s_4 = 6.9029 = 0.832 + 7.70766x_4. \qquad \therefore x_4 = 0.858$ and $h_4 = 2274$ kJ/kg
$\therefore w_T = h_3 - h_4 = 3625 - 2274 = 1351$ kJ/kg

15. (B) $\left.\begin{array}{l} s_3 = 6.9029 \\ P_4 = 4 \end{array}\right\} \therefore h_4 = 3307. \quad q_{\text{in}} = h_3 - h_2 + h_5 - h_4 = 3625 - 251 + 3674 - 3307 = 3741$ kJ/kg

16. (D) $\dot{W}_P = \dot{m}\Delta P/\rho = \dfrac{2400}{60}(10\,000 - 20)/1000 = 400$ kW

17. (C) $P_a = P - P_v = 100 - 3.169 = 96.83. \qquad \therefore \omega = 0.622 \times 3.169/96.83 = 0.0204$

18. (D) $\left.\begin{array}{l} 0.622P_v/P_a = 0.015 \\ 100 = P_v + P_a \end{array}\right\} \qquad \therefore P_v = 2.355. \qquad \therefore \phi = \dfrac{2.355}{3.169} = 0.743$

19. (B) At $\phi = 90\%$ and $T = 5°$C, $\omega = 0.0048$. At $T = 25°$C and $\omega = 0.0048$, $\phi = 25\%$

20. (B) $\Delta\omega = 0.025 - 0.0079 = 0.0171$. $\Delta\text{water} = \rho_{\text{air}}\Delta\omega = \dfrac{100}{0.287 \times 298}(3 \times 10 \times 20) \times 0.0171 = 12$ kg

Final Exam (This exam covers the material of the three-hour exams)

Name____________

1. Select the correct statement for a liquid.

(A) The enthalpy depends on pressure only.
(B) The enthalpy depends on temperature only.
(C) The specific volume depends on pressure only.
(D) The internal energy depends on temperature only.

2. Steam is expanded at constant temperature of 50°C from 100 kPa to 10 kPa. Select the correct diagram.

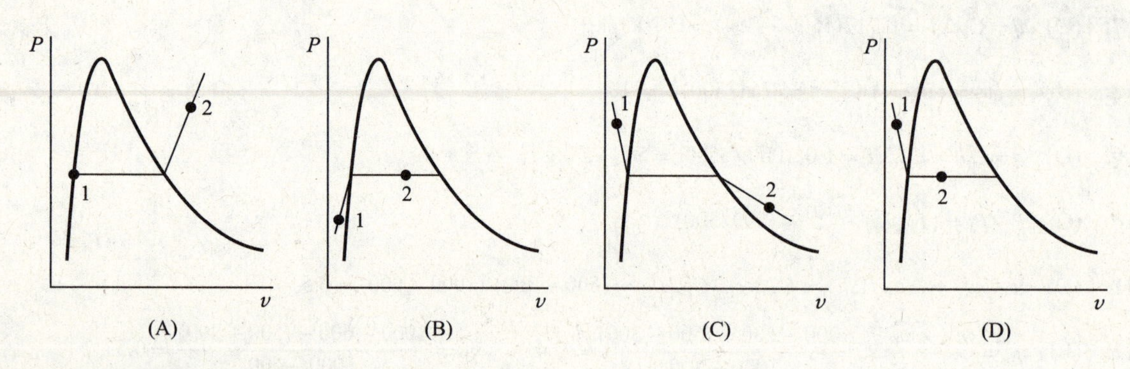

3. In a quasiequilibrium process

(A) the pressure remains constant.
(B) the pressure varies with temperature.
(C) the pressure is everywhere constant at an instant.
(D) the prêssure force results in a positive work input.

4. Which of the following first-law statements is wrong?

(A) The internal energy change equals the work of a system for an adiabatic process.
(B) The heat transfer equals the enthalpy change for an adiabatic process.
(C) The heat transfer equals the quasiequilibrium work of a system for a constant-volume process in which the internal energy remains constant.
(D) The net heat transfer equals the work output for an engine operating on a cycle.

5. Which of the following statements about work for a quasiequilibrium process is wrong?

(A) Work is energy crossing a boundary.
(B) The differential of work is inexact.
(C) Work is the area under a curve on the *P*-*T* diagram.
(D) Work is a path function.

6. The gage pressure at a depth of 20 m in water is nearest:

(A) 100 kPa (B) 200 kPa (C) 300 kPa (D) 400 kPa

7. The force of the water on the 2-m-wide vertical plane area is:
(A) 400 kN (B) 350 kN
(C) 300 kN (D) 250 kN

8. One kilogram of steam in a cylinder requires 320 kJ of heat while $P = \text{const.} = 1$ MPa. Find T_2 if $T_1 = 300\,°\text{C}$.
(A) 480 °C (B) 450 °C (C) 420 °C (D) 380 °C

9. Calculate the magnitude of the work in Question 8.
(A) 85 kJ (B) 72 kJ (C) 66 kJ (D) 59 kJ

10. 200 kg of a solid at 0 °C is added to 42 kg of water at 80 °C. The final temperature is:
(A) 58 °C (B) 52 °C (C) 48 °C (D) 42 °C

11. The pressure of steam at 400 °C and $u = 2949$ kJ/kg is nearest:
(A) 1.9 MPa (B) 1.8 MPa (C) 1.7 MPa (D) 1.6 MPa

12. The enthalpy of steam at 400 kPa and $v = 0.7$ m^3/kg is nearest:
(A) 3230 kJ/kg (B) 3240 kJ/kg (C) 3250 kJ/kg (D) 3260 kJ/kg

13. The heat needed to raise the temperature of 0.1 m^3 of saturated steam from 120 °C to 500 °C in a rigid tank is:
(A) 306 kJ (B) 221 kJ (C) 152 kJ (D) 67 kJ

14. Estimate C_p for steam at 4 MPa and 350 °C.
(A) 2.71 kJ/kg·K (B) 2.53 kJ/kg·K (C) 2.31 kJ/kg·K (D) 2.04 kJ/kg·K

15. A cycle involving a system has 3 processes. Find W_{1-2}.
(A) 50 kJ (B) 40 kJ
(C) 30 kJ (D) 20 kJ

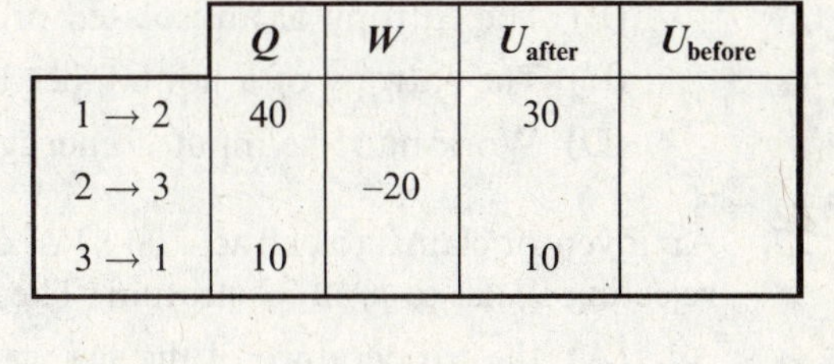

	Q	W	U_{after}	U_{before}
1 → 2	40		30	
2 → 3		–20		
3 → 1	10		10	

16. The gas methane is heated at $P = \text{const.} = 200$ kPa from 0 °C to 300 °C. The heat required is:
(A) 766 kJ/kg (B) 731 kJ/kg
(C) 692 kJ/kg (D) 676 kJ/kg

17. Find Q_{2-3} for the cycle shown.
(A) – 500 kJ/kg (B) – 358 kJ/kg
(C) 500 kJ/kg (D) 358 kJ/kg

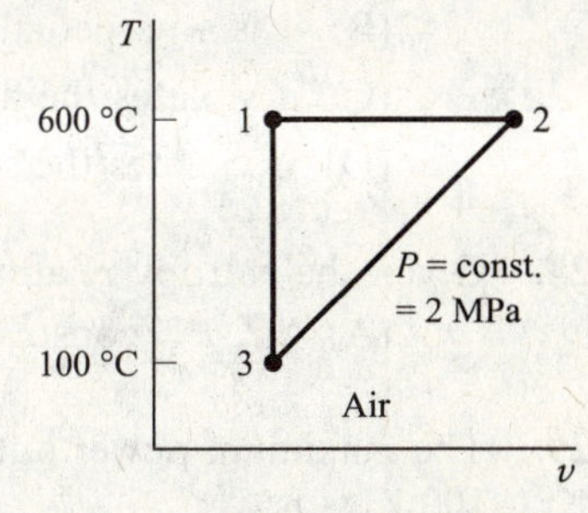

Questions 17, 18, 19

18. Find W_{1-2} for the cycle shown.
(A) 500 kJ/kg (B) 358 kJ/kg
(C) 213 kJ/kg (D) 178 kJ/kg

19. Find Q_{1-2} for the cycle shown.
(A) 500 kJ/kg (B) 358 kJ/kg
(C) 213 kJ/kg (D) 178 kJ/kg

20. R134a enters an expansion valve as saturated liquid at 16 °C and exits at −20 °C. Its quality is:
(A) 0.3 (B) 0.26 (C) 0.22 (D) 0.20

21. The work needed to compress 1 kg/s of air adiabatically in a steady-flow compressor is:
(A) 180 kW (B) 130 kW (C) 90 kW (D) 70 kW

22. 200 kJ/kg of heat is added to air in a steady-flow heat exchanger. The temperature rise is:
(A) 200 °C (B) 220 °C (C) 240 °C (D) 260 °C

23. A nozzle accelerates air from 10 m/s to 800 m/s. The temperature rise is nearest:
(A) 200 °C (B) 240 °C (C) 280 °C (D) 320 °C

24. The net entropy change in the universe during a process is:
(A) zero
(B) equal to or greater than zero
(C) equal to or less than zero
(D) undefined

25. An isentropic process:
(A) is adiabatic and reversible.
(B) is reversible but may not be adiabatic.
(C) is adiabatic but may not be reversible.
(D) is always reversible.

26. Which of the following second-law statements is wrong?
(A) Heat is always rejected from a heat engine.
(B) The entropy of an isolated process must remain constant or increase.
(C) The entropy of a hot copper block decreases as it cools.
(D) Work must be input if energy is transferred from a cold body to a hot body.

27. An inventor claims to extract 100 kJ of energy from 30 °C sea water and generate 7 kJ of heat while rejecting 93 kJ to a 10 °C stratum. The scheme fails because:
(A) the temperature of the sea water is too low.
(B) it is a perpetual-motion machine.
(C) it violates the first law.
(D) it violates the second law.

28. Given the entropy relationship $\Delta s = C_p \ln T_2/T_1$. For which of the following is it incorrect?
(A) Air, $P = \text{const.}$ (B) Water (C) A reservoir (D) Copper

29. The maximum power output of an engine operating between 600 °C and 20 °C if the heat input is 100 kJ/s is:
(A) 83.3 kW (B) 77.3 kW (C) 66.4 kW (D) 57.3 kW

30. Ten kilograms of iron at 300 °C is chilled in a large volume of ice and water. If $C_{\text{iron}} = 0.45$ kJ/kg·K, the net entropy change is nearest:
(A) 0.41 kJ/kg (B) 0.76 kJ/kg (C) 1.20 kJ/kg (D) 1.61 kJ/kg

31. A farmer has a pressurized air tank at 2000 kPa and 20 °C. Estimate the exiting temperature from a valve assuming isentropic flow.
(A) −150 °C (B) −120 °C (C) −90 °C (D) −40 °C

32. Three temperatures in a diesel cycle are $T_1 = 300$ K, $T_2 = 950$ K, $T_3 = 2000$ K. Find the heat input assuming constant specific heats.
(A) 1150 kJ/kg (B) 1050 kJ/kg (C) 950 kJ/kg (D) 850 kJ/kg

33. The compression ratio for the diesel cycle of Question 32 is nearest:
(A) 19.8 (B) 18.4 (C) 17.8 (D) 17.2

34. The temperature T_4 for the diesel cycle of Question 32 is nearest:
(A) 950 K (B) 900 K (C) 850 K (D) 800 K

35. Find q_B in kJ/kg:
(A) 3400 (B) 3200
(C) 2900 (D) 2600

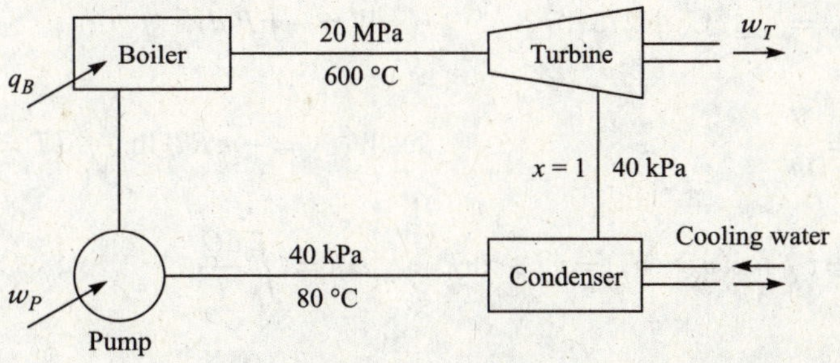

36. Find w_P in kJ/kg:
(A) 20 (B) 30
(C) 40 (D) 50

37. Find w_T in kJ/kg:
(A) 1000 (B) 950
(C) 925 (D) 900

Questions 35–39

38. The efficiency of the turbine is nearest:
(A) 67% (B) 69% (C) 72% (D) 75%

39. The mass flux of the condenser cooling water if its $\Delta T = 30$ °C and $\dot{m}_{steam} = 50$ kg/s is:
(A) 1990 kg/s (B) 1920 kg/s (C) 1840 kg/s (D) 1770 kg/s

40. Using equations, calculate the maximum possible specific humidity ω if $T = 30$ °C. ($P = 100$kPa.)
(A) 0.0273 (B) 0.0274 (C) 0.0275 (D) 0.0276

41. The outside temperature is 35 °C. A wet cloth, attached to the bulb of a thermometer, reads 28 °C with air blowing over it. Estimate the humidity.
(A) 50% (B) 55% (C) 60% (D) 65%

42. Outside air at 10 °C and 100% humidity is heated to 30 °C. How much heat is needed if $\dot{m}_{air} = 10$ kg/s?
(A) 185 kJ/s (B) 190 kJ/s (C) 195 kJ/s (D) 205 kJ/s

Equation of use in the closed-book exam (use this page for your work)

Air: $k = 1.4$, $R = 0.287$ kJ/kg·K, $C_p = 1.00$ kJ/kg·K, $C_v = 0.717$ kJ/kg·K

Water: $\rho = 1000 \text{kg/m}^3$, $C_p = 4.18$ kJ/kg·K

$PV = mRT$ $P = \rho RT$ $Pv = RT$ $P = \rho g \Delta h$

$h = u + Pv$ $\Delta u = C_v \Delta T$ $\Delta h = C_p \Delta T$

$\Delta s = C_v \ln \frac{T_2}{T_1} + R \ln \frac{V_2}{V_1}$ $\Delta s = C_p \ln \frac{T_2}{T_1} - R \ln \frac{P_2}{P_1}$ $\frac{T_2}{T_1} = \left(\frac{P_2}{P_1}\right)^{\frac{k-1}{k}} = \left(\frac{V_1}{V_2}\right)^{k-1}$

$T_{\text{abs}} =$ ___ °C $+ 273$ $W = -\int P\, dV$

$v = \frac{V}{m}$ $W_{1-2} = -mRT \ln \frac{V_2}{V_1}$ $(T = \text{const.})$

$v = v_f + x(v_g - v_f) = v_f + xv_{fg}$ $\Delta S = \int \frac{\delta Q}{T}$

$Q - W = \Delta U + \Delta KE + \Delta PE$

$\rho A \mathcal{V} = \dot{m}$

$$\dot{Q} - \dot{W}_s = \dot{m}\left(h_2 - h_1 + \frac{\mathcal{V}_2^2 - \mathcal{V}_1^2}{2}\right) \quad \text{(gas or vapor)}$$

$$= \dot{m}\left(\frac{P_2 - P_1}{\rho} + u_2 - u_1 + \frac{\mathcal{V}_2^2 - \mathcal{V}_1^2}{2}\right) \quad \text{(liquid)}$$

$\eta_{\max} = \frac{\text{output}}{\text{input}} = 1 - \frac{T_L}{T_H}$ $(\text{COP}_\text{R})_{\max} = \frac{\text{output}}{\text{input}} = \frac{T_L}{T_H - T_L}$

$\phi = \frac{P_v}{P_g}$ $\omega = 0.622 \frac{P_v}{P_g}$ $P = P_v + P_a$

Solution to Final Exam

1. (D)

2. (C)

3. (C)

4. (B)

5. (C)

6. (B) $P = \rho g \Delta h = 1000 \times 9.8 \times 20 = 196\,000$ Pa

7. (B) The pressure varies linearly from top to bottom. Hence, use the average pressure at the center. $\therefore F = P_{avg} A = \rho g h_{center} A = 1000 \times 9.8 \times 3 \times (2 \times 6) = 352\,800$ N

8. (B) $Q = m\Delta H$. $320 = 1(h_2 - 3051)$. $\therefore h_2 = 3371$. $P_2 = 1$ MPa. $\therefore T_2 = 450\,^\circ\text{C}$

9. (B) $W = Pm\Delta v = 1000 \times 1(0.3304 - 0.2579) = 72.5$ kJ

10. (D) $m_s C_s \Delta T_s = m_w C_w \Delta T_w$. $200 \times 0.8(T_2 - 0) = 42 \times 4.18(80 - T_2)$, $T_2 = 41.9\,^\circ\text{C}$

11. (C) Interpolate: $P = \dfrac{2951.3 - 2949}{2951.3 - 2945.2} \times 0.5 + 11.5 = 1.698$ MPa

12. (C) $\dfrac{0.7 - 0.6548}{0.7139 - 0.6548}(3273.4 - 3169.6) + 3169.6 = 3249$ kJ/kg

13. (D) $m = V/v = 0.1/0.892 = 0.112$ kg. $Q = m\Delta v = 0.112(3130 - 2529) = 67.3$ kJ

14. (B) $C_p = \Delta h/\Delta T = (3213.5 - 2960.7)/(400 - 300) = 2.53$ kJ/kg·K

15. (D) From last row, $U_1 = 10$. From first row, $40 - W = 30 - 10$. $\therefore W = 20$ kJ

16. (D) $q = \Delta h = C_p \Delta T = 2.254 \times 300 = 676$ kJ/kg

17. (A) $q = C_p \Delta T = 1.0(100 - 600) = -500$ kJ/kg

18. (C) $v_3/v_2 = T_3/T_2 = 373/873$. $w = RT \ln v_2/v_1 = 0.287 \times 873 \ln 873/373 = 213$ kJ/kg

19. (C) $q = w = 213$ kJ/kg

20. (C) $h_2 = h_1 = 71.69 = 24.26 + 211.05x_2$. $\therefore x_2 = 0.225$

21. (A) $\dot{W} = \dot{m}\Delta h = \dot{m} C_p \Delta T = 1 \times 1.0(200 - 20) = 180$ kW

22. (A) $q = \Delta h = C_p \Delta T$. $200 = 1.0\Delta T$. $\therefore \Delta T = 200\,^\circ\text{C}$

23. (D) $0 = h_1 - h_1 + \dfrac{V_2^2 - V_1^2}{2}$. $1.0 \times \Delta T = \dfrac{10^2 - 800^2}{2 \times 1000}$. $\therefore \Delta T = -320\,^\circ\text{C}$

24. (B)

25. (C)

26. (D)

27. (D) $\eta = \frac{7}{100} = 0.7.$ $\quad \eta_{\max} = 1 - \frac{T_L}{T_H} = 1 - \frac{283}{303} = 0.066.$ Thus, it violates the second law.

28. (C)

29. (C) $\eta_{\max} = 1 - \frac{T_L}{T_H} = 1 - \frac{293}{873} = 0.664 = \frac{\dot{W}}{\dot{Q}_{in}} = \frac{\dot{W}}{100}.$ $\quad \therefore \dot{W} = 66.4 \text{ kW}$

30. (D) $Q = mC_i\Delta T = 10 \times 0.45(300 - 0) = 1350 \text{ kJ}.$ $\quad \Delta S_{net} = 10 \times 0.45 \ln \frac{273}{573} + \frac{1350}{273} = 1.61 \text{ kJ/kg}$

31. (A) $T_2 = T_1\left(\frac{P_2}{P_1}\right)^{\frac{k-1}{k}} = 293\left(\frac{100}{2000}\right)^{0.2857} = 124 \text{ K or } -149\,°\text{C}$

32. (B) $q_{2-3} = \Delta h = C_p\Delta T = 1.0 \times (2000 - 950) = 1050 \text{ kJ/kg}$

33. (C) $\frac{v_1}{v_2} = \left(\frac{T_2}{T_1}\right)^{1/(k-1)} = \left(\frac{950}{300}\right)^{2.5} = 17.8$

34. (C) $\frac{v_3}{v_2} = \frac{T_3}{T_2} = \frac{2000}{950} = 2.11.$ $\quad \frac{v_4}{v_3} = \frac{v_1}{v_2} \times \frac{v_2}{v_3} = \frac{17.8}{2.11} = 8.46.$ $\quad T_4 = T_3\left(\frac{v_3}{v_4}\right)^{k-1} = 2000/8.46^{0.4} = 851 \text{ K}$

35. (B) $q = h_3 - h_2 = 3538 - 335 = 3203 \text{ kJ/kg}$

36. (A) $w_p = \frac{\Delta P}{\rho} = \frac{20\,000 - 40}{1000} = 20 \text{ kJ/kg}$

37. (D) $w_T = h_3 - h_4 = 3538 - 2637 = 901 \text{ kJ/kg}$

38. (B) $\eta = \frac{h_3 - h_4}{h_3 - h_4} = \frac{901}{3538 - 2232} = 0.690$

where $s_3 = s_4 = 6.51 = 1.026 + 6.645x_{4'}.$ $\quad \therefore x_{4'} = 0.825$ and $h_{4'} = 2232$

39. (C) $\dot{m}_s\Delta h_s = \dot{m}_w\Delta h_w.$ $\quad 50(2637 - 335) = \dot{m}_w \times 4.18 \times 15.$ $\quad \therefore \dot{m}_w = 1836 \text{ kg/s}$

40. (D) $\phi = 1.$ $\quad \therefore P_v = P_g = 4.246 \text{ kPa}.$ $\quad \omega = \frac{0.622 \times 4.246}{100 - 4.246} = 0.0276$

41. (C) From the psychrometric chart $\phi = 60\%$.

42. (D) $\dot{Q} = \dot{m}\Delta h = 10(50 - 29.5) = 205$ using $\omega_1 = \omega_2$ on the psychrometric chart.

APPENDIX A

Conversions of Units

Length	Force	Mass	Velocity
1 cm = 0.3937 in	1 lbf = 0.4448×10^6 dyne	1 oz = 28.35 g	1 mph = 1.467 ft/sec
1 m = 3.281 ft	1 dyne = 2.248×10^{-6} lbf	1 lbm = 0.4536 kg	1 mph = 0.8684 knot
1 km = 0.6214 mi	1 kip = 1000 lbf	1 slug = 32.17 lbm	1 ft/sec = 0.3048 m/s
1 in = 2.54 cm	1 N = 0.2248 lbf	1 slug = 14.59 kg	1 km/h = 0.2778 m/s
1 ft = 0.3048 m	1 N = 10^5 dyne	1 kg = 2.205 lbm	1 knot = 1.688 ft/sec
1 mi = 1.609 km			
1 mi = 5280 ft			
1 mi = 1760 yd			

Work and Heat	Power	Pressure	Volume
1 J = 10^7 ergs	1 ho = 550 ft-lb/sec	1 psi = 2.036 in Hg	1 ft^3 = 7.481 gal (U.S.)
1 ft-lbf = 1.356 J	1 hp = 2545 Btu/hr	1 psi = 27.7 in H_2O	1 ft^3 = 0.02832 m^3
1 Cal = 3.088 ft-lb	1 hp = 0.7455 kW	1 atm = 29.92 in Hg	1 gal (U.S.) = 231 in^3
1 Cal = 0.003968 Btu	1 W = 1 J/s	1 atm = 33.93 ft H_2O	1 gal (Brit.) = 1.2 gal (U.S.)
1 Btu = 1055 J	1 W = 1.0×10^7 dyne·cm/s	1 atm = 101.3 kPa	1 L = 10^{-3} m^3
1 Btu = 0.2930 W·hr	1 W = 3.412 Btu/hr	1 atm = 1.0133 bar	1 L = 0.03531 ft^3
1 Btu = 778 ft-lb	1 kW = 1.341 hp	1 atm = 14.7 psi	1 L = 0.2642 gal
1 kWh = 3412 Btu	1 ton = 12,000 Btu/hr	1 in Hg = 0.4912 psi	1 m^3 = 264.2 gal
1 therm = 10^5 Btu	1 ton = 3.517 kW	1 ft H_2O = 0.4331 psi	1 m^3 = 35.31 ft^3
1 quad = 10^{15} Btu		1 psi = 6.895 kPa	1 ft^3 = 28.32 L
		1 kPa = 0.145 psi	1 in^3 = 16.387 Cm^3

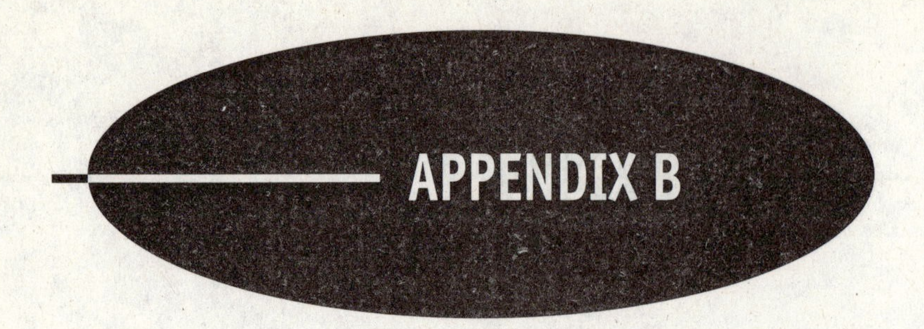

Material Properties

Table B-1 Properties of the U.S. Standard Atmosphere

$P_0 = 101.3$ kPa, $\rho_0 = 1.225$ kg/m^3

Altitude m	Temperature °C	Pressure P/P_0	Density ρ/ρ_0
0	15.2	1.000	1.000
1,000	9.7	0.8870	0.9075
2,000	2.2	0.7846	0.8217
3,000	−4.3	0.6920	0.7423
4,000	−10.8	0.6085	0.6689
5,000	−17.3	0.5334	0.6012
6,000	−23.8	0.4660	0.5389
7,000	−30.3	0.4057	0.4817
8,000	−36.8	0.3519	0.4292
10,000	−49.7	0.2615	0.3376
12,000	−56.3	0.1915	0.2546
14,000	−56.3	0.1399	0.1860
16,000	−56.3	0.1022	0.1359
18,000	−56.3	0.07466	0.09930
20,000	−56.3	0.05457	0.07258
30,000	−46.5	0.01181	0.01503
40,000	−26.6	0.2834×10^{-2}	0.3262×10^{-2}
50,000	−2.3	0.7874×10^{-3}	0.8383×10^{-3}
60,000	−17.2	0.2217×10^{-3}	0.2497×10^{-3}
70,000	−53.3	0.5448×10^{-4}	0.7146×10^{-4}

Table B-1E Properties of the U.S. Standard Atmosphere

$P_0 = 14.7$ psia, $\rho_0 = 0.0763$ kg/ft^3

Altitude ft	Temperature °F	Pressure P/P_0	Density ρ/ρ_0
0	59.0	1.00	1.00
1,000	55.4	0.965	0.975
2,000	51.9	0.930	0.945
5,000	41.2	0.832	0.865
10,000	23.4	0.688	0.743
15,000	5.54	0.564	0.633
20,000	−12.3	0.460	0.536
25,000	−30.1	0.371	0.451
30,000	−48.0	0.297	0.376
35,000	−65.8	0.235	0.311
36,000	−67.6	0.224	0.299
40,000	−67.6	0.185	0.247
50,000	−67.6	0.114	0.153
100,000	−67.6	0.0106	0.0140
110,000	−47.4	0.00657	0.00831
150,000	113.5	0.00142	0.00129
200,000	160.0	0.314×10^{-3}	0.262×10^{-3}
260,000	−28	0.351×10^{-4}	0.422×10^{-4}

Table B-2 Properties of Various Ideal Gases

Gas	Chemical Formula	Molar Mass	R		C_p		C_v		k
			kJ/kg·K	ft-lbf/lbm-°R	kJ/kg·K	Btu/lbm-°R	kJ/kg·K	Btu/lbm-°R	
Air	–	28.97	0.287 0	53.34	1.003	0.240	0.717	0.171	1.400
Argon	Ar	39.95	0.208 1	38.68	0.520	0.1253	0.312	0.0756	1.667
Butane	C_4H_{10}	58.12	0.143 0	26.58	1.716	0.415	1.573	0.381	1.091
Carbon dioxide	CO_2	44.01	0.188 9	35.10	0.842	0.203	0.653	0.158	1.289
Carbon monoxide	CO	28.01	0.296 8	55.16	1.041	0.249	0.744	0.178	1.400
Ethane	C_2H_6	30.07	0.276 5	51.38	1.766	0.427	1.490	0.361	1.186
Ethylene	C_2H_4	28.05	0.296 4	55.07	1.548	0.411	1.252	0.340	1.237
Helium	He	4.00	2.077 0	386.0	5.198	1.25	3.116	0.753	1.667
Hydrogen	H_2	2.02	4.124 2	766.4	14.209	3.43	10.085	2.44	1.409
Methane	CH_4	16.04	0.518 4	96.35	2.254	0.532	1.735	0.403	1.299
Neon	Ne	20.18	0.412 0	76.55	1.020	0.246	0.618	0.1477	1.667
Nitrogen	N_2	28.01	0.296 8	55.15	1.042	0.248	0.745	0.177	1.400
Octane	C_8H_{18}	114.23	0.072 8	13.53	1.711	0.409	1.638	0.392	1.044
Oxygen	O_2	32.00	0.259 8	48.28	0.922	0.219	0.662	0.157	1.393
Propane	C_3H_8	44.10	0.188 6	35.04	1.679	0.407	1.491	0.362	1.126
Steam	H_2O	18.02	0.461 5	85.76	1.872	0.445	1.411	0.335	1.327

Note: C_p, C_v, and k are at 300 K. Also, kJ/kg·K is the same as kJ/kg·°C.

SOURCE: G. J. Van Wylen and R. E. Sonntag, *Fundamentals of Classical Thermodynamics*, Wiley, New York, 1976.

Table B-3 Critical Point Constants

			Temperature		Pressure		Volume		
Substance	Formula	Molar Mass	K	°R	MPa	psia	ft³/lbmol	m³/kmol	Z_{cr}
Air		28.97	133	239	3.77	547	1.41	0.0883	0.30
Ammonia	NH_3	17.03	405.5	729.8	11.28	1636	1.16	0.0724	0.243
Argon	Ar	39.94	151	272	4.86	705	1.20	0.0749	0.290
Benzene	C_6H_6	78.11	562	1012	4.92	714	4.17	0.2603	0.274
Butane	C_4H_{10}	58.12	425.2	765.2	3.80	551	4.08	0.2547	0.274
Carbon dioxide	CO_2	44.01	304.2	547.5	7.39	1070	1.51	0.0943	0.275
Carbon monoxide	CO	28.01	133	240	3.50	507	1.49	0.0930	0.294
Carbon tetrachloride	CCl_4	153.84	556.4	1001.5	4.56	661	4.42	0.2759	0.272
Ethane	C_2H_6	30.07	305.5	549.8	4.88	708	2.37	0.148	0.284
Ethylene	C_2H_4	28.05	282.4	508.3	5.12	742	1.99	0.1242	0.271
Helium	He	4.00	5.3	9.5	0.23	33.2	0.926	0.0578	0.302
Hydrogen	H_2	2.02	33.3	59.9	1.30	188	1.04	0.0649	0.304
Methane	CH_4	16.04	191.1	343.9	4.64	673	1.59	0.0993	0.290
Neon	Ne	20.18	44.5	80.1	2.73	395	0.668	0.0417	0.308
Nitrogen	N_2	28.02	126.2	227.1	3.39	492	1.44	0.0899	0.291
Oxygen	O_2	32.00	154.8	278.6	5.08	736	1.25	0.078	0.308
Propane	C_3H_8	44.09	370.0	665.9	4.26	617	3.20	0.1998	0.277
Propylene	C_3H_6	42.08	365.0	656.9	4.62	670	2.90	0.1810	0.276
R134a	CF_3CH_2F	102.03	374.3	613.7	4.07	596	2.96	0.2478	0.324
Sulfur dioxide	SO_2	64.06	430.7	775.2	7.88	1143	1.95	0.1217	0.269
Water	H_2O	18.02	647.4	1165.3	22.1	3204	0.90	0.0568	0.233

SOURCE: K. A. Kobe and R. E. Lynn, Jr., *Chem. Rev.*, **52**: 117–236 (1953).

Table B-4 Specific Heats of Liquids and Solids

C_p, kJ/kg·°C

Liquids

Substance	State	C_p	Substance	State	C_p
Water	1 atm, 25 °C	4.177	Glycerin	1 atm, 10 °C	2.32
Ammonia	sat., −20 °C	4.52	Bismuth	1 atm, 425 °C	0.144
	sat., 50 °C	5.10	Mercury	1 atm, 10 °C	0.138
Freon 12	sat., −20 °C	0.908	Sodium	1 atm, 95 °C	1.38
	sat., 50 °C	1.02	Propane	1 atm, 0 °C	2.41
Benzene	1 atm, 15 °C	1.80	Ethyl Alcohol	1 atm, 25 °C	2.43

Solids

Substance	T, °C	C_p	Substance	T, °C	C_p
Ice	−11	2.033	Lead	−100	0.118
	−2.2	2.10		0	0.124
Aluminum	−100	0.699		100	0.134
	0	0.870	Copper	−100	0.328
	100	0.941		0	0.381
Iron	20	0.448		100	0.393
Silver	20	0.233			

SOURCE: Kenneth Wark, *Thermodynamics*, 3d ed., McGraw-Hill, New York, 1981.

Table B-4E Specific Heats of Liquids and Solids

C_p, Btu/lbm-°F

Liquids

Substance	State	C_p	Substance	State	C_p
Water	1 atm, 77 °C	1.00	Glycerin	1 atm, 50 °C	0.555
Ammonia	sat., −4 °C	1.08	Bismuth	1 atm, 800 °C	0.0344
	sat., 120 °C	1.22	Mercury	1 atm, 50 °C	0.0330
Freon 12	sat., −4 °C	0.217	Sodium	1 atm, 200 °C	0.330
	sat., 120 °C	0.244	Propane	1 atm, 32 °C	0.577
Benzene	1 atm, 60 °F	0.431	Ethyl Alcohol	1 atm, 77 °C	0.581

Solids

Substance	T, °F	C_p	Substance	T, °F	C_p
Ice	−76	0.392	Silver	−4	0.0557
	−12	0.486	Lead	−150	0.0282
Aluminum	−28	0.402		30	0.0297
	−150	0.167		210	0.0321
	30	0.208	Copper	−150	0.0785
	210	0.225		30	0.0911
Iron	−4	0.107		210	0.0940

SOURCE: Kenneth Wark, *Thermodynamics*, 3d ed., McGraw-Hill, New York, 1981.

Table B-5 Constant-Pressure Specific Heat of Various Ideal Gases

$\theta \equiv T(\text{Kelvin})/100$

Gas	C_p kJ/kmol·K	Range K	Max. Error %
N_2	$39.060 - 512.79\theta^{-1.5} + 1072.78^{-2} - 820.40\theta^{-3}$	300–3500	0.43
O_2	$37.432 + 0.020102\theta^{1.5} - 178.57\theta^{-1.5} + 236.88\theta^{-2}$	300–3500	0.30
H_2	$56.505 - 702.74\theta^{-0.75} + 1165.0\theta^{-1} - 560.70\theta^{-1.5}$	300–3500	0.60
CO	$69.145 - 0.70463\theta^{0.75} - 200.77\theta^{-0.5} + 176.76\theta^{-0.75}$	300–3500	0.42
OH	$81.546 - 59.350\theta^{0.25} + 17.329\theta^{0.75} - 4.2660\theta$	300–3500	0.43
NO	$59.283 - 1.7096\theta^{0.5} - 70.613\theta^{-0.5} + 74.889\theta^{-1.5}$	300–3500	0.34
H_2O	$143.05 - 183.54\theta^{0.25} + 82.751\theta^{0.5} - 3.6989\theta$	300–3500	0.43
CO_2	$-3.7357 + 30.529\theta^{0.5} - 4.1034\theta + 0.024198\theta^{2}$	300–3500	0.19
NO_2	$46.045 + 216.10\theta^{-0.5} - 363.66\theta^{-0.75} + 232.550\theta^{-2}$	300–3500	0.26
CH_4	$-672.87 + 439.74\theta^{0.25} - 24.875\theta^{0.75} + 323.88\theta^{-0.5}$	300–2000	0.15
C_2H_4	$-95.395 + 123.15\theta^{0.5} - 35.641\theta^{0.75} + 182.77\theta^{-3}$	300–2000	0.07

SOURCE: G. J. Van Wylen and R. E. Sonntag, *Fundamentals of Classical Thermodynamics*, Wile

Table B-5E Constant-Pressure Specific Heat of Various Ideal Gases

$\theta \equiv T(\text{Rankine})/180$

Gas	C_p Btu/lbmol-°R	Range °R	Max. Error %
N_2	$9.3355 - 122.56\theta^{-1.5} + 256.38\theta^{-2} - 196.08\theta^{-3}$	540–6300	0.43
O_2	$8.9465 + 4.8044 \times 10^{-3}\theta^{1.5} - 42.679\theta^{-1.5} + 56.615\theta^{-2}$	540–6300	0.30
H_2	$13.505 - 167.96\theta^{-0.75} + 278.44\theta^{-1} - 134.01\theta^{-1.5}$	540–6300	0.60
CO	$16.526 - 0.16841\theta^{0.75} - 47.985\theta^{-0.5} + 42.246\theta^{-0.75}$	540–6300	0.42
OH	$19.490 - 14.185\theta^{0.25} + 4.1418\theta^{0.75} - 1.0196\theta$	540–6300	0.43
NO	$14.169 - 0.40861\theta^{0.5} - 16.877\theta^{-0.5} + 17.899\theta^{-1.5}$	540–6300	0.34
H_2O	$34.190 - 43.868\theta^{0.25} + 19.778\theta^{0.5} - 0.88407\theta$	540–6300	0.43
CO_2	$-0.89286 + 7.2967\theta^{0.5} - 0.98074\theta + 5.7835 \times 10^{-3}\theta^{-2}$	540–6300	0.19
NO_2	$11.005 + 51.650\theta^{0.5} - 86.916\theta^{0.75} + 55.580\theta^{-2}$	540–6300	0.26
CH_4	$-160.82 + 105.10\theta^{0.25} - 5.9452\theta^{0.75} + 77.408\theta^{-0.5}$	540–3600	0.15
C_2H_4	$-22.800 + 29.433\theta^{0.5} - 8.5185\theta^{0.75} + 43.683\theta^{-3}$	540–3600	0.07

SOURCE: G. J. Van Wylen and R. E. Sonntag, *Fundamentals of Classical Thermodynamics*, Wiley, New York, 1976.

Table B-6 Enthalpy of Formation and Enthalpy of Vaporization

25 °C (77 °F), 1 atm

Substance	Formula	$\bar{h}^\circ_f$, kJ/kmol	$\bar{h}_{fg}$, kJ/kmol	$\bar{h}^\circ_f$, Btu/lbmol	$\bar{h}_{fg}$, Btu/lbmol
Carbon	$C(s)$	0		0	
Hydrogen	$H_2(g)$	0		0	
Nitrogen	$N_2(g)$	0		0	
Oxygen	$O_2(g)$	0		0	
Carbon monoxide	$CO(g)$	−110 530		−47,540	
Carbon dioxide	$CO_2(g)$	−393 520		−169,300	
Water	$H_2O(g)$	−241 820		−104,040	
Water	$H_2O(l)$	−285 830	44 010	−122,970	
Hydrogen peroxide	$H_2O_2(g)$	−136 310	61 090	−58,640	26,260
Ammonia	$NH_3(g)$	−46 190		−19,750	
Oxygen	$O(g)$	249 170		+ 107,210	
Hydrogen	$H(g)$	218 000		+ 93,780	
Nitrogen	$N(g)$	472 680		+203,340	
Hydroxyl	$OH(g)$	39 040		+ 16,790	
Methane	$CH_4(g)$	−74 850		−32,210	
Acetylene (Ethyne)	$C_2H_2(g)$	226 730		+ 97,540	
Ethylene (Ethene)	$C_2H_4(g)$	52 280		+ 22,490	
Ethane	$C_2H_6(g)$	−84 680		−36,420	
Propylene (Propene)	$C_3H_6(g)$	20 410		+ 8,790	
Propane	$C_3H_8(g)$	−103 850	15 060	−44,680	6,480
n-Butane	$C_4H_{10}(g)$	−126 150	21 060	−54,270	9,060
n-Pentane	$C_5H_{12}(g)$	−146 440	31 410		
n-Octane	$C_8H_{18}(g)$	−208 450	41 460	−89,680	17,835
Benzene	$C_6H_6(g)$	82 930	33 830	+ 35,680	14,550
Methyl alcohol	$CH_3OH(g)$	−200 890	37 900	−86,540	16,090
Ethyl alcohol	$C_2H_5OH(g)$	−235 310	42 340	−101,230	18,220

SOURCES: JANAF Thermochemical Tables, NSRDS-NBS-37, 1971; *Selected Values of Chemical Thermodynamic Properties*, NBS Technical Note 270–3, 1968; and API Res. Project 44, Carnegie Press, Carnegie Institute of Technology, Pittsburgh, 1953.

Table B-7 Enthalpy of Combustion and Enthalpy of Vaporization

25 °C (77 °F), 1 atm

Substance	Formula	–HHV, kJ/kmol	$\bar{h}_{fg}$, kJ/kmol	–HHV, Btu/lbmol	$\bar{h}_{fg}$, Btu/lbmol
Hydrogen	$H_2(g)$	–285 840		–122,970	
Carbon	$C(s)$	–393 520		–169,290	
Carbon monoxide	$CO(g)$	–282 990		–121,750	
Methane	$CH_4(g)$	–890 360		–383,040	
Acetylene	$C_2H_2(g)$	–1 299 600		–559,120	
Ethylene	$C_2H_4(g)$	–1 410 970		–607,010	
Ethane	$C_2H_6(g)$	–1 559 900		–671,080	
Propylene	$C_3H_6(g)$	–2 058 500		–885,580	
Propane	$C_3H_8(g)$	–2 220 000	15 060	–955,070	6,480
n-Butane	$C_4H_{10}(g)$	–2 877 100	21 060	–1,237,800	9,060
n-Pentane	$C_5H_{12}(g)$	–3 536 100	26 410	–1,521,300	11,360
n-Hexane	$C_6H_{14}(g)$	–4 194 800	31 530	–1,804,600	13,560
n-Heptane	$C_7H_{16}(g)$	–4 853 500	36 520	–2,088,000	15,710
n-Octane	$C_8H_{18}(g)$	–5 512 200	41 460	–2,371,400	17,835
Benzene	$C_6H_6(g)$	–3 301 500	33 830	–1,420,300	14,550
Toluene	$C_7H_8(g)$	–3 947 900	39 920	–1,698,400	17,180
Methyl alcohol	$CH_3OH(g)$	–764 540	37 900	–328,700	16,090
Ethyl alcohol	$C_2H_5OH(g)$	–1 409 300	42 340	–606,280	18,220

Note: Water appears as a liquid in the products of combustion.

SOURCE: Kenneth Wark, *Thermodynamics*, 3d ed., McGraw-Hill, New York, 1981, pp. 834–835, Table A-23M.

Table B-8 Constants for the van der Waals and the Redlich–Kwong Equation of State

Van der Waals equation

	a, kPa·m^6/kg^2	b, m^3/kg	a, lbf-ft^4/lbm^2	b, ft^3/lbm
Air	0.1630	0.00127	870	0.0202
Ammonia	1.468	0.00220	7850	0.0351
Carbon dioxide	0.1883	0.000972	1010	0.0156
Carbon monoxide	0.1880	0.00141	1010	0.0227
Freon 12	0.0718	0.000803	394	0.0132
Helium	0.214	0.00587	1190	0.0959
Hydrogen	6.083	0.0132	32,800	0.212
Methane	0.888	0.00266	4780	0.0427
Nitrogen	0.1747	0.00138	934	0.0221
Oxygen	0.1344	0.000993	720	0.0159
Propane	0.481	0.00204	2580	0.0328
Water	1.703	0.00169	9130	0.0271

Table B-8 (*Continued*)

Redlich–Kwong Equation

	a, kPa·m^6·K$^{1/2}$/kg^2	b, m^3/kg	a, lbf-ft^4-°R$^{1/2}$/lbm^2	b, ft^3/lbm
Air	1.905	0.000878	13,600	0.014
Ammonia	30.0	0.00152	215,000	0.0243
Carbon dioxide	3.33	0.000674	24,000	0.0108
Carbon monoxide	2.20	0.000978	15,900	0.0157
Freon 12	1.43	0.000557	10,500	0.00916
Helium	0.495	0.00407	3,710	0.0665
Hydrogen	35.5	0.00916	257,000	0.147
Methane	12.43	0.00184	89,700	0.0296
Nitrogen	1.99	0.000957	14,300	0.0153
Oxygen	1.69	0.000689	12,200	0.0110
Propane	9.37	0.00141	67,600	0.0228
Water	43.9	0.00117	316,000	0.0188

APPENDIX C

Thermodynamic Properties of Water (Steam Tables)

Table C-1 Properties of Saturated H_2O—Temperature Table

		Volume, m^3/kg		**Energy**, kJ/kg		**Enthalpy**, kJ/kg			**Entropy**, kJ/kg·K		
T, °C	P, MPa	v_f	v_g	u_f	u_g	h_f	h_{fg}	h_g	s_f	s_{fg}	s_g
0.010	0.0006113	0.001000	206.1	0.0	2375.3	0.0	2501.3	2501.3	0.0000	9.1571	9.1571
2	0.0007056	0.001000	179.9	8.4	2378.1	8.4	2496.6	2505.0	0.0305	9.0738	9.1043
5	0.0008721	0.001000	147.1	21.0	2382.2	21.0	2489.5	2510.5	0.0761	8.9505	9.0266
10	0.001228	0.001000	106.4	42.0	2389.2	42.0	2477.7	2519.7	0.1510	8.7506	8.9016
15	0.001705	0.001001	77.93	63.0	2396.0	63.0	2465.9	2528.9	0.2244	8.5578	8.7822
20	0.002338	0.001002	57.79	83.9	2402.9	83.9	2454.2	2538.1	0.2965	8.3715	8.6680
25	0.003169	0.001003	43.36	104.9	2409.8	104.9	2442.3	2547.2	0.3672	8.1916	8.5588
30	0.004246	0.001004	32.90	125.8	2416.6	125.8	2430.4	2556.2	0.4367	8.0174	8.4541
35	0.005628	0.001006	25.22	146.7	2423.4	146.7	2418.6	2565.3	0.5051	7.8488	8.3539
40	0.007383	0.001008	19.52	167.5	2430.1	167.5	2406.8	2574.3	0.5723	7.6855	8.2578
45	0.009593	0.001010	15.26	188.4	2436.8	188.4	2394.8	2583.2	0.6385	7.5271	8.1656
50	0.01235	0.001012	12.03	209.3	2443.5	209.3	2382.8	2592.1	0.7036	7.3735	8.0771
55	0.01576	0.001015	9.569	230.2	2450.1	230.2	2370.7	2600.9	0.7678	7.2243	7.9921
60	0.01994	0.001017	7.671	251.1	2456.6	251.1	2358.5	2609.6	0.8310	7.0794	7.9104
65	0.02503	0.001020	6.197	272.0	2463.1	272.0	2346.2	2618.2	0.8934	6.9384	7.8318
70	0.03119	0.001023	5.042	292.9	2469.5	293.0	2333.8	2626.8	0.9549	6.8012	7.7561
75	0.03858	0.001026	4.131	313.9	2475.9	313.9	2321.4	2635.3	1.0155	6.6678	7.6833
80	0.04739	0.001029	3.407	334.8	2482.2	334.9	2308.8	2643.7	1.0754	6.5376	7.6130
85	0.05783	0.001032	2.828	355.8	2488.4	355.9	2296.0	2651.9	1.1344	6.4109	7.5453
90	0.07013	0.001036	2.361	376.8	2494.5	376.9	2283.2	2660.1	1.1927	6.2872	7.4799
95	0.08455	0.001040	1.982	397.9	2500.6	397.9	2270.2	2668.1	1.2503	6.1664	7.4167

Table C-1 ***(Continued)***

		Volume, m³/kg		**Energy**, kJ/kg		**Enthalpy**, kJ/kg			**Entropy**, kJ/kg·K		
T, °C	P, MPa	v_f	v_g	u_f	u_g	h_f	h_{fg}	h_g	s_f	s_{fg}	s_g
100	0.1013	0.001044	1.673	418.9	2506.5	419.0	2257.0	2676.0	1.3071	6.0486	7.3557
110	0.1433	0.001052	1.210	461.1	2518.1	461.3	2230.2	2691.5	1.4188	5.8207	7.2395
120	0.1985	0.001060	0.8919	503.5	2529.2	503.7	2202.6	2706.3	1.5280	5.6024	7.1304
130	0.2701	0.001070	0.6685	546.0	2539.9	546.3	2174.2	2720.5	1.6348	5.3929	7.0277
140	0.3613	0.001080	0.5089	588.7	2550.0	589.1	2144.8	2733.9	1.7395	5.1912	6.9307
150	0.4758	0.001090	0.3928	631.7	2559.5	632.2	2114.2	2746.4	1.8422	4.9965	6.8387
160	0.6178	0.001102	0.3071	674.9	2568.4	675.5	2082.6	2758.1	1.9431	4.8079	6.7510
170	0.7916	0.001114	0.2428	718.3	2576.5	719.2	2049.5	2768.7	2.0423	4.6249	6.6672
180	1.002	0.001127	0.1941	762.1	2583.7	763.2	2015.0	2778.2	2.1400	4.4466	6.5866
190	1.254	0.001141	0.1565	806.2	2590.0	807.5	1978.8	2786.4	2.2363	4.2724	6.5087
200	1.554	0.001156	0.1274	850.6	2595.3	852.4	1940.8	2793.2	2.3313	4.1018	6.4331
210	1.906	0.001173	0.1044	895.5	2599.4	897.7	1900.8	2798.5	2.4253	3.9340	6.3593
220	2.318	0.001190	0.08620	940.9	2602.4	943.6	1858.5	2802.1	2.5183	3.7686	6.2869
230	2.795	0.001209	0.07159	986.7	2603.9	990.1	1813.9	2804.0	2.6105	3.6050	6.2155
240	3.344	0.001229	0.05977	1033.2	2604.0	1037.3	1766.5	2803.8	2.7021	3.4425	6.1446
250	3.973	0.001251	0.05013	1080.4	2602.4	1085.3	1716.2	2801.5	2.7933	3.2805	6.0738
260	4.688	0.001276	0.04221	1128.4	2599.0	1134.4	1662.5	2796.9	2.8844	3.1184	6.0028
270	5.498	0.001302	0.03565	1177.3	2593.7	1184.5	1605.2	2789.7	2.9757	2.9553	5.9310
280	6.411	0.001332	0.03017	1227.4	2586.1	1236.0	1543.6	2779.6	3.0674	2.7905	5.8579
290	7.436	0.001366	0.02557	1278.9	2576.0	1289.0	1477.2	2766.2	3.1600	2.6230	5.7830
300	8.580	0.001404	0.02168	1332.0	2563.0	1344.0	1405.0	2749.0	3.2540	2.4513	5.7053
310	9.856	0.001447	0.01835	1387.0	2546.4	1401.3	1326.0	2727.3	3.3500	2.2739	5.6239
320	11.27	0.001499	0.01549	1444.6	2525.5	1461.4	1238.7	2700.1	3.4487	2.0883	5.5370
330	12.84	0.001561	0.01300	1505.2	2499.0	1525.3	1140.6	2665.9	3.5514	1.8911	5.4425
340	14.59	0.001638	0.01080	1570.3	2464.6	1594.2	1027.9	2622.1	3.6601	1.6765	5.3366
350	16.51	0.001740	0.008815	1641.8	2418.5	1670.6	893.4	2564.0	3.7784	1.4338	5.2122
360	18.65	0.001892	0.006947	1725.2	2351.6	1760.5	720.7	2481.2	3.9154	1.1382	5.0536
370	21.03	0.002213	0.004931	1844.0	2229.0	1890.5	442.2	2332.7	4.1114	0.6876	4.7990
374.136	22.088	0.003155	0.003155	2029.6	2029.6	2099.3	0.0	2099.3	4.4305	0.0000	4.4305

SOURCES: Keenan, Keyes, Hill, and Moore, *Steam Tables*, Wiley, New York, 1969; G. J. Van Wylen and R. E. Sonntag, *Fundamentals of Classical Thermodynamics*, Wiley, New York, 1973.

Table C-2 · Properties of Saturated H_2O—Pressure Table

		Volume, m^3/kg		Energy, kJ/kg		Enthalpy, kJ/kg			Entropy, kJ/kg·K		
P, MPa	T, °C	v_f	v_g	u_f	u_g	h_f	h_{fg}	h_g	s_f	s_{fg}	s_g
0.000611	0.01	0.001000	206.1	0.0	2375.3	0.0	2501.3	2501.3	0.0000	9.1571	9.1571
0.0008	3.8	0.001000	159.7	15.8	2380.5	15.8	2492.5	2508.3	0.0575	9.0007	9.0582
0.001	7.0	0.001000	129.2	29.3	2385.0	29.3	2484.9	2514.2	0.1059	8.8706	8.9765
0.0012	9.7	0.001000	108.7	40.6	2388.7	40.6	2478.5	2519.1	0.1460	8.7639	8.9099
0.0014	12.0	0.001001	93.92	50.3	2391.9	50.3	2473.1	2523.4	0.1802	8.6736	8.8538
0.0016	14.0	0.001001	82.76	58.9	2394.7	58.9	2468.2	2527.1	0.2101	8.5952	8.8053
0.0018	15.8	0.001001	74.03	66.5	2397.2	66.5	2464.0	2530.5	0.2367	8.5259	8.7626
0.002	17.5	0.001001	67.00	73.5	2399.5	73.5	2460.0	2533.5	0.2606	8.4639	8.7245
0.003	24.1	0.001003	45.67	101.0	2408.5	101.0	2444.5	2545.5	0.3544	8.2240	8.5784
0.004	29.0	0.001004	34.80	121.4	2415.2	121.4	2433.0	2554.4	0.4225	8.0529	8.4754
0.006	36.2	0.001006	23.74	151.5	2424.9	151.5	2415.9	2567.4	0.5208	7.8104	8.3312
0.008	41.5	0.001008	18.10	173.9	2432.1	173.9	2403.1	2577.0	0.5924	7.6371	8.2295
0.01	45.8	0.001010	14.67	191.8	2437.9	191.8	2392.8	2584.6	0.6491	7.5019	8.1510
0.012	49.4	0.001012	12.36	206.9	2442.7	206.9	2384.1	2591.0	0.6961	7.3910	8.0871
0.014	52.6	0.001013	10.69	220.0	2446.9	220.0	2376.6	2596.6	0.7365	7.2968	8.0333
0.016	55.3	0.001015	9.433	231.5	2450.5	231.5	2369.9	2601.4	0.7719	7.2149	7.9868
0.018	57.8	0.001016	8.445	241.9	2453.8	241.9	2363.9	2605.8	0.8034	7.1425	7.9459
0.02	60.1	0.001017	7.649	251.4	2456.7	251.4	2358.3	2609.7	0.8319	7.0774	7.9093
0.03	69.1	0.001022	5.229	289.2	2468.4	289.2	2336.1	2625.3	0.9439	6.8256	7.7695
0.04	75.9	0.001026	3.993	317.5	2477.0	317.6	2319.1	2636.7	1.0260	6.6449	7.6709
0.06	85.9	0.001033	2.732	359.8	2489.6	359.8	2293.7	2653.5	1.1455	6.3873	7.5328
0.08	93.5	0.001039	2.087	391.6	2498.8	391.6	2274.1	2665.7	1.2331	6.2023	7.4354
0.1	99.6	0.001043	1.694	417.3	2506.1	417.4	2258.1	2675.5	1.3029	6.0573	7.3602
0.12	104.8	0.001047	1.428	439.2	2512.1	439.3	2244.2	2683.5	1.3611	5.9378	7.2980
0.14	109.3	0.001051	1.237	458.2	2517.3	458.4	2232.0	2690.4	1.4112	5.8360	7.2472
0.16	113.3	0.001054	1.091	475.2	2521.8	475.3	2221.2	2696.5	1.4553	5.7472	7.2025
0.18	116.9	0.001058	0.9775	490.5	2525.9	490.7	2211.1	2701.8	1.4948	5.6683	7.1631
0.2	120.2	0.001061	0.8857	504.5	2529.5	504.7	2201.9	2706.6	1.5305	5.5975	7.1280
0.3	133.5	0.001073	0.6058	561.1	2543.6	561.5	2163.8	2725.3	1.6722	5.3205	6.9927
0.4	143.6	0.001084	0.4625	604.3	2553.6	604.7	2133.8	2738.5	1.7770	5.1197	6.8967
0.6	158.9	0.001101	0.3157	669.9	2567.4	670.6	2086.2	2756.8	1.9316	4.8293	6.7609
0.8	170.4	0.001115	0.2404	720.2	2576.8	721.1	2048.0	2769.1	2.0466	4.6170	6.6636
1	179.9	0.001127	0.1944	761.7	2583.6	762.8	2015.3	2778.1	2.1391	4.4482	6.5873
1.2	188.0	0.001139	0.1633	797.3	2588.8	798.6	1986.2	2784.8	2.2170	4.3072	6.5242
1.4	195.1	0.001149	0.1408	828.7	2592.8	830.3	1959.7	2790.0	2.2847	4.1854	6.4701
1.6	201.4	0.001159	0.1238	856.9	2596.0	858.8	1935.2	2794.0	2.3446	4.0780	6.4226
1.8	207.2	0.001168	0.1104	882.7	2598.4	884.8	1912.3	2797.1	2.3986	3.9816	6.3802
2	212.4	0.001177	0.09963	906.4	2600.3	908.8	1890.7	2799.5	2.4478	3.8939	6.3417
3	233.9	0.001216	0.06668	1004.8	2604.1	1008.4	1795.7	2804.1	2.6462	3.5416	6.1878
4	250.4	0.001252	0.04978	1082.3	2602.3	1087.3	1714.1	2801.4	2.7970	3.2739	6.0709
6	275.6	0.001319	0.03244	1205.4	2589.7	1213.3	1571.0	2784.3	3.0273	2.8627	5.8900
8	295.1	0.001384	0.02352	1305.6	2569.8	1316.6	1441.4	2758.0	3.2075	2.5365	5.7440
10	311.1	0.001452	0.01803	1393.0	2544.4	1407.6	1317.1	2724.7	3.3603	2.2546	5.6149
12	324.8	0.001527	0.01426	1472.9	2513.7	1491.3	1193.6	2684.9	3.4970	1.9963	5.4933
14	336.8	0.001611	0.01149	1548.6	2476.8	1571.1	1066.5	2637.6	3.6240	1.7486	5.3726
16	347.4	0.001711	0.009307	1622.7	2431.8	1650.0	930.7	2580.7	3.7468	1.4996	5.2464
18	357.1	0.001840	0.007491	1698.9	2374.4	1732.0	777.2	2509.2	3.8722	1.2332	5.1054
20	365.8	0.002036	0.005836	1785.6	2293.2	1826.3	583.7	2410.0	4.0146	0.9135	4.9281
22.088	374.136	0.003155	0.003155	2029.6	2029.6	2099.3	0.0	2099.3	4.4305	0.0000	4.4305

SOURCES: Keenan, Keyes, Hill, and Moore, *Steam Tables*, Wiley, New York, 1969; G. J. Van Wylen and R. E. Sonntag, *Fundamentals of Classical Thermodynamics*, Wiley, New York, 1973.

Table C-3 Superheated Steam (T in °C, v in m³/kg, u and h in kJ/kg, s in kJ/kg·K)

T	v	u	h	s	v	u	h	s	v	u	h	s
	P = 0.010 MPa (45.81 °C)				P = 0.050 MPa (81.33 °C)				P = 0.10 MPa (99.63 °C)			
Sat.	14.674	2437.9	2584.7	8.1502	3.240	2483.9	2645.9	7.5939	1.6940	2506.1	2675.5	7.3594
50	14.869	2443.9	2592.6	8.1749								
100	17.196	2515.5	2687.5	8.4479	3.418	2511.6	2682.5	7.6947	1.6958	2506.7	2676.2	7.3614
150	19.512	2587.9	2783.0	8.6882	3.889	2585.6	2780.1	7.9401	1.9364	2582.8	2776.4	7.6134
200	21.825	2661.3	2879.5	8.9038	4.356	2659.9	2877.7	8.1580	2.172	2658.1	2875.3	7.8343
250	24.136	2736.0	2977.3	9.1002	4.820	2735.0	2976.0	8.3556	2.406	2733.7	2974.3	8.0333
300	26.445	2812.1	3076.5	9.2813	5.284	2811.3	3075.5	8.5373	2.639	2810.4	3074.3	8.2158
400	31.063	2968.9	3279.6	9.6077	6.209	2968.5	3278.9	8.8642	3.103	2967.9	3278.2	8.5435
500	35.679	3132.3	3489.1	9.8978	7.134	3132.0	3488.7	9.1546	3.565	3131.6	3488.1	8.8342
600	40.295	3302.5	3705.4	10.1608	8.057	3302.2	3705.1	9.4178	4.028	3301.9	3704.7	9.0976
700	44.911	3479.6	3928.7	10.4028	8.981	3479.4	3928.5	9.6599	4.490	3479.2	3928.2	9.3398
800	49.526	3663.8	4159.0	10.6281	9.904	3663.6	4158.9	9.8852	4.952	3663.5	4158.6	9.5652
900	54.141	3855.0	4396.4	10.8396	10.828	3854.9	4396.3	10.0967	5.414	3854.8	4396.1	9.7767
1000	58.757	4053.0	4640.6	11.0393	11.751	4052.9	4640.5	10.2964	5.875	4052.8	4640.3	9.9764
1100	63.372	4257.5	4891.2	11.2287	12.674	4257.4	4891.1	10.4859	6.337	4257.3	4891.0	10.1659
1200	67.987	4467.9	5147.8	11.4091	13.597	4467.8	5147.7	10.6662	6.799	4467.7	5147.6	10.3463
1300	72.602	4683.7	5409.7	11.5811	14.521	4683.6	5409.6	10.8382	7.260	4683.5	5409.5	10.5183
	P = 0.20 MPa (120.23 °C)				P = 0.30 MPa (133.55 °C)				P = 0.40 MPa (143.63 °C)			
Sat.	.8857	2529.5	2706.7	7.1272	.6058	2543.6	2725.3	6.9919	.4625	2553.6	2738.6	6.8959
150	.9596	2576.9	2768.8	7.2795	.6339	2570.8	2761.0	7.0778	.4708	2564.5	2752.8	6.9299
200	1.0803	2654.4	2870.5	7.5066	.7163	2650.7	2865.6	7.3115	.5342	2646.8	2860.5	7.1706
250	1.1988	2731.2	2971.0	7.7086	.7964	2728.7	2967.6	7.5166	.5951	2726.1	2964.2	7.3789
300	1.3162	2808.6	3071.8	7.8926	.8753	2806.7	3069.3	7.7022	.6548	2804.8	3066.8	7.5662
400	1.5493	2966.7	3276.6	8.2218	1.0315	2965.6	3275.6	8.0330	.7726	2964.4	3273.4	7.8985
500	1.7814	3130.8	3487.1	8.5133	1.1867	3130.0	3486.0	8.3251	.8893	3129.2	3484.9	8.1913
600	2.013	3301.4	3704.0	8.7770	1.3414	3300.8	3703.2	8.5892	1.0055	3300.2	3702.4	8.4558
700	2.244	3478.8	3927.6	9.0194	1.4957	3478.4	3927.1	8.8319	1.1215	3477.9	3926.5	8.6987
800	2.475	3663.1	4158.2	9.2449	1.6499	3662.9	4157.8	9.0576	1.2372	3662.4	4157.3	8.9244
900	2.706	3854.5	4395.8	9.4566	1.8041	3854.2	4395.4	9.2692	1.3529	3853.9	4395.1	9.1362
1000	2.937	4052.5	4640.0	9.6563	1.9581	4052.3	4639.7	9.4690	1.4685	4052.0	4639.4	9.3360
1100	3.168	4257.0	4890.7	9.8458	2.1121	4256.5	4890.4	9.6585	1.5840	4256.5	4890.2	9.5256
1200	3.399	4467.5	5147.3	10.0262	2.2661	4467.2	5147.1	9.8389	1.6996	4467.0	5146.8	9.7060
1300	3.630	4683.2	5409.3	10.1982	2.4201	4683.0	5409.0	10.0110	1.8151	4682.8	5408.8	9.8780
	P = 0.50 MPa (151.86 °C)				P = 0.60 MPa (158.85 °C)				P = 0.80 MPa (170.43 °C)			
Sat.	.3749	2561.2	2748.7	6.8213	.3157	2567.4	2756.8	6.7600	.2404	2576.8	2769.1	6.6628
200	.4249	2642.9	2855.4	7.0592	.3520	2638.9	2850.1	6.9665	.2608	2630.6	2839.3	6.8158
250	.4744	2723.5	2960.7	7.2709	.3938	2720.9	2957.2	7.1816	.2931	2715.5	2950.0	7.0384
300	.5226	2802.9	3064.2	7.4599	.4344	2801.0	3061.6	7.3724	.3241	2797.2	3056.5	7.2328
350	.5701	2882.6	3167.7	7.6329	.4742	2881.2	3165.7	7.5464	.3544	2878.2	3161.7	7.4089
400	.6173	2963.2	3271.9	7.7938	.5137	2962.1	3270.3	7.7079	.3843	2959.7	3267.1	7.5716
500	.7109	3128.4	3483.9	8.0873	.5920	3127.6	3482.8	8.0021	.4433	3126.0	3480.6	7.8673
600	.8041	3299.6	3701.7	8.3522	.6697	3299.1	3700.9	8.2674	.5018	3297.9	3699.4	8.1333
700	.8969	3477.5	3925.9	8.5952	.7472	3477.0	3925.3	8.5107	.5601	3476.2	3924.2	8.3770
800	.9896	3662.1	4156.9	8.8211	.8245	3661.8	4156.5	8.7367	.6181	3661.1	4155.6	8.6033
900	1.0822	3853.6	4394.7	9.0329	.9017	3853.4	4394.4	8.9486	.6761	3852.8	4393.7	8.8153
1000	1.1747	4051.8	4639.1	9.2328	.9788	4051.5	4638.8	9.1485	.7340	4051.0	4638.2	9.0153
1100	1.2672	4256.3	4889.9	9.4224	1.0559	4256.1	4889.6	9.3381	.7919	4255.6	4889.1	9.2050
1200	1.3596	4466.8	5146.6	9.6029	1.1330	4466.5	5146.3	9.5185	.8497	4466.1	5145.9	9.3855
1300	1.4521	4682.5	5408.6	9.7749	1.2101	4682.3	5408.3	9.6906	.9076	4681.8	5407.9	9.5575

Table C-3 ***(Continued)***

T	v	u	h	s	v	u	h	s	v	u	h	s
	P = 1.00 MPa (179.91 °C)				P = 1.20 MPa (187.99 °C)				P = 1.40 MPa (195.07 °C)			
Sat.	.19 444	2583.6	2778.1	6.5865	.163 33	2588.8	2784.8	6.5233	.140 84	2592.8	2790.0	6.4693
200	.2060	2621.9	2827.9	6.6940	.169 30	2612.8	2815.9	6.5898	.143 02	2603.1	2803.3	6.4975
250	.2327	2709.9	2942.6	6.9247	.192 34	2704.2	2935.0	6.8294	.163 50	2698.3	2927.2	6.7467
300	.2579	2793.2	3051.2	7.1229	.2138	2789.2	3045.8	7.0317	.182 28	2785.2	3040.4	6.9534
350	.2825	2875.2	3157.7	7.3011	.2345	2872.2	3153.6	7.2121	.2003	2869.2	3149.5	7.1360
400	.3066	2957.3	3263.9	7.4651	.2548	2954.9	3260.7	7.3774	.2178	2952.5	3257.5	7.3026
500	.3541	3124.4	3478.5	7.7622	.2946	3122.8	3476.3	7.6759	.2521	3321.1	3474.1	7.6027
600	.4011	3296.8	3697.9	8.0290	.3339	3295.6	3696.3	7.9435	.2860	3294.4	3694.8	7.8710
700	.4478	3475.3	3923.1	8.2731	.3729	3474.4	3922.0	8.1881	.3195	3473.6	3920.8	8.1160
800	.4943	3660.4	4154.7	8.4996	.4118	3659.7	4153.8	8.4148	.3528	3659.0	4153.0	8.8431
900	.5407	3852.2	4392.9	8.7118	.4505	3851.6	4392.2	8.6272	.3861	3851.1	4391.5	8.5556
1000	.5871	4050.5	4637.6	8.9119	.4892	4050.0	4637.0	8.8274	.4192	4049.5	4636.4	8.7559
1100	.6335	4255.1	4888.6	9.1017	.5278	4254.6	4888.0	9.0172	.4524	4254.1	4887.5	8.9457
1200	.6798	4465.6	5145.4	9.2822	.5665	4465.1	5144.9	9.1977	.4855	4464.7	5144.4	9.1262
1300	.7261	4681.3	5407.4	9.4543	.6051	4680.9	5407.0	9.3698	.5186	4680.4	5406.5	9.2984
	P = 1.60 MPa (201.41 °C)				P = 1.80 MPa (207.15 °C)				P = 2.00 MPa (212.42°C)			
Sat.	.123 80	2596.0	2794.0	6.4218	.110 42	2598.4	2797.1	6.3794	.099 63	2600.3	2799.5	6.3409
225	.132 87	2644.7	2857.3	6.5518	.136 73	2636.6	2846.7	6.4808	.103 77	2628.3	2835.8	6.4147
250	.141 84	2692.3	2919.2	6.6732	.124 97	2686.0	2911.0	6.6066	.111 44	2679.6	2902.5	6.5453
300	.158 62	2781.1	3034.8	6.8844	.140 21	2776.9	3029.2	6.8226	.125 47	2772.6	3023.5	6.7664
350	.174 56	2866.1	3145.4	7.0694	.154 57	2863.0	3141.2	7.0100	.138 57	2859.8	3137.0	6.9563
400	.190 05	2950.1	3254.2	7.2374	.168 47	2947.7	3250.9	7.1794	.151 20	2945.2	3247.6	7.1271
500	.2203	3119.5	3472.0	7.5390	.195 50	3117.9	3469.8	7.4825	.175 68	3116.2	3467.6	7.4317
600	.2500	3293.3	3693.2	7.8080	.2220	3292.1	3691.7	7.7523	.199 60	3290.9	3690.1	7.7024
700	.2794	3472.7	3919.7	8.0535	.2482	3471.8	3918.5	7.9983	.2232	3470.9	3917.4	7.9487
800	.3086	3658.3	4152.1	8.2808	.2742	3657.6	4151.2	8.2258	.2467	3657.0	4150.3	8.1765
900	.3377	3850.5	4390.8	8.4935	.3001	3849.9	4390.1	8.4386	.2700	3849.3	4389.4	8.3895
1000	.5668	4049.0	4635.8	8.6938	.3260	4048.5	4635.2	8.6391	.2933	4048.0	4634.6	8.5901
1100	.3958	4253.7	4887.0	8.8837	.3518	4253.2	4886.4	8.8290	.3166	4252.7	4885.9	8.7800
1200	.4248	4464.2	5143.9	9.0643	.3776	4463.7	5143.4	9.0096	.3398	4463.3	5142.9	8.9607
1300	.4538	4679.9	5406.0	9.2364	.4034	4679.5	5405.6	9.1818	.3631	4679.0	5405.1	9.1329
	P = 2.50 MPa (233.99 °C)				P = 3.00 MPa (233.90 °C)				P = 3.50 MPa (242.60 °C)			
Sat.	.079 98	2603.1	2803.1	6.2575	.066 68	2604.1	2804.2	6.1869	.05707	2603.7	2803.4	6.1253
225	.080 27	2605.6	2806.3	6.2639								
250	.087 00	2662.6	2880.1	6.4085	.070 58	2644.0	2855.8	6.2872	.058 72	2623.7	2829.2	6.1749
300	.098 90	2761.6	3008.8	6.6438	.081 14	2750.1	2993.5	6.5390	.068 42	2738.0	2977.5	6.4461
350	.109 76	2851.9	3126.3	6.8403	.090 53	2843.7	3115.3	6.7428	.076 78	2835.3	3104.0	6.6579
400	.120 10	2939.1	3239.3	7.0148	.099 36	2932.8	3230.9	6.9212	.084 53	2926.4	3222.3	6.8405
450	.130 14	3025.5	3350.8	7.1746	.107 87	3020.4	3344.0	7.0834	.091 96	3015.3	3337.2	7.0052
500	.139 98	3112.1	3462.1	7.3234	.116 19	3108.0	3456.5	7.2338	.099 18	3103.0	3450.9	7.1572
600	.159 30	3288.0	3686.3	7.5960	.132 43	3285.0	3682.3	7.5085	.113 24	3282.1	3678.4	7.4339
700	.178 32	3468.7	3914.5	7.8435	.148 38	3466.5	3911.7	7.7571	.126 99	3464.3	3908.8	7.6837
800	.197 16	3655.3	4148.2	8.0720	.164 14	3653.5	4145.9	7.9862	.140 56	3651.8	4143.7	7.9134
900	.215 90	3847.9	4387.6	8.2853	.179 80	3846.5	4385.9	8.1999	.154 02	3845.0	4384.1	8.1276
1000	.2346	4046.7	4633.1	8.4861	.195 41	4045.4	4631.6	8.4009	.167 43	4044.1	4630.1	8.3288
1100	.2532	4251.5	4884.6	8.6762	.210 98	4250.3	4883.3	8.5912	.180 80	4249.2	4881.9	8.5192
1200	.2718	4462.1	5141.7	8.8569	.226 52	4460.9	5140.5	8.7720	.194 15	4459.8	5139.3	8.7000
1300	.2905	4677.8	5404.0	9.0291	.242 06	4676.6	5402.8	8.9442	.207 49	4675.5	5401.7	8.8723

Table C-3 (*Continued*)

T	v	u	h	s	v	u	h	s	v	u	h	s
	$P = 4.0$ MPa (250.40 °C)				$P = 4.5$ MPa (257.49 °C)				$P = 5.0$ MPa (263.99 °C)			
Sat.	.049 78	2602.3	2801.4	6.0701	.044 06	2600.1	2798.3	6.0198	.039 44	2597.1	2794.3	5.9734
275	.054 57	2667.9	2886.2	6.2285	.047 30	2650.3	2863.2	6.1401	.041 41	2631.3	2838.3	6.0544
300	.058 84	2725.3	2960.7	6.3615	.051 35	2712.0	2943.1	6.2828	.045 32	2698.0	2924.5	6.2084
350	.066 45	2826.7	3092.5	6.5821	.058 40	2817.8	3080.6	6.5131	.051 94	2808.7	3068.4	6.4493
400	.073 41	2919.9	3213.6	6.7690	.064 75	2913.3	3204.7	6.7047	.057 81	2906.6	3195.7	6.6459
450	.080 02	3010.2	3330.3	6.9363	.070 74	3005.0	3323.3	6.8746	.063 30	2999.7	3316.2	6.8186
500	.086 43	3099.5	3445.3	7.0901	.076 51	3095.3	3439.6	7.0301	.068 57	3091.0	3433.8	6.9759
600	.098 85	3279.1	3674.4	7.3688	.087 65	3276.0	3670.5	7.3110	.078 69	3273.0	3666.5	7.2589
700	.110 95	3462.1	3905.9	7.6198	.098 47	3459.9	3903.0	7.5631	.088 49	3457.6	3900.1	7.5122
800	.122 87	3650.0	4141.5	7.8502	.109 11	3648.3	4139.3	7.7942	.098 11	3646.6	4137.1	7.7440
900	.134 69	3843.6	4382.3	8.0647	.119 65	3842.2	4380.6	8.0091	.107 62	3840.7	4378.8	7.9593
1000	.146 45	4042.9	4628.7	8.2662	.130 13	4041.6	4627.2	8.2108	.117 07	4040.4	4625.7	8.1612
1100	.158 17	4248.0	4880.6	8.4567	.140 56	4246.8	4879.3	8.4015	.126 48	4245.6	4878.0	8.3520
1200	.169 87	4458.6	5138.1	8.6376	.150 98	4457.5	5136.9	8.5825	.135 87	4456.3	5135.7	8.5331
1300	.181 56	4674.3	5400.5	8.8100	.161 39	4673.1	5399.4	8.7549	.145 26	4672.0	5398.2	8.7055
	$P = 6.0$ MPa (275.64 °C)				$P = 7.0$ MPa (285.88 °C)				$P = 8.0$ MPa (295.06 °C)			
Sat.	.032 44	2589.7	2784.3	5.8892	.027 37	2580.5	2772.1	5.8133	.023 52	2569.8	2758.0	5.7432
300	.036 16	2667.2	2884.2	6.0674	.029 47	2632.2	2838.4	3.9305	.024 26	2590.9	2785.0	5.7906
350	.042 23	2789.6	3043.0	6.3335	.035 24	2769.4	3016.0	6.2283	.029 95	2747.7	2987.3	6.1301
400	.047 39	2892.9	3177.2	6.5408	.039 93	2878.6	3158.1	6.4478	.034 32	2863.8	3138.3	6.3634
450	.052 14	2988.9	3301.8	6.7193	.044 16	2978.0	3287.1	6.6327	.038 17	2966.7	3272.0	6.5551
500	.056 65	3082.2	3422.2	6.8803	.048 14	3073.4	3410.3	6.7975	.041 75	3064.3	3398.3	6.7240
550	.061 01	3174.6	3540.6	7.0288	.051 95	3167.2	3530.9	6.9486	.045 16	3159.8	3521.0	6.8778
600	.065 25	3266.9	3658.4	7.1677	.055 65	3260.7	3650.3	7.0894	.048 45	3254.4	3642.0	7.0206
700	.073 52	3453.1	3894.2	7.4234	.062 83	3448.5	3888.3	7.3476	.054 81	3443.9	3882.4	7.2812
800	.081 60	3643.1	4132.7	7.6566	.069 81	3639.5	4128.2	7.5822	.060 97	3636.0	4123.8	7.5173
900	.089 58	3837.8	4375.3	7.8727	.076 69	3835.0	4371.8	7.7991	.067 02	3832.1	4368.3	7.7351
1000	.097 49	4037.8	4622.7	8.0751	.083 50	4035.3	4619.8	8.0020	.073 01	4032.8	4616.9	7.9384
1100	.105 36	4243.3	4875.4	8.2661	.090 27	4240.9	4872.8	8.1933	.078 96	4238.6	4870.3	8.1300
1200	.113 21	4454.0	5133.3	8.4474	.097 03	4451.7	5130.9	8.3747	.084 89	4449.5	5128.5	8.3115
1300	.121 06	4669.6	5396.0	8.6199	.103 77	4667.3	5393.7	8.5473	.090 80	4665.0	5391.5	8.4842
	$P = 9.0$ MPa (303.40 °C)				$P = 10.0$ MPa (311.06 °C)				$P = 12.5$ MPa (327.89 °C)			
Sat.	.020 48	2557.8	2742.1	5.6772	.018 026	2544.4	2724.7	5.6141	.013 495	2505.1	2673.8	5.4624
325	.023 27	2646.6	2856.0	5.8712	.019 861	2610.4	2809.1	5.7568				
350	.025 80	2724.4	2956.6	6.0361	.022 42	2699.4	2923.4	5.9443	.016 126	2624.6	2826.2	5.7118
400	.029 93	2848.4	3117.8	6.2854	.026 41	2832.4	3096.5	6.2120	.020 00	2789.3	3039.3	6.0417
450	.033 50	2955.2	3256.6	6.4844	.029 75	2943.4	3240.9	6.4190	.022 99	2912.5	3199.8	6.2719
500	.036 77	3055.2	3336.1	6.6576	.032 79	3045.8	3373.7	6.5966	.025 60	3021.7	3341.8	6.4618
550	.039 87	3152.2	3511.0	6.8142	.035 64	3144.6	3500.9	6.7561	.028 01	3125.0	3475.2	6.6290
600	.042 85	3248.1	3633.7	6.9589	.038 37	3241.7	3625.3	6.9029	.030 29	3225.4	3604.0	6.7810
650	.045 74	3343.6	3755.3	7.0943	.041 01	3338.2	3748.2	7.0398	.032 48	3324.4	3730.4	6.9218
700	.048 57	3439.3	3876.5	7.2221	.043 58	3434.7	3870.5	7.1687	.034 60	3422.9	3855.3	7.0536
800	.054 09	3632.5	4119.3	7.4596	.048 59	3628.9	4114.8	7.4077	.038 69	3620.0	4103.6	7.2965
900	.059 50	3829.2	4364.3	7.6783	.053 49	3826.3	4361.2	7.6272	.042 67	3819.1	4352.5	7.5182
1000	.064 85	4030.3	4614.0	7.8821	.058 32	4027.8	4611.0	7.8315	.046 58	4021.6	4603.8	7.7237
1100	.070 16	4236.3	4867.7	8.0740	.063 12	4234.0	4865.1	8.0237	.050 45	4228.2	4858.8	7.9165
1200	.075 44	4447.2	5126.2	8.2556	.067 89	4444.9	5123.8	8.2055	.054 30	4439.3	5118.0	8.0987
1300	.080 72	4662.7	5389.2	8.4284	.072 65	4460.5	5387.0	8.3783	.058 13	4654.8	5381.4	8.2717

Table C-3 (*Continued*)

T	v	u	h	s	v	u	h	s	v	u	h	s
	P = 15.0 MPa (342.24°C)				P = 17.5 MPa (354.75°C)				P = 20.0 MPa (365.81°C)			
Sat.	.010 337	2455.5	2610.5	5.3098	.0079 20	2390.2	2528.8	5.1419	.005 834	2293.0	2409.7	4.9269
350	.011 470	2520.4	2692.4	5.4421								
400	.015 649	2740.7	2975.5	5.8811	.012 447	2685.0	2902.9	5.7213	.009 942	2619.3	2818.1	5.5540
450	.018 445	2879.5	3156.2	6.1404	.015 174	2844.2	3109.7	6.0184	.012 695	2806.2	3060.1	5.9017
500	.020 80	2996.6	3308.6	6.3443	.017 358	2970.3	3274.1	6.2383	.014 768	2942.9	3238.2	6.1401
550	.022 93	3104.7	3448.6	6.5199	.019 288	3083.9	3421.4	6.4230	.016 555	3062.4	3393.5	6.3348
600	.024 91	3208.6	3582.3	6.6776	.021 06	3191.5	3560.2	6.5866	.018 178	3174.0	3537.6	6.5048
650	.026 80	3310.3	3712.3	6.8224	.022 74	3296.0	3693.9	6.7357	.019 693	3281.4	3675.3	6.6582
700	.028 61	3410.9	3840.1	6.9572	.024 34	3398.7	3824.6	6.8736	.021 13	3386.4	3809.0	6.7993
800	.032 10	3610.9	4092.4	7.2040	.027 38	3601.8	4081.1	7.1244	.023 85	3592.7	4069.7	7.0544
900	.035 46	3811.9	4343.8	7.4279	.030 31	3804.7	4335.1	7.3507	.026 45	3797.5	4326.4	7.2830
1000	.038 75	4015.4	4596.6	7.6348	.033 16	4009.3	4589.5	7.5589	.028 97	4003.1	4582.5	7.4925
1100	.042 00	4222.6	4852.6	7.8283	.035 97	4216.9	4846.4	7.7531	.031 45	4211.3	4840.2	7.6874
1200	.045 23	4433.8	5112.3	8.0108	.038 76	4428.3	5106.6	7.9360	.033 91	4422.8	5101.0	7.8707
1300	.048 45	4649.1	5376.0	8.1840	.041 54	4643.5	5370.5	8.1093	.036 36	4638.0	5365.1	8.0442
	P = 25.0 MPa				P = 30.0 MPa				P = 40.0 MPa			
375	.001 9731	1798.7	1848.0	4.0320	.001 789 2	1737.8	1791.5	3.9305	.001 640 7	1677.1	1742.8	3.8290
400	.006 004	2430.1	2580.2	5.1418	.002 790	2067.4	2151.1	4.4728	.001 907 7	1854.6	1930.9	4.1135
425	.007 881	2609.2	2806.3	5.4723	.005 303	2455.1	2614.2	5.1504	.002 532	2096.9	2198.1	4.5029
450	.009 162	2720.7	2949.7	5.6744	.006 735	2619.3	2821.4	5.4424	.003 693	2365.1	2512.8	4.9459
500	.011 123	2884.3	3162.4	5.9592	.008 678	2820.7	3081.1	5.7905	.005 622	2678.4	2903.3	5.4700
550	.012 724	3017.5	3335.6	6.1765	.010 168	2970.3	3275.4	6.0342	.006 984	2869.7	3149.1	5.7785
600	.014 137	3137.9	3491.4	6.3602	.011 446	3100.5	3443.9	6.2331	.008 094	3022.6	3346.4	6.0114
650	.015 433	3251.6	3637.4	6.5229	.012 596	3221.0	3598.9	6.4058	.009 063	3158.0	3520.6	6.2054
700	.016 646	3361.3	3777.5	6.6707	.013 661	3335.8	3745.6	6.5606	.009 941	3283.6	3681.2	6.3750
800	.018 912	3574.3	4047.1	6.9345	.015 623	3555.5	4024.2	6.8332	.011 523	3517.8	3978.7	6.6662
900	.021 045	3783.0	4309.1	7.1680	.017 448	3768.5	4291.9	7.0718	.012 962	3739.4	4257.9	6.9150
1000	.023 10	3990.9	4568.5	7.3802	.019 196	3978.8	4554.7	7.2867	.014 324	3954.6	4527.6	7.1356
1100	.025 12	4200.2	4828.2	7.5765	.020 903	4189.2	4816.3	7.4845	.015 642	4167.4	4793.1	7.3364
1200	.027 11	4412.0	5089.9	7.7605	.022 589	4401.3	5079.0	7.6692	.016 940	4380.1	5057.7	7.5224
1300	.029 10	4626.9	5354.4	7.9342	.024 266	4616.0	5344.0	7.8432	.018 229	4594.3	5323.5	7.6969

SOURCES: Keenan, Keyes, Hill, and Moore, *Steam Tables*, Wiley, New York, 1969; G. J. Van Wylen and R. E. Sonntag, *Fundamentals of Classical Thermodynamics*, Wiley, New York, 1973.

Table C-4 Compressed Liquid

	$P = 5$ MPa (263.99°C)				$P = 10$ MPa (311.06°C)				$P = 15$ MPa (342.42°C)			
T	v	u	h	s	v	u	h	s	v	u	h	s
0	0.000 997 7	0.04	5.04	0.0001	0.000 995 2	0.09	10.04	0.0002	0.000 992 8	0.15	15.05	0.0004
20	0.000 999 5	83.65	88.65	0.2956	0.000 997 2	83.36	93.33	0.2945	0.000 995 0	83.06	97.99	0.2934
40	0.001 005 6	166.95	171.97	0.5705	0.001 003 4	166.35	176.38	0.5686	0.001 001 3	165.76	180.78	0.5666
60	0.001 014 9	250.23	255.30	0.8285	0.001 012 7	249.36	259.49	0.8258	0.001 010 5	248.51	263.67	0.8232
80	0.001 026 8	333.72	338.85	1.0720	0.001 024 5	332.59	342.83	1.0688	0.001 022 2	331.48	346.81	1.0656
100	0.001 041 0	417.52	422.72	1.3030	0.001 038 5	416.12	426.50	1.2992	0.001 036 1	414.74	430.28	1.2955
120	0.001 057 6	501.80	507.09	1.5233	0.001 054 9	500.08	510.64	1.5189	0.001 052 2	498.40	514.19	1.5145
140	0.001 076 8	586.76	592.15	1.7343	0.001 073 7	584.68	595.42	1.7292	0.001 070 7	582.66	598.72	1.7242
160	0.001 098 8	672.62	678.12	1.9375	0.001 095 3	670.13	681.08	1.9317	0.001 091 8	667.71	684.09	1.9260
180	0.001 124 0	759.63	765.25	2.1341	0.001 119 9	756.65	767.84	2.1275	0.001 115 9	753.76	770.50	2.1210
200	0.001 153 0	848.1	853.9	2.3255	0.001 148 0	844.5	856.0	2.3178	0.001 143 3	841.0	858.2	2.3104
220	0.001 186 6	938.4	944.4	2.5128	0.001 180 5	934.1	945.9	2.5039	0.001 174 8	929.9	947.5	2.4953
240	0.001 226 4	1031.4	1037.5	2.6979	0.001 218 7	1026.0	1038.1	2.6872	0.001 211 4	1020.8	1039.0	2.6771
260	0.001 274 9	1127.9	1134.3	2.8830	0.001 264 5	1121.1	1133.7	2.8699	0.001 255 0	1114.6	1133.4	2.8576

	$P = 20$ MPa (365.81°C)				$P = 30$ MPa				$P = 50$ MPa			
T	v	u	h	s	v	u	h	s	v	u	h	s
0	0.000 990 4	0.19	20.01	0.0004	0.000 985 6	0.25	29.82	0.0001	0.000 976 6	0.20	49.03	0.0014
20	0.000 992 8	82.77	102.62	0.2923	0.000 988 6	82.17	111.84	0.2899	0.000 980 4	81.00	130.02	0.2848
40	0.000 999 2	165.17	185.16	0.5646	0.000 995 1	164.04	193.89	0.5607	0.000 987 2	161.86	211.21	0.5527
60	0.001 008 4	247.68	267.85	0.8206	0.001 004 2	246.06	276.19	0.8154	0.000 996 2	242.98	292.79	0.8052
80	0.001 019 9	330.40	350.80	1.0624	0.001 015 6	328.30	358.77	1.0561	0.001 007 3	324.34	374.70	1.0440
100	0.001 033 7	413.39	434.06	1.2917	0.001 029 0	410.78	441.66	1.2844	0.001 020 1	405.88	456.89	1.2703
120	0.001 049 6	496.76	517.76	1.5102	0.001 044 5	493.59	524.93	1.5018	0.001 034 8	487.65	539.39	1.4857
140	0.001 067 8	580.69	602.04	1.7193	0.001 062 1	576.88	608.75	1.7098	0.001 051 5	569.77	622.35	1.6915
160	0.001 088 5	665.35	687.12	1.9204	0.001 082 1	660.82	693.28	1.9096	0.001 070 3	652.41	705.92	1.8891
180	0.001 112 0	750.95	773.20	2.1147	0.001 104 7	745.59	778.73	2.1024	0.001 091 2	735.69	790.25	2.0794
200	0.001 138 8	837.7	860.5	2.3031	0.001 130 2	831.4	865.3	2.2893	0.001 114 6	819.7	875.5	2.2634
240	0.001 204 6	1016.0	1040.0	2.6674	0.001 192 0	1006.9	1042.6	2.6490	0.001 170 2	990.7	1049.2	2.6158
280	0.001 296 5	1204.7	1230.6	3.0248	0.001 275 5	1190.7	1229.0	2.9986	0.001 241 5	1167.2	1229.3	2.9537
320	0.001 443 7	1415.7	1444.6	3.3979	0.001 399 7	1390.7	1432.7	3.3539	0.001 338 8	1353.3	1420.2	3.2868
360	0.001 822 6	1702.8	1739.3	3.8772	0.001 626 5	1626.6	1675.4	3.7494	0.001 483 8	1556.0	1630.2	3.6291

SOURCES: Keenan, Keyes, Hill, and Moore, *Steam Tables*, Wiley, New York, 1969; G. J. Van Wylen and R. E. Sonntag, *Fundamentals of Classical Thermodynamics*, Wiley, New York, 1973.

Table C-5 Saturated Solid—Vapor

		Volume, m³/kg		**Energy**, kJ/kg			**Enthalpy**, kJ/kg			**Entropy**, kJ/kg·K		
		Sat. Solid	**Sat. Vapor**	**Sat. Solid**	**Subl.**	**Sat. Vapor**	**Sat. Solid**	**Subl.**	**Sat. Vapor**	**Sat. Solid**	**Subl.**	**Sat. Vapor**
T, °C	P, kPa	$v_i \times 10^3$	v_g	u_i	u_{ig}	u_g	h_i	h_{ig}	h_g	s_i	s_{ig}	s_g
0.01	0.6113	1.0908	206.1	−333.40	2708.7	2375.3	−333.40	2834.8	2501.4	−1.221	10.378	9.156
0	0.6108	1.0908	206.3	−333.43	2708.8	2375.3	−333.43	2834.8	2501.3	−1.221	10.378	9.157
−2	0.5176	1.0904	241.7	−337.62	2710.2	2372.6	−337.62	2835.3	2497.7	−1.237	10.456	9.219
−4	0.4375	1.0901	283.8	−341.78	2711.6	2369.8	−341.78	2835.7	2494.0	−1.253	10.536	9.283
−6	0.3689	1.0898	334.2	−345.91	2712.9	2367.0	−345.91	2836.2	2490.3	−1.268	10.616	9.348
−8	0.3102	1.0894	394.4	−350.02	2714.2	2364.2	−350.02	2836.6	2486.6	−1.284	10.698	9.414
−10	0.2602	1.0891	466.7	−354.09	2715.5	2361.4	−354.09	3837.0	2482.9	−1.299	10.781	9.481
−12	0.2176	1.0888	553.7	−358.14	2716.8	2358.7	−358.14	2837.3	2479.2	−1.315	10.865	9.550
−14	0.1815	1.0884	658.8	−362.15	2718.0	2355.9	−362.15	2837.6	2475.5	−1.331	10.950	9.619
−16	0.1510	1.0881	786.0	−366.14	2719.2	2353.1	−366.14	2837.9	2471.8	−1.346	11.036	9.690
−20	0.1035	1.0874	1128.6	−374.03	2721.6	2347.5	−374.03	2838.4	2464.3	−1.377	11.212	9.835
−24	0.0701	1.0868	1640.1	−381.80	2723.7	2342.0	−381.80	2838.7	2456.9	−1.408	11.394	9.985
−28	0.0469	1.0861	2413.7	−389.45	2725.8	2336.4	−389.45	2839.0	2449.5	−1.439	11.580	10.141
−32	0.0309	1.0854	3600	−396.98	2727.8	2330.8	−396.98	2839.1	2442.1	−1.471	11.773	10.303
−36	0.0201	1.0848	5444	−404.40	2729.6	2325.2	−404.40	2839.1	2434.7	−1.501	11.972	10.470
−40	0.0129	1.0841	8354	−411.70	2731.3	2319.6	−411.70	2838.9	2427.2	−1.532	12.176	10.644

SOURCES: Keenan, Keyes, Hill, and Moore, *Steam Tables*, Wiley, New York, 1969; G. J. Van Wylen and R. E. Sonntag, *Fundamentals of Classical Thermodynamics*, Wiley, New York, 1973.

Table C-1E Properties of Saturated H_2O—Temperature Table

		Volume, ft³/lbm		**Energy**, Btu/lbm			**Enthalpy**, Btu/lbm			**Entropy**, Btu/lbm-°R		
Temp, T, °F	**Press.** P, psia	**Sat. Liquid** v_f	**Sat. Vapor** v_g	**Sat. Liquid** u_f	**Evap.** u_{fg}	**Sat. Vapor** u_g	**Sat. Liquid** h_f	**Evap.** h_{fg}	**Sat. Vapor** h_g	**Sat. Liquid** s_f	**Evap.** s_{fg}	**Sat. Vapor** s_g
32.018	0.08866	0.016022	3302	0.00	1021.2	1021.2	0.01	1075.4	1075.4	0.00000	2.1869	2.1869
35	0.09992	0.016021	2948	2.99	1019.2	1022.2	3.00	1073.7	1076.7	0.00607	2.1704	2.1764
40	0.12166	0.016020	2445	8.02	1015.8	1023.9	8.02	1070.9	1078.9	0.01617	2.1430	2.1592
45	0.14748	0.016021	2037	13.04	1012.5	1025.5	13.04	1068.1	1081.1	0.02618	2.1162	2.1423
50	0.17803	0.016024	1704.2	18.06	1009.1	1027.2	18.06	1065.2	1083.3	0.03607	2.0899	2.1259
60	0.2563	0.016035	1206.9	28.08	1002.4	1030.4	28.08	1059.6	1087.7	0.05555	2.0388	2.0943
70	0.3632	0.016051	867.7	38.09	995.6	1033.7	38.09	1054.0	1092.0	0.07463	1.9896	2.0642
80	0.5073	0.016073	632.8	48.08	988.9	1037.0	48.09	1048.3	1096.4	0.09332	1.9423	2.0356
90	0.6988	0.016099	467.7	58.07	982.2	1040.2	58.07	1042.7	1100.7	0.11165	1.8966	2.0083
100	0.9503	0.016130	350.0	68.04	975.4	1043.5	68.05	1037.0	1105.0	0.12963	1.8526	1.9822
110	1.2763	0.016166	265.1	78.02	968.7	1046.7	78.02	1031.3	1109.3	0.14730	1.8101	1.9574
120	1.6945	0.016205	203.0	87.99	961.9	1049.9	88.00	1025.5	1113.5	0.16465	1.7690	1.9336
130	2.225	0.016247	157.17	97.97	955.1	1053.0	97.98	1019.8	1117.8	0.18172	1.7292	1.9109
140	2.892	0.016293	122.88	107.95	948.2	1056.2	107.96	1014.0	1121.9	0.19851	1.6907	1.8892
150	3.722	0.016343	96.99	117.95	941.3	1059.3	117.96	1008.1	1126.1	0.21503	1.6533	1.8684
160	4.745	0.016395	77.23	127.94	934.4	1062.3	127.96	1002.2	1130.1	0.23130	1.6171	1.8484
170	5.996	0.016450	62.02	137.95	927.4	1065.4	137.97	996.2	1134.2	0.24732	1.5819	1.8293
180	7.515	0.016509	50.20	147.97	920.4	1068.3	147.99	990.2	1138.2	0.26311	1.5478	1.8109
190	9.343	0.016570	40.95	158.00	913.3	1071.3	158.03	984.1	1142.1	0.27866	1.5146	1.7932
200	11.529	0.016634	33.63	168.04	906.2	1074.2	168.07	977.9	1145.9	0.29400	1.4822	1.7762
210	14.125	0.016702	27.82	178.10	898.9	1077.0	178.14	971.6	1149.7	0.30913	1.4508	1.7599
212	14.698	0.016716	26.80	180.11	897.5	1077.6	180.16	970.3	1150.5	0.31213	1.4446	1.7567
220	17.188	0.016772	23.15	188.17	891.7	1079.8	188.22	965.3	1153.5	0.32406	1.4201	1.7441
230	20.78	0.016845	19.386	198.26	884.3	1082.6	198.32	958.8	1157.1	0.33880	1.3091	1.7289
240	24.97	0.016922	16.327	208.36	876.9	1085.3	208.44	952.3	1160.7	0.35335	1.3609	1.7143
250	29.82	0.017001	13.826	218.49	869.4	1087.9	218.59	945.6	1164.2	0.36772	1.3324	1.7001
260	35.42	0.017084	11.768	228.64	861.8	1090.5	228.76	938.8	1167.6	0.38193	1.3044	1.6864
270	41.85	0.017170	10.066	238.82	854.1	1093.0	238.95	932.0	1170.9	0.39597	1.2771	1.6731
280	49.18	0.017259	8.650	249.02	846.3	1095.4	249.18	924.9	1174.1	0.40986	1.2504	1.6602
290	57.53	0.017352	7.467	259.25	838.5	1097.7	259.44	917.8	1177.2	0.42360	1.2241	1.6477
300	66.98	0.017448	6.472	269.52	830.5	1100.0	269.73	910.4	1180.2	0.43720	1.1984	1.6356
320	89.60	0.017652	4.919	290.14	814.1	1104.2	290.43	895.3	1185.8	0.46400	1.1483	1.6123
340	117.93	0.017872	3.792	310.91	797.1	1108.0	311.30	879.5	1190.8	0.49031	1.0997	1.5901
360	152.92	0.018108	2.961	331.84	779.6	1111.4	332.35	862.9	1195.2	0.51617	1.0526	1.5688
380	195.60	0.018363	2.339	352.95	761.4	1114.3	353.62	845.4	1199.0	0.54163	1.0067	1.5483
400	247.1	0.018638	1.8661	374.27	742.4	1116.6	375.12	826.8	1202.0	0.56672	0.9617	1.5284
420	308.5	0.018936	1.5024	395.81	722.5	1118.3	396.89	807.2	1204.1	0.59152	0.9175	1.5091
440	381.2	0.019260	1.2192	417.62	701.7	1119.3	418.98	786.3	1205.3	0.61605	0.8740	1.4900
460	466.3	0.019614	0.9961	439.7	679.8	1119.6	441.4	764.1	1205.5	0.6404	0.8308	1.4712
480	565.5	0.020002	0.8187	462.2	656.7	1118.9	464.3	740.3	1204.6	0.6646	0.7878	1.4524
500	680.0	0.02043	0.6761	485.1	632.3	1117.4	487.7	714.8	1202.5	0.6888	0.7448	1.4335
520	811.4	0.02091	0.5605	508.5	606.2	1114.8	511.7	687.3	1198.9	0.7130	0.7015	1.4145
540	961.5	0.02145	0.4658	532.6	578.4	1111.0	536.4	657.5	1193.8	0.7374	0.6576	1.3950
560	1131.8	0.02207	0.3877	557.4	548.4	1105.8	562.0	625.0	1187.0	0.7620	0.6129	1.3749
580	1324.3	0.02278	0.3225	583.1	515.9	1098.9	588.6	589.3	1178.0	0.7872	0.5668	1.3540
600	1541.0	0.02363	0.2677	609.9	480.1	1090.0	616.7	549.7	1166.4	0.8130	0.5187	1.3317
620	1784.4	0.02465	0.2209	638.3	440.2	1078.5	646.4	505.0	1151.4	0.8398	0.4677	1.3075
640	2057.1	0.02593	0.1805	668.7	394.5	1063.2	678.6	453.4	1131.9	0.8681	0.4122	1.2803
660	2362	0.02767	0.14459	702.3	340.0	1042.3	714.4	391.1	1105.5	0.8990	0.3493	1.2483
680	2705	0.03032	0.11127	741.7	269.3	1011.0	756.9	309.8	1066.7	0.9350	0.2718	1.2068
700	3090	0.03666	0.07438	801.7	145.9	947.7	822.7	167.5	990.2	0.9902	0.1444	1.1346
705.44	3204	0.05053	0.05053	872.6	0	872.6	902.5	0	902.5	1.0580	0	1.0580

SOURCE: Keenan, Keyes, Hill, and Moore, *Steam Tables*, Wiley, New York, 1969.

Table C-2E Properties of Saturated H_2O—Pressure Table

		Volume, ft^3/lbm		**Energy**, Btu/lbm			**Enthalpy**, Btu/lbm			**Entropy**, Btu/lbm-°R		
Press. P, psia	**Temp.** T, °F	**Sat. Liquid** v_f	**Sat. Vapor** v_g	**Sat. Liquid** u_f	**Evap.** u_{fg}	**Sat. Vapor** u_g	**Sat. Liquid** h_f	**Evap.** h_{fg}	**Sat. Vapor** h_g	**Sat. Liquid** s_f	**Evap.** s_{fg}	**Sat. Vapor** s_g
1.0	101.70	0.016136	333.6	69.74	974.3	1044.0	69.74	1036.0	1105.8	0.13266	1.8453	1.9779
2.0	126.04	0.016230	173.75	94.02	957.8	1051.8	94.02	1022.1	1116.1	0.17499	1.7448	1.9198
3.0	141.43	0.016300	118.72	109.38	947.2	1056.6	109.39	1013.1	1122.5	0.20089	1.6852	1.8861
4.0	152.93	0.016358	90.64	120.88	939.3	1060.2	120.89	1006.4	1127.3	0.21983	1.6426	1.8624
5.0	162.21	0.016407	73.53	130.15	932.9	1063.0	130.17	1000.9	1131.0	0.23486	1.6093	1.8441
6.0	170.03	0.016451	61.98	137.98	927.4	1065.4	138.00	996.2	1134.2	0.24736	1.5819	1.8292
8.0	182.84	0.016526	47.35	150.81	918.4	1069.2	150.84	988.4	1139.3	0.26754	1.5383	1.8058
10	193.19	0.016590	38.42	161.20	911.0	1072.2	161.23	982.1	1143.3	0.28358	1.5041	1.7877
14.696	211.99	0.016715	26.80	180.10	897.5	1077.6	180.15	970.4	1150.5	0.31212	1.4446	1.7567
15	213.03	0.016723	26.29	181.14	896.8	1077.9	181.19	969.7	1150.9	0.31367	1.4414	1.7551
20	227.96	0.016830	20.09	196.19	885.8	1082.0	196.26	960.1	1156.4	0.33580	1.3962	1.7320
25	240.08	0.016922	16.306	208.44	876.9	1085.3	208.52	952.2	1160.7	0.35345	1.3607	1.7142
30	250.34	0.017004	13.748	218.84	869.2	1088.0	218.93	945.4	1164.3	0.36821	1.3314	1.6996
35	259.30	0.017073	11.900	227.93	862.4	1090.3	228.04	939.3	1167.4	0.38093	1.3064	1.6873
40	267.26	0.017146	10.501	236.03	856.2	1092.3	236.16	933.8	1170.0	0.39214	1.2845	1.6767
45	274.46	0.017209	9.403	243.37	850.7	1094.0	243.51	928.8	1172.3	0.40218	1.2651	1.6673
50	281.03	0.017269	8.518	250.08	845.5	1095.6	250.24	924.2	1174.4	0.41129	1.2476	1.6589
55	287.10	0.017325	7.789	256.28	840.8	1097.0	256.46	919.9	1176.3	0.41963	1.2317	1.6513
60	292.73	0.017378	7.177	262.06	836.3	1098.3	262.25	915.8	1178.0	0.42733	1.2170	1.6444
65	298.00	0.017429	6.657	267.46	832.1	1099.5	267.67	911.9	1179.6	0.43450	1.2035	1.6380
70	302.96	0.017478	6.209	272.56	828.1	1100.6	272.79	908.3	1181.0	0.44120	1.1909	1.6321
75	307.63	0.017524	5.818	277.37	824.3	1101.6	277.61	904.8	1182.4	0.44749	1.790	1.6265
80	312.07	0.017570	5.474	281.95	820.6	1102.6	282.21	901.4	1183.6	0.45344	1.1679	1.6214
85	316.29	0.017613	5.170	286.30	817.1	1103.5	286.58	898.2	1184.8	0.45907	1.1574	1.6165
90	320.31	0.017655	4.898	290.46	813.8	1104.3	290.76	895.1	1185.9	0.46442	1.1475	1.6119
95	324.16	0.017696	4.654	294.45	810.6	1105.0	294.76	892.1	1186.9	0.46952	1.1380	1.6076
100	327.86	0.017736	4.434	298.28	807.5	1105.8	298.61	889.2	1187.8	0.47439	1.1290	1.6034
110	334.82	0.017813	4.051	305.52	801.6	1107.1	305.88	883.7	1189.6	0.48355	1.1122	1.5957
120	341.30	0.017886	3.730	312.27	796.0	1108.3	312.67	878.5	1191.1	0.49201	1.0966	1.5886
130	347.37	0.017957	3.457	318.61	790.7	1109.4	319.04	873.5	1192.5	0.49989	1.0822	1.5821
140	353.08	0.018024	3.221	324.58	785.7	1110.3	325.05	868.7	1193.8	0.50727	1.0688	1.5761
150	358.48	0.018089	3.016	330.24	781.0	1111.2	330.75	864.2	1194.9	0.51422	1.0562	1.5704
160	363.60	0.018152	2.836	335.63	776.4	1112.0	336.16	859.8	1196.0	0.52078	1.0443	1.5651
170	368.47	0.018214	2.676	340.76	772.0	1112.7	341.33	355.6	1196.9	0.52700	1.0330	1.5600
180	373.13	0.018273	2.533	345.68	767.7	1113.4	346.29	851.5	1197.8	0.53292	1.0223	1.5553
190	337.59	0.018331	2.405	350.39	763.6	1114.0	351.04	847.5	1198.6	0.53857	1.0122	1.5507
200	381.86	0.018387	2.289	354.9	759.6	1114.6	355.6	843.7	1199.3	0.5440	1.0025	1.5664
300	417.43	0.018896	1.5442	393.0	725.1	1118.2	394.1	809.8	1203.9	0.5883	0.9232	1.5115
400	444.70	0.019340	1.1620	422.8	696.7	1119.5	424.2	781.2	1205.5	0.6218	0.8638	1.4856
500	467.13	0.019748	0.9283	447.7	671.7	1119.4	449.5	755.8	1205.3	0.6490	0.8154	1.4645
600	486.33	0.02013	0.7702	469.4	649.1	1118.6	471.7	732.4	1204.1	0.6723	0.7742	1.4464
700	503.23	0.02051	0.6558	488.9	628.2	1117.0	491.5	710.5	1202.0	0.6927	0.7378	1.4305
800	518.36	0.02087	0.5691	506.6	608.4	1115.0	509.7	689.6	1199.3	0.7110	0.7050	1.4160
900	532.12	0.02123	0.5009	523.0	589.6	1112.6	526.6	669.5	1196.0	0.7277	0.6750	1.4027
1000	544.75	0.02159	0.4459	538.4	571.5	1109.9	542.4	650.0	1192.4	0.7432	0.6471	1.3903
1200	567.37	0.02232	0.3623	566.7	536.8	1103.5	571.7	612.3	1183.9	0.7712	0.5961	1.3673
1400	587.25	0.02307	0.3016	592.7	503.3	1096.0	598.6	575.5	1174.1	0.7964	0.5497	1.3461
1600	605.06	0.02386	0.2552	616.9	470.5	1087.4	624.0	538.9	1162.9	0.8196	0.5062	1.3258
1800	621.21	0.02472	0.2183	640.0	437.6	1077.7	648.3	502.1	1150.4	0.8414	0.4645	1.3060
2000	636.00	0.02565	0.18813	662.4	404.2	1066.6	671.9	464.4	1136.3	0.8623	0.4238	1.2861
2500	668.31	0.02860	0.13059	717.7	313.4	1031.0	730.9	360.5	1091.4	0.9131	0.3196	1.2327
3000	695.52	0.03431	0.08404	783.4	185.4	968.8	802.5	213.0	1015.5	0.9732	0.1843	1.1575
3203.6	705.44	0.05053	0.05053	872.6	0	872.6	902.5	0	902.5	1.0580	0	1.0580

SOURCE: Keenan, Keyes, Hill, and Moore, *Steam Tables*, Wiley, New York, 1969.

Table C-3E Properties of Superheated Steam

°F	*v*	*u*	*h*	*s*	*v*	*u*	*h*	*s*	*v*	*u*	*h*	*s*
	P = 1.0 psia (101.70 °F)				P = 5.0 psia (162.21 °F)				P = 10.0 psia (193.19 °F)			
Sat.	333.6	1044.0	1105.8	1.9779	73.53	1063.0	1131.0	1.8441	38.42	1072.2	1143.3	1.7877
200	392.5	1077.5	1150.1	2.0508	78.15	1076.3	1148.6	1.8715	38.85	1074.7	1146.6	1.7927
240	416.4	1091.2	1168.3	2.0775	83.00	1090.3	1167.1	1.8987	41.32	1089.0	1165.5	1.8205
280	440.3	1105.0	1186.5	2.1028	87.83	1104.3	1185.5	1.9244	43.77	1103.3	1184.3	1.8467
320	464.2	1118.9	1204.8	2.1269	92.64	1118.3	1204.0	1.9487	46.20	1117.6	1203.1	1.8714
360	488.1	1132.5	1223.2	2.1500	97.45	1132.4	1222.6	1.9719	48.62	1131.8	1221.8	1.8948
400	511.9	1147.0	1241.8	2.1720	102.24	1146.6	1241.2	1.9941	51.03	1146.1	1240.5	1.9171
500	571.5	1182.8	1288.5	2.2235	114.20	1182.5	1288.2	2.0458	57.04	1182.2	1287.7	1.9690
600	631.1	1219.3	1336.1	2.2706	126.15	1219.1	1335.8	2.0930	63.03	1218.9	1335.5	2.0164
700	690.7	1256.7	1384.5	2.3142	138.08	1256.5	1384.3	2.1367	69.01	1256.3	1384.0	2.0601
800	750.3	1294.9	1433.7	2.3550	150.01	1294.7	1433.5	2.1775	74.98	1294.6	1433.3	2.1009
1000	869.5	1373.9	1534.8	2.4294	173.86	1373.9	1534.7	2.2520	86.91	1373.8	1534.6	2.1755
	P = 14.696 psia (211.99 °F)				P = 20 psia (277.96 °F)				P = 40 psia (267.26 °F)			
Sat.	26.80	1077.6	1150.5	1.7567	20.09	1082.0	1156.4	1.7320	10.501	1092.3	1170.0	1.6767
240	28.00	1087.9	1164.0	1.7764	20.47	1086.5	1162.3	1.7405				
280	29.69	1102.4	1183.1	1.8030	21.73	1101.4	1181.8	1.7676	10.711	1097.3	1176.6	1.6857
320	31.36	1116.8	1202.1	1.8280	22.98	1116.0	1201.1	1.7930	11.360	1112.8	1196.9	1.7124
360	33.02	1131.2	1221.0	1.8516	24.21	1130.6	1220.1	1.8168	11.996	1128.0	1216.8	1.7373
400	34.67	1145.6	1239.9	1.8741	25.43	1145.1	1239.2	1.8395	12.623	1143.0	1236.4	1.7606
500	38.77	1181.8	1287.3	1.9263	28.46	1181.5	1286.3	1.8919	14.164	1180.1	1284.9	1.8140
600	42.86	1218.6	1335.2	1.9737	31.47	1218.4	1334.8	1.9395	15.685	1217.3	1333.4	1.8621
700	46.93	1256.1	1383.8	2.0175	34.47	1255.9	1383.5	1.9834	17.196	1255.1	1382.4	1.9063
800	51.00	1294.4	1433.1	2.0584	37.46	1294.3	1432.9	2.0243	18.701	1293.7	1432.1	1.9474
1000	59.13	1373.7	1534.5	2.1330	43.44	1373.5	1534.3	2.0989	21.70	1373.1	1533.8	2.0223
1200	67.25	1465	1639.3	2.2003	49.41	1456.4	1639.2	2.1663	24.69	1456.1	1638.9	2.0897
	P = 60 psia (292.73 °F)				P = 80 psia (312.07 °F)				P = 100 psia (327.86 °F)			
Sat.	7.177	1098.3	1178.0	1.6444	5.474	1102.6	1183.6	1.6214	4.434	1105.8	1187.8	1.6034
320	7.485	1109.5	1192.6	1.6634	5.544	1106.0	1188.0	1.6271				
360	7.924	1125.3	1213.3	1.6893	5.886	1122.5	1209.7	1.6541	4.662	1119.7	1205.9	1.6259
400	8.353	1140.3	1233.5	1.7134	6.217	1138.5	1230.6	1.6790	4.934	1136.2	1227.5	1.6517
500	9.399	1178.6	1283.0	1.7678	7.017	1177.2	1281.1	1.7346	5.587	1175.7	1279.1	1.7085
600	10.425	1216.3	1332.1	1.8165	7.794	1215.3	1330.7	1.7838	6.216	1214.2	1329.3	1.7582
700	11.440	1254.4	1381.4	1.8609	8.561	1253.6	1380.3	1.8285	6.834	1252.8	1379.2	1.8033
800	12.448	1293.0	1431.2	1.9022	9.321	1292.4	1430.4	1.8700	7.445	1291.8	1429.6	1.8449
1000	14.454	1372.7	1533.2	1.9773	10.831	1372.3	1532.6	1.9453	8.657	1371.9	1532.1	1.9204
1200	16.452	1455.8	1638.5	2.0448	12.333	1455.5	1638.1	2.0130	9.861	1455.2	1637.7	1.9882
1400	18.445	1542.5	1747.3	2.1067	13.830	1542.3	1747.0	2.0749	11.060	1542.0	1746.7	2.0502
1600	20.44	1632.8	1859.7	2.1641	15.324	1632.6	1859.5	2.1323	12.257	1632.4	1859.3	2.1076
	P = 120 psia (341.30 °F)				P = 140 psia (353.08 °F)				P = 160 psia (363.60 °F)			
Sat.	3.730	1108.3	1191.1	1.5886	3.221	1110.3	1193.8	1.5761	2.836	1112.0	1196.0	1.5651
360	3.844	1116.7	1202.0	1.6021	3.259	1113.5	1198.0	1.5812				
400	4.079	1133.8	1224.4	1.6288	3.466	1131.4	1221.2	1.6088	3.007	1128.8	1217.8	1.5911
450	4.360	1154.3	1251.2	1.6590	3.713	1152.4	1248.6	1.6399	3.228	1150.5	1246.1	1.6230
500	4.633	1174.2	1277.1	1.6868	3.952	1172.7	1275.1	1.6682	3.440	1171.2	1273.0	1.6518
600	5.164	1213.2	1327.8	1.7371	4.412	1212.1	1326.4	1.7191	3.848	1211.1	1325.0	1.7034
700	5.682	1252.0	1378.2	1.7825	4.860	1251.2	1377.1	1.7648	4.243	1250.4	1376.0	1.7494
800	6.195	1291.2	1428.7	1.8243	5.301	1290.5	1427.9	1.8068	4.631	1289.9	1427.0	1.7916
1000	7.208	1371.5	1531.5	1.9000	6.173	1371.0	1531.0	1.8827	5.397	1370.6	1530.4	1.8677
1200	8.213	1454.9	1637.3	1.9679	7.036	1454.6	1636.9	1.9507	6.154	1454.3	1636.5	1.9358
1400	9.214	1541.8	1746.4	2.0300	7.895	1541.6	1746.1	2.0129	6.906	1541.4	1745.9	1.9980
1600	10.212	1632.3	1859.0	2.0875	8.752	1632.1	1858.8	2.0704	7.656	1631.9	1858.6	2.0556

Table C-3E (*Continued*)

°F	*v*	*u*	*h*	*s*	*v*	*u*	*h*	*s*	*v*	*u*	*h*	*s*
	P = 180 psia (373.13 °F)				P = 200 psia (381.68 °F)				P = 300 psia (417.43 °F)			
Sat.	2.533	1113.4	1197.8	1.5553	2.289	1114.6	1199.3	1.5464	1.5442	1118.2	1203.9	1.5115
400	2.648	1126.2	1214.4	1.5749	2.361	1123.5	1210.8	1.5600				
450	2.850	1148.5	1243.4	1.6078	2.548	1146.4	1240.7	1.5938	1.6361	1135.4	1226.2	1.5363
500	3.042	1169.6	1270.9	1.6372	2.724	1168.0	1268.8	1.6239	1.7662	1159.5	1257.5	1.5701
600	3.409	1210.0	1323.5	1.6893	3.058	1208.9	1322.1	1.6767	2.004	1203.2	1314.5	1.6266
700	3.763	1249.6	1374.9	1.7357	3.379	1248.8	1373.8	1.7234	2.227	1244.6	1368.3	1.6751
800	4.110	1289.3	1426.2	1.7781	3.693	1288.6	1425.3	1.7660	2.442	1285.4	1421.0	1.7187
900	4.453	1329.4	1477.7	1.8175	4.003	1328.9	1477.1	1.8055	2.653	1328.3	1473.6	1.7589
1000	4.793	1370.2	1529.8	1.8545	4.310	1369.8	1529.3	1.8425	2.860	1367.7	1526.5	1.7964
1200	5.467	1454.0	1636.1	1.9227	4.918	1453.7	1635.7	1.9109	3.270	1452.2	1633.8	1.8653
1400	6.137	1541.2	1745.6	1.9849	5.521	1540.9	1745.3	1.9732	3.675	1539.8	1743.8	1.9279
1600	6.804	1631.7	1858.4	2.0425	6.123	1631.6	1858.2	2.0308	4.078	1630.7	1857.0	1.9857
	P = 400 psia (447.70 °F)				P = 500 psia (467.13 °F)				P = 600 psia (486.33 °F)			
Sat.	1.1620	1119.5	1205.5	1.4856	0.9283	1119.4	1205.3	1.4645	0.7702	1118.6	1204.1	1.4464
500	1.2843	1150.1	1245.2	1.5282	0.9924	1139.7	1231.5	1.4923	0.7947	1128.0	1216.2	1.4592
550	1.3833	1174.6	1277.0	1.5605	1.0792	1166.7	1266.6	1.5279	0.8749	1158.2	1255.4	1.4990
600	1.4760	1197.3	1306.6	1.5892	1.1583	1191.1	1298.3	1.5585	0.9456	1184.5	1289.5	1.5320
700	1.6503	1240.4	1362.5	1.6397	1.3040	1236.0	1356.7	1.6112	1.0727	1231.5	1350.6	1.5872
800	1.8163	1282.1	1416.6	1.6844	1.4407	1278.8	1412.1	1.6571	1.1900	1275.4	1407.6	1.6343
900	1.9776	1323.7	1470.1	1.7252	1.5723	1321.0	1466.5	1.6987	1.3021	1318.4	1462.9	1.6766
1000	2.136	1365.5	1523.6	1.7632	1.7008	1363.3	1520.7	1.7371	1.4108	1361.2	1517.8	1.7155
1100					1.8271	1406.0	1575.1	1.7731	1.5173	1404.2	1572.7	1.7519
1200	2.446	1450.7	1631.8	1.8327	1.9518	1449.2	1629.8	1.8072	1.6222	1447.7	1627.8	1.7861
1400	2.752	1538.7	1742.4	1.8956	2.198	1537.6	1741.0	1.8704	1.8289	1536.5	1739.5	1.8497
1600	3.055	1629.8	1855.9	1.9535	2.442	1628.9	1854.8	1.9285	2.033	1628.0	1853.7	1.9080
	P = 800 psia (518.36 °F)				P = 1000 psia (544.75 °F)				P = 2000 psia (636.00 °F)			
550	0.6154	1138.8	1229.9	1.4469	0.4534	1114.8	1198.7	1.3966				
600	0.6776	1170.1	1270.4	1.4861	0.5140	1153.7	1248.8	1.4450				
650	0.7324	1197.2	1305.6	1.5186	0.5637	1184.7	1289.1	1.4822	0.2057	1091.1	1167.2	1.3141
700	0.7829	1222.1	1338.0	1.5471	0.6080	1212.0	1324.6	1.5135	0.2487	1147.7	1239.8	1.3782
800	0.8764	1268.5	1398.2	1.5969	0.6878	1261.2	1388.5	1.5664	0.3071	1220.1	1333.8	1.4562
900	0.9640	1312.9	1455.6	1.6408	0.7610	1307.3	1488.1	1.6120	0.3534	1276.8	1407.6	1.5126
1000	1.0482	1356.7	1511.9	1.6807	0.8305	1352.2	1505.9	1.6530	0.3945	1328.1	1474.1	1.5598
1100	1.1300	1400.5	1567.8	1.7178	0.8976	1396.8	1562.9	1.6908	0.4325	1377.2	1537.2	1.6017
1200	1.2102	1444.6	1623.8	1.7526	0.9630	1441.5	1619.7	1.7261	0.4685	1425.2	1598.6	1.6393
1400	1.3674	1534.2	1736.6	1.8167	1.0905	1531.9	1733.7	1.7909	0.5368	1520.2	1718.8	1.7082
1600	1.5218	1626.2	1851.5	1.8754	1.2152	1624.4	1849.3	1.8499	0.6020	1615.4	1838.2	1.7692
	P = 3000 psia (695.52 °F)				P = 4000 psia				P = 5000 psia			
650					0.02447	657.7	675.8	0.8574	0.02377	648.0	670.0	0.8482
700	0.09771	1003.9	1058.1	1.1944	0.02867	742.1	763.4	0.9345	0.02676	721.8	746.6	0.9156
750	0.14831	1114.7	1197.1	1.3122	0.06331	960.7	1007.5	1.1395	0.03364	821.4	852.6	1.0049
800	0.17572	1167.6	1265.2	1.3675	0.10522	1095.0	1172.9	1.2740	0.05932	987.2	1042.1	1.1583
850	0.19731	1207.7	1317.2	1.4080	0.12833	1156.5	1251.5	1.3352	0.08556	1092.7	1171.9	1.2956
900	0.2160	1241.8	1361.7	1.4414	0.14622	1201.5	1309.7	1.3789	0.10385	1155.1	1251.1	1.3190
1000	0.2485	1301.7	1439.6	1.4967	0.17520	1272.9	1402.6	1.4449	0.13120	1242.0	1363.4	1.3988
1100	0.2772	1356.2	1510.1	1.5434	0.19954	1333.9	1481.6	1.4973	0.15302	1310.6	1452.2	1.4577
1200	0.3036	1408.0	1576.6	1.5848	0.2213	1390.1	1553.9	1.5423	0.17199	1371.6	1530.8	1.5066
1300					0.2414	1443.7	1622.4	1.5823	0.18918	1428.6	1603.7	1.5493
1400	0.3524	1508.1	1703.7	1.6571	0.2603	1495.7	1688.4	1.6188	0.20517	1483.2	1673.0	1.5876
1600	0.3978	1606.3	1827.1	1.7201	0.2959	1597.1	1816.1	1.6841	0.2348	1587.9	1805.2	1.6551

SOURCE: Keenan, Keyes, Hill, and Moore, *Steam Tables*, Wiley, New York, 1969.

Table C-4E Compressed Liquid (T in °F, v in ft^3/lbm, u and h in Btu/lbm, s in Btu/lbm-°R)

°F	v	u	h	s	v	u	h	s	v	u	h	s
	P = 500 psia (467.13 °F)				P = 1000 psia (544.75 °F)				P = 1500 psia (596.39 °F)			
Sat.	0.019748	447.70	449.53	0.64904	0.021591	538.39	542.38	0.74320	0.023461	604.97	611.48	0.80824
32	0.015994	0.00	1.49	0.00000	0.015967	0.03	2.99	0.00005	0.015939	0.05	4.47	0.00007
50	0.015998	18.02	19.50	0.03599	0.015972	17.99	20.94	0.03592	0.015946	17.95	22.38	0.03584
100	0.016106	67.87	69.36	0.12932	0.016082	67.70	70.68	0.12901	0.016058	67.53	71.99	0.12870
150	0.016318	117.66	119.17	0.21457	0.016293	117.38	120.40	0.21410	0.016268	117.10	121.62	0.21364
200	0.016608	167.65	169.19	0.29341	0.016580	167.26	170.32	0.29281	0.016554	166.87	171.46	0.29221
250	0.016972	217.99	219.56	0.36702	0.016941	217.47	220.61	0.36628	0.016910	216.96	221.65	0.36554
300	0.017416	268.92	270.53	0.43641	0.017379	268.24	271.46	0.43552	0.017343	267.58	272.39	0.43463
350	0.017954	320.71	322.37	0.50249	0.017909	319.83	323.15	0.50140	0.017865	318.98	323.94	0.50034
400	0.018608	373.68	375.40	0.56604	0.018550	372.55	375.98	0.56472	0.018493	371.45	376.59	0.56343
450	0.019420	428.40	430.19	0.62798	0.019340	426.89	430.47	0.62632	0.019264	425.44	430.79	0.62470
500					0.02036	483.8	487.5	0.6874	0.02024	481.8	487.4	0.6853
550									0.02158	542.1	548.1	0.7469
	P = 2000 psia (636.00 °F)				P = 3000 psia (695.52 °F)				P = 5000 psia			
Sat.	0.025649	662.40	671.89	0.86227	0.034310	783.45	802.50	0.97320				
32	0.015912	0.06	5.95	0.00008	0.015859	0.09	8.90	0.00009	0.015755	0.11	14.70	−0.00001
50	0.015920	17.91	23.81	0.03575	0.015870	17.84	26.65	0.03555	0.015773	17.67	32.26	0.03508
100	0.016034	67.37	73.30	0.12839	0.015987	67.04	75.91	0.12777	0.015897	66.40	81.11	0.12651
200	0.016527	166.49	172.60	0.29162	0.016476	165.74	174.89	0.29046	0.016376	164.32	179.47	0.28818
300	0.017308	266.93	273.33	0.43376	0.017240	265.66	275.23	0.43205	0.017110	263.25	279.08	0.42875
400	0.018439	370.38	377.21	0.56216	0.018334	368.32	378.50	0.55970	0.018141	364.47	381.25	0.55506
450	0.019191	424.04	431.14	0.62313	0.019053	421.36	431.93	0.62011	0.018803	416.44	433.84	0.61451
500	0.02014	479.8	487.3	0.6832	0.019944	476.2	487.3	0.6794	0.019603	469.8	487.9	0.6724
560	0.02172	551.8	559.8	0.7565	0.021382	546.2	558.0	0.7508	0.020835	536.7	556.0	0.7411
600	0.02330	605.4	614.0	0.8086	0.02274	597.0	609.6	0.8004	0.02191	584.0	604.2	0.7876
640					0.02475	654.3	668.0	0.8545	0.02334	634.6	656.2	0.8357
680					0.02879	728.4	744.3	0.9226	0.02535	690.6	714.1	0.8873
700									0.02676	721.8	746.6	0.9156

SOURCE: Keenan, Keyes, Hill, and Moore, *Steam Tables*, Wiley, New York, 1969.

Table C-5E Saturated Solid—Vapor

		Volume, ft³/lbm		**Energy**, Btu/lbm			**Enthalpy**, Btu/lbm			**Entropy**, Btu/lbm-°R		
		Sat. Solid	**Sat. Vapor**	**Sat. Solid**	**Subl.**	**Sat. Vapor**	**Sat. Solid**	**Subl.**	**Sat. Vapor**	**Sat. Solid**	**Subl.**	**Sat. Vapor**
T, °F	P, psia	v_i	$v_g \times 10^{-3}$	u_i	u_{ig}	u_g	h_i	h_{ig}	h_g	s_i	s_{ig}	s_g
32.018	0.0887	0.01747	3.302	−143.34	1164.6	1021.2	−143.34	1218.7	1075.4	−0.292	2.479	2.187
32	0.0886	0.01747	3.305	−143.35	1164.6	1021.2	−143.35	1218.7	1075.4	−0.292	2.479	2.187
30	0.0808	0.01747	3.607	−144.35	1164.9	1020.5	−144.35	1218.9	1074.5	−0.294	2.489	2.195
25	0.0641	0.01746	4.506	−146.84	1165.7	1018.9	−146.84	1219.1	1072.3	−0.299	2.515	2.216
20	0.0505	0.01745	5.655	−149.31	1166.5	1017.2	−149.31	1219.4	1070.1	−0.304	2.542	2.238
15	0.0396	0.01745	7.13	−151.75	1167.3	1015.5	−151.75	1219.7	1067.9	−0.309	2.569	2.260
10	0.0309	0.01744	9.04	−154.17	1168.1	1013.9	−154.17	1219.9	1065.7	−0.314	2.597	2.283
5	0.0240	0.01743	11.52	−156.56	1168.8	1012.2	−156.56	1220.1	1063.5	−0.320	2.626	2.306
0	0.0185	0.01743	14.77	−158.93	1169.5	1010.6	−158.93	1220.2	1061.2	−0.325	2.655	2.330
−5	0.0142	0.01742	19.03	−161.27	1170.2	1008.9	−161.27	1220.3	1059.0	−0.330	2.684	2.354
−10	0.0109	0.01741	24.66	−163.59	1170.9	1007.3	−163.59	1220.4	1056.8	−0.335	2.714	2.379
−15	0.0082	0.01740	32.2	−165.89	1171.5	1005.6	−165.89	1220.5	1054.6	−0.340	2.745	2.405
−20	0.0062	0.01740	42.2	−168.16	1172.1	1003.9	−168.16	1220.6	1052.4	−0.345	2.776	2.431
−25	0.0046	0.01739	55.7	−170.40	1172.7	1002.3	−170.40	1220.6	1050.2	−0.351	2.808	2.457
−30	0.0035	0.01738	74.1	−172.63	1173.2	1000.6	−172.63	1220.6	1048.0	−0.356	2.841	2.485
−35	0.0026	0.01737	99.2	−174.82	1173.8	998.9	−174.82	1220.6	1045.8	−0.361	2.874	2.513
−40	0.0019	0.01737	133.8	−177.00	1174.3	997.3	−177.00	1220.6	1043.6	−0.366	2.908	2.542

SOURCE: Keenan, Keyes, Hill, and Moore, *Steam Tables*, Wiley, New York, 1969.

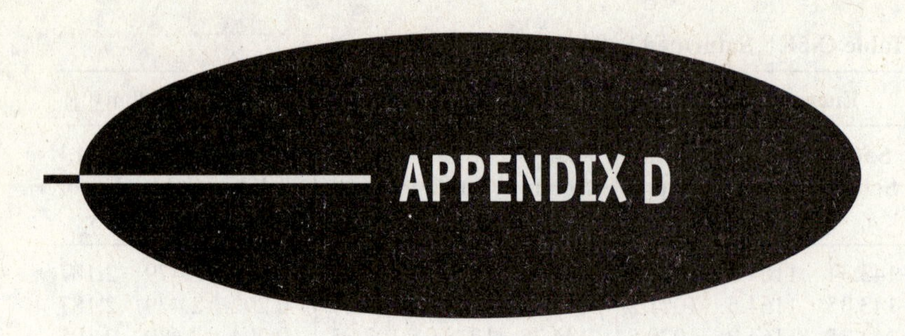

APPENDIX D

Thermodynamic Properties of R134a

Table D-1 Saturated R134a—Temperature Table

		Specific Volume m³/kg		Internal Energy kJ/kg		Enthalpy kJ/kg			Entropy kJ/kg·K	
Temp. °C	**Pressure** kPa	**Sat. Liquid** $v_f \times 10^3$	**Sat. Vapor** v_g	**Sat. Liquid** u_f	**Sat. Vapor** u_g	**Sat. Liquid** h_f	**Evap.** h_{fg}	**Sat. Vapor** h_g	**Sat. Liquid** s_f	**Sat. Vapor** s_g
–40	51.64	0.7055	0.3569	–0.04	204.45	0.00	222.88	222.88	0.0000	0.9560
–36	63.32	0.7113	0.2947	4.68	206.73	4.73	220.67	225.40	0.0201	0.9506
–32	77.04	0.7172	0.2451	9.47	209.01	9.52	218.37	227.90	0.0401	0.9456
–28	93.05	0.7233	0.2052	14.31	211.29	14.37	216.01	230.38	0.0600	0.9411
–26	101.99	0.7265	0.1882	16.75	212.43	16.82	214.80	231.62	0.0699	0.9390
–24	111.60	0.7296	0.1728	19.21	213.57	19.29	213.57	232.85	0.0798	0.9370
–22	121.92	0.7328	0.1590	21.68	214.70	21.77	212.32	234.08	0.0897	0.9351
–20	132.99	0.7361	0.1464	24.17	215.84	24.26	211.05	235.31	0.0996	0.9332
–18	144.83	0.7395	0.1350	26.67	216.97	26.77	209.76	236.53	0.1094	0.9315
–16	157.48	0.7428	0.1247	29.18	218.10	29.30	208.45	237.74	0.1192	0.9298
–12	185.40	0.7498	0.1068	34.25	220.36	34.39	205.77	240.15	0.1388	0.9267
–8	217.04	0.7569	0.0919	39.38	222.60	39.54	203.00	242.54	0.1583	0.9239
–4	252.74	0.7644	0.0794	44.56	224.84	44.75	200.15	244.90	0.1777	0.9213
0	292.82	0.7721	0.0689	49.79	227.06	50.02	197.21	247.23	0.1970	0.9190
4	337.65	0.7801	0.0600	55.08	229.27	55.35	194.19	249.53	0.2162	0.9169
8	387.56	0.7884	0.0525	60.43	231.46	60.73	191.07	251.80	0.2354	0.9150
12	442.94	0.7971	0.0460	65.83	233.63	66.18	187.85	254.03	0.2545	0.9132
16	504.16	0.8062	0.0405	71.29	235.78	71.69	184.52	256.22	0.2735	0.9116
20	571.60	0.8157	0.0358	76.80	237.91	77.26	181.09	258.36	0.2924	0.9102
24	645.66	0.8257	0.0317	82.37	240.01	82.90	177.55	260.45	0.3113	0.9089
26	685.30	0.8309	0.0298	85.18	241.05	85.75	175.73	261.48	0.3208	0.9082
28	726.75	0.8362	0.0281	88.00	242.08	88.61	173.89	262.50	0.3302	0.9076
30	770.06	0.8417	0.0265	90.84	243.10	91.49	172.00	263.50	0.3396	0.9070
32	815.28	0.8473	0.0250	93.70	244.12	94.39	170.09	264.48	0.3490	0.9064
34	862.47	0.8530	0.0236	96.58	245.12	97.31	168.14	265.45	0.3584	0.9058
36	911.68	0.8590	0.0223	99.47	246.11	100.25	166.15	266.40	0.3678	0.9053
38	962.98	0.8651	0.0210	102.38	247.09	103.21	164.12	267.33	0.3772	0.9047

		Specific Volume m³/kg		Internal Energy kJ/kg		Enthalpy kJ/kg			Entropy kJ/kg·K	
Temp. °C	Pressure kPa	Sat. Liquid $v_f \times 10^3$	Sat. Vapor v_g	Sat. Liquid u_f	Sat. Vapor u_g	Sat. Liquid h_f	Evap. h_{fg}	Sat. Vapor h_g	Sat. Liquid s_f	Sat. Vapor s_g
40	1016.4	0.8714	0.0199	105.30	248.06	106.19	162.05	268.24	0.3866	0.9041
42	1072.0	0.8780	0.0188	108.25	249.02	109.19	159.94	269.14	0.3960	0.9035
44	1129.9	0.8847	0.0177	111.22	249.96	112.22	157.79	270.01	0.4054	0.9030
48	1252.6	0.8989	0.0159	117.22	251.79	118.35	153.33	271.68	0.4243	0.9017
52	1385.1	0.9142	0.0142	123.31	253.55	124.58	148.66	273.24	0.4432	0.9004
56	1527.8	0.9308	0.0127	129.51	255.23	130.93	143.75	274.68	0.4622	0.8990
60	1681.3	0.9488	0.0114	135.82	256.81	137.42	138.57	275.99	0.4814	0.8973
70	2116.2	1.0027	0.0086	152.22	260.15	154.34	124.08	278.43	0.5302	0.8918
80	2632.4	1.0766	0.0064	169.88	262.14	172.71	106.41	279.12	0.5814	0.8827
90	3243.5	1.1949	0.0046	189.82	261.34	193.69	82.63	276.32	0.6380	0.8655
100	3974.2	1.5443	0.0027	218.60	248.49	224.74	34.40	259.13	0.7196	0.8117

SOURCE: Table D-1 through D-3 are based on equations from D. P. Wilson and R. S. Basu, "Thermodynamic Properties of a New Stratospherically Safe Working Fluid—Refrigerant 134a," *ASHRAE Trans.*, Vol. 94, Pt. 2, 1988, pp. 2095–2118.

Table D-2 Saturated R134a—Pressure Table

		Specific Volume m³/kg		Internal Energy kJ/kg		Enthalpy kJ/kg			Entropy kJ/kg·K	
Pressure kPa	Temp. °C	Sat. Liquid $v_f \times 10^3$	Sat. Vapor v_g	Sat. Liquid u_f	Sat. Vapor u_g	Sat. Liquid h_f	Evap. h_{fg}	Sat. Vapor h_g	Sat. Liquid s_f	Sat. Vapor s_g
60	–37.07	0.7097	0.3100	3.14	206.12	3.46	221.27	224.72	0.0147	0.9520
80	–31.21	0.7184	0.2366	10.41	209.46	10.47	217.92	228.39	0.0440	0.9447
100	–26.43	0.7258	0.1917	16.22	212.18	16.29	215.06	231.35	0.0678	0.9395
120	–22.36	0.7323	0.1614	21.23	214.50	21.32	212.54	233.86	0.0879	0.9354
140	–18.80	0.7381	0.1395	25.66	216.52	25.77	210.27	236.04	0.1055	0.9322
160	–15.62	0.7435	0.1229	29.66	218.32	29.78	208.19	237.97	0.1211	0.9295
180	–12.73	0.7485	0.1098	33.31	219.94	33.45	206.26	239.71	0.1352	0.9273
200	–10.09	0.7532	0.0993	36.69	221.43	36.84	204.46	241.30	0.1481	0.9253
240	–5.37	0.7618	0.0834	42.77	224.07	42.95	201.14	244.09	0.1710	0.9222
280	–1.23	0.7697	0.0719	48.18	226.38	48.39	198.13	246.52	0.1911	0.9197
320	2.48	0.7770	0.0632	53.06	228.43	53.31	195.35	248.66	0.2089	0.9177
360	5.84	0.7839	0.0564	57.54	230.28	57.82	192.76	250.58	0.2251	0.9160
400	8.93	0.7904	0.0509	61.69	231.97	62.00	190.32	252.32	0.2399	0.9145
500	15.74	0.8056	0.0409	70.93	235.64	71.33	184.74	256.07	0.2723	0.9117
600	21.58	0.8196	0.0341	78.99	238.74	79.48	179.71	259.19	0.2999	0.9097
700	26.72	0.8328	0.0292	86.19	241.42	86.78	175.07	261.85	0.3242	0.9080
800	31.33	0.8454	0.0255	92.75	243.78	93.42	170.73	264.15	0.3459	0.9066
900	35.53	0.8576	0.0226	98.79	245.88	99.56	166.62	266.18	0.3656	0.9054
1000	39.39	0.8695	0.0202	104.42	247.77	105.29	162.68	267.97	0.3838	0.9043
1200	46.32	0.8928	0.0166	114.69	251.03	115.76	155.23	270.99	0.4164	0.9023
1400	52.43	0.9159	0.0140	123.98	253.74	125.26	148.14	273.40	0.4453	0.9003
1600	57.92	0.9392	0.0121	132.52	256.00	134.02	141.31	275.33	0.4714	0.8982
1800	62.91	0.9631	0.0105	140.49	257.88	142.22	134.60	276.83	0.4954	0.8959
2000	67.49	0.9878	0.0093	148.02	259.41	149.99	127.95	277.94	0.5178	0.8934
2500	77.59	1.0562	0.0069	165.48	261.84	168.12	111.06	279.17	0.5687	0.8854
3000	86.22	1.1416	0.0053	181.88	262.16	185.30	92.71	278.01	0.6156	0.8735

Table D-3 Superheated R134a

T, °C	*v*, m³/kg	*u*, kJ/kg	*h*, kJ/kg	*s*, kJ/kg·K	*v*, m³/kg	*u*, kJ/kg	*h*, kJ/kg	*s*, kJ/kg·K
	$P = 0.06$ MPa (–37.07 °C)				$P = 0.10$ MPa (–26.43 °C)			
Sat.	0.31003	206.12	224.72	0.9520	0.19170	212.18	231.35	0.9395
–20	0.33536	217.86	237.98	1.0062	0.19770	216.77	236.54	0.9602
–10	0.34992	224.97	245.96	1.0371	0.20686	224.01	244.70	0.9918
0	0.36433	232.24	254.10	1.0675	0.21587	231.41	252.99	1.0227
10	0.37861	239.69	262.41	1.0973	0.22473	238.96	261.43	1.0531
20	0.39279	247.32	270.89	1.1267	0.23349	246.67	270.02	1.0829
30	0.40688	255.12	279.53	1.1557	0.24216	254.54	278.76	1.1122
40	0.42091	263.10	288.35	1.1844	0.25076	262.58	287.66	1.1411
50	0.43487	271.25	297.34	1.2126	0.25930	270.79	296.72	1.1696
60	0.44879	279.58	306.51	1.2405	0.26779	279.16	305.94	1.1977
70	0.46266	288.08	315.84	1.2681	0.27623	287.70	315.32	1.2254
80	0.47650	296.75	325.34	1.2954	0.28464	296.40	324.87	1.2528
90	0.49031	305.58	335.00	1.3224	0.29302	305.27	334.57	1.2799
	$P = 0.14$ MPa (–18.80 °C)				$P = 0.18$ MPa (–12.73 °C)			
Sat.	0.13945	216.52	236.04	0.9322	0.10983	219.94	239.71	0.9273
–10	0.14519	223.03	243.40	0.9606	0.11135	222.02	242.06	0.9362
0	0.15219	230.55	251.86	0.9922	0.11678	229.67	250.69	0.9684
10	0.15875	238.21	260.43	1.0230	0.12207	237.44	259.41	0.9998
20	0.16520	246.01	269.13	1.0532	0.12723	245.33	268.23	1.0304
30	0.17155	253.96	277.97	1.0828	0.13230	253.36	277.17	1.0604
40	0.17783	262.06	286.96	1.1120	0.13730	261.53	286.24	1.0898
50	0.18404	270.32	296.09	1.1407	0.14222	269.85	295.45	1.1187
60	0.19020	278.74	305.37	1.1690	0.14710	278.31	304.79	1.1472
70	0.19633	287.32	314.80	1.1969	0.15193	286.93	314.28	1.1753
80	0.20241	296.06	324.39	1.2244	0.15672	295.71	323.92	1.2030
90	0.20846	304.95	334.14	1.2516	0.16148	304.63	333.70	1.2303
100	0.21449	314.01	344.04	1.2785	0.16622	313.72	343.63	1.2573
	$P = 0.20$ MPa (–10.09 °C)				$P = 0.24$ MPa (–5.37 °C)			
Sat.	0.09933	221.43	241.30	0.9253	0.08343	224.07	244.09	0.9222
–10	0.09938	221.50	241.38	0.9256				
0	0.10438	229.23	250.10	0.9582	0.08574	228.31	248.89	0.9399
10	0.10922	237.05	258.89	0.9898	0.08993	236.26	257.84	0.9721
20	0.11394	244.99	267.78	1.0206	0.09399	244.30	266.85	1.0034
30	0.11856	253.06	276.77	1.0508	0.09794	252.45	275.95	1.0339
40	0.12311	261.26	285.88	1.0804	0.10181	260.72	285.16	1.0637
50	0.12758	269.61	295.12	1.1094	0.10562	269.12	294.47	1.0930
60	0.13201	278.10	304.50	1.1380	0.10937	277.67	303.91	1.1218
70	0.13639	286.74	314.02	1.1661	0.11307	286.35	313.49	1.1501
80	0.14073	295.53	323.68	1.1939	0.11674	295.18	323.19	1.1780
90	0.14504	304.47	333.48	1.2212	0.12037	304.15	333.04	1.2055
100	0.14932	313.57	343.43	1.2483	0.12398	313.27	343.03	1.2326

Table D-3 (*Continued*)

T, °C	v, m³/kg	u, kJ/kg	h, kJ/kg	s, kJ/kg·K	v, m³/kg	u, kJ/kg	h, kJ/kg	s, kJ/kg·K
	P = 0.28 MPa (−1.23 °C)				P = 0.32 MPa (2.48 °C)			
Sat.	0.07193	226.38	246.52	0.9197	0.06322	228.43	248.66	0.917
0	0.07240	227.37	247.64	0.9238				
10	0.07613	235.44	256.76	0.9566	0.06576	234.61	255.65	0.942
20	0.07972	243.59	265.91	0.9883	0.06901	242.87	264.95	0.974
30	0.08320	251.83	275.12	1.0192	0.07214	251.19	274.28	1.006
40	0.08660	260.17	284.42	1.0494	0.07518	259.61	283.67	1.036
50	0.08992	268.64	293.81	1.0789	0.07815	268.14	293.15	1.066
60	0.09319	277.23	303.32	1.1079	0.08106	276.79	302.72	1.095
70	0.09641	285.96	312.95	1.1364	0.08392	285.56	312.41	1.124
80	0.09960	294.82	322.71	1.1644	0.08674	294.46	322.22	1.152
90	0.10275	303.83	332.60	1.1920	0.08953	303.50	332.15	1.180
100	0.10587	312.98	342.62	1.2193	0.09229	312.68	342.21	1.207
110	0.10897	322.27	352.78	1.2461	0.09503	322.00	352.40	1.234
120	0.11205	331.71	363.08	1.2727	0.09774	331.45	362.73	1.261
	P = 0.40 MPa (8.93 °C)				P = 0.50 MPa (15.74 °C)			
Sat.	0.05089	231.97	252.32	0.9145	0.04086	235.64	256.07	0.911
10	0.05119	232.87	253.35	0.9182				
20	0.05397	241.37	262.96	0.9515	0.04188	239.40	260.34	0.926
30	0.05662	249.89	272.54	0.9837	0.04416	248.20	270.28	0.959
40	0.05917	258.47	282.14	1.0148	0.04633	256.99	280.16	0.991
50	0.06164	267.13	291.79	1.0452	0.04842	265.83	290.04	1.022
60	0.06405	275.89	301.51	1.0748	0.05043	274.73	299.95	1.053
70	0.06641	284.75	311.32	1.1038	0.05240	283.72	309.92	1.082
80	0.06873	293.73	321.23	1.1322	0.05432	292.80	319.96	1.111
90	0.07102	302.84	331.25	1.1602	0.05620	302.00	330.10	1.139
100	0.07327	312.07	341.38	1.1878	0.05805	311.31	340.33	1.167
110	0.07550	321.44	351.64	1.2149	0.05988	320.74	350.68	1.194
120	0.07771	330.94	362.03	1.2417	0.06168	330.30	361.14	1.221
130	0.07991	340.58	372.54	1.2681	0.06347	339.98	371.72	1.248
140	0.08208	350.35	383.18	1.2941	0.06524	349.79	382.42	1.274
	P = 0.60 MPa (21.58 °C)				P = 0.70 MPa (26.72 °C)			
Sat.	0.03408	238.74	259.19	0.9097	0.02918	241.42	261.85	0.9080
30	0.03581	246.41	267.89	0.9388	0.02979	244.51	265.37	0.9197
40	0.03774	255.45	278.09	0.9719	0.03157	253.83	275.93	0.9539
50	0.03958	264.48	288.23	1.0037	0.03324	263.08	286.35	0.9867
60	0.04134	273.54	298.35	1.0346	0.03482	272.31	296.69	1.0182
70	0.04304	282.66	308.48	1.0645	0.03634	281.57	307.01	1.0487
80	0.04469	291.86	318.67	1.0938	0.03781	290.88	317.35	1.0784
90	0.04631	301.14	328.93	1.1225	0.03924	300.27	327.74	1.1074
100	0.04790	310.53	339.27	1.1505	0.04064	309.74	338.19	1.1358
110	0.04946	320.03	349.70	1.1781	0.04201	319.31	348.71	1.1637
120	0.05099	329.64	360.24	1.2053	0.04335	328.98	359.33	1.1910
130	0.05251	339.38	370.88	1.2320	0.04468	338.76	370.04	1.2179
140	0.05402	349.23	381.64	1.2584	0.04599	348.66	380.86	1.2444
150	0.05550	359.21	392.52	1.2844	0.04729	358.68	391.79	1.2706
160	0.05698	369.32	403.51	1.3100	0.04857	368.82	402.82	1.2963

Table D-3 ***(Continued)***

T, °C	*v*, m³/kg	*u*, kJ/kg	*h*, kJ/kg	*s*, kJ/kg·K	*v*, m³/kg	*u*, kJ/kg	*h*, kJ/kg	*s*, kJ/kg·K
	P = 0.80 MPa (31.33 °C)				*P* = 0.90 MPa (35.53 °C)			
Sat.	0.02547	243.78	264.15	0.9066	0.02255	245.88	266.18	0.9054
40	0.02691	252.13	273.66	0.9374	0.02325	250.32	271.25	0.9217
50	0.02846	261.62	284.39	0.9711	0.02472	260.09	282.34	0.9566
60	0.02992	271.04	294.98	1.0034	0.02609	269.72	293.21	0.9897
70	0.03131	280.45	305.50	1.0345	0.02738	279.30	303.94	1.0214
80	0.03264	289.89	316.00	1.0647	0.02861	288.87	314.62	1.0521
90	0.03393	299.37	326.52	1.0940	0.02980	298.46	325.28	1.0819
100	0.03519	308.93	337.08	1.1227	0.03095	308.11	335.96	1.1109
110	0.03642	318.57	347.71	1.1508	0.03207	317.82	346.68	1.1392
120	0.03762	328.31	358.40	1.1784	0.03316	327.62	357.47	1.1670
130	0.03881	338.14	369.19	1.2055	0.03423	337.52	368.33	1.1943
140	0.03997	348.09	380.07	1.2321	0.03529	347.51	379.27	1.2211
150	0.04113	358.15	391.05	1.2584	0.03633	357.61	390.31	1.2475
160	0.04227	368.32	402.14	1.2843	0.03736	367.82	401.44	1.2735
170	0.04340	378.61	413.33	1.3098	0.03838	378.14	412.68	1.2992
180	0.04452	389.02	424.63	1.3351	0.03939	388.57	424.02	1.3245
	P = 1.00 MPa (39.39 °C)				*P* = 1.20 MPa (46.32 °C)			
Sat.	0.02020	247.77	267.97	0.9043	0.01663	251.03	270.99	0.9023
40	0.02029	248.39	268.68	0.9066				
50	0.02171	258.48	280.19	0.9428	0.01712	254.98	275.52	0.9164
60	0.02301	268.35	291.36	0.9768	0.01835	265.42	287.44	0.9527
70	0.02423	278.11	302.34	1.0093	0.01947	275.59	298.96	0.9868
80	0.02538	287.82	313.20	1.0405	0.02051	285.62	310.24	1.0192
90	0.02649	297.53	324.01	1.0707	0.02150	295.59	321.39	1.0503
100	0.02755	307.27	334.82	1.1000	0.02244	305.54	332.47	1.0804
110	0.02858	317.06	345.65	1.1286	0.02335	315.50	343.52	1.1096
120	0.02959	326.93	356.52	1.1567	0.02423	325.51	354.58	1.1381
130	0.03058	336.88	367.46	1.1841	0.02508	335.58	365.68	1.1660
140	0.03154	346.92	378.46	1.2111	0.02592	345.73	376.83	1.1933
150	0.03250	357.06	389.56	1.2376	0.02674	355.95	388.04	1.2201
160	0.03344	367.31	400.74	1.2638	0.02754	366.27	399.33	1.2465
170	0.03436	377.66	412.02	1.2895	0.02834	376.69	410.70	1.2724
180	0.03528	388.12	423.40	1.3149	0.02912	387.21	422.16	1.2980
	P = 1.40 MPa (52.43 °C)				*P* = 1.60 MPa (57.92 °C)			
Sat.	0.01405	253.74	273.40	0.9003	0.01208	256.00	275.33	0.8982
60	0.01495	262.17	283.10	0.9297	0.01233	258.48	278.20	0.9069
70	0.01603	272.87	295.31	0.9658	0.01340	269.89	291.33	0.9457
80	0.01701	283.29	307.10	0.9997	0.01435	280.78	303.74	0.9813
90	0.01792	293.55	318.63	1.0319	0.01521	291.39	315.72	1.0148
100	0.01878	303.73	330.02	1.0628	0.01601	301.84	327.46	1.0467
110	0.01960	313.88	341.32	1.0927	0.01677	312.20	339.04	1.0773
120	0.02039	324.05	352.59	1.1218	0.01750	322.53	350.53	1.1069
130	0.02115	334.25	363.86	1.1501	0.01820	332.87	361.99	1.1357
140	0.02189	344.50	375.15	1.1777	0.01887	343.24	373.44	1.1638
150	0.02262	354.82	386.49	1.2048	0.01953	353.66	384.91	1.1912
160	0.02333	365.22	397.89	1.2315	0.02017	364.15	396.43	1.2181
170	0.02403	375.71	409.36	1.2576	0.02080	374.71	407.99	1.2445
180	0.02472	386.29	420.90	1.2834	0.02142	385.35	419.62	1.2704
190	0.02541	396.96	432.53	1.3088	0.02203	396.08	431.33	1.2960
200	0.02608	407.73	444.24	1.3338	0.02263	406.90	443.11	1.3212

Table D-1E Properties of Saturated R134a—Temperature Table

		Specific Volume ft^3/lbm		Internal Energy Btu/lbm		Enthalpy Btu/lbm			Entropy Btu/lbm-°R	
Temp. °F	Pressure psia	Sat. Liquid v_f	Sat. Vapor v_g	Sat. Liquid u_f	Sat. Vapor u_g	Sat. Liquid h_f	Evap. h_{fg}	Sat. Vapor h_g	Sat. Liquid s_f	Sat. Vapor s_g
–40	7.490	0.01130	5.7173	–0.02	87.90	0.00	95.82	95.82	0.0000	0.2283
–30	9.920	0.01143	4.3911	2.81	89.26	2.83	94.49	97.32	0.0067	0.2266
–20	12.949	0.01156	3.4173	5.69	90.62	5.71	93.10	98.81	0.0133	0.2250
–15	14.718	0.01163	3.0286	7.14	91.30	7.17	92.38	99.55	0.0166	0.2243
–10	16.674	0.01170	2.6918	8.61	91.98	8.65	91.64	100.29	0.0199	0.2236
–5	18.831	0.01178	2.3992	10.09	92.66	10.13	90.89	101.02	0.0231	0.2230
0	21.203	0.01185	2.1440	11.58	93.33	11.63	90.12	101.75	0.0264	0.2224
5	23.805	0.01193	1.9208	13.09	94.01	13.14	89.33	102.47	0.0296	0.2219
10	26.651	0.01200	1.7251	14.60	94.68	14.66	88.53	103.19	0.0329	0.2214
15	29.756	0.01208	1.5529	16.13	95.35	16.20	87.71	103.90	0.0361	0.2209
20	33.137	0.01216	1.4009	17.67	96.02	17.74	86.87	104.61	0.0393	0.2205
25	36.809	0.01225	1.2666	19.22	96.69	19.30	86.02	105.32	0.0426	0.2200
30	40.788	0.01233	1.1474	20.78	97.35	20.87	85.14	106.01	0.0458	0.2196
40	49.738	0.01251	0.9470	23.94	98.67	24.05	83.34	107.39	0.0522	0.2189
50	60.125	0.01270	0.7871	27.14	99.98	27.28	81.46	108.74	0.0585	0.2183
60	72.092	0.01290	0.6584	30.39	101.27	30.56	79.49	110.05	0.0648	0.2178
70	85.788	0.01311	0.5538	33.68	102.54	33.89	77.44	111.33	0.0711	0.2173
80	101.37	0.01334	0.4682	37.02	103.78	37.27	75.29	112.56	0.0774	0.2169
85	109.92	0.01346	0.4312	38.72	104.39	38.99	74.17	113.16	0.0805	0.2167
90	118.99	0.01358	0.3975	40.42	105.00	40.72	73.03	113.75	0.0836	0.2165
95	128.62	0.01371	0.3668	42.14	105.60	42.47	71.86	114.33	0.0867	0.2163
100	138.83	0.01385	0.3388	43.87	106.18	44.23	70.66	114.89	0.0898	0.2161
105	149.63	0.01399	0.3131	45.62	106.76	46.01	69.42	115.43	0.0930	0.2159
110	161.04	0.01414	0.2896	47.39	107.33	47.81	68.15	115.96	0.0961	0.2157
115	173.10	0.01429	0.2680	49.17	107.88	49.63	66.84	116.47	0.0992	0.2155
120	185.82	0.01445	0.2481	50.97	108.42	51.47	65.48	116.95	0.1023	0.2153
140	243.86	0.01520	0.1827	58.39	110.41	59.08	59.57	118.65	0.1150	0.2143
160	314.63	0.01617	0.1341	66.26	111.97	67.20	52.58	119.78	0.1280	0.2128
180	400.22	0.01758	0.0964	74.83	112.77	76.13	43.78	119.91	0.1417	0.2101
200	503.52	0.02014	0.0647	84.90	111.66	86.77	30.92	117.69	0.1575	0.2044
210	563.51	0.02329	0.0476	91.84	108.48	94.27	19.18	113.45	0.1684	0.1971

SOURCE: Table D-1E through D-3E are based on equations from D. P. Wilson and R. S. Basu, "Thermodynamic Properties of a New Stratospherically Safe Working Fluid—Refrigerant 134a," *ASHRAE Trans.*, Vol. 94, Pt. 2, 1988, pp. 2095–2118.

Table D-2E Properties of Saturated R134a—Pressure Table

		Specific Volume ft³/lbm		Internal Energy Btu/lbm		Enthalpy Btu/lbm			Entropy Btu/lbm- °R	
Pressure psia	**Temp** °F	**Sat. Liquid** v_f	**Sat. Vapor** v_g	**Sat. Liquid** u_f	**Sat. Vapor** u_g	**Sat. Liquid** h_f	**Evap.** h_{fg}	**Sat. Vapor** h_g	**Sat. Liquid** s_f	**Sat. Vapor** s_g
5	–53.48	0.01113	8.3508	–3.74	86.07	–3.73	97.53	93.79	–0.0090	0.2311
10	–29.71	0.01143	4.3581	2.89	89.30	2.91	94.45	97.37	0.0068	0.2265
15	–14.25	0.01164	2.9747	7.36	91.40	7.40	92.27	99.66	0.0171	0.2242
20	–2.48	0.01181	2.2661	10.84	93.00	10.89	90.50	101.39	0.0248	0.2227
30	15.38	0.01209	1.5408	16.24	95.40	16.31	87.65	103.96	0.0364	0.2209
40	29.04	0.01232	1.1692	20.48	97.23	20.57	85.31	105.88	0.0452	0.2197
50	40.27	0.01252	0.9422	24.02	98.71	24.14	83.29	107.43	0.0523	0.2189
60	49.89	0.01270	0.7887	27.10	99.96	27.24	81.48	108.72	0.0584	0.2183
70	58.35	0.01286	0.6778	29.85	101.05	30.01	79.82	109.83	0.0638	0.2179
80	65.93	0.01302	0.5938	32.33	102.02	32.53	78.28	110.81	0.0686	0.2175
90	72.83	0.01317	0.5278	34.62	102.89	34.84	76.84	111.68	0.0729	0.2172
100	79.17	0.01332	0.4747	36.75	103.68	36.99	75.47	112.46	0.0768	0.2169
120	90.54	0.01360	0.3941	40.61	105.06	40.91	72.91	113.82	0.0839	0.2165
140	100.56	0.01386	0.3358	44.07	106.25	44.43	70.52	114.95	0.0902	0.2161
160	109.56	0.01412	0.2916	47.23	107.28	47.65	68.26	115.91	0.0958	0.2157
180	117.74	0.01438	0.2569	50.16	108.18	50.64	66.10	116.74	0.1009	0.2154
200	125.28	0.01463	0.2288	52.90	108.98	53.44	64.01	117.44	0.1057	0.2151
220	132.27	0.01489	0.2056	55.48	109.68	56.09	61.96	118.05	0.1101	0.2147
240	138.79	0.01515	0.1861	57.93	110.30	58.61	59.96	118.56	0.1142	0.2144
260	144.92	0.01541	0.1695	60.28	110.84	61.02	57.97	118.99	0.1181	0.2140
280	150.70	0.01568	0.1550	62.53	111.31	63.34	56.00	119.35	0.1219	0.2136
300	156.17	0.01596	0.1424	64.71	111.72	65.59	54.03	119.62	0.1254	0.2132
350	168.72	0.01671	0.1166	69.88	112.45	70.97	49.03	120.00	0.1338	0.2118
400	179.95	0.01758	0.0965	74.81	112.77	76.11	43.80	119.91	0.1417	0.2102
450	190.12	0.01863	0.0800	79.63	112.60	81.18	38.08	119.26	0.1493	0.2079
500	199.38	0.02002	0.0657	84.54	111.76	86.39	31.44	117.83	0.1570	0.2047

Table D-3E Superheated R134a Vapor

T, °F	v, ft^3/lbm	u, Btu/lbm	h, Btu/lbm	s, Btu/lbm-°R	v, ft^3/lbm	u, Btu/lbm	h, Btu/lbm	s, Btu/lbm-°R
	P = 10 psia (–29.71 °F)				P = 15 psia (–14.25 °F)			
Sat.	4.3581	89.30	97.37	0.2265	2.9747	91.40	99.66	0.2242
–20	4.4718	90.89	99.17	0.2307				
0	4.7026	94.24	102.94	0.2391	3.0893	93.84	102.42	0.2303
20	4.9297	97.67	106.79	0.2472	3.2468	97.33	106.34	0.2386
40	5.1539	101.19	110.72	0.2553	3.4012	100.89	110.33	0.2468
60	5.3758	104.80	114.74	0.2632	3.5533	104.54	114.40	0.2548
80	5.5959	108.50	118.85	0.2709	3.7034	108.28	118.56	0.2626
100	5.8145	112.29	123.05	0.2786	3.8520	112.10	122.79	0.2703
120	6.0318	116.18	127.34	0.2861	3.9993	116.01	127.11	0.2779
140	6.2482	120.16	131.72	0.2935	4.1456	120.00	131.51	0.2854
160	6.4638	124.23	136.19	0.3009	4.2911	124.09	136.00	0.2927
180	6.6786	128.38	140.74	0.3081	4.4359	128.26	140.57	0.3000
200	6.8929	132.63	145.39	0.3152	4.5801	132.52	145.23	0.3072
	P = 20 psia (–2.48 °F)				P = 30 psia (15.38 °F)			
Sat.	2.2661	93.00	101.39	0.2227	1.5408	95.40	103.96	0.2209
0	2.2816	93.43	101.88	0.2238				
20	2.4046	96.98	105.88	0.2323	1.5611	96.26	104.92	0.2229
40	2.5244	100.59	109.94	0.2406	1.6465	99.98	109.12	0.2315
60	2.6416	104.28	114.06	0.2487	1.7293	103.75	113.35	0.2398
80	2.7569	108.05	118.25	0.2566	1.8098	107.59	117.63	0.2478
100	2.8705	111.90	122.52	0.2644	1.8887	111.49	121.98	0.2558
120	2.9829	115.83	126.87	0.2720	1.9662	115.47	126.39	0.2635
140	3.0942	119.85	131.30	0.2795	2.0426	119.53	130.87	0.2711
160	3.2047	123.95	135.81	0.2869	2.1181	123.66	135.42	0.2786
180	3.3144	128.13	140.40	0.2922	2.1929	127.88	140.05	0.2859
200	3.4236	132.40	145.07	0.3014	2.2671	132.17	144.76	0.2932
220	3.5323	136.76	149.83	0.3085	2.3407	136.55	149.54	0.3003
	P = 40 psia (29.04 °F)				P = 50 psia (40.27 °F)			
Sat.	1.1692	97.23	105.88	0.2197	0.9422	98.71	107.43	0.2189
40	1.2065	99.33	108.26	0.2245				
60	1.2723	103.20	112.62	0.2331	0.9974	102.62	111.85	0.2276
80	1.3357	107.11	117.00	0.2414	1.0508	106.62	116.34	0.2361
100	1.3973	111.08	121.42	0.2494	1.1022	110.65	120.85	0.2443
120	1.4575	115.11	125.90	0.2573	1.1520	114.74	125.39	0.2523
140	1.5165	119.21	130.43	0.2650	1.2007	118.88	129.99	0.2601
160	1.5746	123.38	135.03	0.2725	1.2484	123.08	134.64	0.2677
180	1.6319	127.62	139.70	0.2799	1.2953	127.36	139.34	0.2752
200	1.6887	131.94	144.44	0.2872	1.3415	131.71	144.12	0.2825
220	1.7449	136.34	149.25	0.2944	1.3873	136.12	148.96	0.2897
240	1.8006	140.81	154.14	0.3015	1.4326	140.61	153.87	0.2969
260	1.8561	145.36	159.10	0.3085	1.4775	145.18	158.85	0.3039
280	1.9112	149.98	164.13	0.3154	1.5221	149.82	163.90	0.3108

Table D-3E (*Continued*)

T, °F	v, ft^3/lbm	u, Btu/lbm	h, Btu/lbm	s, Btu/lbm-°R	v, ft^3/lbm	u, Btu/lbm	h, Btu/lbm	s, Btu/lbm-°R
	P = 60 psia (49.89 °F)				P = 70 psia (58.35 °F)			
Sat.	0.7887	99.96	108.72	0.2183	0.6778	101.05	109.83	0.2179
60	0.8135	102.03	111.06	0.2229	0.6814	101.40	110.23	0.2186
80	0.8604	106.11	115.66	0.2316	0.7239	105.58	114.96	0.2276
100	0.9051	110.21	120.26	0.2399	0.7640	109.76	119.66	0.2361
120	0.9482	114.35	124.88	0.2480	0.8023	113.96	124.36	0.2444
140	0.9900	118.54	129.53	0.2559	0.8393	118.20	129.07	0.2524
160	1.0308	122.79	134.23	0.2636	0.8752	122.49	133.82	0.2601
180	1.0707	127.10	138.98	0.2712	0.9103	126.83	138.62	0.2678
200	1.1100	131.47	143.79	0.2786	0.9446	131.23	143.46	0.2752
220	1.1488	135.91	148.66	0.2859	0.9784	135.69	148.36	0.2825
240	1.1871	140.42	153.60	0.2930	1.0118	140.22	153.33	0.2897
260	1.2251	145.00	158.60	0.3001	1.0448	144.82	158.35	0.2968
280	1.2627	149.65	163.67	0.3070	1.0774	149.48	163.44	0.3038
300	1.3001	154.38	168.81	0.3139	1.1098	154.22	168.60	0.3107
	P = 80 psia (65.93 °F)				P = 90 psia (72.83 °F)			
Sat.	0.5938	102.02	110.81	0.2175	0.5278	102.89	111.68	0.2172
80	0.6211	105.03	114.23	0.2239	0.5408	104.46	113.47	0.2205
100	0.6579	109.30	119.04	0.2327	0.5751	108.82	118.39	0.2295
120	0.6927	113.56	123.82	0.2411	0.6073	113.15	123.27	0.2380
140	0.7261	117.85	128.60	0.2492	0.6380	117.50	128.12	0.2463
160	0.7584	122.18	133.41	0.2570	0.6675	121.87	132.98	0.2542
180	0.7898	126.55	138.25	0.2647	0.6961	126.28	137.87	0.2620
200	0.8205	130.98	143.13	0.2722	0.7239	130.73	142.79	0.2696
220	0.8506	135.47	148.06	0.2796	0.7512	135.25	147.76	0.2770
240	0.8803	140.02	153.05	0.2868	0.7779	139.82	152.77	0.2843
260	0.9095	144.63	158.10	0.2940	0.8043	144.45	157.84	0.2914
280	0.9384	149.32	163.21	0.3010	0.8303	149.15	162.97	0.2984
300	0.9671	154.06	168.38	0.3079	0.8561	153.91	168.16	0.3054
320	0.9955	158.88	173.62	0.3147	0.8816	158.73	173.42	0.3122
	P = 100 psia (79.17 °F)				P = 120 psia (90.54 °F)			
Sat.	0.4747	103.68	112.46	0.2169	0.3941	105.06	113.82	0.2165
80	0.4761	103.87	112.68	0.2173				
100	0.5086	108.32	117.73	0.2265	0.4080	107.26	116.32	0.2210
120	0.5388	112.73	122.70	0.2352	0.4355	111.84	121.52	0.2301
140	0.5674	117.13	127.63	0.2436	0.4610	116.37	126.61	0.2387
160	0.5947	121.55	132.55	0.2517	0.4852	120.89	131.66	0.2470
180	0.6210	125.99	137.49	0.2595	0.5082	125.42	136.70	0.2550
200	0.6466	130.48	142.45	0.2671	0.5305	129.97	141.75	0.2628
220	0.6716	135.02	147.45	0.2746	0.5520	134.56	146.82	0.2704
240	0.6960	139.61	152.49	0.2819	0.5731	139.20	151.92	0.2778
260	0.7201	144.26	157.59	0.2891	0.5937	143.89	157.07	0.2850
280	0.7438	148.98	162.74	0.2962	0.6140	148.63	162.26	0.2921
300	0.7672	153.75	167.95	0.3031	0.6339	153.43	167.51	0.2991
320	0.7904	158.59	173.21	0.3099	0.6537	158.29	172.81	0.3060

Table D-3E (*Continued*)

T, °F	v, ft³/lbm	u, Btu/lbm	h, Btu/lbm	s, Btu/lbm-°R	v, ft³/lbm	u, Btu/lbm	h, Btu/lbm	s, Btu/lbm-°R
	P = 140 psia (100.6 °F)				P = 160 psia (109.6 °F)			
Sat.	0.3358	106.25	114.95	0.2161	0.2916	107.28	115.91	0.2157
120	0.3610	110.90	120.25	0.2254	0.3044	109.88	118.89	0.2209
140	0.3846	115.58	125.54	0.2344	0.3269	114.73	124.41	0.2303
160	0.4066	120.21	130.74	0.2429	0.3474	119.49	129.78	0.2391
180	0.4274	124.82	135.89	0.2511	0.3666	124.20	135.06	0.2475
200	0.4474	129.44	141.03	0.2590	0.3849	128.90	140.29	0.2555
220	0.4666	134.09	146.18	0.2667	0.4023	133.61	145.52	0.2633
240	0.4852	138.77	151.34	0.2742	0.4192	138.34	150.75	0.2709
260	0.5034	143.50	156.54	0.2815	0.4356	143.11	156.00	0.2783
280	0.5212	148.28	161.78	0.2887	0.4516	147.92	161.29	0.2856
300	0.5387	153.11	167.06	0.2957	0.4672	152.78	166.61	0.2927
320	0.5559	157.99	172.39	0.3026	0.4826	157.69	171.98	0.2996
340	0.5730	162.93	177.78	0.3094	0.4978	162.65	177.39	0.3065
360	0.5898	167.93	183.21	0.3162	0.5128	167.67	182.85	0.3132
	P = 180 psia (117.7 °F)				P = 200 psia (125.3 °F)			
Sat.	0.2569	108.18	116.74	0.2154	0.2288	108.98	117.44	0.2151
120	0.2595	108.77	117.41	0.2166				
140	0.2814	113.83	123.21	0.2264	0.2446	112.87	121.92	0.2226
160	0.3011	118.74	128.77	0.2355	0.2636	117.94	127.70	0.2321
180	0.3191	123.56	134.19	0.2441	0.2809	122.88	133.28	0.2410
200	0.3361	128.34	139.53	0.2524	0.2970	127.76	138.75	0.2494
220	0.3523	133.11	144.84	0.2603	0.3121	132.60	144.15	0.2575
240	0.3678	137.90	150.15	0.2680	0.3266	137.44	149.53	0.2653
260	0.3828	142.71	155.46	0.2755	0.3405	142.30	154.90	0.2728
280	0.3974	147.55	160.79	0.2828	0.3540	147.18	160.28	0.2802
300	0.4116	152.44	166.15	0.2899	0.3671	152.10	165.69	0.2874
320	0.4256	157.38	171.55	0.2969	0.3799	157.07	171.13	0.2945
340	0.4393	162.36	177.00	0.3038	0.3926	162.07	176.60	0.3014
360	0.4529	167.40	182.49	0.3106	0.4050	167.13	182.12	0.3082
	P = 300 psia (156.2 °F)				P = 400 psia (179.9 °F)			
Sat.	0.1424	111.72	119.62	0.2132	0.0965	112.77	119.91	0.2102
160	0.1462	112.95	121.07	0.2155				
180	0.1633	118.93	128.00	0.2265	0.0965	112.79	119.93	0.2102
200	0.1777	124.47	134.34	0.2363	0.1143	120.14	128.60	0.2235
220	0.1905	129.79	140.36	0.2453	0.1275	126.35	135.79	0.2343
240	0.2021	134.99	146.21	0.2537	0.1386	132.12	142.38	0.2438
260	0.2130	140.12	151.95	0.2618	0.1484	137.65	148.64	0.2527
280	0.2234	145.23	157.63	0.2696	0.1575	143.06	154.72	0.2610
300	0.2333	150.33	163.28	0.2772	0.1660	148.39	160.67	0.2689
320	0.2428	155.44	168.92	0.2845	0.1740	153.69	166.57	0.2766
340	0.2521	160.57	174.56	0.2916	0.1816	158.97	172.42	0.2840
360	0.2611	165.74	180.23	0.2986	0.1890	164.26	178.26	0.2912
380	0.2699	170.94	185.92	0.3055	0.1962	169.57	184.09	0.2983
400	0.2786	176.18	191.64	0.3122	0.2032	174.90	189.94	0.3051

Ideal-Gas Tables

Table E-1 Properties of Air

T, K	h, kJ/kg	P_r	u, kJ/kg	v_r	$s°$, kJ/kg·K	T, K	h, kJ/kg	P_r	u, kJ/kg	v_r	$s°$, kJ/kg·K
200	199.97	0.3363	142.56	1707	1.29559	780	800.03	43.35	576.12	51.64	2.69013
220	219.97	0.4690	156.82	1346	1.39105	820	843.98	52.49	608.59	44.84	2.74504
240	240.02	0.6355	171.13	1084	1.47824	860	888.27	63.09	641.40	39.12	2.79783
260	260.09	0.8405	185.45	887.8	1.55848	900	932.93	75.29	674.58	34.31	2.84856
280	280.13	1.0889	199.75	738.0	1.63279	940	977.92	89.28	708.08	30.22	2.89748
290	290.16	1.2311	206.91	676.1	1.66802	980	1023.25	105.2	741.98	26.73	2.94468
300	300.19	1.3860	214.07	621.2	1.70203	1020	1068.89	123.4	776.10	23.72	2.99034
310	310.24	1.5546	221.25	572.3	1.73498	1060	1114.86	143.9	810.62	21.14	3.03449
320	320.29	1.7375	228.43	528.6	1.76690	1100	1161.07	167.1	845.33	18.896	3.07732
340	340.42	2.149	242.82	454.1	1.82790	1140	1207.57	193.1	880.35	16.946	3.11883
360	360.58	2.626	257.24	393.4	1.88543	1180	1254.34	222.2	915.57	15.241	3.15916
380	380.77	3.176	271.69	343.4	1.94001	1220	1301.31	254.7	951.09	13.747	3.19834
400	400.98	3.806	286.16	301.6	1.99194	1260	1348.55	290.8	986.90	12.435	3.23638
420	421.26	4.522	300.69	266.6	2.04142	1300	1395.97	330.9	1022.82	11.275	3.27345
440	441.61	5.332	315.30	236.8	2.08870	1340	1443.60	375.3	1058.94	10.247	3.30959
460	462.02	6.245	329.97	211.4	2.13407	1380	1491.44	424.2	1095.26	9.337	3.34474
480	482.49	7.268	344.70	189.5	2.17760	1420	1539.44	478.0	1131.77	8.526	3.37901
500	503.02	8.411	359.49	170.6	2.21952	1460	1587.63	537.1	1168.49	7.801	3.41247
520	523.63	9.684	374.36	154.1	2.25997	1500	1635.97	601.9	1205.41	7.152	3.44516
540	544.35	11.10	389.34	139.7	2.29906	1540	1684.51	672.8	1242.43	6.569	3.47712
560	565.17	12.66	404.42	127.0	2.33685	1580	1733.17	750.0	1279.65	6.046	3.50829
580	586.04	14.38	419.55	115.7	2.37348	1620	1782.00	834.1	1316.96	5.574	3.53879
600	607.02	16.28	434.78	105.8	2.40902	1660	1830.96	925.6	1354.48	5.147	3.56867
620	628.07	18.36	450.09	96.92	2.44356	1700	1880.1	1025	1392.7	4.761	3.5979
640	649.22	20.65	465.05	88.99	2.47716	1800	2003.3	1310	1487.2	3.944	3.6684
660	670.47	23.13	481.01	81.89	2.50985	1900	2127.4	1655	1582.6	3.295	3.7354
680	691.82	25.85	496.62	75.50	2.54175	2000	2252.1	2068	1678.7	2.776	3.7994
700	713.27	28.80	512.33	69.76	2.57277	2100	2377.4	2559	1775.3	2.356	3.8605
720	734.82	32.02	528.14	64.53	2.60319	2200	2503.2	3138	1872.4	2.012	3.9191
740	756.44	35.50	544.02	59.82	2.63280						

SOURCE: J. H. Keenan and J. Kaye, *Gas Tables*, Wiley, New York, 1945.

Table E-2 Molar Properties of Nitrogen, N_2

$\bar{h}^\circ_f = 0$ kJ/kmol

T, K	$\bar{h}$, kJ/kmol	$\bar{u}$, kJ/kmol	$\bar{s}$,° kJ/kmol·K	T, K	$\bar{h}$, kJ/kmol	$\bar{u}$, kJ/kmol	$\bar{s}$,° kJ/kmol·K
0	0	0	0	1000	30 129	21 815	228.057
220	6 391	4 562	182.639	1020	30 784	22 304	228.706
240	6 975	4 979	185.180	1040	31 442	22 795	229.344
260	7 558	5 396	187.514	1060	32 101	23 288	229.973
280	8 141	5 813	189.673	1080	32 762	23 782	230.591
298	8 669	6 190	191.502	1100	33 426	24 280	231.199
300	8 723	6 229	191.682	1120	34 092	24 780	231.799
320	9 306	6 645	193.562	1140	34 760	25 282	232.391
340	9 888	7 061	195.328	1160	35 430	25 786	232.973
360	10 471	7 478	196.995	1180	36 104	26 291	233.549
380	11 055	7 895	198.572	1200	36 777	26 799	234.115
400	11 640	8 314	200.071	1240	38 129	27 819	235.223
420	12 225	8 733	201.499	1260	38 807	28 331	235.766
440	12 811	9 153	202.863	1280	39 488	28 845	236.302
460	13 399	9 574	204.170	1300	40 170	29 361	236.831
480	13 988	9 997	205.424	1320	40 853	29 878	237.353
500	14 581	10 423	206.630	1340	41 539	30 398	237.867
520	15 172	10 848	207.792	1360	42 227	30 919	238.376
540	15 766	11 277	208.914	1380	42 915	31 441	238.878
560	16 363	11 707	209.999	1400	43 605	31 964	239.375
580	16 962	12 139	211.049	1440	44 988	33 014	240.350
600	17 563	12 574	212.066	1480	46 377	34 071	241.301
620	18 166	13 011	213.055	1520	47 771	35 133	242.228
640	18 772	13 450	214.018	1560	49 168	36 197	243.137
660	19 380	13 892	214.954	1600	50 571	37 268	244.028
680	19 991	14 337	215.866	1700	54 099	39 965	246.166
700	20 604	14 784	216.756	1800	57 651	42 685	248.195
720	21 220	15 234	217.624	1900	61 220	45 423	250.128
740	21 839	15 686	218.472	2000	64 810	48 181	251.969
760	22 460	16 141	219.301	2100	68 417	50 957	253.726
780	23 085	16 599	220.113	2200	72 040	53 749	255.412
800	23 714	17 061	220.907	2300	75 676	56 553	257.02
820	24 342	17 524	221.684	2400	79 320	59 366	258.580
840	24 974	17 990	222.447	2500	82 981	62 195	260.073
860	25 610	18 459	223.194	2600	86 650	65 033	261.512
880	26 248	18 931	223.927	2700	90 328	67 880	262.902
900	26 890	19 407	224.647	2800	94 014	70 734	264.241
920	27 532	19 883	225.353	2900	97 705	73 593	265.538
940	28 178	20 362	226.047	3000	101 407	76 464	266.793
960	28 826	20 844	226.728	3100	105 115	79 341	268.007
980	29 476	21 328	227.398	3200	108 830	82 224	269.186

SOURCE: JANAF Thermochemical Tables, NSRDS-NBS-37, 1971.

Table E-3 Molar Properties of Oxygen, O_2

$\bar{h}^\circ_f = 0$ kJ/kmol

T	$\bar{h}$	$\bar{u}$	$\bar{s}^\circ$	T	$\bar{h}$	$\bar{u}$	$\bar{s}^\circ$
0	0	0	0	1020	32 088	23 607	244.164
220	6 404	4 575	196.171	1040	32 789	24 142	244.844
240	6 984	4 989	198.696	1060	33 490	24 677	245.513
260	7 566	5 405	201.027	1080	34 194	25 214	246.171
280	8 150	5 822	203.191	1100	34 899	25 753	246.818
298	8 682	6 203	205.033	1120	35 606	26 294	247.454
300	8 736	6 242	205.213	1140	36 314	26 836	248.081
320	9 325	6 664	207.112	1160	37 023	27 379	248.698
340	9 916	7 090	208.904	1180	37 734	27 923	249.307
360	10 511	7 518	210.604	1200	38 447	28 469	249.906
380	11 109	7 949	212.222	1220	39 162	29 018	250.497
400	11 711	8 384	213.765	1240	39 877	29 568	251.079
420	12 314	8 822	215.241	1260	40 594	30 118	251.653
440	12 923	9 264	216.656	1280	41 312	30 670	252.219
460	13 535	9 710	218.016	1300	42 033	31 224	252.776
480	14 151	10 160	219.326	1320	42 753	31 778	253.325
500	14 770	10 614	220.589	1340	43 475	32 334	253.868
520	15 395	11 071	221.812	1360	44 198	32 891	254.404
540	16 022	11 533	222.997	1380	44 923	33 449	254.932
560	16 654	11 998	224.146	1400	45 648	34 008	255.454
580	17 290	12 467	225.262	1440	47 102	35 129	256.475
600	17 929	12 940	226.346	1480	48 561	36 256	257.474
620	18 572	13 417	227.400	1520	50 024	37 387	258.450
640	19 219	13 898	228.429	1540	50 756	37 952	258.928
660	19 870	14 383	229.430	1560	51 490	38 520	259.402
680	20 524	14 871	230.405	1600	52 961	39 658	260.333
700	21 184	15 364	231.358	1700	56 652	42 517	262.571
720	21 845	15 859	232.291	1800	60 371	45 405	264.701
740	22 510	16 357	233.201	1900	64 116	48 319	266.722
760	23 178	16 859	234.091	2000	67 881	51 253	268.655
780	23 850	17 364	234.960	2100	71 668	54 208	270.504
800	24 523	17 872	235.810	2200	75 484	57 192	272.278
820	25 199	18 382	236.644	2300	79 316	60 193	273.981
840	25 877	18 893	237.462	2400	83 174	63 219	275.625
860	26 559	19 408	238.264	2500	87 057	66 271	277.207
880	27 242	19 925	239.051	2600	90 956	69 339	278.738
900	27 928	20 445	239.823	2700	94 881	72 433	280.219
920	28 616	20 967	240.580	2800	98 826	75 546	281.654
940	29 306	21 491	241.323	2900	102 793	78 682	283.048
960	29 999	22 017	242.052	3000	106 780	81 837	284.399
980	30 692	22 544	242.768	3100	110 784	85 009	285.713
1000	31 389	23 075	243.471	3200	114 809	88 203	286.989

SOURCE: JANAF Thermochemical Tables, NSRDS-NBS-37, 1971.

Table E-4 Molar Properties of Carbon Dioxide, CO_2

$\bar{h}^\circ_f = -393\,520$ kJ/kmol

T	$\bar{h}$	$\bar{u}$	$\bar{s}^\circ$	T	$\bar{h}$	$\bar{u}$	$\bar{s}^\circ$
0	0	0	0	1020	43 859	35 378	270.293
220	6 601	4 772	202.966	1040	44 953	36 306	271.354
240	7 280	5 285	205.920	1060	46 051	37 238	272.400
260	7 979	5 817	208.717	1080	47 153	38 174	273.430
280	8 697	6 369	211.376	1100	48 258	39 112	274.445
298	9 364	6 885	213.685	1120	49 369	40 057	275.444
300	9 431	6 939	213.915	1140	50 484	41 006	276.430
320	10 186	7 526	216.351	1160	51 602	41 957	277.403
340	10 959	8 131	218.694	1180	52 724	42 913	278.361
360	11 748	8 752	220.948	1200	53 848	43 871	279.307
380	12 552	9 392	223.122	1220	54 977	44 834	280.238
400	13 372	10 046	225.225	1240	56 108	45 799	281.158
420	14 206	10 714	227.258	1260	57 244	46 768	282.066
440	15 054	11 393	229.230	1280	58 381	47 739	282.962
460	15 916	12 091	231.144	1300	59 522	48 713	283.847
480	16 791	12 800	233.004	1320	60 666	49 691	284.722
500	17 678	13 521	234.814	1340	61 813	50 672	285.586
520	18 576	14 253	236.575	1360	62 963	51 656	286.439
540	19 485	14 996	238.292	1380	64 116	52 643	287.283
560	20 407	15 751	239.962	1400	65 271	53 631	288.106
580	21 337	16 515	241.602	1440	67 586	55 614	289.743
600	22 280	17 291	243.199	1480	69 911	57 606	291.333
620	23 231	18 076	244.758	1520	72 246	59 609	292.888
640	24 190	18 869	246.282	1560	74 590	61 620	294.411
660	25 160	19 672	247.773	1600	76 944	63 741	295.901
680	26 138	20 484	249.233	1700	82 856	68 721	299.482
700	27 125	21 305	250.663	1800	88 806	73 840	302.884
720	28 121	22 134	252.065	1900	94 793	78 996	306.122
740	29 124	22 972	253.439	2000	100 804	84 185	309.210
760	30 135	23 817	254.787	2100	106 864	89 404	312.160
780	31 154	24 669	256.110	2200	112 939	94 648	314.988
800	32 179	25 527	257.408	2300	119 035	99 912	317.695
820	33 212	26 394	258.682	2400	125 152	105 197	320.302
840	34 251	27 267	259.934	2500	131 290	110 504	322.308
860	35 296	28 125	261.164	2600	137 449	115 832	325.222
880	36 347	29 031	262.371	2700	143 620	121 172	327.549
900	37 405	29 922	263.559	2800	149 808	126 528	329.800
920	38 467	30 818	264.728	2900	156 009	131 898	331.975
940	39 535	31 719	265.877	3000	162 226	137 283	334.084
960	40 607	32 625	267.007	3100	168 456	142 681	336.126
980	41 685	33 537	268.119	3200	174 695	148 089	338.109
1000	42 769	34 455	269.215				

SOURCE: JANAF Thermochemical Tables, NSRDS-NBS-37, 1971.

Table E-5 Molar Properties of Carbon Monoxide, CO

$\bar{h}_f^\circ = -110\,530$ kJ/kmol

T	$\bar{h}$	$\bar{u}$	$\bar{s}^\circ$	T	$\bar{h}$	$\bar{u}$	$\bar{s}^\circ$
0	0	0	0	1040	31 688	23 041	235.728
220	6 391	4 562	188.683	1060	32 357	23 544	236.364
240	6 975	4 979	191.221	1080	33 029	24 049	236.992
260	7 558	5 396	193.554	1100	33 702	24 557	237.609
280	8 140	5 812	195.713	1120	34 377	25 065	238.217
300	8 723	6 229	197.723	1140	35 054	25 575	238.817
320	9 306	6 645	199.603	1160	35 733	26 088	239.407
340	9 889	7 062	201.371	1180	36 406	26 602	239.989
360	10 473	7 480	203.040	1200	37 095	27 118	240.663
380	11 058	7 899	204.622	1220	37 780	27 637	241.128
400	11 644	8 319	206.125	1240	38 466	28 426	241.686
420	12 232	8 740	207.549	1260	39 154	28 678	242.236
440	12 821	9 163	208.929	1280	39 844	29 201	242.780
460	13 412	9 587	210.243	1300	40 534	29 725	243.316
480	14 005	10 014	211.504	1320	41 226	30 251	243.844
500	14 600	10 443	212.719	1340	41 919	30 778	244.366
520	15 197	10 874	213.890	1360	42 613	31 306	244.880
540	15 797	11 307	215.020	1380	43 309	31 836	245.388
560	16 399	11 743	216.115	1400	44 007	32 367	245.889
580	17 003	12 181	217.175	1440	45 408	33 434	246.876
600	17 611	12 622	218.204	1480	46 813	34 508	247.839
620	18 221	13 066	219.205	1520	48 222	35 584	248.778
640	18 833	13 512	220.179	1560	49 635	36 665	249.695
660	19 449	13 962	221.127	1600	51 053	37 750	250.592
680	20 068	14 414	222.052	1700	54 609	40 474	252.751
700	20 690	14 870	222.953	1800	58 191	43 225	254.797
720	21 315	15 328	223.833	1900	61 794	45 997	256.743
740	21 943	15 789	224.692	2000	65 408	48 780	258.600
760	22 573	16 255	225.533	2100	69 044	51 584	260.370
780	23 208	16 723	226.357	2200	72 688	54 396	262.065
800	23 844	17 193	227.162	2300	76 345	57 222	263.692
820	24 483	17 665	227.952	2400	80 015	60 060	265.253
840	25 124	18 140	228.724	2500	83 692	62 906	266.755
860	25 768	18 617	229.482	2600	87 383	65 766	268.202
880	26 415	19 099	230.227	2700	91 077	68 628	269.596
900	27 066	19 583	230.957	2800	94 784	71 504	270.943
920	27 719	20 070	231.674	2900	98 495	74 383	272.249
940	28 375	20 559	232.379	3000	102 210	77 267	273.508
960	29 033	21 051	233.072	3100	105 939	80 164	274.730
980	29 693	21 545	233.752	3150	107 802	81 612	275.326
1000	30 355	22 041	234.421	3200	109 667	83 061	275.914
1020	31 020	22 540	235.079				

SOURCE: JANAF Thermochemical Tables, NSRDS-NBS-37, 1971.

Table E-6 Molar Properties of Water, H_2O

$\bar{h}_f^\circ = -241\,810$ kJ/kmol

T	$\bar{h}$	$\bar{u}$	$\bar{s}^\circ$	T	$\bar{h}$	$\bar{u}$	$\bar{s}^\circ$
0	0	0	0	1020	36 709	28 228	233.415
220	7 295	5 466	178.576	1040	37 542	28 895	234.223
240	7 961	5 965	181.471	1060	38 380	29 567	235.020
260	8 627	6 466	184.139	1080	39 223	30 243	235.806
280	9 296	6 968	186.616	1100	40 071	30 925	236.584
298	9 904	7 425	188.720	1120	40 923	31 611	237.352
300	9 966	7 472	188.928	1140	41 780	32 301	238.110
320	10 639	7 978	191.098	1160	42 642	32 997	238.859
340	11 314	8 487	193.144	1180	43 509	33 698	239.600
360	11 992	8 998	195.081	1200	44 380	34 403	240.333
380	12 672	9 513	196.920	1220	45 256	35 112	241.057
400	13 356	10 030	198.673	1240	46 137	35 827	241.773
420	14 043	10 551	200.350	1260	47 022	36 546	242.482
440	14 734	11 075	201.955	1280	47 912	37 270	243.183
460	15 428	11 603	203.497	1300	48 807	38 000	243.877
480	16 126	12 135	204.982	1320	49 707	38 732	244.564
500	16 828	12 671	206.413	1340	50 612	39 470	245.243
520	17 534	13 211	207.799	1360	51 521	40 213	245.915
540	18 245	13 755	209.139	1400	53 351	41 711	247.241
560	18 959	14 303	210.440	1440	55 198	43 226	248.543
580	19 678	14 856	211.702	1480	57 062	44 756	249.820
600	20 402	15 413	212.920	1520	58 942	46 304	251.074
620	21 130	15 975	214.122	1560	60 838	47 868	252.305
640	21 862	16 541	215.285	1600	62 748	49 445	253.513
660	22 600	17 112	216.419	1700	67 589	53 455	256.450
680	23 342	17 688	217.527	1800	72 513	57 547	259.262
700	24 088	18 268	218.610	1900	77 517	61 720	261.969
720	24 840	18 854	219.668	2000	82 593	65 965	264.571
740	25 597	19 444	220.707	2100	87 735	70 275	267.081
760	26 358	20 039	221.720	2200	92 940	74 649	269.500
780	27 125	20 639	222.717	2300	98 199	79 076	271.839
800	27 896	21 245	223.693	2400	103 508	83 553	274.098
820	28 672	21 855	224.651	2500	108 868	88 082	276.286
840	29 454	22 470	225.592	2600	114 273	92 656	278.407
860	30 240	23 090	226.517	2700	119 717	97 269	280.462
880	31 032	23 715	227.426	2800	125 198	101 917	282.453
900	31 828	24 345	228.321	2900	130 717	106 605	284.390
920	32 629	24 980	229.202	3000	136 264	111 321	286.273
940	33 436	25 621	230.070	3100	141 846	116 072	288.102
960	34 247	26 265	230.924	3150	144 648	118 458	288.9
980	35 061	26 913	231.767	3200	147 457	120 851	289.884
1000	35 882	27 568	232.597	3250	150 250	123 250	290.7

SOURCE: JANAF Thermochemical Tables, NSRDS-NBS-37, 1971.

Table E-1E Properties of Air

T, °R	h, Btu/lbm	P_r	u, Btu/lbm	v_r	s,° Btu/lbm-°R
400	95.53	0.4858	68.11	305.0	0.52890
440	105.11	0.6776	74.93	240.6	0.55172
480	114.69	0.9182	81.77	193.65	0.57255
520	124.27	1.2147	88.62	158.58	0.59173
537	128.10	1.3593	91.53	146.34	0.59945
540	129.06	1.3860	92.04	144.32	0.60078
560	133.86	1.5742	95.47	131.78	0.60950
580	138.66	1.7800	98.90	120.70	0.61793
600	143.47	2.005	102.34	110.88	0.62607
620	148.28	2.249	105.78	102.12	0.63395
640	153.09	2.514	109.21	94.30	0.64159
660	157.92	2.801	112.67	87.27	0.64902
680	162.73	3.111	116.12	80.96	0.65621
700	167.56	3.446	119.58	75.25	0.66321
720	172.39	3.806	123.04	70.07	0.67002
740	177.23	4.193	126.51	65.38	0.67665
760	182.08	4.607	129.99	61.10	0.68312
780	186.94	5.051	133.47	57.20	0.68942
800	191.81	5.526	136.97	53.63	0.69558
820	196.69	6.033	140.47	50.35	0.70160
840	201.56	6.573	143.98	47.34	0.70747
860	206.46	7.149	147.50	44.57	0.71323
880	211.35	7.761	151.02	42.01	0.71886
900	216.26	8.411	154.57	39.64	0.72438
920	221.18	9.102	158.12	37.44	0.72979
940	226.11	9.834	161.68	35.41	0.73509
960	231.06	10.610	165.26	33.52	0.74030
980	236.02	11.430	168.83	31.76	0.74540
1000	240.98	12.298	172.43	30.12	0.75042
1020	245.97	13.215	176.04	28.59	0.75536
1040	250.95	14.182	179.66	27.17	0.76019
1060	255.96	15.203	183.29	25.82	0.76496
1080	260.97	16.278	186.93	24.58	0.76964
1100	265.99	17.413	190.58	23.40	0.77426
1120	271.03	18.604	194.25	22.30	0.77880
1160	281.14	21.18	201.63	20.293	0.78767
1200	291.30	24.01	209.05	18.514	0.79628
1240	301.52	27.13	216.53	16.932	0.80466
1280	311.79	30.55	244.05	15.518	0.81280
1320	322.11	34.31	231.63	14.253	0.82075
1360	332.48	38.41	239.25	13.118	0.82848
1400	342.90	42.88	246.93	12.095	0.83604
1440	353.37	47.75	254.66	11.172	0.84341
1480	363.89	53.04	262.44	10.336	0.85062
1520	374.47	58.78	270.26	9.578	0.85767
1560	385.08	65.00	278.13	8.890	0.86456
1600	395.74	71.73	286.06	8.263	0.87130
1640	406.45	78.99	294.03	7.691	0.87791
1680	417.20	86.82	302.04	7.168	0.88439
1720	428.00	95.24	310.09	6.690	0.89074
1760	438.83	104.30	318.18	6.251	0.89697
1800	449.71	114.03	326.32	5.847	0.90308

Table E-1E (*Continued*)

T, °R	h, Btu/lbm	P_r	u, Btu/lbm	v_r	$s°$, Btu/lbm-°R
1900	477.09	141.51	346.85	4.974	0.91788
2000	504.71	174.00	367.61	4.258	0.93205
2200	560.59	256.6	409.78	3.176	0.95868
2400	617.22	367.6	452.70	2.419	0.98331
2600	674.49	513.5	496.26	1.8756	1.00623
2800	732.33	702.0	540.40	1.4775	1.02767
3000	790.68	941.4	585.04	1.1803	1.04779

SOURCE: J. H. Keenan and J. Kaye, *Gas Tables*, Wiley, New York, 1945.

Table E-2E Molar Properties of Nitrogen, N_2

$\bar{h}_f^\circ = 0$ Btu/lbmol

T, °R	$\bar{h}$, Btu/lbmol	$\bar{u}$, Btu/lbmol	$\bar{s}°$, Btu/lbmol-°R	T, °R	$\bar{h}$, Btu/lbmol	$\bar{u}$, Btu/lbmol	$\bar{s}°$, Btu/lbmol-°R
300	2082.0	1486.2	41.695	1100	7695.0	5510.5	50.783
320	2221.0	1585.5	42.143	1120	7839.3	5615.2	50.912
340	2360.0	1684.8	42.564	1160	8129.0	5825.4	51.167
400	2777.0	1982.6	43.694	1200	8420.0	6037.0	51.413
440	3055.1	2181.3	44.357	1240	8712.6	6250.1	51.653
480	3333.1	2379.9	44.962	1280	9006.4	6464.5	51.887
520	3611.3	2578.6	45.519	1320	9301.8	6680.4	52.114
537	3729.5	2663.1	45.743	1360	9598.6	6897.8	52.335
540	3750.3	2678.0	45.781	1400	9896.9	7116.7	52.551
560	3889.5	2777.4	46.034	1440	10196.6	7337.0	52.763
580	4028.7	2876.9	46.278	1480	10497.8	7558.7	52.969
600	4167.9	2976.4	46.514	1520	10800.4	7781.9	53.171
620	4307.1	3075.9	46.742	1560	11104.3	8006.4	53.369
640	4446.4	3175.5	46.964	1600	11409.7	8232.3	53.561
660	4585.8	3275.2	47.178	1640	11716.4	8459.6	53.751
680	4725.3	3374.9	47.386	1680	12024.3	8688.1	53.936
700	4864.9	3474.8	47.588	1720	12333.7	8918.0	54.118
720	5004.5	3574.7	47.785	1760	12644.3	9149.2	54.297
740	5144.3	3674.7	47.977	1800	12956.3	9381.7	54.472
760	5284.1	3774.9	48.164	1900	13741.6	9968.4	54.896
780	5424.2	3875.2	48.345	2000	14534.4	10562.6	55.303
800	5564.4	3975.7	48.522	2200	16139.8	11770.9	56.068
820	5704.7	4076.3	48.696	2400	17767.9	13001.8	56.777
840	5845.3	4177.1	48.865	2600	19415.8	14252.5	57.436
860	5985.9	4278.1	49.031	2800	21081.1	15520.6	58.053
880	6126.9	4379.4	49.193	3000	22761.5	16803.9	58.632
900	6268.1	4480.8	49.352	3100	23606.8	17450.6	58.910
920	6409.6	4582.6	49.507	3200	24455.0	18100.2	59.179
940	6551.2	4684.5	49.659	3300	25306.0	18752.7	59.442
960	6693.1	4786.7	49.808	3400	26159.7	19407.7	59.697
980	6835.4	4889.3	49.955	3600	27874.4	20725.3	60.186
1000	6977.9	4992.0	50.099	3700	28735.1	21387.4	60.422
1020	7120.7	5095.1	50.241	3800	29597.9	22051.6	60.562
1040	7263.8	5198.5	50.380	3900	30462.8	22717.9	60.877
1060	7407.2	5302.2	50.516	5300	42728.3	32203.2	63.563
1080	7551.0	5406.2	50.651	5380	43436.0	32752.1	63.695

SOURCE: J. H. Keenan and J. Kaye, *Gas Tables*, Wiley, New York, 1945.

Table E-3E Molar Properties of Oxygen, O_2

$\bar{h}_f^\circ = 0$ Btu/lbmol

T, °R	$\bar{h}$	$\bar{u}$	$\bar{s}^\circ$	T, °R	$\bar{h}$	$\bar{u}$	$\bar{s}^\circ$
300	2073.5	1477.8	44.927	1280	9254.6	6712.7	55.386
320	2212.6	1577.1	45.375	1320	9571.6	6950.2	55.630
340	2351.7	1676.5	45.797	1360	9890.2	7189.4	55.867
400	2769.1	1974.8	46.927	1400	10210.4	7430.1	56.099
420	2908.3	2074.3	47.267	1440	10532.0	7672.4	56.326
440	3047.5	2173.8	47.591	1480	10855.1	7916.0	56.547
480	3326.5	2373.3	48.198	1520	11179.6	8161.1	56.763
520	3606.1	2573.4	48.757	1560	11505.4	8407.4	56.975
537	3725.1	2658.7	48.982	1600	11832.5	8655.1	57.182
540	3746.2	2673.8	49.021	1640	12160.9	8904.1	57.385
560	3886.6	2774.5	49.276	1680	12490.4	9154.1	57.582
580	4027.3	2875.5	49.522	1720	12821.1	9405.4	57.777
600	4168.3	2976.8	49.762	1760	13153.0	9657.9	57.968
620	4309.7	3078.4	49.993	1800	13485.8	9911.2	58.155
640	4451.4	3180.4	50.218	1900	14322.1	10549.0	58.607
660	4593.5	3282.9	50.437	2000	15164.0	11192.3	59.039
680	4736.2	3385.8	50.650	2200	16862.6	12493.7	59.848
700	4879.3	3489.2	50.858	2400	18579.2	13813.1	60.594
720	5022.9	3593.1	51.059	2600	20311.4	15148.1	61.287
740	5167.0	3697.4	51.257	2800	22057.8	16497.4	61.934
760	5311.4	3802.2	51.450	3000	23817.7	17860.1	62.540
780	5456.4	3907.5	51.638	3100	24702.5	18546.3	62.831
800	5602.0	4013.3	51.821	3200	25590.5	19235.7	63.113
820	5748.1	4119.7	52.002	3300	26481.6	19928.2	63.386
840	5894.8	4226.6	52.179	3400	27375.9	20623.9	63.654
860	6041.9	4334.1	52.352	3600	29173.9	22024.8	64.168
880	6189.6	4442.0	52.522	3700	30077.5	22729.8	64.415
900	6337.9	4550.6	52.688	3800	30984.1	23437.8	64.657
920	6486.7	4659.7	52.852	3900	31893.6	24148.7	64.893
940	6636.1	4769.4	53.012	4100	33721.6	25579.5	65.350
960	6786.0	4879.5	53.170	4200	34639.9	26299.2	65.571
980	6936.4	4990.3	53.326	4300	35561.1	27021.9	65.788
1000	7087.5	5101.6	53.477	4400	36485.0	27747.2	66.000
1020	7238.9	5213.3	53.628	4600	38341.4	29206.4	66.413
1040	7391.0	5325.7	53.775	4700	39273.6	29940.0	66.613
1060	7543.6	5438.6	53.921	4800	40208.6	30676.4	66.809
1080	7697.8	5552.1	54.064	4900	41146.1	31415.3	67.003
1100	7850.4	5665.9	54.204	5100	43029.1	32901.2	67.380
1120	8004.5	5780.3	54.343	5200	43974.3	33647.9	67.562
1160	8314.2	6010.6	54.614	5300	44922.2	34397.1	67.743
1200	8625.8	6242.8	54.879	5380	45682.1	34998.1	67.885
1240	8939.4	6476.9	55.136				

SOURCE: J. H. Keenan and J. Kaye, *Gas Tables*, Wiley, New York, 1945.

Table E-4E Molar Properties of Carbon Dioxide, CO_2

$\bar{h}_f^\circ = -169{,}300$ Btu/lbmol

T, °R	$\bar{h}$	$\bar{u}$	$\bar{s}^\circ$	T, °R	$\bar{h}$	$\bar{u}$	$\bar{s}^\circ$
300	2108.2	1512.4	46.353	1320	12376.4	9755.0	60.412
340	2407.3	1732.1	47.289	1340	12617.0	9955.9	60.593
380	2716.4	1961.8	48.148	1360	12858.5	10157.7	60.772
420	3035.7	2201.7	48.947	1380	13101.0	10360.5	60.949
460	3365.7	2452.2	49.698	1400	13344.7	10564.5	61.124
480	3534.7	2581.5	50.058	1420	13589.1	10769.2	61.298
500	3706.2	2713.3	50.408	1440	13834.5	10974.8	61.469
520	3880.3	2847.7	50.750	1460	14080.8	11181.4	61.639
537	4030.2	2963.8	51.032	1480	14328.0	11388.9	61.808
540	4056.8	2984.4	51.082	1500	14576.0	11597.2	61.974
580	4417.2	3265.4	51.726	1520	14824.9	11806.4	62.138
600	4600.9	3409.4	52.038	1540	15074.7	12016.5	62.302
620	4786.8	3555.6	52.343	1560	15325.3	12227.3	62.464
640	4974.9	3704.0	52.641	1580	15576.7	12439.0	62.624
660	5165.2	3854.6	52.934	1600	15829.0	12651.6	62.783
680	5357.6	4007.2	53.225	1620	16081.9	12864.8	62.939
700	5552.0	4161.9	53.503	1640	16335.7	13078.9	63.095
720	5748.4	4318.6	53.780	1660	16590.2	13293.7	63.250
740	5946.8	4477.3	54.051	1700	17101.4	13725.4	63.555
760	6147.0	4637.9	54.319	1800	18391.5	14816.9	64.292
780	6349.1	4800.1	54.582	1900	19697.8	15924.7	64.999
800	6552.9	4964.2	54.839	2000	21018.7	17046.9	65.676
820	6758.3	5129.9	55.093	2100	22352.7	18182.4	66.327
840	6965.7	5297.6	55.343	2200	23699.0	19330.1	66.953
860	7174.7	5466.9	55.589	2300	25056.3	20488.8	67.557
880	7385.3	5637.7	55.831	2400	26424.0	21657.9	68.139
900	7597.6	5810.3	56.070	2500	27801.2	22836.5	68.702
920	7811.4	5984.4	56.305	2600	29187.1	24023.8	69.245
940	8026.8	6160.1	56.536	2700	30581.2	25219.4	69.771
960	8243.8	6337.4	56.765	2800	31982.8	26422.4	70.282
980	8462.2	6516.1	56.990	2900	33391.5	27632.5	70.776
1000	8682.1	6696.2	57.212	3000	34806.6	28849.0	71.255
1020	8903.4	6877.8	57.432	3100	36227.9	30071.7	71.722
1040	9126.2	7060.9	57.647	3200	37654.7	31299.9	72.175
1060	9350.3	7245.3	57.861	3300	39086.7	32533.3	72.616
1080	9575.8	7431.1	58.072	3400	40523.6	33771.6	73.045
1100	9802.6	7618.1	58.281	3500	41965.2	35014.7	73.462
1120	10030.6	7806.4	58.485	3600	43411.0	36261.9	73.870
1140	10260.1	7996.2	58.689	3700	44860.6	37512.9	74.267
1160	10490.6	8187.0	58.889	3800	46314.0	38767.7	74.655
1180	10722.3	8379.0	59.088	3900	47771.0	40026.1	75.033
1200	10955.3	8572.3	59.283	4000	49231.4	41287.9	75.404
1220	11189.4	8766.6	59.477	4200	52162.0	43821.4	76.119
1240	11424.6	8962.1	59.668	4400	55105.1	46367.3	76.803
1260	11661.0	9158.8	59.858	4600	58059.7	48924.7	77.460
1280	11898.4	9356.5	60.044	4800	61024.9	51492.7	78.091
1300	12136.9	9555.3	60.229	5000	64000.0	54070.6	78.698

SOURCE: J. H. Keenan and J. Kaye, *Gas Tables*, Wiley, New York, 1945.

Table E-5E Molar Properties of Carbon Monoxide, CO

$\bar{h}_f^\circ = -47{,}550$ Btu/lbmol

T, °R	$\bar{h}$	$\bar{u}$	$\bar{s}^\circ$	T, °R	$\bar{h}$	$\bar{u}$	$\bar{s}^\circ$
300	2081.9	1486.1	43.223	1400	9948.1	7167.9	54.129
340	2359.9	1684.7	44.093	1420	10100.0	7280.1	54.237
380	2637.9	1883.3	44.866	1460	10404.8	7505.4	54.448
420	2916.0	2081.9	45.563	1500	10711.1	7732.3	54.655
460	3194.0	2280.5	46.194	1520	10864.9	7846.4	54.757
500	3472.1	2479.2	46.775	1540	11019.0	7960.8	54.858
520	3611.2	2578.6	47.048	1560	11173.4	8075.4	54.958
537	3729.5	2663.1	47.272	1580	11328.2	8190.5	55.056
540	3750.3	2677.9	47.310	1600	11483.4	8306.0	55.154
580	4028.7	2876.9	47.807	1620	11638.9	8421.8	55.251
620	4307.4	3076.2	48.272	1640	11794.7	8537.9	55.347
660	4586.5	3275.8	48.709	1660	11950.9	8654.4	55.441
700	4866.0	3475.9	49.120	1700	12264.3	8888.3	55.628
720	5006.1	3576.3	49.317	1800	13053.2	9478.6	56.078
740	5146.4	3676.9	49.509	1900	13849.8	10076.6	56.509
760	5286.8	3775.5	49.697	2000	14653.2	10681.5	56.922
780	5427.4	3878.4	49.880	2100	15463.3	11293.0	57.317
800	5568.2	3979.5	50.058	2200	16279.4	11910.5	57.696
820	5709.4	4081.0	50.232	2300	17101.0	12533.5	58.062
840	5850.7	4182.6	50.402	2400	17927.4	13161.3	58.414
860	5992.3	4284.5	50.569	2500	18758.8	13794.1	58.754
880	6134.2	4386.6	50.732	2600	19594.3	14431.0	59.081
900	6276.4	4489.1	50.892	2700	20434.0	15072.2	59.398
920	6419.0	4592.0	51.048	2800	21277.2	15716.8	59.705
940	6561.7	4695.0	51.202	2900	22123.8	16364.8	60.002
960	6704.9	4798.5	51.353	3000	22973.4	17015.8	60.290
980	6848.4	4902.3	51.501	3100	23826.0	17669.8	60.569
1000	6992.2	5006.3	51.646	3200	24681.2	18326.4	60.841
1020	7136.4	5110.8	51.788	3300	25539.0	18985.6	61.105
1040	7281.0	5215.7	51.929	3400	26399.3	19647.3	61.362
1060	7425.9	5320.9	52.067	3500	27261.8	20311.2	61.612
1080	7571.1	5426.4	52.203	3600	28126.6	20977.5	61.855
1100	7716.8	5532.3	52.337	3700	28993.5	21645.8	62.093
1120	7862.9	5638.7	52.468	3800	29862.3	22316.0	62.325
1140	8009.2	5745.4	52.598	3900	30732.9	22988.0	62.551
1160	8156.1	5852.5	52.726	4000	31605.2	23661.7	62.772
1180	8303.3	5960.0	52.852	4100	32479.1	24337.0	62.988
1200	8450.8	6067.8	52.976	4200	33354.4	25013.8	63.198
1220	8598.8	6176.0	53.098	4300	34231.2	25692.0	63.405
1240	8747.2	6284.7	53.218	4400	35109.2	26371.4	63.607
1260	8896.0	6393.8	53.337	4600	36869.3	27734.3	63.998
1280	9045.0	6503.1	53.455	4700	37751.0	28417.4	64.188
1300	9194.6	6613.0	53.571	5000	40402.7	30473.4	64.735
1320	9344.6	6723.2	53.685	5100	41288.6	31160.7	64.910
1340	9494.8	6833.7	53.799	5200	42175.5	31849.0	65.082
1360	9645.5	6944.7	53.910	5300	43063.2	32538.1	65.252
1380	9796.6	7056.1	54.021	5380	43774.1	33090.1	65.385

SOURCE: J. H. Keenan and J. Kaye, *Gas Tables*, Wiley, New York, 1945.

Table E-6E Molar Properties of Water Vapor, H_2O

$\bar{h}_f^\circ = -104{,}040$ Btu/lbmol

T, °R	$\bar{h}$, Btu/lbmol	$\bar{u}$, Btu/lbmol	$\bar{s}^\circ$, Btu/lbmol-°R	T, °R	$\bar{h}$, Btu/lbmol	$\bar{u}$, Btu/lbmol	$\bar{s}^\circ$, Btu/lbmol-°R
300	2,367.6	1,771.8	40.439	1300	10,714.5	8,132.9	52.494
340	2,686.0	2,010.8	41.435	1340	11,076.6	8,415.5	52.768
380	3,004.4	2,249.8	42.320	1380	11,441.4	8,700.9	53.037
420	3,323.2	2,489.1	43.117	1420	11,808.8	8,988.9	53.299
460	3,642.3	2,728.8	43.841	1460	12,178.8	9,279.4	53.556
500	3,962.0	2,969.1	44.508	1500	12,551.4	9,572.7	53.808
537	4,258.0	3,191.9	45.079	1600	13,494.4	10,317.6	54.418
540	4,282.4	3,210.0	45.124	1700	14,455.4	11,079.4	54.999
580	4,603.7	3,451.9	45.696	1800	15,433.0	11,858.4	55.559
620	4,926.1	3,694.9	46.235	1900	16,428	12,654	56.097
660	5,250.0	3,939.3	46.741	2100	18,467	14,297	57.119
700	5,575.4	4,185.3	47.219	2300	20,571	16,003	58.077
740	5,902.6	4,433.1	47.673	2500	22,735	17,771	58.980
780	6,231.7	4,682.7	48.106	2700	24,957	19,595	59.837
820	6,562.6	4,934.2	48.520	2900	27,231	21,472	60.650
860	6,895.6	5,187.8	48.916	3100	29,553	23,397	61.426
900	7,230.9	5,443.6	49.298	3300	31,918	25,365	62.167
940	7,568.4	5,701.7	49.665	3500	34,324	27,373	62.876
980	7,908.2	5,962.0	50.019	3700	36,765	29,418	63.557
1020	8,250.4	6,224.8	50.360	3900	39,240	31,495	64.210
1060	8,595.0	6,490.0	50.693	4100	41,745	33,603	64.839
1100	8,942.0	6,757.5	51.013	4300	44,278	35,739	65.444
1140	9,291.4	7,027.5	51.325	4500	46,836	37,900	66.028
1180	9,643.4	7,300.1	51.630	4700	49,417	40,083	66.591
1220	9,998.0	7,575.2	51.925	4900	52,019	42,288	67.135
1260	10,354.9	7,852.7	52.212	5000	53,327	43,398	67.401

SOURCE: J. H. Keenan and J. Kaye, *Gas Tables*, Wiley, New York, 1945.

Psychrometric Charts

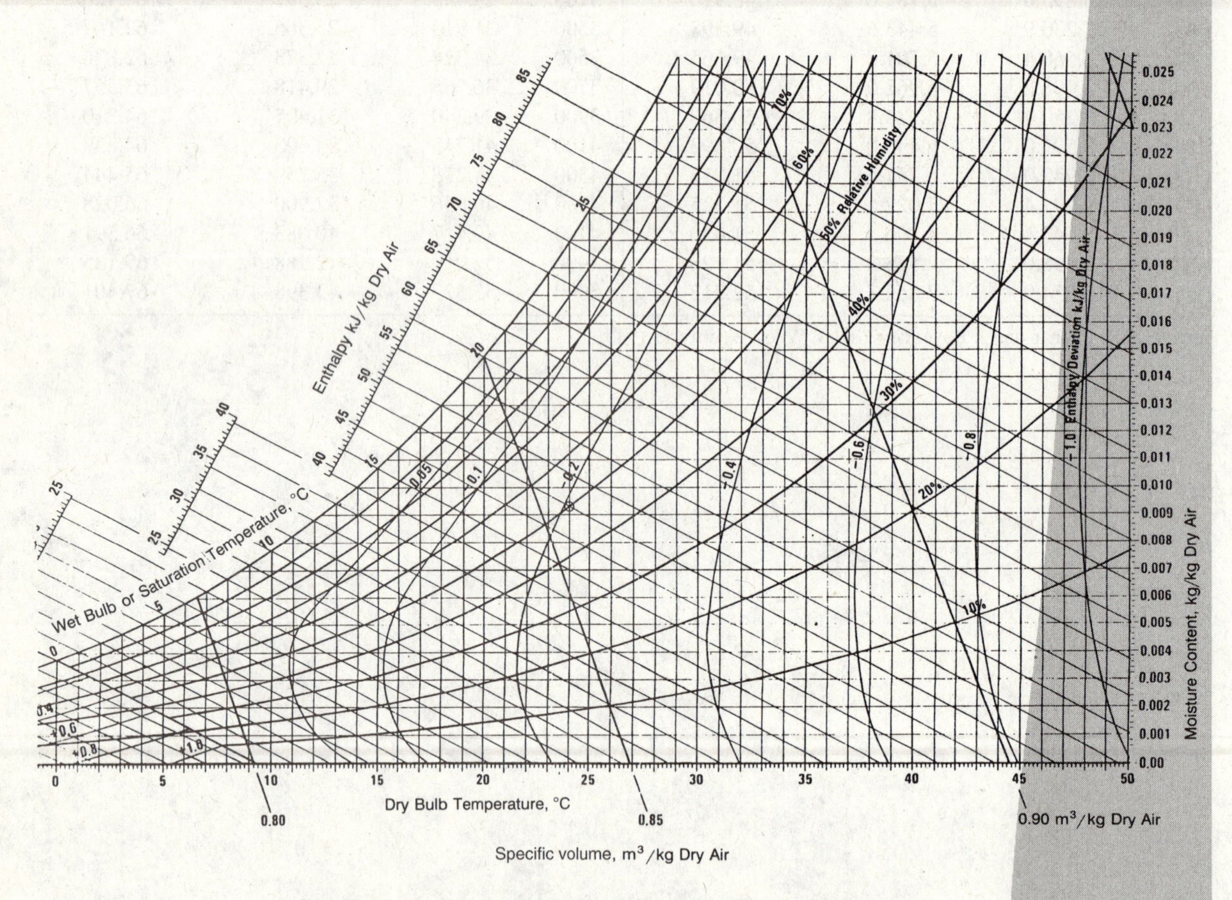

Fig. F-1 Psychrometric Chart, $P = 1$ atm. (Carrier Corporation.)

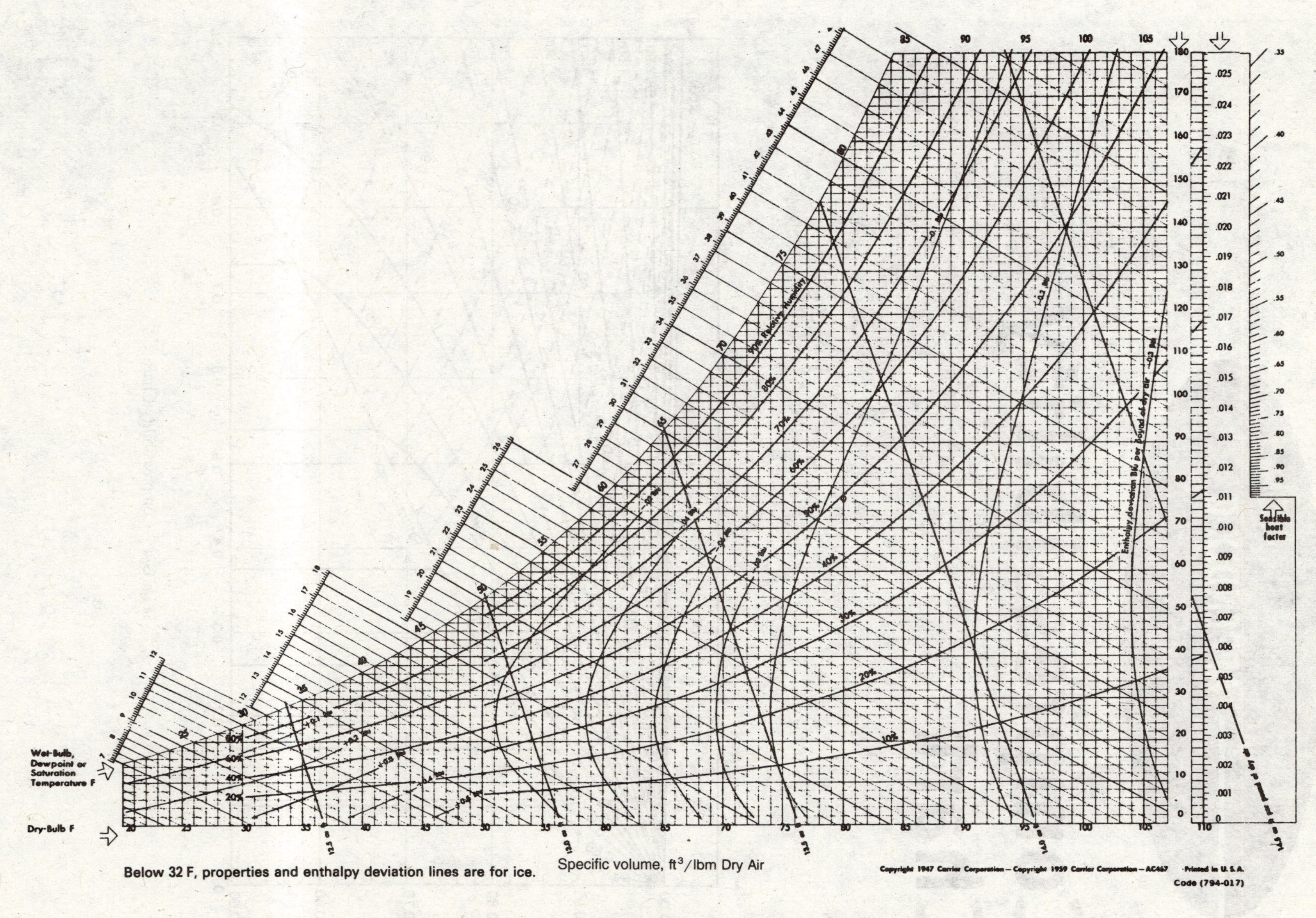

Fig. F-1E Psychrometric Chart, $P = 1$ atm. (Carrier Corporation.)

Compressibility Chart

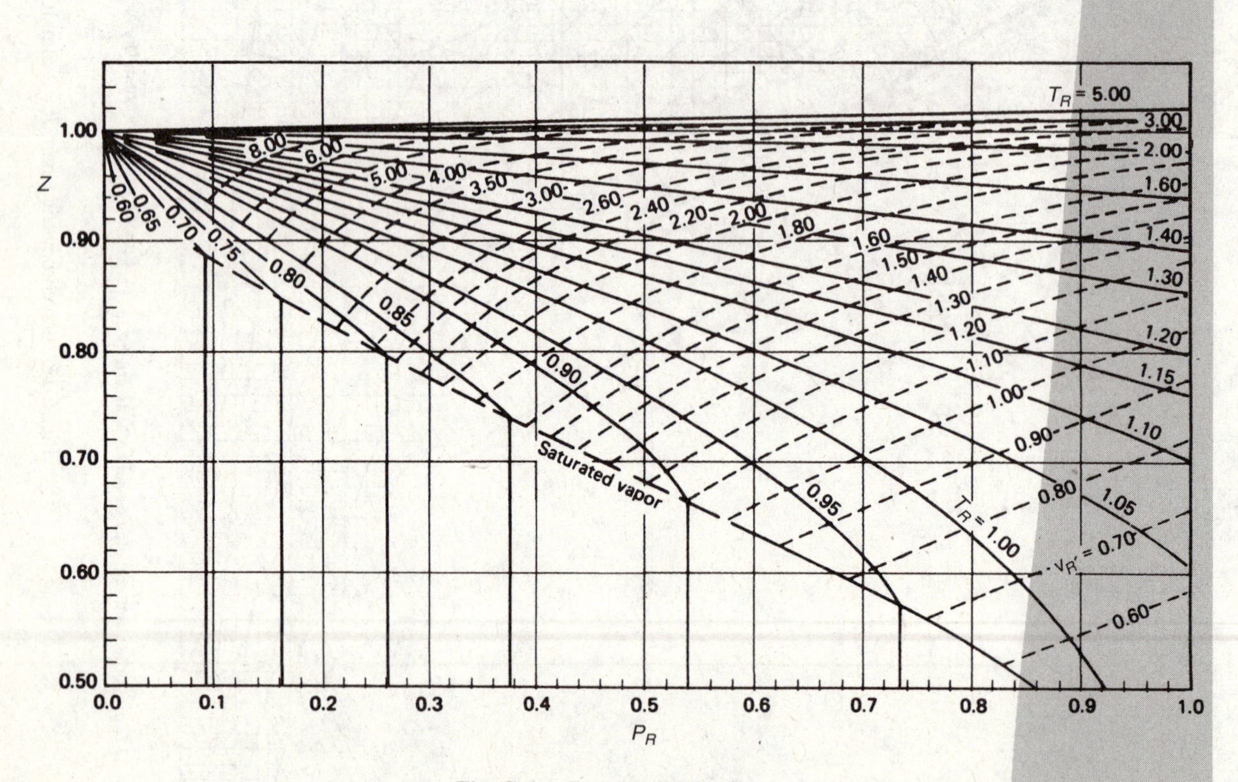

Fig. G-1 Compressibility Chart.

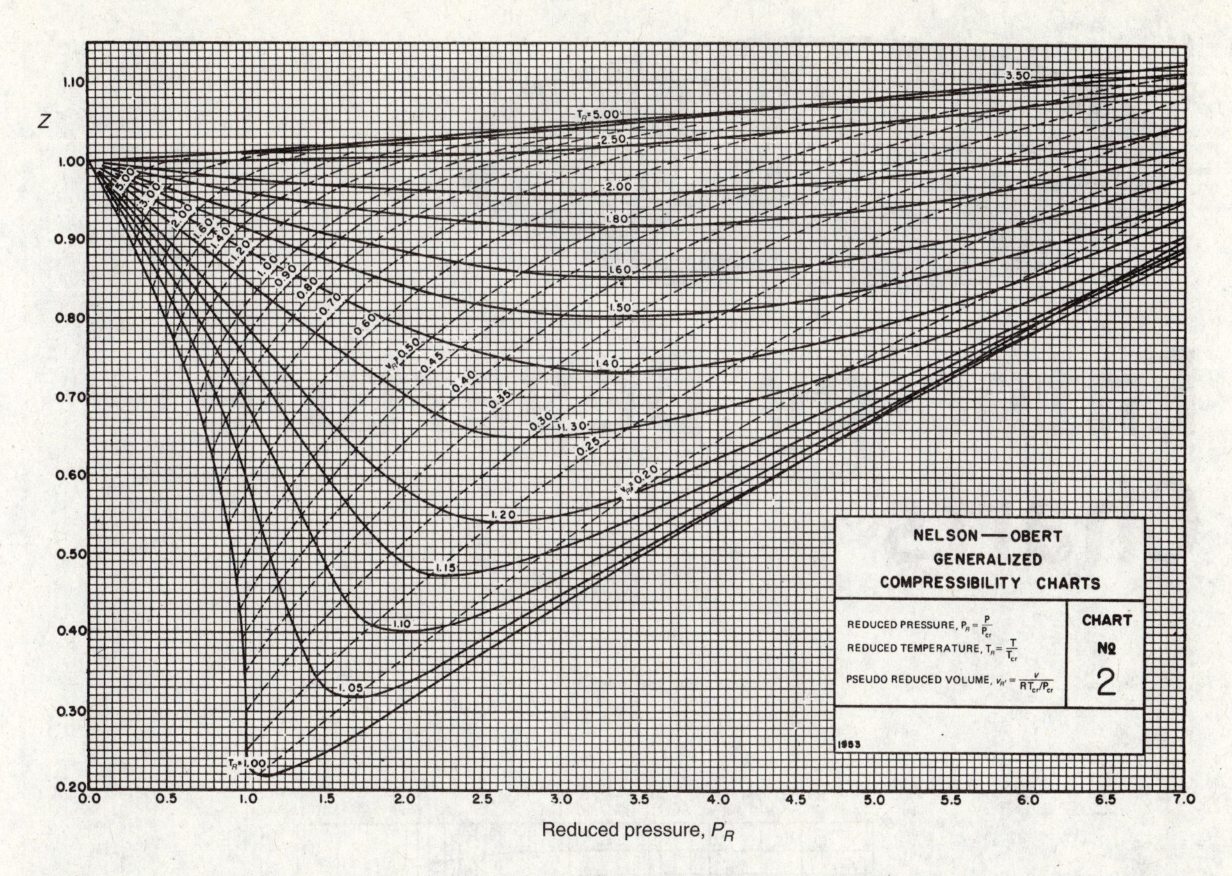

Fig. G-2 Compressibility Chart (*continued*). [V. M. Faires, *Problems on Thermodynamics*, Macmillan, New York, 1962. Data from L. C. Nelson and E. F. Obert, Generalized Compressibility Charts, *Chem. Eng.* **61**: 203 (1954).]

Enthalpy Departure Charts

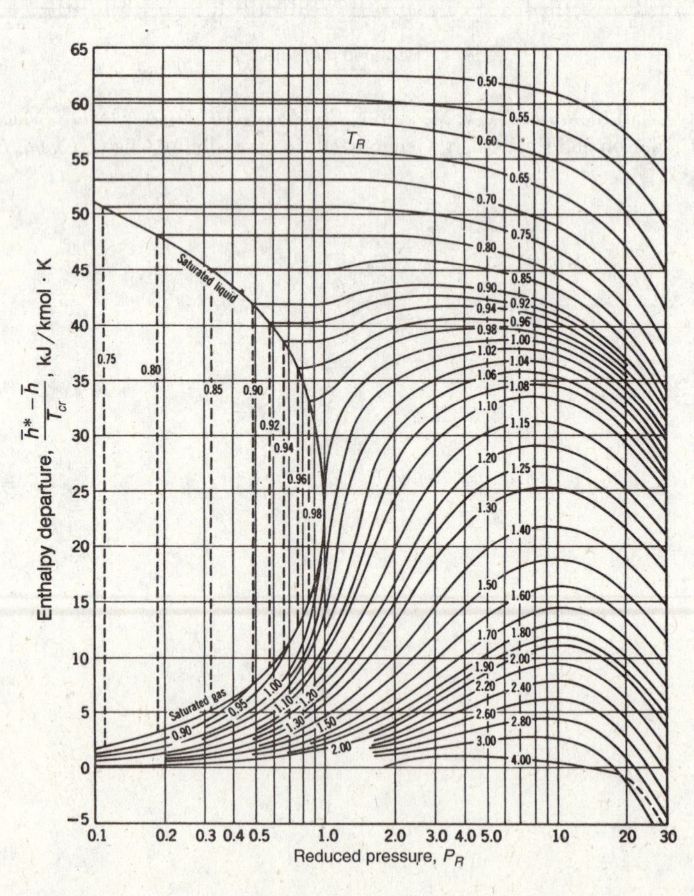

Fig. H-1 Enthalpy Departure Chart. [G. J. Van Wylen and R. E. Sonntag, *Fundamentals of Classical Thermodynamics*, 3d ed., Wiley, New York.]

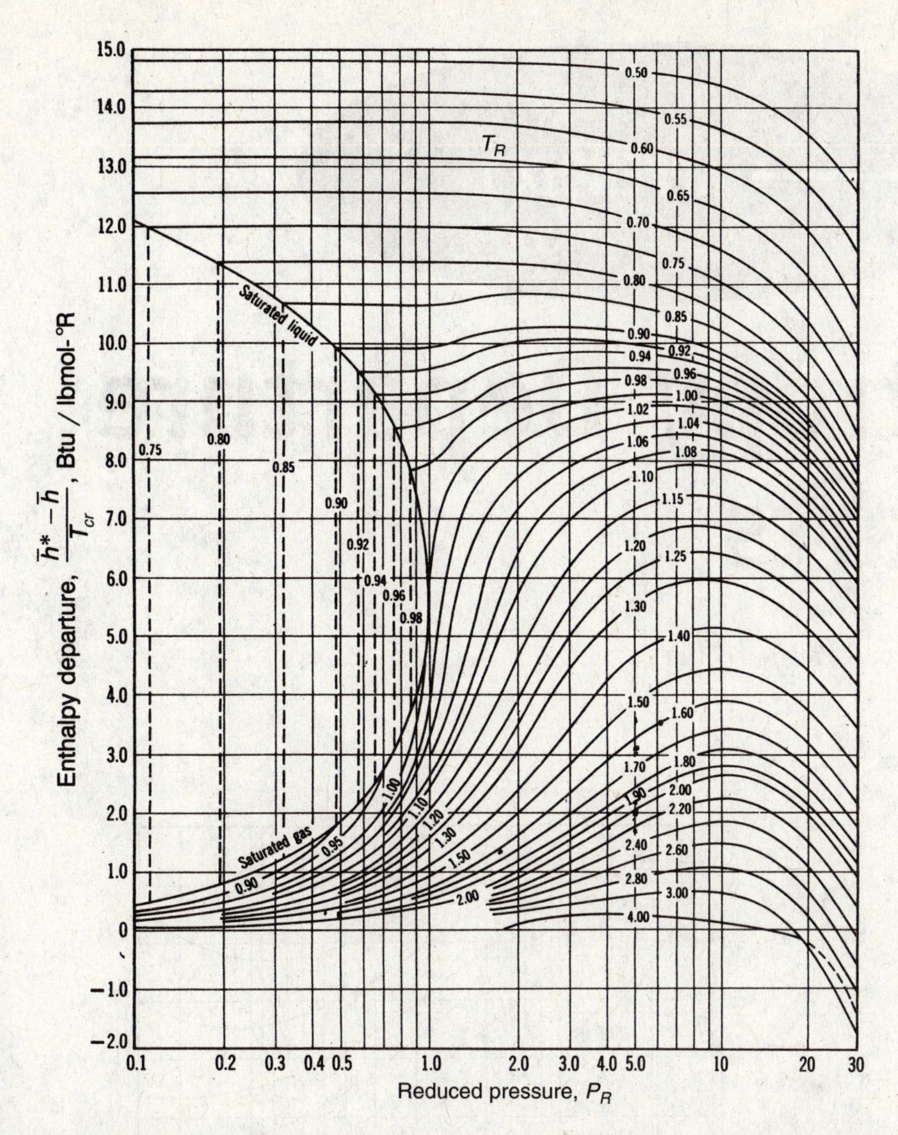

Fig. H-1E Enthalpy Departure Chart. [G. J. Van Wylen and R. E. Sonntag, *Fundamentals of Classical Thermodynamics*, 3d ed., Wiley, New York.]

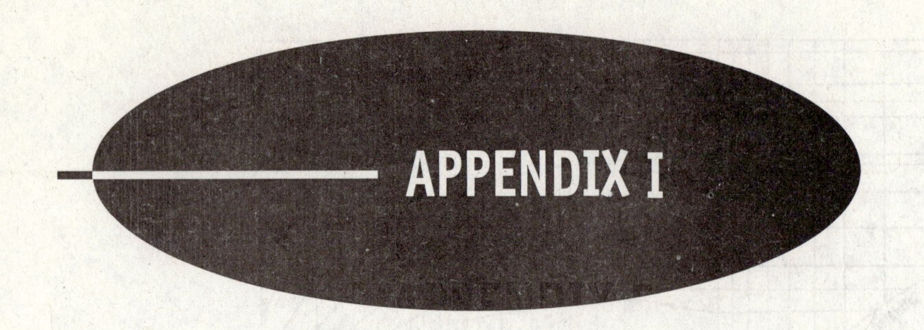

APPENDIX I

Entropy Departure Charts

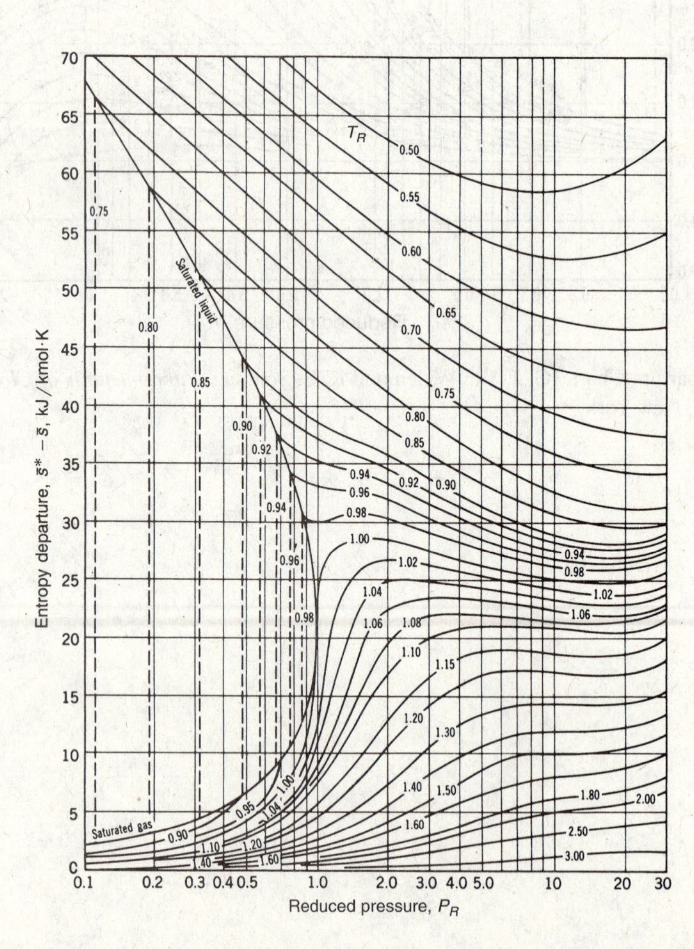

Fig. I-1 [G. J. Van Wylen and R. E. Sonntag, *Fundamentals of Classical Thermodynamics*, 3d ed., Wiley, New York.]

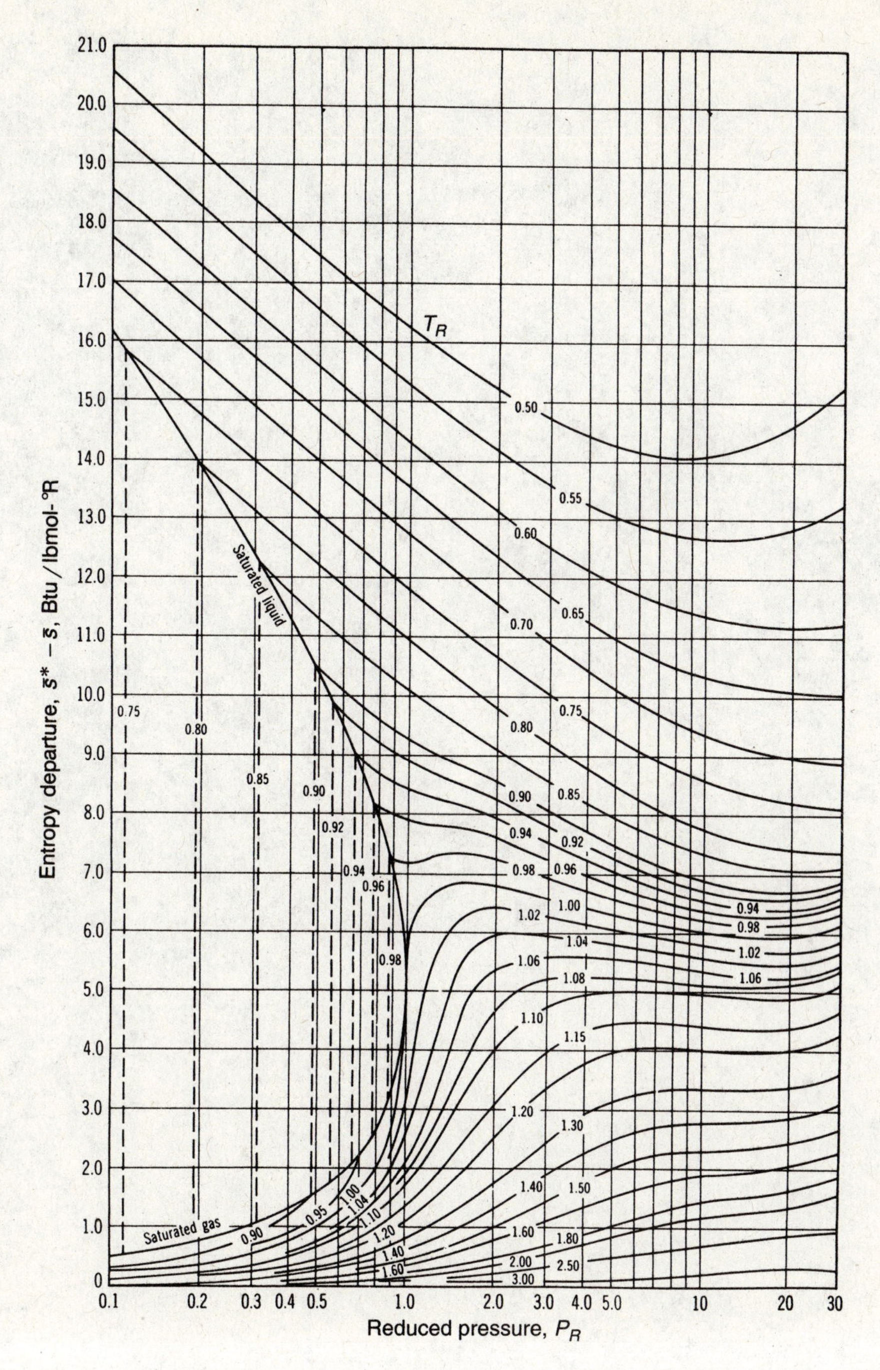

Fig. I-1E [G. J. Van Wylen and R. E. Sonntag, *Fundamentals of Classical Thermodynamics*, 3d ed., Wiley, New York.]

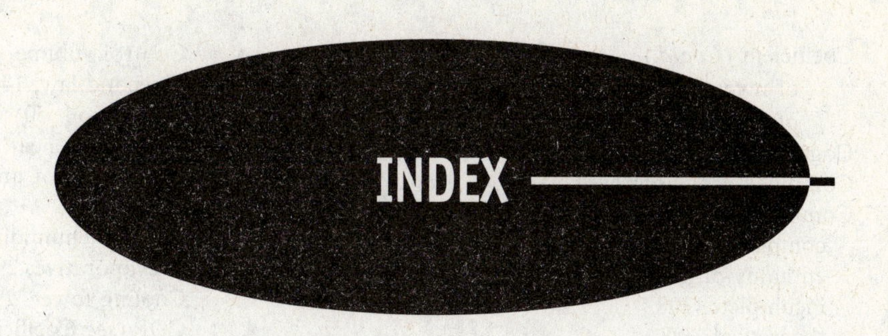
INDEX